Developments in Economic Geology, 9

GEOCHEMICAL EXPLORATION 1976

Association of Exploration Geochemists
Special Volume No. 6

Further titles in this series

1. *I.L. ELLIOTT and W.K. FLETCHER*
GEOCHEMICAL EXPLORATION 1974

2. *P.M.D. BRADSHAW*
CONCEPTUAL MODELS IN EXPLORATION GEOCHEMISTRY
The Canadian Cordillera and Canadian Shield

3. *G.J.S. GOVETT and M.H. GOVETT*
WORLD MINERALS SUPPLIES
Assessment and Perspective

4. R.T. SHUEY
SEMICONDUCTING ORE MINERALS

5. *J.S. SUMNER*
PRINCIPLES OF INDUCED POLARIZATION FOR GEOPHYSICAL EXPLORATION

6. *R.A. RICH, H.D. HOLLAND and U. PETERSEN*
HYDROTHERMAL URANIUM DEPOSITS

7. *J.G. MORSE*
NUCLEAR METHODS IN MINERAL EXPLORATION AND PRODUCTION

8. *M. KUZVART and M. BÖHMER*
PROSPECTING AND EXPLORATION FOR MINERAL DEPOSITS

Developments in Economic Geology, 9

GEOCHEMICAL EXPLORATION 1976

Proceedings of the Sixth International Geochemical Exploration Symposium held in Sydney, N.S.W., Australia, August 1976 during the 25th International Geological Congress

Edited by

C.R.M. BUTT

Division of Mineralogy, Commonwealth Scientific and Industrial Research Organization, Wembley, Western Australia

and

I.G.P. WILDING

Australian Anglo American Limited, Perth, Western Australia

Association of Exploration Geochemists
Special Volume No. 6

Reprinted from the Journal of Geochemical Exploration, Volume 8, No. 1/2

ELSEVIER SCIENTIFIC PUBLISHING COMPANY
Amsterdam — Oxford — New York 1977

ELSEVIER SCIENTIFIC PUBLISHING COMPANY
335 Jan van Galenstraat
P.O. Box 211, Amsterdam, The Netherlands

Distributors for the United States and Canada:

ELSEVIER NORTH-HOLLAND INC.
52, Vanderbilt Avenue
New York, N.Y. 10017

Library of Congress Cataloging in Publication Data

International Geochemical Exploration Symposium, 6th, Sydney, 1976.
Geochemical exploration, 1976.

(Developments in economic geology ; 9) (Special volume - Association of Exploration Geochemists ; no. 6)
"Reprinted from the Journal of geochemical exploration, volume 8, no. 1/2."
Bibliography: p.
1. Geochemical prospecting--Congresses. I. Butt, C. R. M. II. Wilding, I. G. P. III. Association of Exploration Geochemists. IV. Title. V. Series. VI. Series: Association of Exploration Geochemists. Special volume - Association of Exploration Geochemists ; no. 6.

TN270.I63 1976 622'.1 77-16444
ISBN 0-444-41653-6

ISBN 0-444-41653-6 (Vol. 9)
ISBN 0-444-41250-6 (Series)

Printed in The Netherlands

FOREWORD

This is the first occasion on which the papers presented at an International Geochemical Exploration Symposium have been published in the Journal of Geochemical Exploration. We hope that, by this means, the papers presented at the Symposium will be available to the widest possible audience.

Thirty of the papers read at the Symposium are presented here; of the remainder, some are published elsewhere and others are not yet available. Extended abstracts of all papers selected for presentation have, however, been published in the Abstracts (Volume II, Section 10B) of the 25th International Geological Congress, 1976. The manuscripts were reviewed by many people and we wish to extend our thanks to them for the care which was taken in making such reviews and for the constructive nature of their comments. We also wish to acknowledge the cooperation of several authors who generously reduced the lengths of their papers to meet the space limitations set for this Proceedings issue.

The Proceedings of the workshop were not formally recorded, in order to promote freer discussion. However, the contributions of the lead speakers have been summarized in Newsletter No. 21 of the Association of Exploration Geochemists (January 1977).

C.R.M. BUTT
I.G.P. WILDING

LIST OF CONTRIBUTORS

SYMPOSIUM COMMITTEES

Organizing Committee

J.F. Gilfillan	*J.F. Gilfillan and Associates Pty. Ltd.*
C.R.M. Butt	*CSIRO Division of Mineralogy*
R. Cox	*Placer Exploration Ltd.*
J.J. Hobbs	*New South Wales Geological Survey*
N.J. Marshall	*Australian Selection Pty. Ltd.*
A.L. Mather	*Geochemical and Mineralogical Laboratories Pty. Ltd.*
W.R. Ryall	*CSIRO Division of Mineralogy*
G.F. Taylor	*CSIRO Division of Mineralogy*
J.R. Wilmshurst	*CSIRO Division of Mineralogy*

Technical Programme Committee

C.R.M. Butt	*CSIRO Division of Mineralogy*
R. Davy	*Geological Survey, Western Australia*
R.H. Mazzucchelli	*Western Mining Corporation Ltd.*
R.E. Smith	*CSIRO Division of Mineralogy*
I.G.P. Wilding	*Australian Anglo American Ltd.*

CONTENTS

INTRODUCTION

The Sixth International Geochemical Exploration Symposium was held in August 1976 in Sydney, New South Wales, Australia, as part of the 25th International Geological Congress. The technical programme of the Symposium was organized by the Association of Exploration Geochemists, within the general format of the Congress. The Association assumed responsibility for Congress Section 10B by including it within the Symposium programme. It was most appropriate that the Symposium be held in Australia, since Australia has the third highest number of Association members, after the U.S.A. and Canada, and is a country in which geochemical techniques of exploration are widely used and have a considerable degree of importance. Bearing in mind the high cost of travel to and within Australia, the meeting was linked with the IGC to encourage greater participation by geochemists from overseas but, as a result, had to compete with the numerous other technical sessions for the available audience. In the event, the Congress was attended by 2300 delegates of whom up to 250 attended individual Symposium sessions. Incidental to this link with IGC, the Association was relieved of the task of arranging accommodation, field excursions etc.

The Symposium was opened by Mr. I.E. Newnham, Director of the Minerals Research Laboratories of the Commonwealth Scientific and Industrial Research Organization (CSIRO). The technical programme commenced with an address by Professor G.J.S. Govett, University of New Brunswick, Canada, the President of the Association of Exploration Geochemists. There followed 38 papers, mainly concerned with exploration for mineral deposits in deeply weathered terrain and for energy resources, with the use of primary dispersion in exploration and with miscellaneous topics, including analytical techniques. A final half-day session was devoted to a workshop entitled "Threshold — fact or fiction".

On behalf of the Association, we would like to thank all those who presented papers, for the success of the meeting was due to their efforts. Special thanks are also due to the lead speakers at the workshop — Dr. A.L. Mather, Mr. N.J. Marshall, Dr. M.A. Chaffee and Dr. D.H. McKenzie — and to all those who contributed to discussions. Regrettably, a number of contributors withdrew from the Symposium at a late stage — a problem which beset the whole Congress. We are indebted to Dr. T. Wignall and Dr. S.H.U. Bowie who agreed to give papers at very short notice to fill gaps in the programme. Daily Organizers and the Chairmen of Sessions are thanked for their assistance with the smooth running of the meeting. We would also like to thank all those who submitted synopses of papers; constraints of time and space prevented the inclusion of many interesting contributions.

The organization and the smooth running of such a Symposium is due to the efforts of many people for a period in excess of two years. While the task was not eased by the necessity of having the Organizing Committee in Sydney and a Technical Programme Committee in Perth, the cooperation of all concerned to overcome this problem is gratefully acknowledged. On behalf of the Association, we would also like to thank members of both Committees for their enthusiasm in carrying out the numerous tasks that ensured the success of the meeting. We benefitted greatly from the ready cooperation of the IGC Organizing Committee and its staff, particularly the Secretary-General, Mr. A. Renwick, who afforded us every assistance. The organization of the meeting was only made possible by the generous donation of time and facilities from the employers of the Committee members. We wish to acknowledge the following organizations who made staff available for reception and publication sales duties through the Congress: Amoco Minerals Australia Ltd., Australian Oil and Gas Corporation Ltd., Australian Selection Pty. Ltd., CSIRO, Geological Survey of N.S.W., Getty Oil Development Ltd., J.F. Gilfillan and Associates Pty. Ltd., Pancontinental Mining Ltd. and Placer Exploration Ltd. We would also like to record our thanks to the Secretarial staff of the CSIRO Division of Mineralogy, particularly Mrs. Cheryl Harris who, with great speed and accuracy, typed much of the correspondence and other paperwork and retyped many of the extended abstracts and complete manuscripts prior to their submission to the publishers.

The Seventh International Geochemical Exploration Symposium is to be held in Golden, Colorado, U.S.A. in April 1978 and we look forward to another stimulating and enjoyable meeting there.

J.F. GILFILLAN
Chairman, Organizing Committee

C.R.M. BUTT
Convenor, Section 10B, 25th IGC

Journal of Geochemical Exploration, 8 (1977) 3–16

Presidential Address

WORLD MINERAL SUPPLIES — THE ROLE OF EXPLORATION GEOCHEMISTRY

G.J.S. GOVETT*

President, Association of Exploration Geochemists

INTRODUCTION

I am immensely pleased to have this opportunity to welcome you on behalf of the Association of Exploration Geochemists to the Sixth International Geochemical Exploration Symposium. It is the first of these biennial symposia to be held in the southern hemisphere and is a remarkable demonstration of the international strength of the Association.

The choice of an appropriate theme for this address was not easy; it was clear that mild exhortations and philosophical musings would not be adequate without "disguising" them by offering something more substantial. The topic I have chosen — mineral supplies and geochemistry — is very much a reflection of my own interests, but I hope it may have broad appeal since the adequacy of mineral supplies is becoming the concern of everyone — and should be of special concern to exploration geochemists.

The world's literature is full of papers and articles that debate the adequacy of mineral resources to sustain the present living standards in industrialized countries and that argue the possibilities of resources being available to permit the numerically less developed majority of the world to attain similar standards. Much of the argument and disagreement stems from the failure of many of the protagonists to distinguish clearly between mineral *reserves* and mineral *resources*.

Mineral reserves are ephemeral figures of the moment — an estimate, of varying reliability, of the tons (pounds, ounces, grams) of a particular commodity that is identified and measured in the ground and which can be removed economically and processed at today's costs and prices and with today's technology. Reserves are dynamic figures that change constantly in response to price, technology, development work and exploration success. Mineral resources, on the other hand, include not only reserves, but also include identified mineral deposits that are not now economically workable (by virtue of grade, mineralogy, lack of appropriate technology, current prices), plus all unknown mineral deposits yet to be discovered. Mineral resources — even if we do not know their magnitude — are conceptually fixed

* Present address: Department of Geology, University of New Brunswick, Fredericton, N.B., Canada.

by past geological events. Fundamentally, the absolute limit of resources of, say, copper is the crustal abundance of copper in the earth's crust; if we had the appropriate technology available and could afford the cost, the resources of copper would be virtually limitless. The dynamic relation between resources and reserves as a function of price, grade, technology and exploration success is shown schematically in Fig. 1.

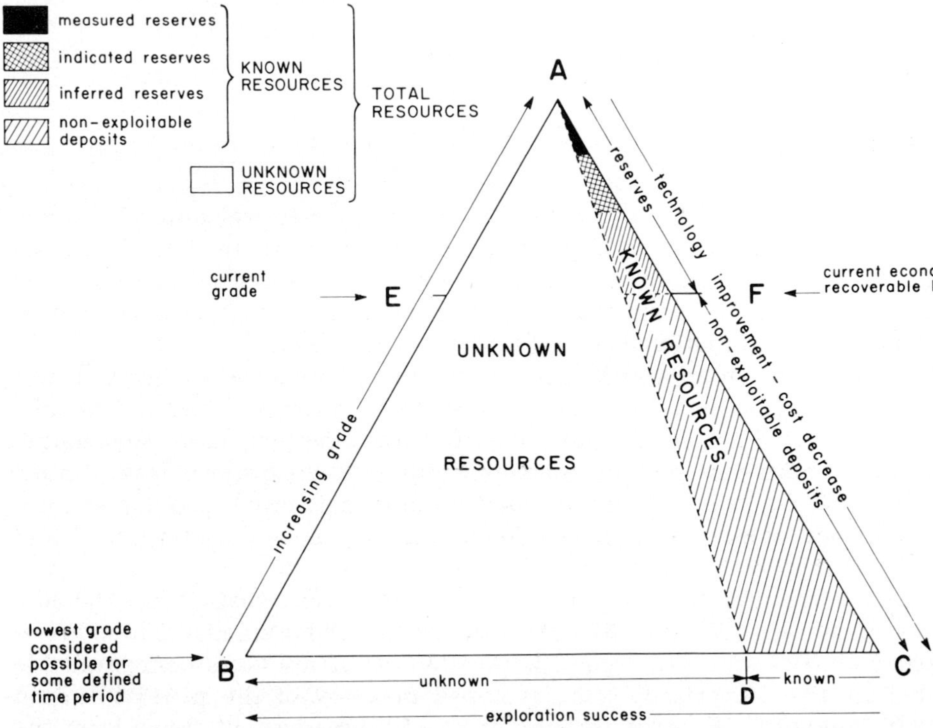

Fig. 1. Schematic illustration showing the dynamic relation between reserves and resources. (From Govett and Govett, 1974; originally published in Resources Policy, vol. 1, p. 53.)

Some encouragement for the future is given by the historical trend of copper production (Fig. 2).The grade has decreased along an approximate negative exponential curve (i.e., the low-grade copper deposits mined today and now regarded as reserves would, even at the beginning of this century, have been classified — if known — as resources). Until recently, cost has also decreased. A related trend of utmost importance is that, although production is steadily increasing, reserves are keeping pace.

The debate on the adequacy of world mineral resources should now properly be a debate about the adequacy of mineral exploration technology to find mineral deposits, and the adequacy of mining and mineral processing technology to convert mineral deposits into orebodies from which raw ma-

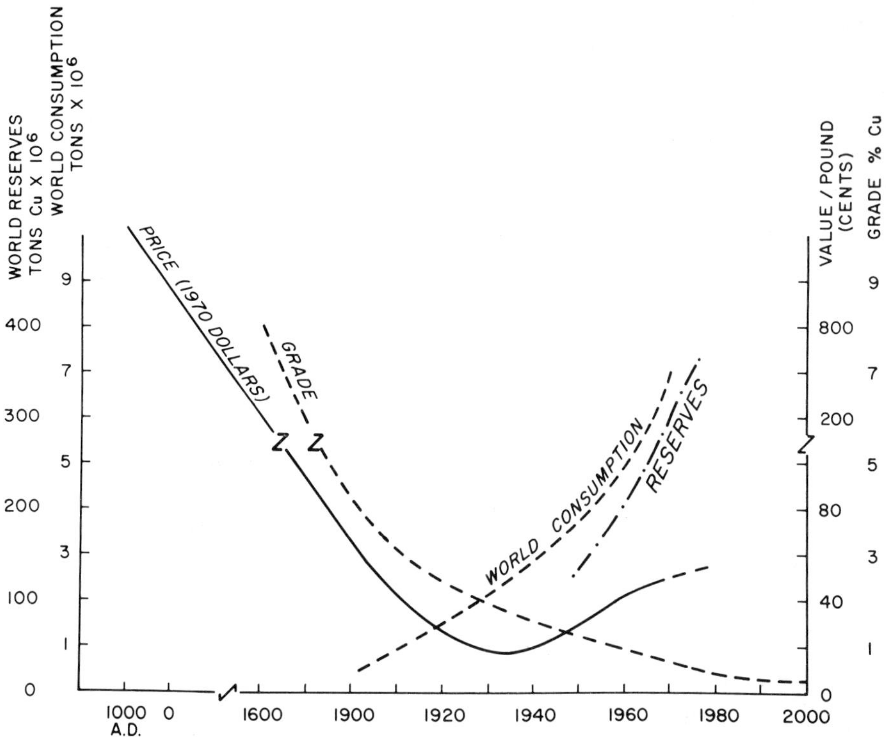

Fig. 2. Historical trend of reserves, mineable grade, price, and world consumption of copper. (From McAllister, 1976.)

terials can be derived economically. Our concern as exploration geochemists is with the first problem — finding mineral deposits. In the remainder of this address, I propose to examine the magnitude of the problem and to discuss the contribution that exploration geochemists may reasonably be expected to make.

THE NATURE OF THE PROBLEM

The magnitude of world demand for minerals in the year 2000 can be estimated with a reasonable degree of confidence (always provided that no major catastrophe of global proportions occurs in the next 25 years). Assuming that little is done to slow the rate of world population growth — and there is no evidence of this happening despite the dire prognosis for the future if the current rate of increase is not slowed — and given the legitimate striving for development in the countries of Asia, Africa, and Latin America, world mineral demand is conservatively expected to increase at an annual rate between 3.6 and 5.5%. What this means in real terms is that if the very conservative figure of 4% growth per year is used, world mineral supplies will have to double in 18 years and increase four-fold in 36 years.

These figures assume that the less developed countries will not be able to achieve their development goals in the next few decades — an assumption which may be incorrect given the increasing political leverage of the "Third World" and the vocal dedication to measures to increase either the size of the "pie" or the less developed countries' share of it. However, if the less developed countries could increase their per capita mineral consumption in the next 30 years to the same levels that now prevail in the U.S.A., annual world mineral production in the year 2000 would have to increase 30 times. Whatever assumptions are made about the global rate of growth of mineral demand, it is clear that pressure will be exerted on some of the world's present conventional mineral reserves — most particularly copper, tin, zinc, lead, tungsten, molybdenum, titanium and, of course, petroleum.

At present, a very large share of the world's industrially important minerals are supplied by five developed countries — the U.S.S.R., the U.S.A., Canada, Australia and South Africa. The five countries produce more than 75% of the world's uranium, vanadium and molybdenum; more than 50% of the ilmenite, nickel, phosphate, manganese, iron ore, potash, lead, chromium, zinc and silver; and more than one-quarter of the copper, tungsten and petroleum. Given this pattern of production, it is not surprising that these five countries also dominate world reserves (see Fig. 3). Inevitably, there are arguments concerning the degree to which this pattern of reserve distribution is a function of where most modern mineral exploration has taken place. This will not be discussed here, but it is perhaps significant that for those minerals which the U.S. Geological Survey has published an

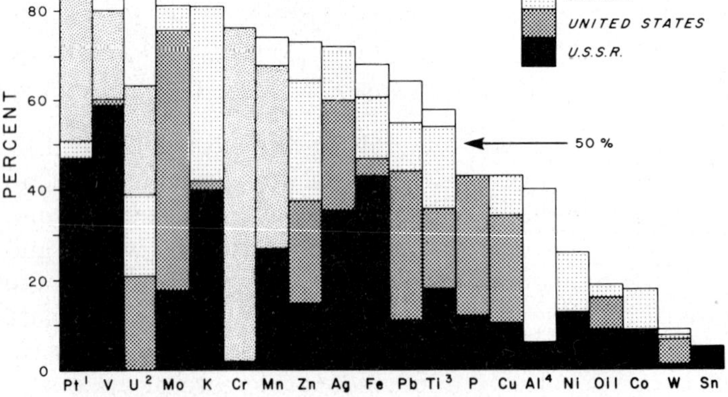

Fig. 3. Percentage share of world reserves of 20 minerals in five developed countries. (After M.H. Govett, 1976.)

assessment for speculative (unknown) resources, the distribution pattern is little changed. Whatever the distribution of *conventional* mineral deposits that emerges, it seems to be reasonably certain that the present major producers will maintain their position over the next 25 years. Three-quarters of the exploration activity outside of the communist countries has been concentrated in the U.S.A., Canada, Australia and South Africa in the past decade; unless this preference for investment in known and politically "safe" countries changes, the pattern will persist.

If one accepts the premise that the dominant proportion of the world's mineral production over the next two decades will continue to be in the present "big five" producer countries, then there are two inevitable conclusions: the average grade of deposits will decrease — which is a problem for mineral extraction technology; and deposits will become increasingly difficult to find — which is the problem of those engaged in mineral exploration.

THE EXPLORATION GEOCHEMISTRY PROBLEM

The magnitude of the problem of finding new mineral deposits in a relatively intensively explored country such as Canada is illustrated in Fig. 4. In the period 1951—1969, the probability of finding a new orebody decreased ten times; total exploration expenditures increased four-fold in this period. It is estimated (Martin et al., 1976) that while between 1971 and 1995 exploration expenditures must increase three times, the value of the metal found and produced will also increase three times; exploration expenditures are expected to remain 4—5% of production value.

The increasing cost and difficulty of finding new mineral deposits in Canada may reasonably be attributed to the decreasing number of deposits having a surface or near-surface expression remaining to be found. This is indicated in the changing pattern of discovery techniques: 85% of producing mines in 1955 had been found by surface prospecting; in the period 1956—1968 only 22% of new discoveries were found by surface prospecting, while 52% were attributed to geophysical techniques (Derry, 1968; 1970). This is a clear reflection of an increasing dependence of finding hidden and buried deposits — a trend that may be expected to intensify and to demand exploration techniques that can detect mineralization at ever-increasing depths, particularly in intensively prospected parts of the world.

Some considerable success has already been achieved in various areas where it has been demonstrated that anomalous haloes of a variety of metals can be detected in rocks for distances from hundreds of metres to the kilometre scale. The technique of measuring volatile constituents at the surface has an obvious attraction, since it is presumed that the constituents are able to travel large distances, even from blind deposits. Most work has probably been done on mercury in rocks, soils and soil gas; other volatile constituents, such as fluorine and iodine, have been tried, but remarkably little work has been done on this promising technique. An exciting new approach, particu-

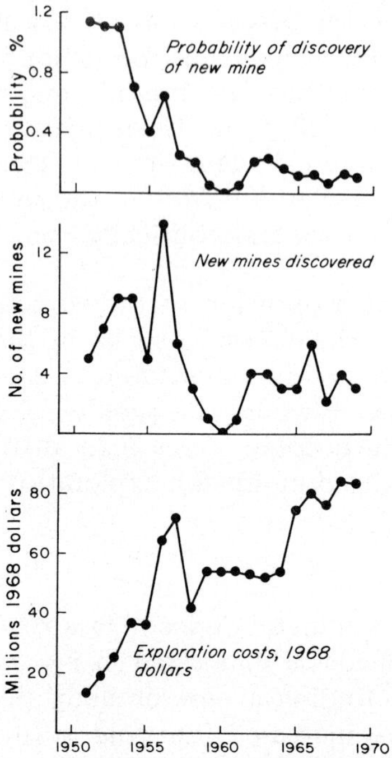

Fig. 4. Exploration costs, numbers of new mines discovered and probability of discovery of new mines, calculated on the basis of an average cost of $30,000 for each exploration try in Canada. All dollars standardized to 1968 dollars. (From Govett, 1976a, based on data from Roscoe, 1971.)

larly in heavily forested regions, is the airborne measurement of organo-metallic particulate matter in the atmosphere by the AIRTRACE^R system of Barringer Research Ltd. (Toronto).

A second mineral exploration problem relevant to the future supply of minerals is the need for improved regional reconnaissance techniques. This is particularly important for the areas of the world that have not yet been intensively prospected; these areas are generally remote from transportation facilities and are frequently physically inhospitable. Therefore, exploration costs are usually high. Low sample density drainage surveys have shown promise for locating broad, regional targets; perhaps the concept has not been pursued with the vigour it deserves. Stream sediment surveys with sample densities of the order of one sample per 200 km^2 have been shown to be successful in Central and West Africa. Similarly, the validity of widely spaced (one sample every 5 km) lake samples has been demonstrated on the Canadian Shield.

Efforts to use rock geochemistry on a regional scale to allow discrimination between potentially ore-bearing and barren plutons and the recognition of particular geographic zones of a lithologic horizon as potentially ore-bearing have been pursued with increasing interest in the past ten years. The results so far have not been uniformly successful. Various approaches have been tried. These include whole rock analysis, distribution of metals in biotites and feldspars in granites, and the distribution of water-soluble halogens in granitic rocks.

NEW APPROACHES TO THE EXPLORATION PROBLEM

The early development of exploration geochemistry has been chronicled by Hawkes (1976), who points out that the earliest reference to the use of systematic measurement of chemical properties of earth materials for exploration in the literature is that by Flerov in 1935; the first important discoveries (at least outside of the U.S.S.R.) using geochemistry were made a mere 25 years ago. Two decades ago, my Ph.D. thesis on soil and stream geochemistry had fewer than two dozen direct literature citations on the subject which, even allowing for my lack of diligence, is in sharp contrast to the more than 1000 references in English derived from a computer search of the past seven years of earth science publications; 88 of these were specifically on the use of rock geochemistry, a field essentially unknown 20 years ago.

Growth in this newest branch of earth science has, then, been phenomenal but progress has been dominantly of an empirical nature — the inevitable and essential first phase of development of a science. A vast amount of data is now available awaiting synthesis and development of hypotheses regarding *processes* of dispersion. This should not be regarded as a leisurely academic pursuit, but as an urgent necessity that will allow development of new techniques to meet the exploration challenges of the next quarter century — notwithstanding the claims of many geochemists, which have been shown to be quite valid in the past, that understanding fundamental processes is not important providing a technique that can be shown to work. Some illustrative examples of the potential usefulness to exploration of determining processes of dispersion will now be considered.

Outside of the U.S.S.R., the sampling and analysis of bedrock as an exploration technique has only been seriously considered for less than 10 years. The reasons for the reluctance to use rock geochemistry are several: while wall-rock anomalies have been recognized since at least the beginning of this century, their extent has generally been regarded as too limited to be of any great significance for exploration; a second reason was the generally unquestioned acceptance of the opinion that, except for leakage haloes, anomalies would not be found in barren, post-mineralization cap-rocks. The discovery of extensive haloes in post-mineralization age basalts overlying cupriferous pyrite deposits in Cyprus in 1967, together with the more or less contemporaneous general recognition and better understanding of the processes of formation of volcanic-sedimentary stratabound sulphide deposits evident in the geological literature, encouraged the belief that processes that are essentially secondary could give rise to useful anomalous haloes in barren cap rocks — at least in a volcanic-sedimentary environment.

The detection of useful geochemical haloes in rock and the mechanism of their formation has been a major preoccupation of the Exploration Geochemistry Group at the University of New Brunswick. Several hypotheses have been proposed to explain these anomalies, including electrochemical dispersion and upward-moving water of compaction; similarly, consideration of the likely effects of hydrothermal fumarolic activity on the pre-ore host rocks, and its effect upon the environment of deposition in a restricted marine basin, provides a hypothetical basis for the presence of extensive haloes in footwall rocks of volcanic-sedimentary massive sulphide deposits. Whether any of the proposed mechanisms are proved to be entirely correct

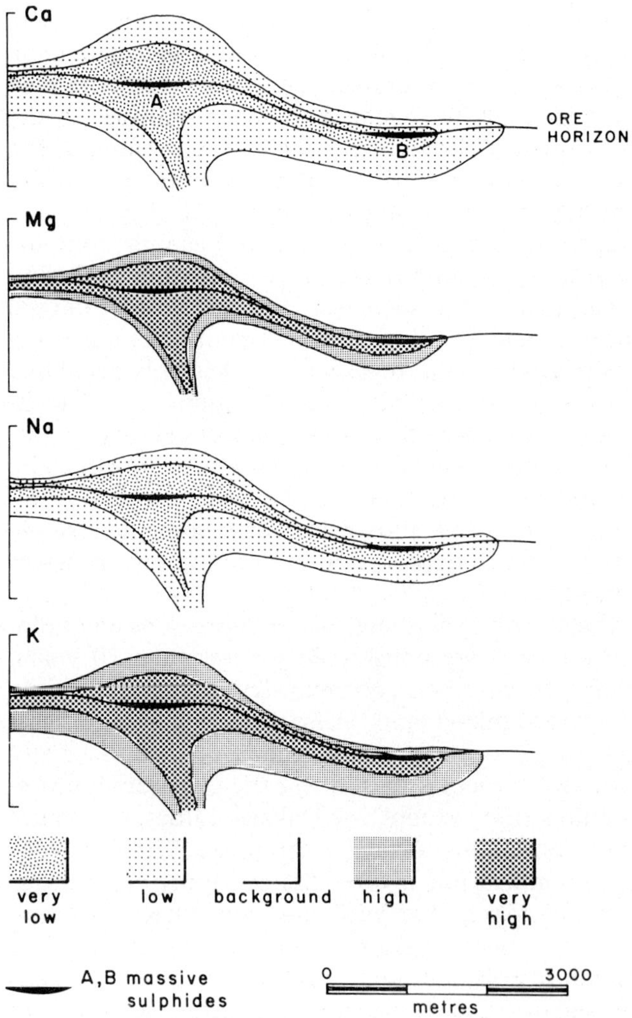

Fig. 5. Schematic illustration of dispersion of calcium, magnesium, sodium and potassium in rocks around volcanic-sedimentary massive sulphide deposits. Sulphide deposit *A* is assumed to overlie source area of base metal sulphides. (Derived from Wahl, 1977.)

is largely irrelevant in the context of practical geochemical exploration. They provide plausible mechanisms that predict extensive dispersion in pre-ore and post-ore rocks associated with volcanic-sedimentary massive sulphide deposits and therefore encourage the development of geochemical techniques that rely on these premises. Accumulated data from studies in Cyprus, New Brunswick, Tasmania and Turkey now indicate that it is probable that all massive sulphide deposits of volcanic-sedimentary origin are likely to have similar haloes of hundreds of metres in extent for one or more of the following combinations of elements: Ca-Mg, Na-K, Fe-Mn, Pb-Zn, Cu-Zn, Ni-Co.

A generalized schematic illustration of characteristic major element haloes to be expected around volcanic-sedimentary massive sulphide deposits is shown in Fig. 5. The extent and distribution of the haloes in the footwall are consistent with Sato's (1972, 1973) model for the genesis of these types of deposits. Sato's model postulates a dense brine source (deposit *A*, Fig. 5) which may flow from the source area to a nearby depression (deposit *B*, Fig. 5). These differences are also reflected in the hanging-wall anomalies which, in terms of recognizable single element haloes, are much more restricted in deposit B-type situations; however, multielement haloes extending hundreds of metres into the stratigraphic hanging wall of B-type deposits have been recognized through multivariate statistical analysis (Whitehead and Govett, 1974).

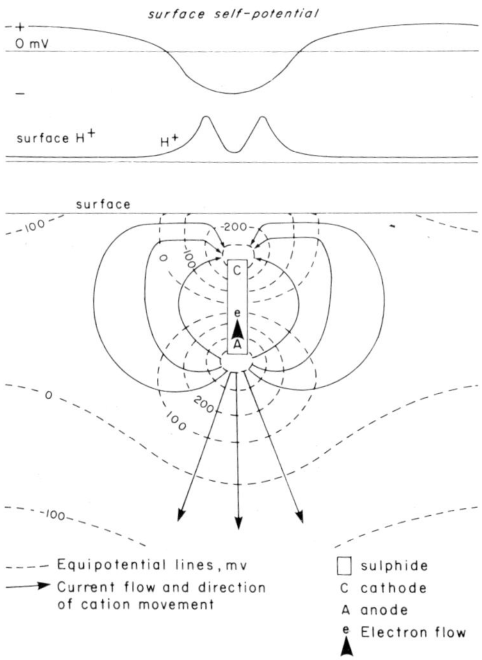

Fig. 6. Theoretical electrochemical model for development of self-potential anomalies around a sulphide conductor, and theoretical distribution of H^+ in surface soils due to electrochemical processes. (Derived from models proposed by Sato and Mooney, 1960; Govett, 1973, 1976b; Bølviken and Logn, 1975; Govett et al., 1976.)

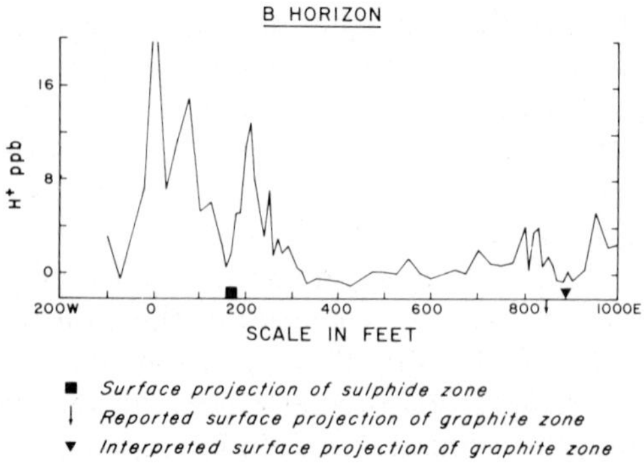

Fig. 7. Distribution of H^+ in soil slurry (1.0 g in 100 ml H_2O), White Lake, Flin Flon, Manitoba. Vertical sulphide zone (Zn-Cu) at a depth of 100 ft (30 m); overburden 25 ft (7.5 m) glacial till with 4 ft (1.2 m) glacio-fluvial clay. (From Govett, 1976b.)

Deeply buried sulphide deposits beneath thick transported overburden (e.g., glacial drift in Canada, especially where there is varved clay) and sulphide bodies permanently below the water table, are generally regarded as poor risks for the use of soil geochemistry; the geochemical response by conventional techniques is usually very weak to non-existent. In glaciated terrain, a means of overcoming this problem has been deep sampling of basal till. A theoretical electrochemical model of dispersion (Govett, 1973; Bølviken and Logn, 1975) provided a hypothetical mechanism for dispersion of elements to surface soils even from deeply buried and blind sulphide deposits beneath the water table, provided that there is electrical conductivity between the sulphide and the surface. The H^+ ion, and, to a lesser extent, the OH^- ion, by virtue of their availability and mobility, should give the most clearly defined response to electrical potential gradients. As shown in Fig. 6, this should result in twin "rabbit-ear" peaks of high H^+ concentration in soils over a deeply buried sulphide deposit where electrochemical dispersion is the dominant mechanism.

Some recent work over a sulphide zone in Canada at a depth of 100 ft (30 m) beneath 25 ft (7.5 m) of glacial overburden that includes an "impermeable" varved clay horizon shows precisely the theoretically predicted "rabbit-ear" H^+ anomaly (Fig. 7). If the proposed mechanism is correct, then any deeply buried sulphide deposit that is an electronic conductor and has electrical conductivity to the surface should be detectable through characteristic patterns for H^+.

THE PROBLEM OF POLICY

Inevitably, the greatest long-term problem in mineral exploration is the detection of deeply buried deposits; the techniques for locating them will have to be largely geophysical and geochemical, albeit guided by improved geological understanding and interpretation. This is a challenge that exploration geochemists must accept. Although there is a constant need for new data from carefully executed case studies, and empirical innovation must continue to contribute to progress, I firmly believe that exploration geochemistry can only continue to develop and provide new techniques through improved understanding of the processes of element dispersion. This demands more — much more — fundamental research than is currently being done. Exploration geochemistry, more than most disciplines, is interdisciplinary in nature. It has much to gain, particularly in research on processes, from a more determined effort to enlist the knowledge of the chemist, the soil scientist, the botanist and the forester.

To identify a problem and suggest a general solution, as I have been doing here, is easy. However, something rather more organized is necessary than exhortations to special groups such as this if noteworthy progress is to be made. Of the three chief components of modern applied science — government, industry and the university — the first two have a direct, although different, interest in ensuring the continued discovery of exploitable mineral deposits. All three have roles to play in developing the appropriate techniques; considerably more progress, and efficient use of available human and material resources, would result from a greater degree of coordination of effort between the three. Although all are capable of research activities and all could mount geochemical surveys, each normally has a somewhat different objective that leads to a different approach — and each has a different *optimum* capability. To achieve the necessary degree of collaboration will require a change in many individual attitudes. The common attitude of the university academic that the geochemical work of industry is unscientific is matched by industry's view that the academic is divorced from reality — while the government scientist regards both with an aloof tolerance in many cases. Worse than this is the too prevalent view in universities that exploration geochemistry is not a respectable academic pursuit and in industry that research is not a defensible investment.

Given the fundamental dependence of modern industrialized society upon mineral raw materials, it is astonishing how little support research in the earth sciences generally receives relative to other sciences. For example, the National Research Council of Canada awards for 1976 in the earth sciences average $9846 to 435 research workers, while in biology the average award was $ 10,625 to each of 1224 workers; in chemistry it was $ 15,027 to each of 542, and in engineering it was $ 10,193 to each of 1345. Thus, not only is the total amount of support to earth sciences as a whole low in absolute terms, the amount awarded to individual research workers in the earth sciences is, on average, the lowest in the general field of science and engineering. The assumption seems to be that somehow the geologist, on his own, will provide the raw materials — and support goes to research in engineering and chemistry on the processing and fabrication of the raw materials.

Even more astonishing, indeed alarming, is the very small proportion of total earth science expenditure on exploration geochemistry research, especially when compared to the relatively large expenditures made by industry (and by many governments) on geochemical exploration surveys to find mineral resources. Although scarcely credible, it seems there is not a single Department of Applied or Exploration Geochemistry in the U.S.A., Canada, Australia or South Africa - four of the "big five" mineral producers; indeed, in the English-speaking world I know of only one applied geochemistry department (Professor Webb's group at Imperial College, London). Given this background, the achievements of exploration geochemistry are remarkable and a tribute to all engaged in the subject.

The situation must be changed, especially in this era when the mining industry is under threat in a number of countries. National science policies must be formulated that encourage the development of research in priority areas — and I would identify mineral exploration as a priority area — without diminishing freedom of individual choice and initiative. In terms of our interest — exploration geochemistry — universities must encourage the establishment of teaching and research departments in exploration geochemistry; governments must increase research support and provide appropriate tax relief to companies that are interested in conducting or supporting research. In an applied subject, the participation of industry is essential; industry should be in the forefront in identifying problems requiring solutions, and universities and governments should be willing to work with industry — and together — to assist in providing solutions to the problem.

Progress demands improved communication between exploration geochemists in the different estates; perhaps the Association of Exploration Geochemists has an important role to play here as an umbrella organization. Let me quote to you from Article I of our Constitution and Bylaws:

> "The objects and purposes [of A.E.G.] are . . . to foster the common scientific interests of exploration geochemists and to facilitate the acquirement of professional knowledge and information relevant to the science of exploration geochemistry and to promote the interchange thereof among its members, to encourage research and development of geochemical exploration methods in universities . . ."

I would consider it appropriate for members of the Association of Exploration Geochemists to form national committees with representatives from industry, government and the universities to consider how best to implement the objectives of the Association in their own country, especially in the context of better using our science to ensure continued discovery of vital mineral resources. Australia, with the impressive growth of its mineral industry (notwithstanding its recent period of gloom), its wide variety of geological conditions, its mineral wealth and its range of climatic conditions, might appropriately form the first such national committee.

In concluding, I should like to state that I do not believe that the world has any fear of an inherent inadequacy of mineral resources; there are likely to be major problems in converting these resources into reserves unless the available *human* resources in the earth sciences are more wisely used and a greater proportion of *financial* resources is devoted to research in mineral exploration. The exploration geochemist has a vital role — indeed, a moral duty — in assuring the world its raw materials. There is a full programme of exciting exploration geochemistry papers to follow that will, I'm sure, demonstrate that exploration geochemists are well-prepared to accept this challenge. I wish you a very successful meeting.

REFERENCES

Bølviken, B. and Logn, Ø., 1975.. An electrochemical model for element distribution around sulphide bodies. In: I.L. Elliott and W.K. Fletcher (Editors), Geochemical Exploration 1974. Elsevier, Amsterdam, pp. 631—648.

Derry, D.R., 1968. Man and minerals . . . exploration. Can. Inst. Min. Metall. Bull., 61 (670): 200—205.

Derry, D.R., 1970. Exploration expenditure, discovery rate and methods. Can. Inst. Min. Metall. Bull., 63(695): 362—366.

Govett, G.J.S., 1973. Differential secondary dispersion in transported soils and post-mineralization rocks: an electrochemical interpretation. In: M.J. Jones (Editor), Geochemical Exploration 1972. Institution of Mining and Metallurgy, London, pp. 81—91.

Govett, G.J.S., 1976a. The development of geochemical exploration methods and techniques. In: G.J.S. Govett and M.H. Govett (Editors), World Mineral Supplies — Assessment and Perspective. Elsevier, Amsterdam, pp. 343—370.

Govett, G.J.S., 1976b. Detection of deeply buried and blind sulphide deposits by measurement of H^+ and conductivity of closely spaced surface soil samples. J. Geochem. Explor., 6: 359—382.

Govett, G.J.S. and Govett, M.H., 1974. The concept and measurement of mineral reserves and resources. Resour. Policy, 1: 46—55.

Govett, G.J.S., Goodfellow, W.D. and Whitehead, R.E.S., 1976. Experimental aqueous dispersion of elements around sulfides. Econ. Geol., 71: 925—940.

Govett, M.H., 1976. Geographic concentration of world mineral supplies, production and consumption. In: G.J.S. Govett and M.H. Govett (Editors), World Mineral Supplies — Assessment and Perspective. Elsevier, Amsterdam, pp. 99—145.

Hawkes, H.E., 1976. The early days of exploration geochemistry. In: G.J.S. Govett (Editor), Exploration Geochemistry in the Appalachians. J. Geochem. Explor., Spec. Issue, 6: 1—11.

Martin, H.L., Cranstone, D.A. and Zwartendyk, J., 1976. Metal mining in Canada to the year 2000. In: G.J.S. Govett and M.H. Govett (Editors), The Canadian Minerals Industry. Resour. Policy, Spec. Issue, 2: 11—24.

McAllister, A.L., 1976. Price, technology, and ore reserves. In: G.J.S. Govett and M.H. Govett (Editors), World Mineral Supplies — Assessment and Perspective. Elsevier, Amsterdam, pp. 37—63.

Roscoe, W.E., 1971. Probability of an exploration discovery in Canada. Can. Inst. Min. Metall. Bull., 64(707): 134—137.

Sato, T., 1972. Behavior of ore-forming solutions in sea water. Mining Geol., 22: 31—42.

Sato, T., 1973. A chloride complex model for Kuroko mineralization. Geochem., J. 7: 240—270.

Sato, M. and Mooney, H.M., 1960. The electrochemical mechanism of sulfide self-potentials. Geophysics, 25: 226—249.

Wahl, J.L., 1977. Rock geochemistry and ore genesis at Heath Steele and Key Anacon massive sulphide deposits. Ph.D. Thesis, University of New Brunswick, Fredericton, N.B.

Whitehead, R.E.S. and Govett, G.J.S., 1974. Exploration rock geochemistry — detection of trace element haloes at Heath Steele Mines (N.B., Canada) by discriminant analysis. J. Geochem. Explor., 3: 371—386.

Journal of Geochemical Exploration, 8 (1977) 17—29

APPLICATION OF GOLD COMPOSITIONAL ANALYSES TO MINERAL EXPLORATION IN THE UNITED STATES

J.C. ANTWEILER and W.L. CAMPBELL

U.S. Geological Survey, Denver Federal Center, Denver, Colo. 80225 (U.S.A.)

(Revised version accepted December 30, 1976)

ABSTRACT

Antweiler, J.C. and Campbell, W.L., 1977. Application of gold compositional analyses to mineral exploration in the United States. J. Geochem. Explor., 8: 17—29.

Native gold is a mineral composed of Au, Ag and Cu in solid solution and it usually contains one or more trace metals as lattice impurities, as mineral inclusions, in grain boundaries or in surface coatings. Alloy proportions of Au, Ag and Cu, together with certain other elements, can be thought of as constituting a gold "signature". Gold is associated with a great variety of ore deposits and has characteristic signatures for each of several types of ore deposits.

Signatures for gold derived from igneous-metamorphic, hypothermal, mesothermal and epithermal deposits reflect conditions of ore formation by their content of Ag, Cu and characteristic associated elements. At higher temperatures of ore formation, gold has low Ag and high Cu content, and Bi and Pb are the most abundant trace elements. But at lower temperatures of ore formation, Ag is high, Cu is low, and Pb is the most abundant trace element. The same trend in gold signatures is observable in gold mining districts, such as Central City, Colorado, where zoning as shown by mineral assemblages indicates ore deposition at progressively lower temperatures as the distance from a central high-temperature zone increases.

The signatures of gold may be useful in searching for porphyry Cu deposits. Signatures from Butte (Montana), Mineral Park (Arizona) and Cala Abajo (Puerto Rico), on the basis of limited sampling, are similar and distinctive. They are characterized by a similar assemblage of trace elements and are relatively high in both Ag and Cu.

Another application of gold compositional data is in tracing placer gold to its bedrock source. For example, the Ag content of placer gold in the Tarryall district of Colorado differed from that of nearly all of the bedrock sources of gold found by early prospectors. However, one lightly prospected area peripheral to the Tertiary quartz monzonite stock at Montgomery Gulch contains gold with a Ag content similar to that of the placer gold. This area is the most likely source of the gold in the productive placers and may be a potential exploration target.

Gold signatures may be useful in prospecting for metals other than gold. Several metals of low crustal abundance — notably Sn, W, Mo and the Pt group metals — are detected in analyses of some gold samples and may indicate economic deposits of these metals.

INTRODUCTION

Mining camps resulting from gold discoveries were often the vanguard of permanent settlement in the western United States. Even though the production of gold from some of the boom camps that mushroomed out of the gold discoveries was later dwarfed by the production of other metals such as Cu at Butte, Montana, or Ag at Coeur d'Alene, Idaho, it was the discovery of gold that led to great hidden mineral wealth. Recently, gold has been suggested as a pathfinder element for porphyry Cu deposits (Learned and Boisson, 1973). Our objective in this paper is to point out that the composition of gold may be a useful aid in looking for ore deposits.

The mineral, native gold, is a natural alloy that is nearly always composed of Au, Ag and Cu in solid solution, plus one or more of at least 22 other elements that may occur in surface or grain boundary coatings, in mineral inclusions, in lattice impurities, or in solid solution (Antweiler and Sutton, 1970). Previous studies (summarized by Gay, 1963, and Jones and Fleischer, 1969) have shown that the composition of native gold varies greatly, not only from one deposit to another, but even within grains from the same deposit (Head, 1935; Eales, 1961; Stumpfl and Clark, 1965; Desborough, 1970). The concentration of Ag and other elements in gold varies so greatly that divergent and sometimes opposing conclusions have been reached by different workers who have investigated similar aspects of gold composition. The net effect is that the potential of gold compositional studies has not been fully realized.

SAMPLING AND ANALYTICAL METHODS

Most of the samples of gold discussed below were obtained by panning alluvium or oxidized ore obtained from veins, mine dumps or prospects. Some were purchased (three from Butte, Montana, one from Hahns Peak, Colorado), and a few were donated (Little Jonny Mine, Colorado, by R.R. Redington; Dixie Mine, Colorado, by William Janes; three samples from Round Mountain, Nevada, by Copper Range Company). Procedures other than panning to recover or purify gold, such as amalgamation, acid-leaching or roasting, were avoided because such procedures alter the composition of gold (Campbell et al., 1973). Foreign minerals were removed by handpicking under a binocular microscope.

The samples were analyzed by a six-step semi-quantitative spectrographic procedure described by Mosier (1975). Supplemental information was obtained by electron microprobe analyses using the methods and equipment described by Desborough (1970).

To minimize the difficulties caused by natural variations in gold composition, five or more analyses were made to determine a signature except in the study of the porphyry Cu deposits, no more gold was available for that study, but more analyses are desirable. Our experience has shown that five analyses are usually adequate to establish confidence in *averages* of the individual determinations. If values for Ag and Cu are widely divergent, more analyses should be made.

The Au content was determined empirically by subtracting from 100 the combined percentages for Ag and Cu. Such an empirical determination was found to be satisfactory. Moreover, it is justifiable, because the mineral, native gold, is usually a solid solution, or alloy, composed of the three elements Au, Ag and Cu. Other elements in gold occur as contaminants in the form of lattice impurities, grain boundary and surface coatings, or discrete mineral inclusions and are not usually an intrinsic part of the alloy. Electron microprobe analyses verify in spot analyses that Au + Ag + Cu = 100%.

GOLD SIGNATURE

We designate the term gold signature as a summary expressed in averages of the important data obtained by analysis of gold; thus the signature consists of percentages and ratios of Au, Ag and Cu; and the presence of certain trace elements that occur consistently in replicate analyses. The trace elements that we include are As, Bi, Cd, Co, Cr, Mo, Ni, Pb, Pd, Pt, Sb, Sn, Te, W and Zn. We do not include Al, Ca, Fe, Mg, Mn, Si or Ti because those elements occur in almost all samples as contaminants.

GOLD SIGNATURES IN RELATION TO CONDITIONS OF ORE DEPOSITION

Gold was collected from several different types of ore deposits to determine the effect of diverse origins and geologic setting on the signature. In Table I hydrothermal deposits that can be confidently classified according to temperature-pressure-depth relationships show that Au and Cu are highest when ore was deposited at high temperature and at great depth as in the Diana mine from the Atlantic City district, Wyoming. Ag, on the other hand, is highest in the epithermal deposit, and decreases as temperature and pressure increase. Arsenic is the most abundant trace element in the signature of the hypothermal deposit, and Bi, Ni and Pb are integral parts of the signature. Bi takes over as the most abundant trace element in the Little Jonny, but gives way to Pb in the Seafoam Mine, and disappears entirely from the signature of Hahns Peak gold as conditions of lower temperature and pressure prevail.

In Table II, signatures of gold are presented from four deposits whose classification is somewhat ambiguous. The Cable Mine gold was deposited with magnetite and manganese minerals. Field evidence suggests that ore metals were deposited by metasomatic reactions when hot hypogene fluids penetrated limestone. Gold from this mine in the early days is said to have contained as little as 0.25% Ag, but our samples contain considerably more than that. Nevertheless, the signature is distinctive — very low Ag, high Cu, and an interesting suite of accessory elements.

The King Solomon Mine also represents a high-temperature deposit, but

TABLE I

Gold signatures from gold ore deposits, I

Location	Number of samples	Number of analyses	Au(%)	Ag(%)	Cu(%)	Au/Ag	Au/Cu	Characteristic trace elements (in order of abundance)
Diana mine, Wyoming (hypothermal)	1	6	92.5	7.4	0.05	12.5	1850	As, Bi, Ni, Pb
Little Jonny Mine, Colorado (upper mesothermal)	1	5	89.6	10.3	0.025	8.7	3600	Bi, Pb
Seafoam Mine, Idaho (lower mesothermal)	5	27	77.6	22.3	0.0165	3.5	4700	Pb, Bi, Sb
Hahns Peak region, Colorado (epithermal)	3	10	66.6	33.3	0.0017	2.0	40,000	Pb, Sb

TABLE II

Gold signatures from gold ore deposits, II

Location	Number of samples	Number of analyses	Au(%)	Ag(%)	Cu(%)	Au/Ag	Au/Cu	Characteristic trace elements (in order of abundance)
Cable Mine, Montana (pyrometasomatic)	3	16	97.5	2.3	0.23	42.4	425	Bi, Pb, Te, Zn, Sb
King Solomon Mine, Colorado (contact metamorphic)	1	6	89.1	10.9	0.056	8.3	1600	Bi, Pb, Zn, Te
Round Mountain District, Nevada (epithermal)	5	15	75.5	24.4	0.0027	3.1	28,000	Pb, Sn, Sb, As
Dixie Mine, Colorado (supergene enrichment of mesothermal veins)	1	18	80.7	19.2	0.0012	4.2	67,000	Pb

one which was probably formed near the surface. A quartz monzonite intrusive penetrated sedimentary rocks consisting largely of sandstone and shale with lenses of limestone. Ore metals were deposited in the sediment (dry rocks) near the intrusive contact, particularly in the limestone lenses. The gold signature supports a high-temperature origin, and suggests similarities to as well as differences from the Cable deposit.

Epithermal gold-bearing quartz veins in rhyolite were mined at Round Mountain, Nevada. The gold signature is unusual in that Sn and As are included. A possible explanation is that two periods of mineralization occurred, the second of which introduced the gold and may have remobilized Sn and As from older veins in granite.

The fourth mine in this group, the Dixie, is characterized by supergene deposition of gold that occurred in hypogene veins. Mineralogically, the Dixie represents a position between mesothermal and epithermal deposits. Electron microprobe analyses show that Ag in some grains ranges from 12 to 40%, with the low values often on the outer edges of the grains.

Gold composition and zoning of hydrothermal ore deposits

Knowledge of zoning patterns around hydrothermal ore deposits can be useful in exploration. In an ideal case, concentric zones of different minerals surround the magmatic source. In the usual case, zones overlap because of many factors; these might include discontinuous surges of ore fluids, porosity, permeability and composition of enclosing rock, configuration of the magma chamber, and the existence of multiple magma chambers.

Variation of the Ag content of gold was observed to correlate with zoning patterns more than 50 years ago (Pryor, 1923). We have attempted to correlate gold signatures with zoning patterns in the Central City gold mining district, Gilpin County, Colorado (Sims et al., 1963) as shown in Fig. 1.

A number of our samples show that the Ag and Cu content of gold samples are related to the zonal position from which they came. Most of the mines of the Central City district with large production records are located near the boundary of the central and intermediate zone. The Ag content of gold from these mines ranges from 7.7 to 16.1% and averages about 9.3%; Cu ranges from 235 to 750 ppm with an average of about 390 ppm. Samples on the outer edge of the intermediate zone or in the peripheral zone, however, range up to as high as 50% Ag, and down to as little as 40 ppm Cu. The average Ag content is about twice that of the samples from the highly productive zone, whereas the average Cu content is less than half.

SIGNATURES OF GOLD IN PORPHYRY COPPER DEPOSITS

Although gold is commonly recovered as a by-product of large copper-mining enterprises, the tenor of gold in Cu ore from porphyry deposits is often so low that the gold-bearing mineral species are not known. For exam-

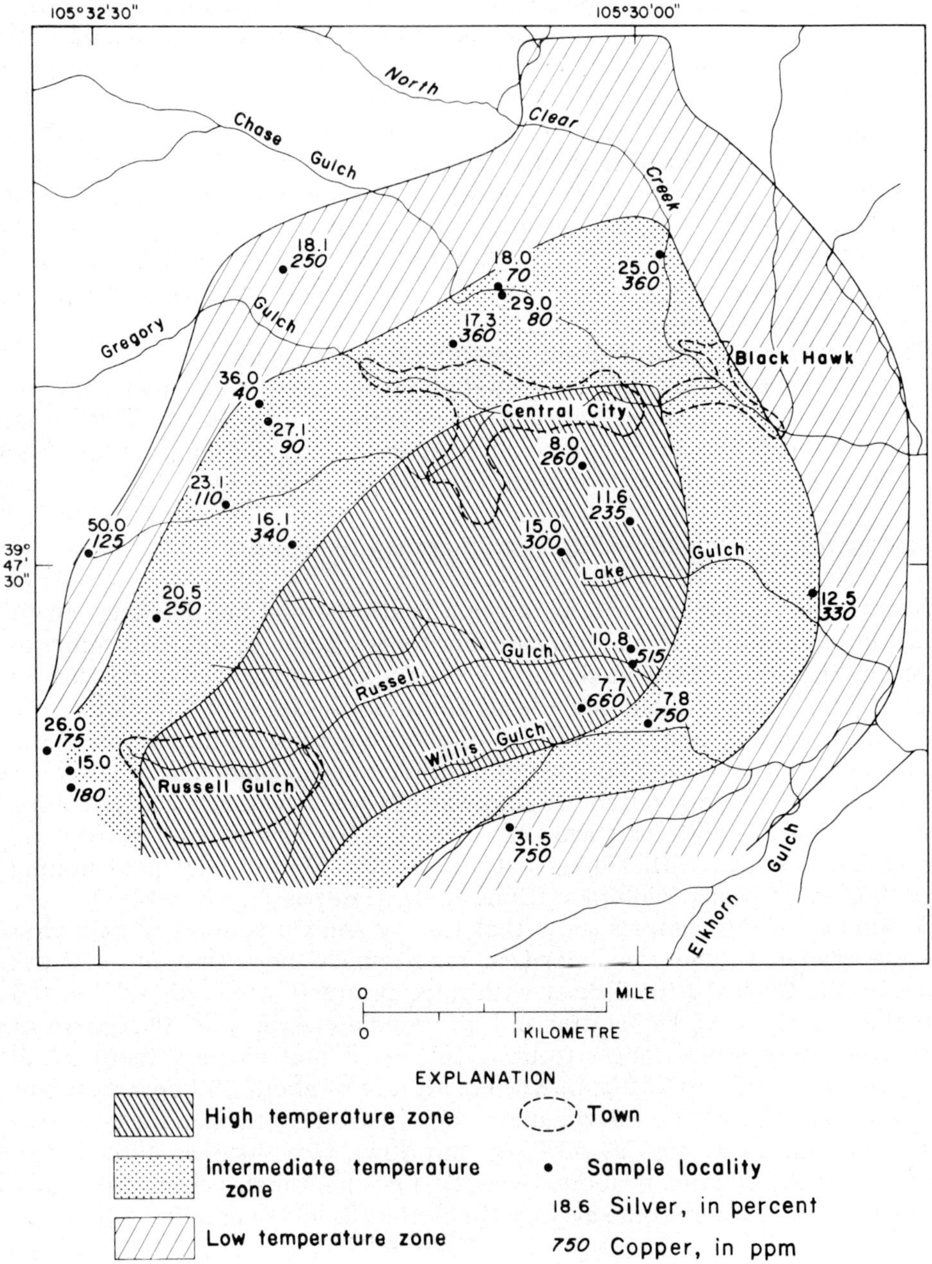

Fig. 1. Map of Central City district showing zoning of mineral deposits, and content of Ag and Cu in samples of gold at various localities. (Adapted from Sims et al., 1963.)

ple, geologists at Butte, Montana and Mineral Park, Arizona were unable to furnish us with gold-bearing specimens and were uncertain as to the mode of occurrence of gold in the Cu ore. Our gold from Butte came from two sources: purchases of gold said to have been recovered in early-day placer mining, and gold that we obtained by panning material from stream drainages and from dumps of inactive mines. We have attempted to locate these samples in Fig. 2, which is an adaptation of the diagram proposed by Meyer et al. (1968), to show their relation to the three-dimensional zoning pattern developed by Anaconda Company geologists. This diagram shows that the peripheral and intermediate zones are several thousand feet away from the central zone of disseminated Cu ore.

Localization of gold in the lower-temperature zones of a porphyry Cu deposit, of course, is not unusual, but the signature of gold from Butte and the other deposits in Table III, is unusual when compared with the signatures in Tables I and II. The Ag content of Butte gold is similar to that in gold from the epithermal deposit at Round Mountain, Nevada, but the Cu content is the same as that in gold from the hypothermal Diana Mine. Moreover, the trace element suite is different.

Sampling at Butte was of a reconnaissance nature, but there and at Mineral Park, Arizona, porphyry Cu was even more limited. Localities where we obtained gold are plotted on the zoning pattern map of Dings (1951) and modified by Eidel et al. (1968) (Fig. 3). Very little gold occurs in the Cu-Mo zone near the Mesozoic (?) Ithaca Peak Granite. However, gold does occur at a distance of a few kilometres in the lower-temperature Zn-Pb-Ag and Ag-Au zones, and has a signature similar to that at Butte (Table III).

Our information on the Cala Abajo, Puerto Rico, porphyry Cu deposit is based on the analysis of only one sample of gold. We recovered this gold by

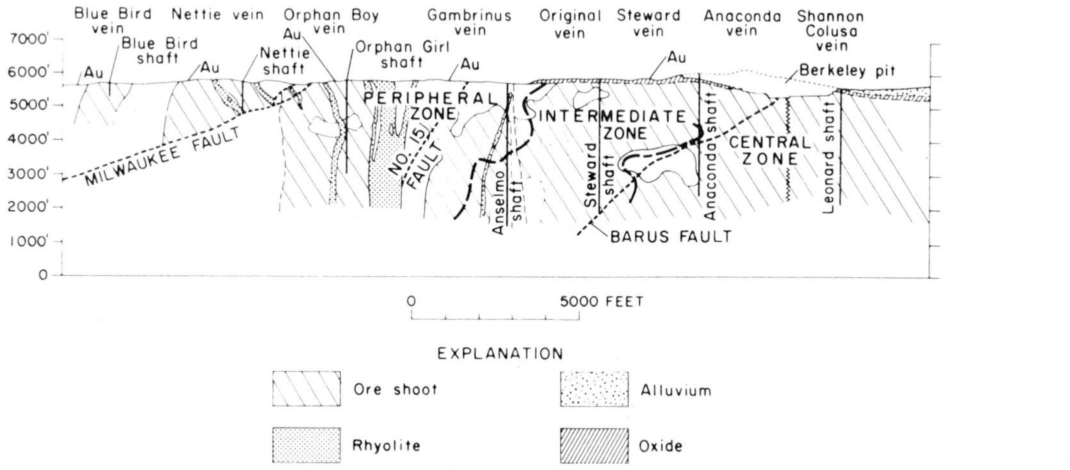

Fig. 2. Longitudinal section of the Anaconda vein at Butte, Montana, showing ore shoots and approximate positions of mineral zones. Approximate localities where gold was collected are designated by symbol Au. (Adapted from Meyer et al., 1968.)

TABLE III

Gold signatures from porphyry Cu deposits

Location	Number of samples	Number of analyses	Au(%)	Ag(%)	Cu(%)	Au/Ag	Au/Cu	Characteristic trace elements (in order of abundance)
Butte, Montana	5	11	74.0	25.9	0.05	2.9	1500	Pb, Bi, Sb, Zn, Sn, Cr, Ni
Mineral Park, Arizona	5	5	70.3	29.6	0.04	2.4	1800	Pb, Bi, Sb, Zn, As, Te
Cala Abajo, Puerto Rico	1	1	74.8	25.1	0.15	3.0	500	Pb, Bi, Sb, Zn, As, Te

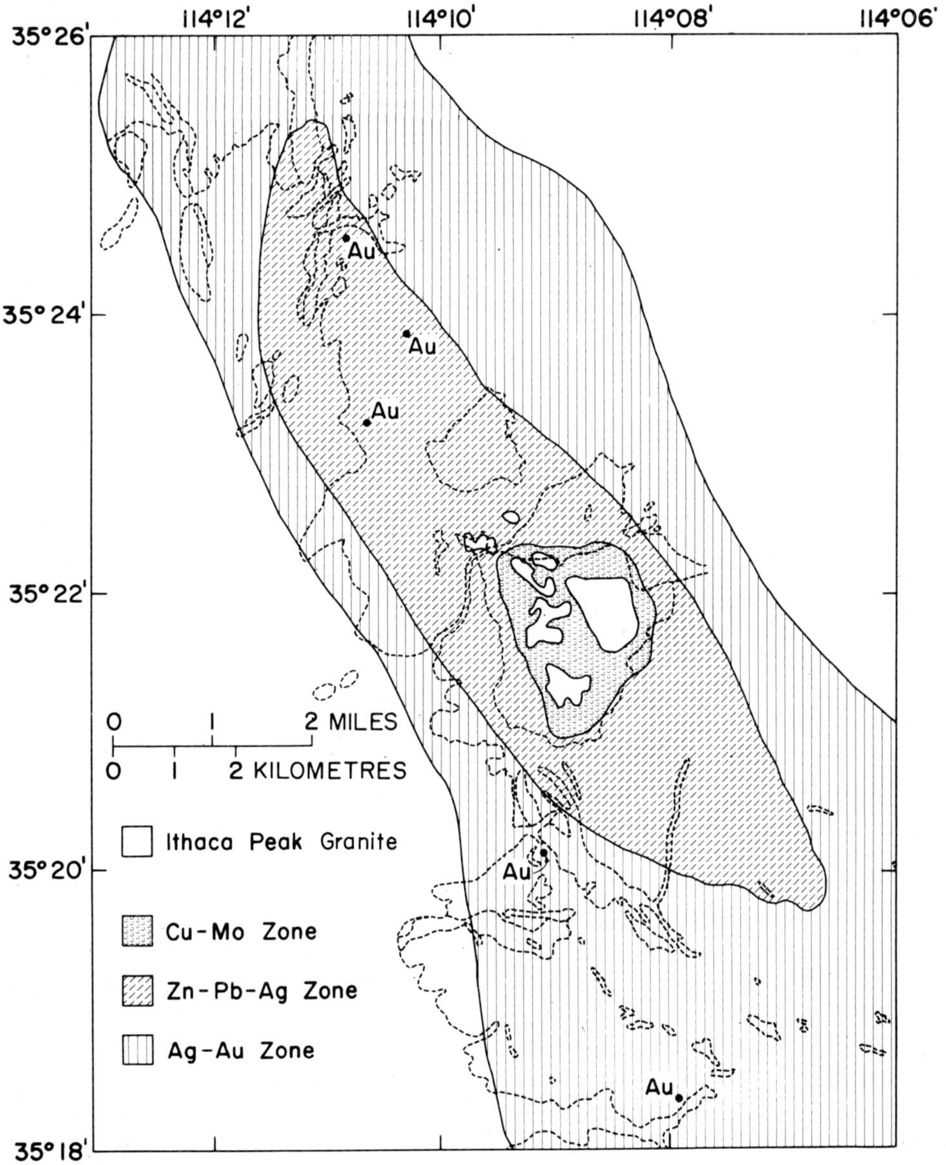

Fig. 3. Map showing zoning around the Mineral Park, Arizona, porphyry Cu deposit and locations where samples of gold were obtained. (Adapted from Eidel et al., 1968, and Dings, 1951.)

panning weathered soil (collected by R.E. Learned) overlying the disseminated Cu deposit. We have not yet had the opportunity to collect additional gold in Puerto Rico, but the similarity of the gold signature in the Cala Abajo deposit with signatures from Butte and Mineral Park prompt us to include its signature here.

Comparison of gold signatures from these three deposits, although based on seventeen samples, indicates that an exciting application of gold signatures may be in the search for porphyry Cu deposits. A much more exhaustive sampling program is warranted to: (1) compare more fully gold signatures for island arc and continental porphyry Cu deposits, (2) determine whether gold signatures change significantly with regard to the spatial distribution of gold around a porphyry deposit, (3) determine whether gold from a porphyry Mo deposit also has a distinctive signature.

GOLD COMPOSITION AS A GUIDE TO THE BEDROCK SOURCE OF PLACER GOLD

Another application of gold compositional data is that of tracing placer gold to its bedrock source. The Tarryall mining district, Park County, Colorado, affords an example. Placer gold was discovered along Tarryall Creek in 1859. Placers were mined along the stream and its tributaries for several years, but prospectors have hunted in vain for the bedrock source. Unfortunately, rock debris from perennial landsliding and three episodes of Pleistocene glaciation (Singewald, 1947) obscures nearly all the bedrock except on the steep mountain slopes. These slopes consist of sedimentary rocks of Pennsylvanian and Mississippian age which are metamorphosed and hydrothermally altered in the vicinity of numerous sills and dykes of porphyry. Prospectors explored many tiny gold-bearing veins in these rocks with pits, tunnels and shafts, but they were not able to produce significant amounts of gold.

Production records compiled by the U.S. Bureau of Mines show that gold from Tarryall placers averaged 9.1% Ag. Our samples of placer gold as shown in Fig. 4 carry from 4.3 to 19.9% Ag, but gold from the areas where production was greatest has 9—10% Ag. The evidence thus suggests that a potentially productive bedrock source will have gold whose Ag concentration is 9—10%. When Singewald (1942) mapped the geology of the area, he discovered a few small outcrops of quartz monzonite in the transported rock debris in Montgomery Gulch. These outcrops, together with poorly exposed sills of similar composition on the north slope of Mt. Silverheels, caused him to infer the existence of a stock whose approximate outline is shown in Fig. 5. Singewald also outlined the area of intense alteration characterized by epidote, magnetite, pyrite and garnet in the Palaeozoic and Mesozoic sedimentary rocks.

Samples of gold were collected from many of the old prospects and a wide range of Ag was found in these samples as shown in Fig. 5. In addition, particular attention was devoted to areas where the stock was evidently only

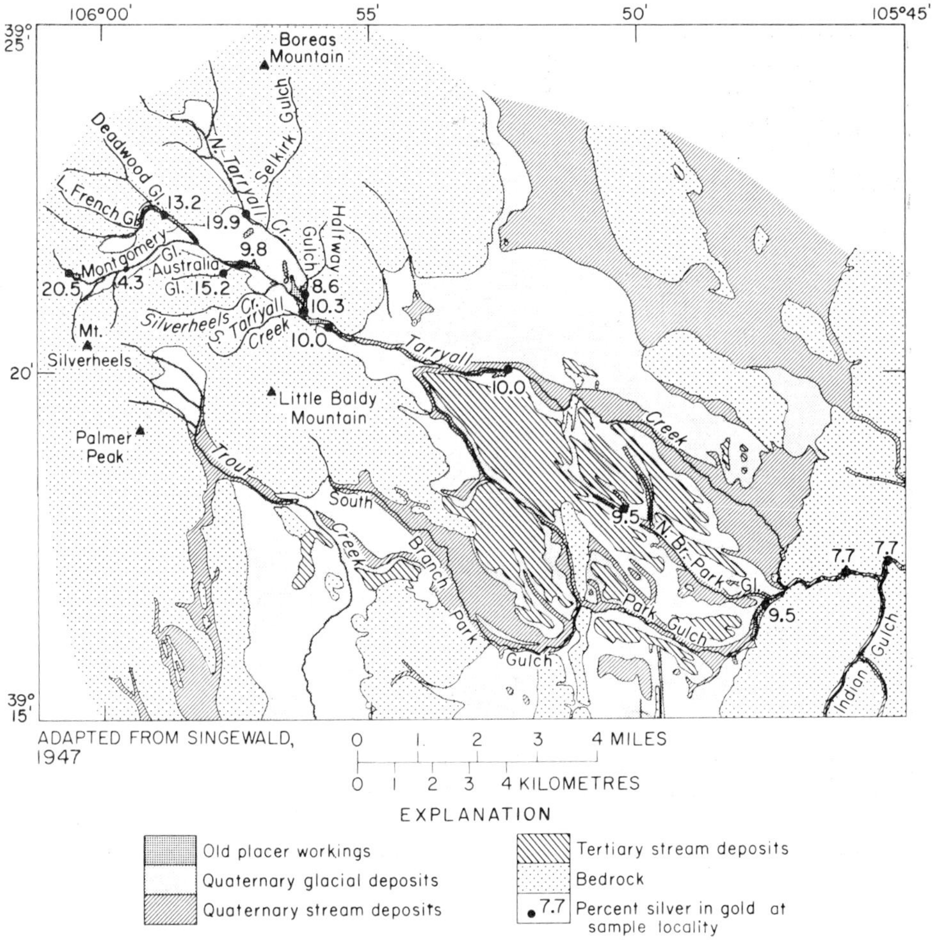

Fig. 4. Map showing Ag content of placer gold, placer workings and their related physiographic features, Park County, Colorado.

thinly veneered or had been recently exposed by new road construction. Several mineralized areas that apparently had been little prospected yielded gold of the composition found in the productive placers. The most promising areas for further exploration are those where gold containing 7—15% Ag can be obtained. Several such areas are enclosed by the ellipse that extends approximately from French Creek to the steep slopes of Mt. Silverheels. These areas for the most part overlie or are peripheral to the stock.

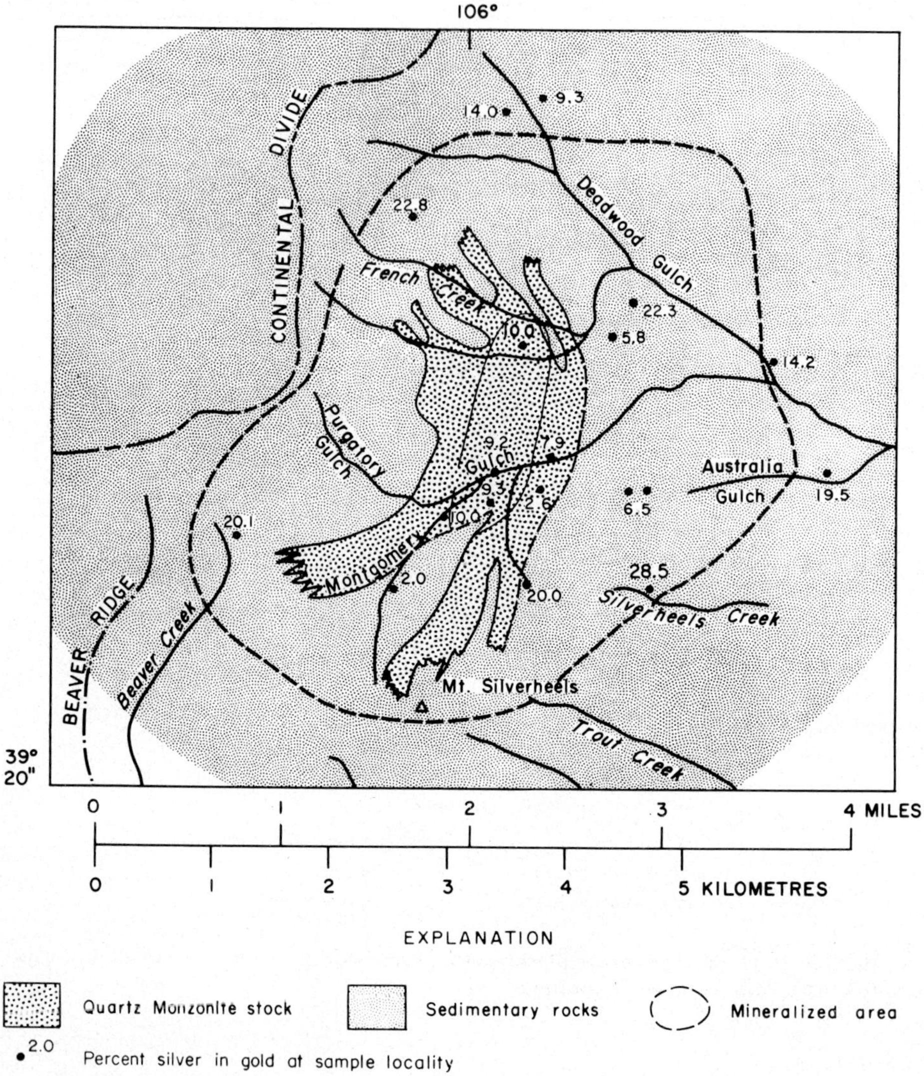

Fig. 5. Map showing Ag content of bedrock gold related to quartz monzonite intrusive, Park County, Colorado. (Adapted from Singewald, 1947.)

GOLD SIGNATURES AS A GUIDE TO OTHER METALS

Gold signatures may also be useful in searching for deposits of metals other than gold. The detection of elements such as Mo, Pt, Sn or W suggests that deposits of those metals may occur near the bedrock source of the gold. Such a possibility was proposed more than 30 years ago by Warren and Thompson (1944) and is still valid. Pt group metals, for example, were found in gold from the New Rambler platinum mine in Wyoming. Also, Sn, W and Mo were detected in gold panned downstream from the Urad molybdenum mine in Colorado. A discovery of Mo, Pt, Sn or W could result from recognition of such elements in the gold signature.

REFERENCES

Antweiler, J.C. and Sutton, A.L., Jr., 1970. Spectrochemical analyses of native gold samples. U.S. Geol. Surv. Rep., USGS-GD-70-003, 28 pp. (available from U.S. Department of Commerce National Technical Information Service, Springfield, Va. 22151 as Rep. PBI-94809).

Campbell, W.L., Mosier, E.L. and Antweiler, J.C., 1973. Effects of laboratory treatments on silver and other elements in native gold. U.S. Geol. Surv. J. Res., 1: 211—220.

Desborough, G.A., 1970. Silver depletion indicated by microanalysis of gold from placer occurrences, western United States. Econ. Geol., 65: 304—311.

Dings, M.G., 1951. The Wallapai mining district, Cerbat Mountains, Mohave County, Arizona. U.S. Geol. Surv., Bull., 978-E: 123—163.

Eales, H.V., 1961. Fineness of gold in some Southern Rhodesian gold mines. Trans. Inst. Min. Metall., 71: 49—73.

Eidel, J.J., Frost, J.E. and Clippinger, D.M., 1968. Copper-molybdenum mineralization at Mineral Park, Mohave County, Arizona. In: J.D. Ridge (Editor), Ore Deposits of the United States, 1933—1967. American Institute of Mining, Metallurgical and Petroleum Engineers, New York, N.Y., pp. 1258—1281.

Gay, 1963. A review of the geochemical characteristics of gold in ore deposits. Witwatersrand Univ. Econ. Geol. Res. Unit Inf. Circ., 12: 70 pp.

Head, R.E., 1935. Form and occurrence of gold in pyrite from a metallurgical standpoint: coated gold. Can. Min. J., 56: 517—521.

Jones, R.S. and Fleischer, M., 1969. Gold in minerals and the composition of native gold. U.S. Geol. Surv., Circ., 712: 17 pp.

Learned, R.E. and Boisson, R., 1973. Gold — a useful pathfinder element in the search for porphyry copper deposits in Puerto Rico. In: M.J. Jones (Editor), Geochemical Exploration, 1972. Institution of Mining and Metallurgy, London, pp. 93—103.

Meyer, C., Shea, E.P., Goddard, C.C., Jr. and Staff of the Anaconda Company, Butte, Montana, 1968. Ore deposits at Butte, Montana. In: J.D. Ridge (Editor), Ore Deposits of the United States, 1933—1967. American Institute of Mining, Metallurgical, and Petroleum Engineers, New York, N.Y., pp. 1373—1416.

Mosier, E.L., 1975. Use of emission spectroscopy for the semiquantitative analysis of trace elements and silver in native gold. In: F.N. Ward (Editor), New and Refined Methods of Trace Analys Is Useful in Geochemical Exploration. U.S. Geol. Surv. Bull., 1408: 97—105.

Pryor, T., 1923. The underground geology of the Kolar gold field. Trans. Inst. Min. Metall., 33: 95—135.

Sims, P.K., Drake, A.A., Jr. and Tooker, E.W., 1963. Economic geology of the Central City district, Gilpin County, Colorado. U.S. Geol. Surv., Prof. Paper, 359: 231 pp.

Singewald, Q.D., 1942. Stratigraphy, structure, and mineralization in the Beaver-Tarryall area, Park County, Colorado. U.S. Geol. Surv., Bull., 928-A: 1—44.

Singewald, Q.D., 1947. Gold placers and their geologic environment in northwestern Park County, Colorado. U.S. Geol. Surv., Bull., 955-D: 103—172.

Stumpfl, E.F. and Clark, A.M., 1965. Electron-probe microanalysis of gold-platinoid concentrates from south west Borneo. Inst. Min. Metall. Bull., 74(708) pt. 14, pp. 933—946.

Warren, H.V. and Thompson, R.M., 1944. Minor elements in gold. Econ. Geol., 39: 457—471.

Journal of Geochemical Exploration, 8 (1977) 31-48

TELLURIUM, A GUIDE TO MINERAL DEPOSITS

J.R. WATTERSON, G.B. GOTT, G.J. NEUERBURG, H.W. LAKIN and J.B. CATHRALL

U.S. Geological Survey, Denver Federal Center, Denver, Colo. 80225 (U.S.A.)

(Revised version accepted February 8, 1977)

ABSTRACT

Watterson, J.R., Gott, G.B., Neuerburg, G.J., Lakin, H.W. and Cathrall, J.B., 1977. Tellurium, a guide to mineral deposits. J. Geochem. Explor., 8: 31—48.

Te dispersion patterns are useful in exploring for different types of mineral deposits and in providing additional information about known ore deposits. The Te content of rocks is given for five mining districts in the western United States: Coeur d'Alene, Idaho; Robinson, near Ely, Nevada; Montezuma, Colorado; Crater Creek area, Colorado; Cripple Creek, Colorado. Many of the analyses were obtained by use of a new analytical method sensitive to 0.001 ppm Te.

The principal ore deposits in the Coeur d'Alene district, Idaho, are Pb-Zn-Ag replacement veins in Precambrian rocks of the Belt Supergroup. Te dispersion patterns show the outlines of the original mineral belts, the effects of intrusive events, the location of ore deposits, the displacements caused by post-ore faulting, and the borders of the 780-km^2 district.

The disseminated porphyry Cu deposits of the Robinson mining district, Nevada, are associated with Cretaceous quartz monzonite stocks that have intruded Palaeozoic carbonate rocks. Te is present in rock samples in concentrations as high as 10,000 ppm and forms a halo around the areas containing the Cu deposits.

The alteration zones in the porphyry Mo district near Montezuma, Colorado, are developed around several small Tertiary intrusions occurring along a regional shear zone. Te haloes reflect the locations of porphyry intrusives, individual deposits and their ore shoots, and the pattern and intensity of adjacent alteration. The Te content of soils over the Montezuma stock is higher than, and varies independently from, the Te content of adjacent outcrops. Soils generally contain more Te than adjacent outcropping rocks. Soil may collect gaseous Te compounds from mineral deposits.

The Crater Creek area is a northwestern extension of the Summitville mining district, Colorado. Te dispersion patterns radiate out from exposed Cu-Pb-Zn veins, from an outcrop of molybdenite stockwork veins and from associated iron-stained altered rock. Te haloes intensify exponentially with proximity to known ore and suggest the presence of Summitville-type chimney deposits.

Most of the gold- and silver-telluride ore in the Cripple Creek district, Colorado, is found in fracture fillings within a volcanic subsidence basin. Haloes of Au, Ag and Te all define the mineralized portions of the fissure veins.

INTRODUCTION

Te is one of the trace elements that have been found to occur in distinctive zones relative to some ore deposits. The zonal patterns vary depending on the depositional environments under which the ore deposits were formed. As a result of geochemical studies of five mining districts in the western United States, Te dispersion patterns have been found to define closely the gold telluride fissure veins in the Cripple Creek district of Colorado; to form a nearly continuous halo peripheral to the porphyry Cu ores in the Robinson district near Ely in east-central Nevada; to define a concentric zone around monzonite stocks that has produced 85% of the Ag, Pb and Zn ores of the Coeur d'Alene district in northern Idaho; and to reflect and extend beyond the alteration patterns associated with the ores of the Montezuma district and the Crater Creek area in Colorado.

GEOCHEMISTRY OF TELLURIUM

The hypogene chemistry of Te reflects the large size of the telluride ion and its extreme rarity in comparison to the sulphide ion. Because the telluride ion is 20% larger than the sulphide ion (2.21 Å vs. 1.84 Å), it does not substitute isomorphically in sulphide minerals to any appreciable extent (Czamanske and Hall, 1975). There are approximately 1.2×10^6 S atoms for each Te atom in the earth's crust (Goldschmidt, 1958). Similarly large concentration differences are probably present in ore-forming solutions and would also tend to prevent appreciable isomorphism between S and Te.

The bonding characteristics of the large telluride ion help to explain why telluride minerals form late in a mineral series, after the bulk of the S has been removed from solution to form sulphide minerals. Shcherbina and Lebedev (in Vlasov, 1966, p. 584) showed that Te and S have opposite affinities for the main telluride-forming metals (Fig. 1). These observations agree, as a trend, with the time sequence postulated for the deposition of sulphide and telluride minerals in a number of telluride vein deposits in Boulder County, Colorado (Kelly and Goddard, 1969, p. 129) and with estimations of the metal zoning orders in the mining districts considered in this report. A study by Lovering et al. (1966) confirms the association of these metals with Te in jasperoid samples.

In the zone of weathering, Te is usually quite immobile. The oxidation of

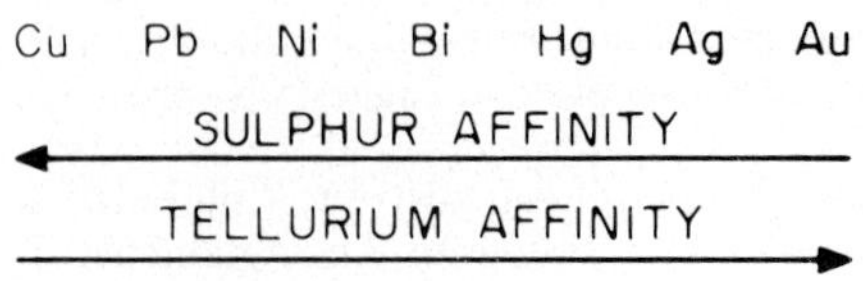

Fig. 1. Order of increasing affinity (arrow) of main telluride-forming metals for S and Te. (Modified from Vlasov, 1966, p. 584.)

pyrite-rich rocks and subsequent production of gossans, or leached cappings, usually prevent Te from migrating appreciably in the zone of weathering, because under the acid conditions of such an oxidizing environment, ions such as TeO_3^{2-} may substitute for OH^- ions; thus TeO_3^{2-} ions can become incorporated in hydrous iron oxides, presumably in the form of basic ferric tellurites (Oldfield et al., 1974). Emmonsite, $Fe_2(TeO_3)_3 \cdot 2H_2O$, and poughite, $Fe_2(TeO_3)_2SO_4 \cdot 3H_2O$, may be minerals formed by this process. In a pyrite-free environment, the tellurite ion could conceivably migrate in the groundwater, but pyrite is almost always present in areas where Te is significantly enriched.

Because sulphates are more soluble than tellurites, tellurites become gradually enriched relative to sulphates in the zone of weathering (Vlasov, 1966).

Te is often widely dispersed in hypogene mineralizing processes. In the supergene zone, however, Te is relatively immobile. These facts eminently qualify Te as an element useful to geochemical exploration programmes.

TELLURIUM DISPERSION PATTERNS AROUND SELECTED ORE DEPOSITS

Coeur d'Alene district

The Coeur d'Alene district is the most productive Ag mining district in the United States. It covers about 780 km^2 in the panhandle of northern Idaho. Within the east-trending district are a series of northwest-trending mineral belts. The host rocks are quartzites, siltites and argillites of the Precambrian Belt Supergroup. The principal ore deposits are replacement veins, and the principal ore minerals are galena, tetrahedrite, and sphalerite. Both geologic evidence (Hobbs et al., 1965) and geochemical evidence (Gott and Cathrall, 1977) indicate that post-ore movement along several of the major east-trending faults in the district has resulted in the offset of some of the mineral belts by as much as 26 km (Gott and Botbol, 1973).

The present-day Te dispersion patterns and their relationships to ore deposits are shown in Fig. 2, and the reconstructed or pre-faulting patterns are shown in Fig. 3. Before post-ore faulting, the largest ore deposits lay in an arc around the southeastern perimeter of the Cretaceous quartz monzonite Gem-Dago Peak stocks. In the reconstructed geochemical patterns, huge haloes of Te, Sb, As, Pb, S and Cd surround these stocks. The shape of the unfaulted Te halo surrounding the reconstructed stock is that of an inverted cone (Fig. 4); 85% of the Coeur d'Alene ore has come from the area of this halo (Fig. 5). Te dispersion patterns also appear to correlate with the location of a pluton inferred by aeromagnetic data to be approximately 1000 m beneath the surface in the southeastern part of the district (M. Dean Kleinkopf, oral communication, 1976; compare the distribution of Te in Figs. 2 or 3 with the stippled area of Fig. 6). Te dispersion patterns define the pre--intrusive mineral belts. A low-level Te halo also appears to surround the entire 780-km^2 district (Figs. 2 and 3).

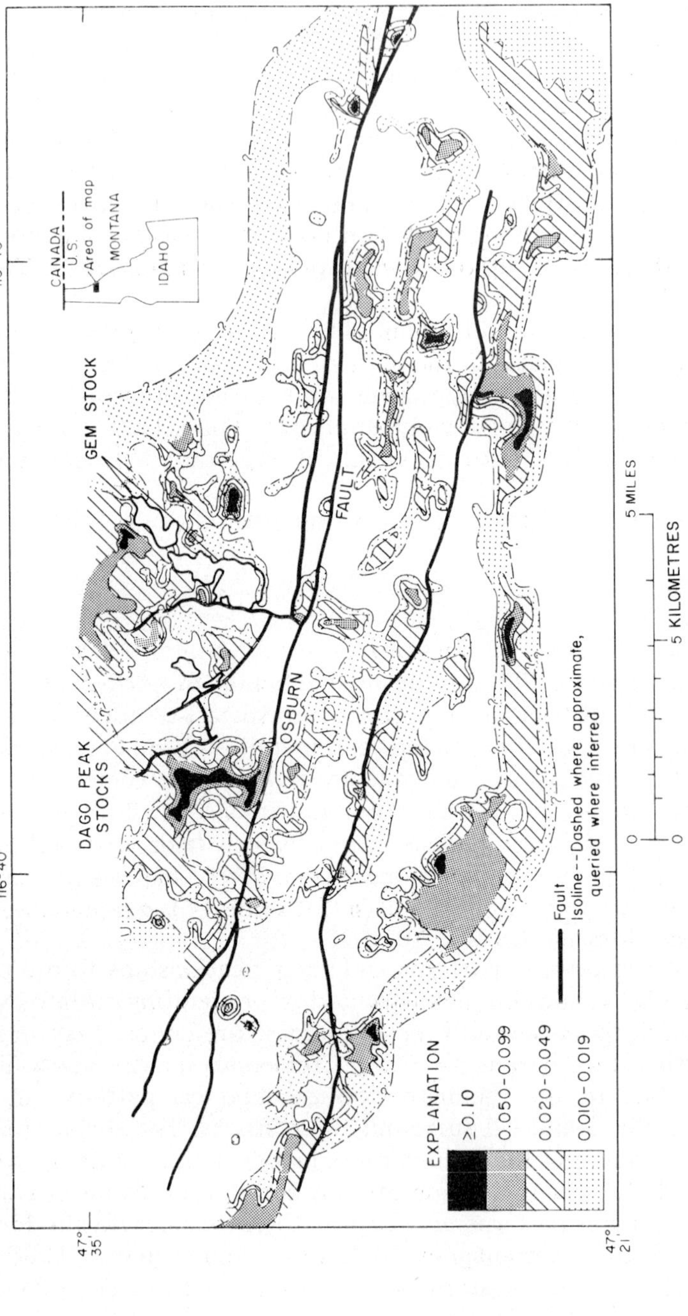

Fig. 2. Distribution of Te (in ppm) in rocks of the present land surface in the Coeur d'Alene mining district, Idaho.

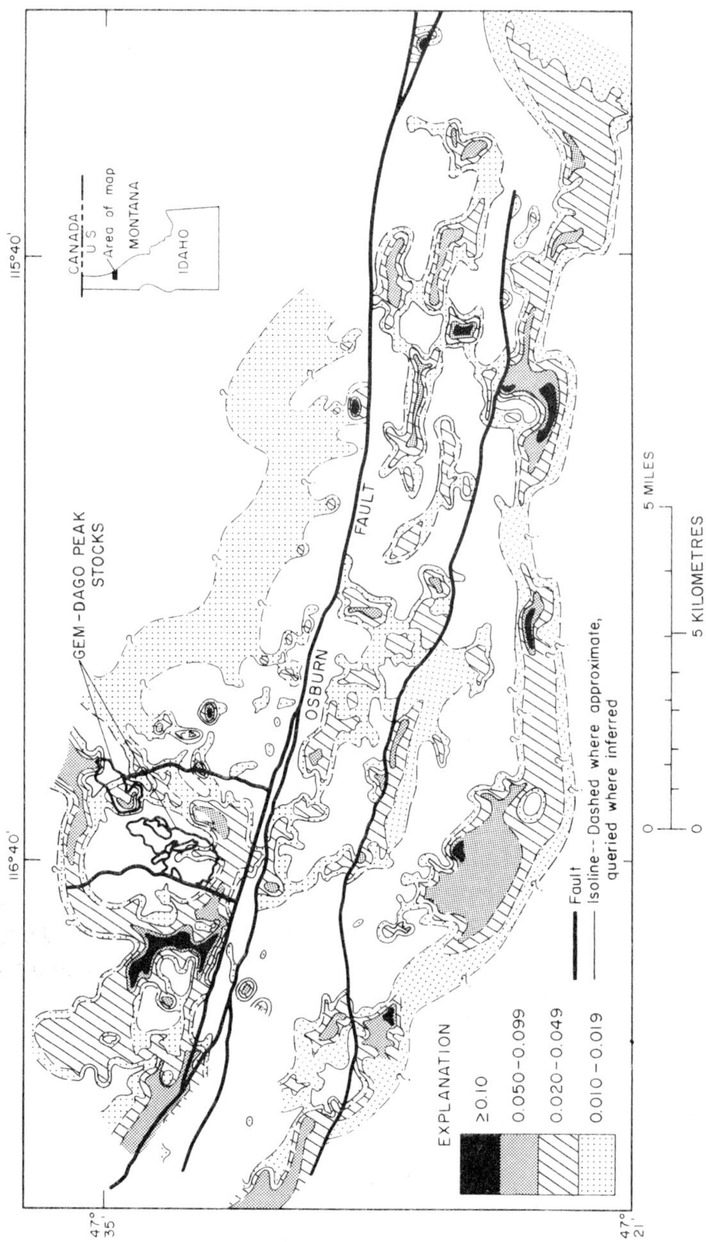

Fig. 3. Distribution of Te (in ppm) in Coeur d'Alene rocks before post-ore faulting.

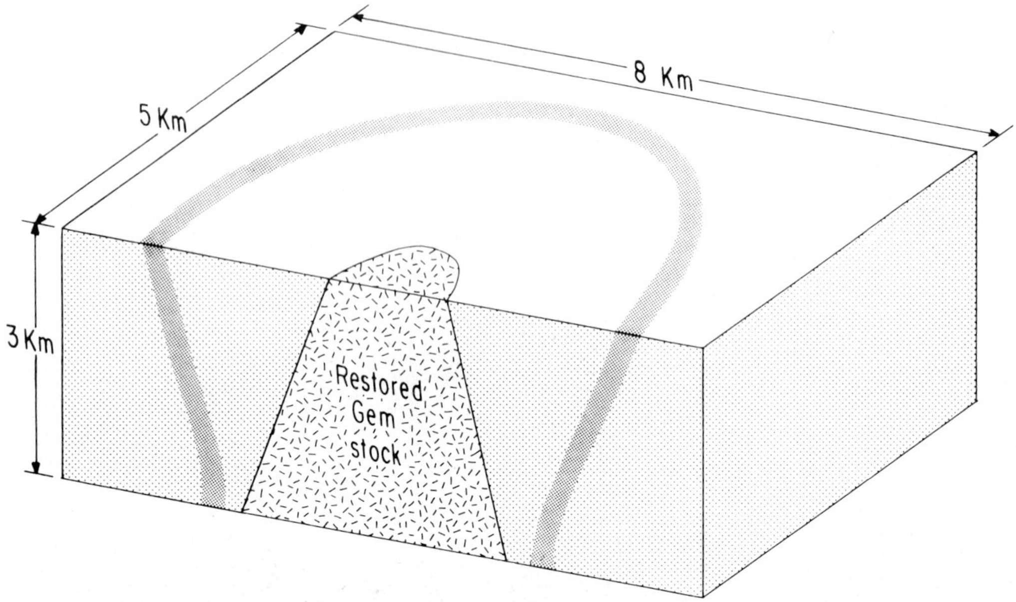

Fig. 4. Simplified sketch of the restored Gem stock and its associated Te halo in the Coeur d'Alene district before post-ore faulting.

Geochemical and geological evidence have led to the conclusion that two mineralizing events have affected the distribution and shape of the Te dispersion patterns. First, an early event resulted in the emplacement of the major deposits in northwest-trending mineral belts. Second, a later event, related to the intrusion of the monzonite stocks, resulted in the remobilization of Te and other volatile elements and their redistribution in the form of haloes around the stocks. Thus the position, shape and size of the Te dispersion patterns around the monzonite stocks have been largely influenced by the heat source centred on the invading magma.

Robinson district

The productive part of the Robinson district, the most important copper-producing district in Nevada, consists of a belt of disseminated porphyry Cu deposits extending about 13 km westward from the town of Ely in east-central Nevada (Fig. 7); another belt of mineralized rock extends 4 km northward from Ely (Gott and McCarthy, 1966). Early mining production was from chalcocite-enriched chalcopyrite ore, but in later years has been from primary chalcopyrite-pyrite ore. Au, Ag, Pt, Pd, Bi, Hg and Pb are elements which combine with, and are associated spatially with, Te in this district.

Fig. 5. Relation of Te halo (in ppm) to the restored Gem stock and to 85% of the recovered Coeur d'Alene ore as shown by underground mine workings.

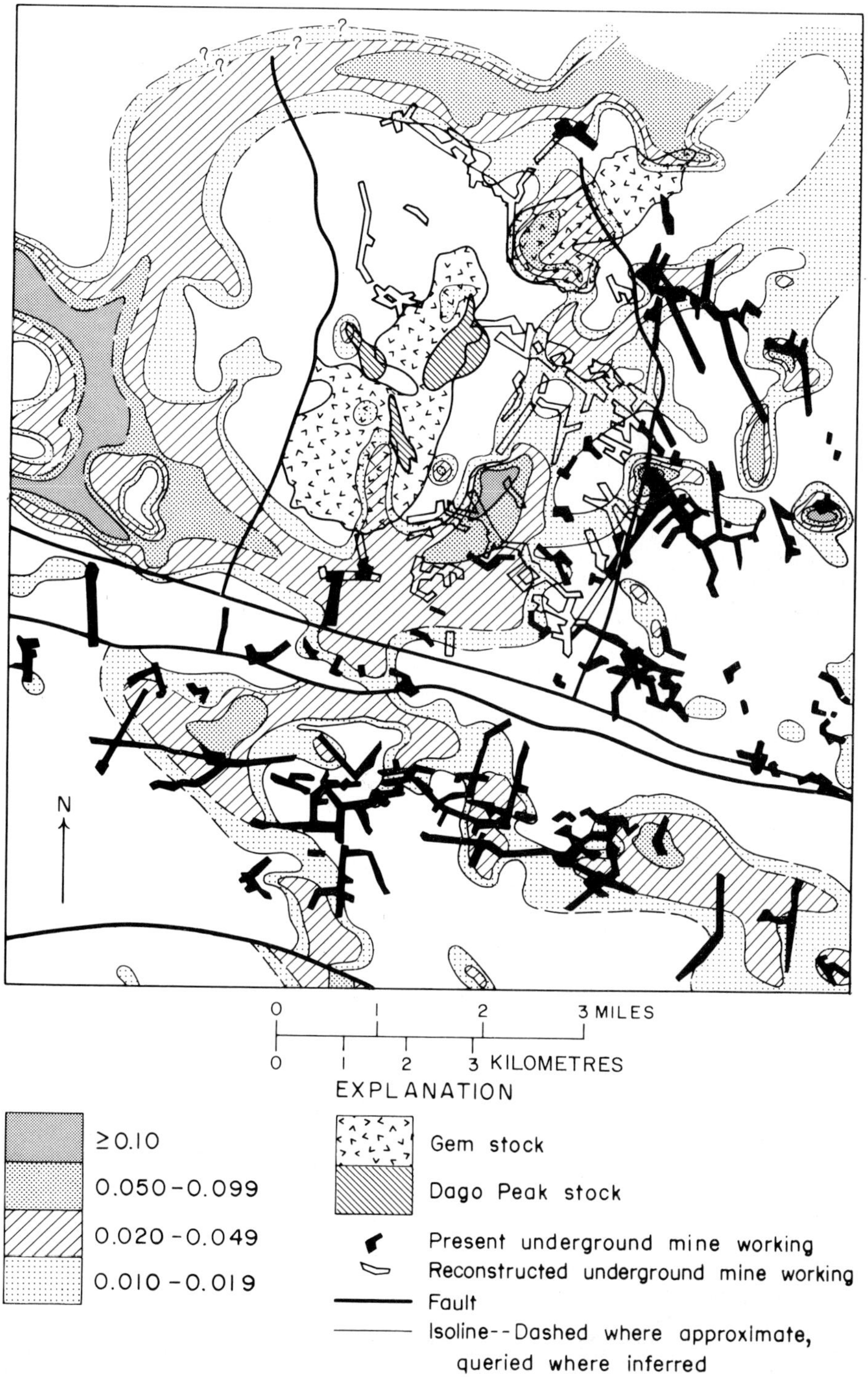
?
N
0 1 2 3 MILES
0 1 2 3 KILOMETRES
EXPLANATION
≥0.10
0.050-0.099
0.020-0.049
0.010-0.019
Gem stock
Dago Peak stock
Present underground mine working
Reconstructed underground mine working
Fault
Isoline--Dashed where approximate, queried where inferred

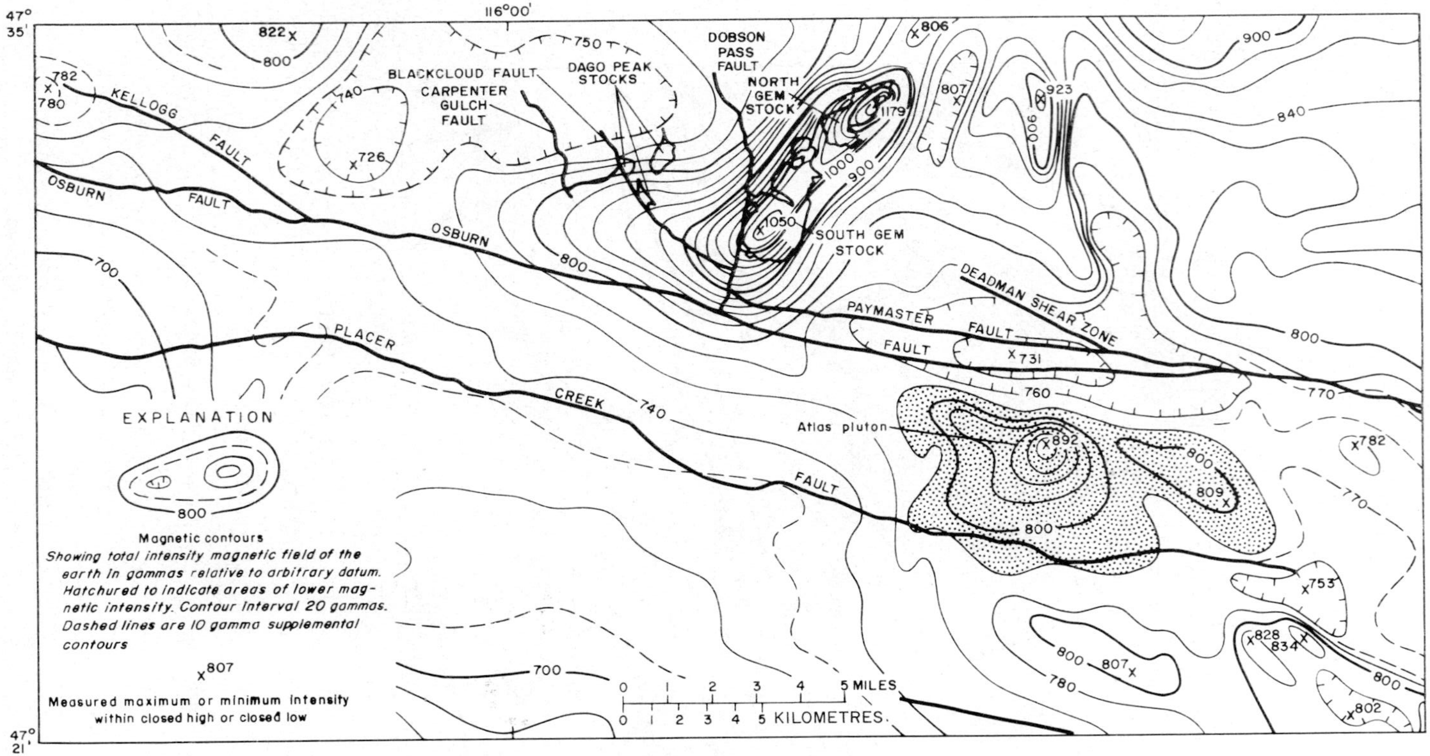

Fig. 6. Aeromagnetic map of the Coeur d'Alene district. Stippled area represents the Atlas pluton, approximately 1000 m from surface.

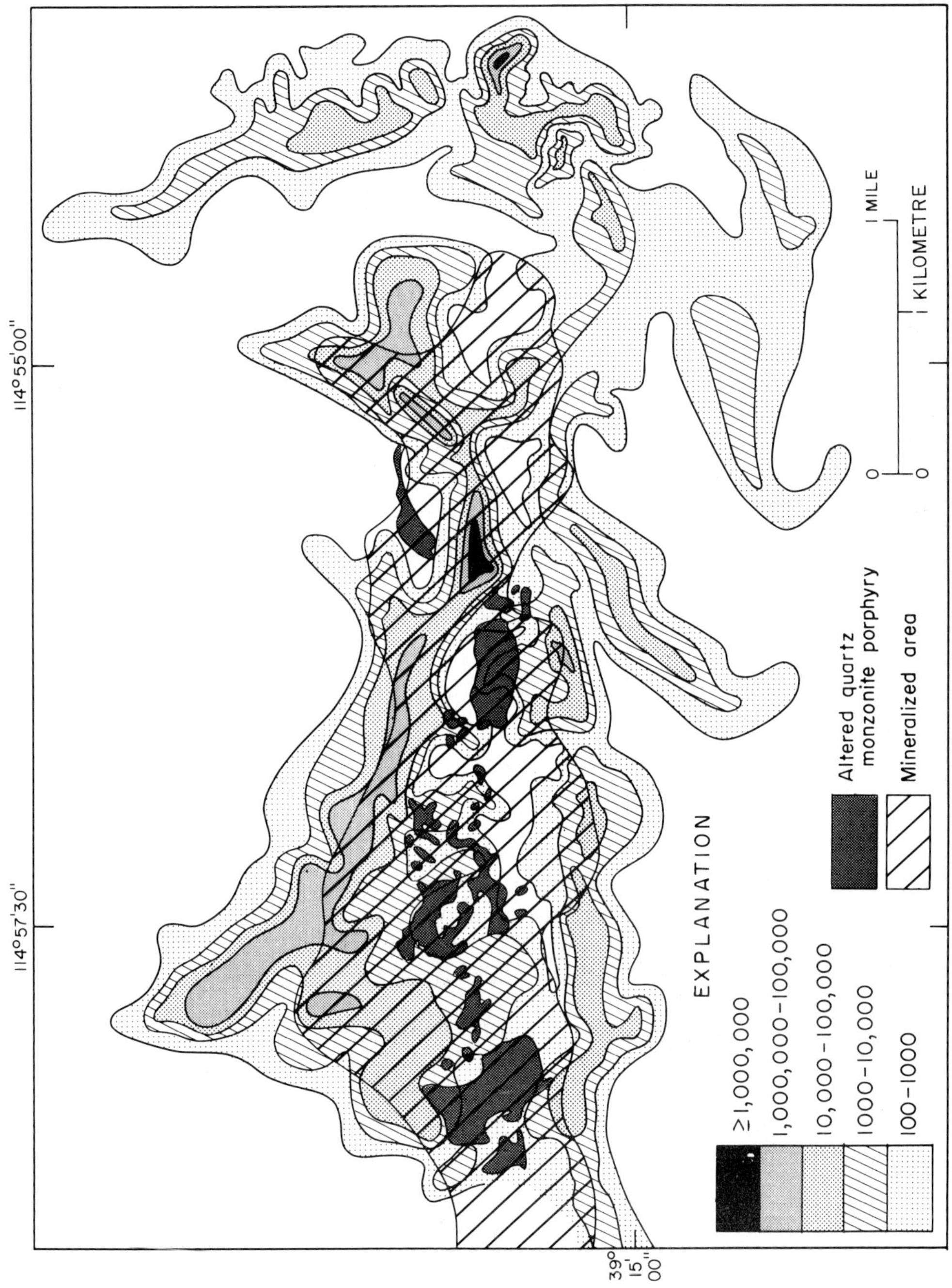

Fig. 7. Heavily mineralized area of the Robinson district, Nevada; superimposed on isolines of the Te dispersion pattern (in ppm).

Rocks within the district are predominantly a sequence of Palaeozoic sedimentary rocks, principally limestones and dolomites, with subordinate amounts of shale and quartzite. This sedimentary sequence is probably 4600—6000 m thick and has been intruded by Tertiary monzonite porphyry stocks along an east-west-trending line. The most intense mineralization is within the porphyry and in adjacent altered zones in the sedimentary rocks (Fig. 7). Parts of the stocks were intensely altered and, in extensive bordering areas, limestone was silicified and replaced with sulphide minerals. Jasperoid and gossan cover an area of about 40 km^2 in the central portion of the zone of highly altered rocks (Fig. 7). The highest concentrations of Te are found near the outer edge of this mineralized zone (Fig. 7). Zoning from the centre outward appears to be in the order: S, Cu, Fe, Zn, Cd, Te, Sb, As, Pb, Hg, Ag and Mn. Cu, Pb, Hg and Ag, the elements listed by Shcherbina (in Vlasov, 1966, p. 584), are in the order of both decreasing chalcophile character and increasing affinity for Te (Fig. 1). On the basis of a median Te value of 2 ppm for all gossans sampled, we estimate that the 60 km^3 of altered rock in the Robinson district may contain as much Te as 60,000 km^3 of average crustal material and may represent a significant "reserve" of Te.

Montezuma district

The Montezuma district (Fig. 8), a part of the Colorado mineral belt, is located just south of Loveland Pass in central Colorado. The district contains stocks and dykes of quartz monzonite porphyry, vein deposits, and altered rocks that originated as products of the venting of a Tertiary batholith through a major Precambrian zone of weakness into a complex of Precambrian gneisses, amphibolites, granites, pegmatites, and Cretaceous black shales. An area of about 130 km^2 of altered rock is localized by, and spreads westward from, the intersection of the Montezuma stock and the Montezuma shear zone (Fig. 8). Mining on a small scale has centred on Ag-Pb-Zn veins developed around several small intrusives along the shear zone. The district may also contain a deep and structurally complex porphyry Mo deposit (Neuerburg et al., 1974).

The relation between the distribution of Te and the distribution of mineralized rock is shown in Fig. 8. Te dispersion patterns reflect, and are equally as complex as, the "mottled" alteration patterns characteristic of this district (Neuerburg et al., 1974). Te dispersion patterns reflect the pervasive alteration that spreads out westward from the Montezuma shear zone; they also delineate the envelopes of intensely altered rock around veins.

The highest Te values are found in altered wall rock associated with Ag-Pb-Zn veins. Although the Te content of rocks decreases exponentially away from these centres of mineralization, Te is detectable throughout the area, even in apparently unaltered rocks several hundred metres beyond visible signs of alteration. Te enriched in apparently unaltered rocks, best observed in our studies in the Coeur d'Alene and Montezuma districts, may be present

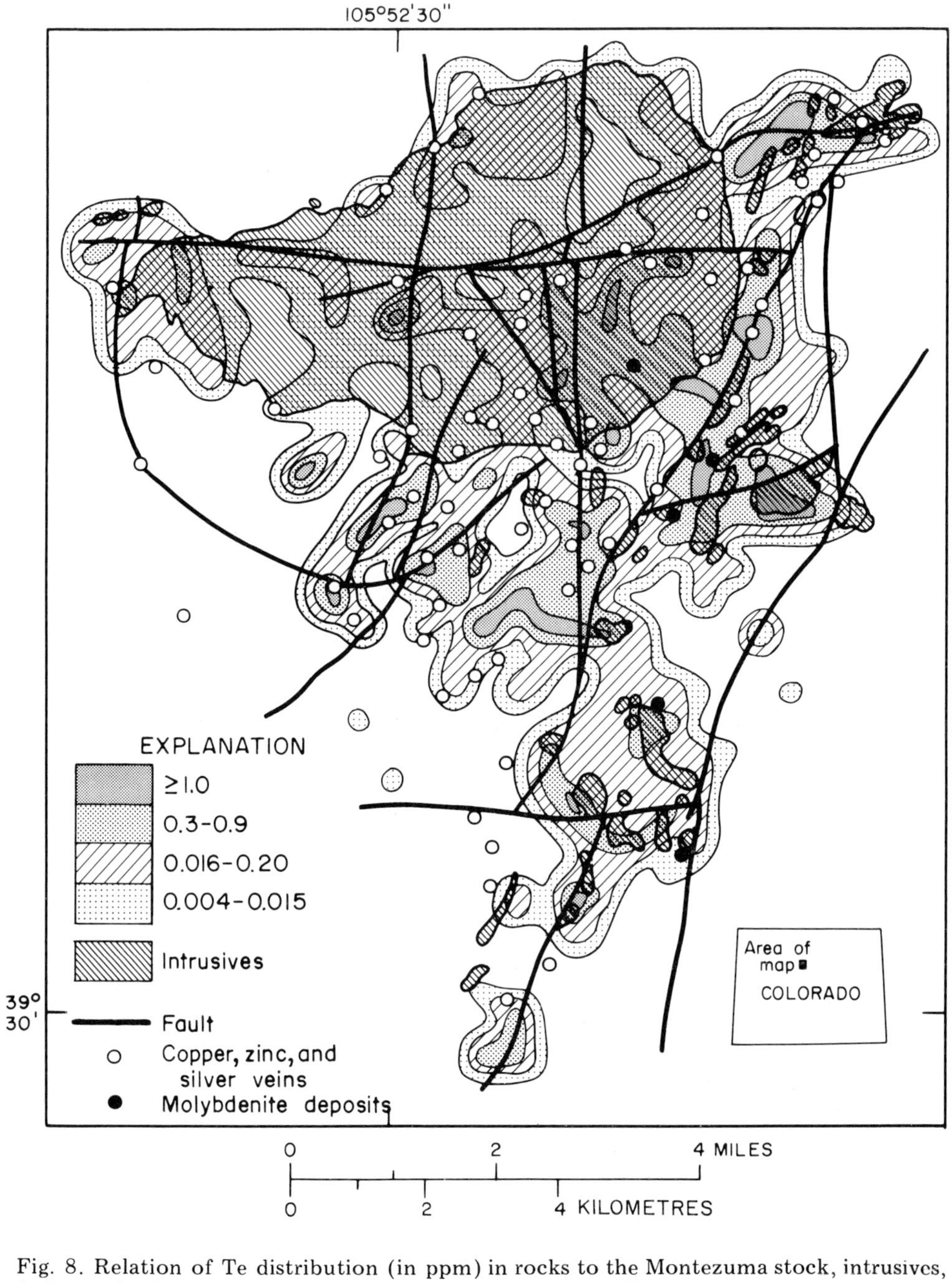

Fig. 8. Relation of Te distribution (in ppm) in rocks to the Montezuma stock, intrusives, faults, and vein deposits of the Montezuma district, Colorado.

as adsorbed Te, a phenomenon discussed for other elements by Griffitts and Nakagawa (1960).

Over the centre of the Montezuma stock, the Te content of soils (Fig. 9) is as much as 100 times greater than that of adjacent rock outcrops; the soil Te patterns do not resemble the rock Te patterns or the distribution of known veins (Watterson and Neuerburg, 1975), but do reflect the intersection of two major structural elements in the stock (compare Fig. 8 with Fig. 9). On the basis of these concentration differences and the results of a labo-

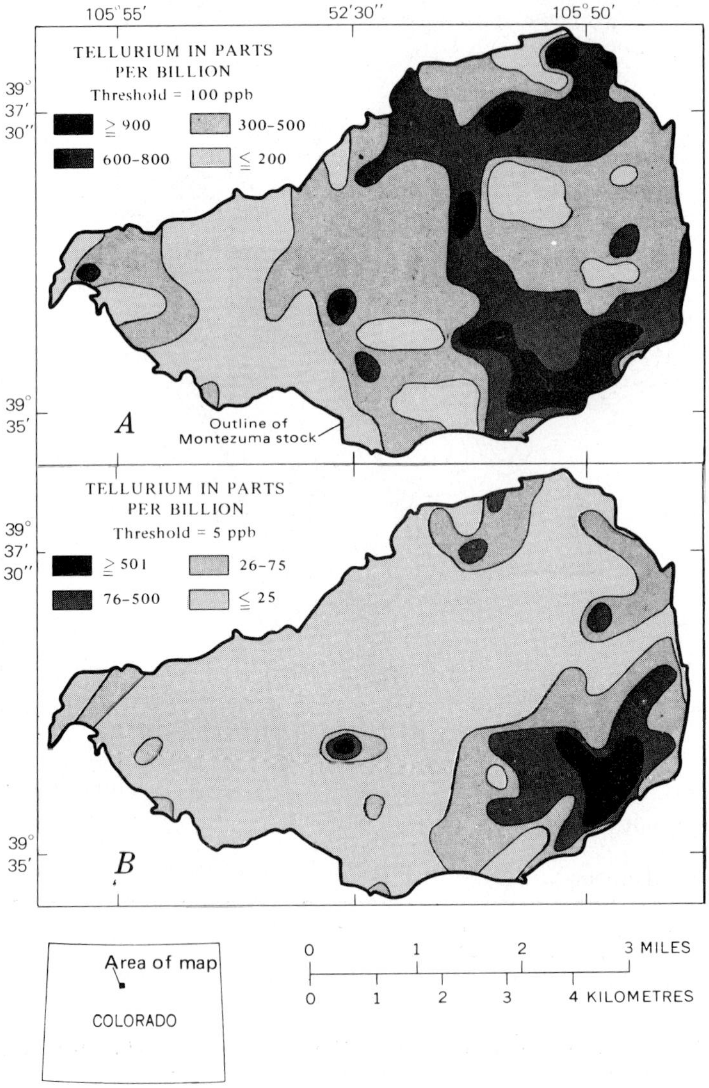

Fig. 9. Comparison of the Te content of soils (A) and rocks (B) over the Montezuma stock.

ratory experiment (U.S. Geological Survey, 1971), we think that these soils may be accumulating gaseous Te compounds emanating from underlying mineral deposits. Where soils were sampled in the districts studied, they generally were found to contain more Te than adjacent outcropping rocks, perhaps because soils accumulate gaseous Te compounds.

Crater Creek area

The U.S. Geological Survey has recently completed a geochemical study of the Crater Creek area (Fig. 10) adjacent to the Summitville mining district

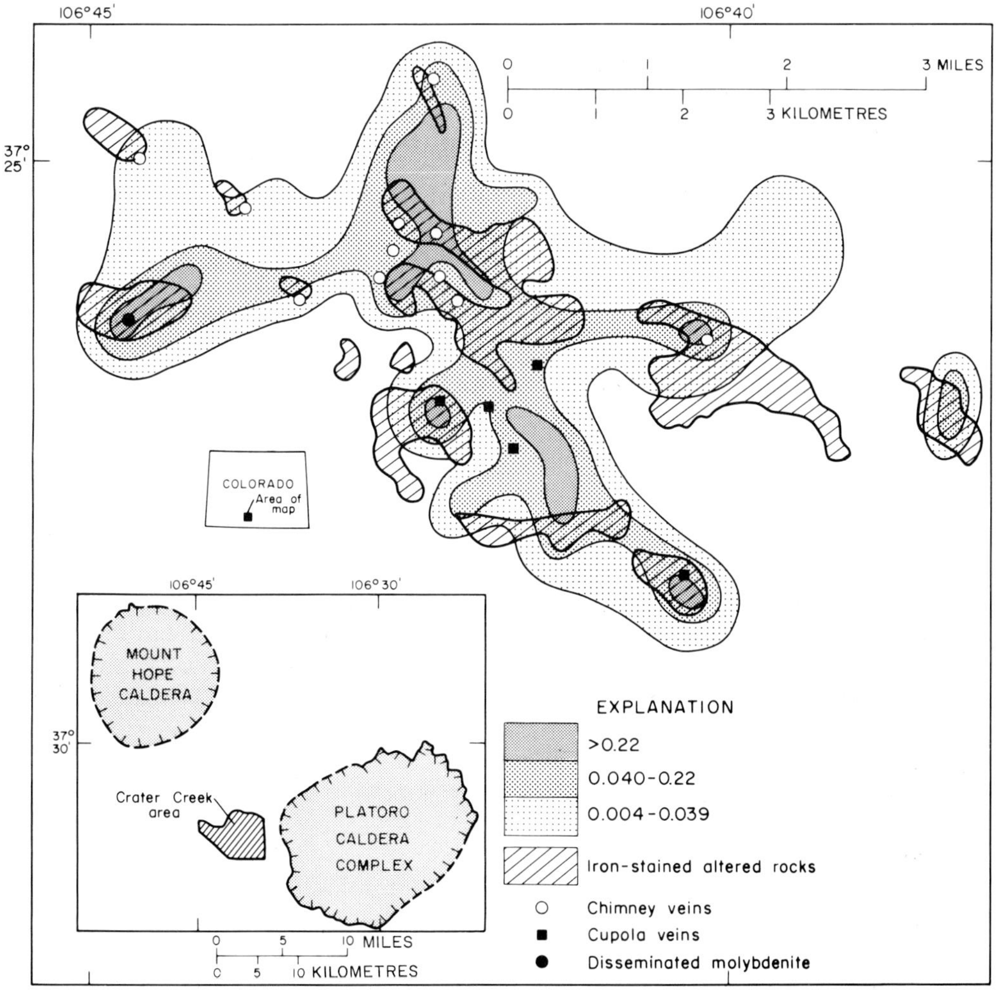

Fig. 10. Distribution of Te (in ppm) in rocks related to alteration and vein deposits of the Crater Creek area, Colorado. The Te dispersion pattern shown approximately fills the area shown in the inset map. (Modified from Lipman, 1975.)

in southern Colorado. The Summitville district, associated with the Platoro caldera complex, is in the San Juan volcanic field at the southwestern end of the Colorado mineral belt (location map, Fig. 10); it has produced Cu, Au, Pb and Ag, mostly from resistant pipes and vein-like masses of vuggy quartz and quartz-alunite rock (Steven and Ratté, 1960). The Crater Creek area, just to the west, is centred on four Tertiary monzonite stocks that have intruded a graben connecting the Platoro caldera with the Mount Hope caldera to the northwest (Fig. 10). Small stocks, dykes, iron-stained and intensely altered rocks, Cu-Pb-Zn veins exposed in mines and prospects, and an outcrop of molybdenite stockwork veins (Fig. 10) mark the Crater Creek area as an extension of the more conspicuous Summitville deposits.

A feature common to both the Crater Creek area and the Summitville district is a rapidly diminishing intensity of alteration across relatively narrow alteration haloes. Alteration gradients in the Crater Creek area, however, are steeper than in the Summitville district. The compactness of the Te dispersion patterns reflects this alteration trend and indicates to us the probable presence of chimney-type deposits similar to, but deeper than, those found near Summitville.

The relation of the Te dispersion pattern to the exposed molybdenite stockwork veins, to the vein deposits, and to areas of iron-stained, altered rock is shown in Fig. 10. As in the Montezuma district, the Te content of rocks decreases exponentially away from the centres of mineralization, but continues at detectable levels throughout the area, even in apparently unaltered rocks.

Cripple Creek district

The Cripple Creek district, in central Colorado, has produced about 21 million ounces of Au since its discovery in 1891 (Gott et al., 1969). The Au deposits are largely confined to an elliptical subsidence basin (Fig. 11) about 6 km long and 3 km wide. The basin is filled with fractured and brecciated Tertiary volcanic rocks and surrounded by Precambrian granite, gneiss and schist. Au and Ag are the only metals of present economic significance; the average ratio of Au/Ag is about 10:1.

The principal ore minerals in the unoxidized zone are gold tellurides; small amounts of native Au occur in the oxidized ores. The Au occurs in narrow, nearly parallel, steeply inclined fissures. Te defines the distribution of Au throughout the zones of complex fissures.

DISCUSSION AND CONCLUSION

In order to depict a trend that seems apparent in the dispersion patterns of Te in four of these districts, we have assigned each of the four deposits a specific erosional level in an hypothetical porphyry metal systems model (Fig. 12). According to this scheme, the Robinson district, the Montezuma

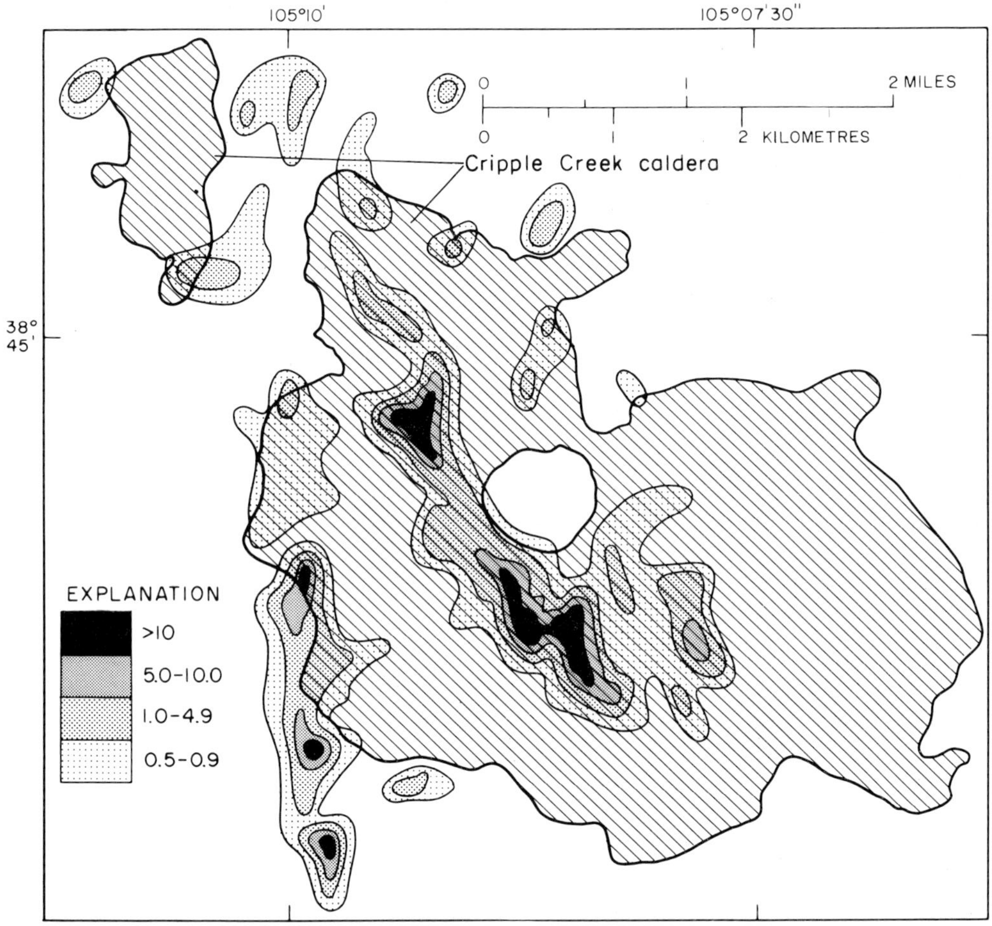

Fig. 11. Distribution of Te (in ppm) in rocks in the Cripple Creek district, Colorado. Note that the analytical cut-off is relatively high.

district, the Crater Creek area and the Cripple Creek district occupy successively higher erosional levels in the model. The trend seems to be for Te to be more broadly dispersed at the lower erosional levels. We conclude that Te may therefore be more useful in searching for disseminated metal deposits such as those of the Robinson and Montezuma districts than in searching for vein deposits of the Crater Creek and Cripple Creek type.

In seeming contradiction to this trend is the apparently strong relation between Te distribution and the location of deep veins in the Coeur d'Alene district. Also in need of explanation is the difference in the apparent shape of the Te halo surrounding the Coeur d'Alene stocks and that surrounding the vertical axis of our model. In the former, the Te halo is in the shape of an inverted cone, and that of the latter is in the shape of a regular cone.

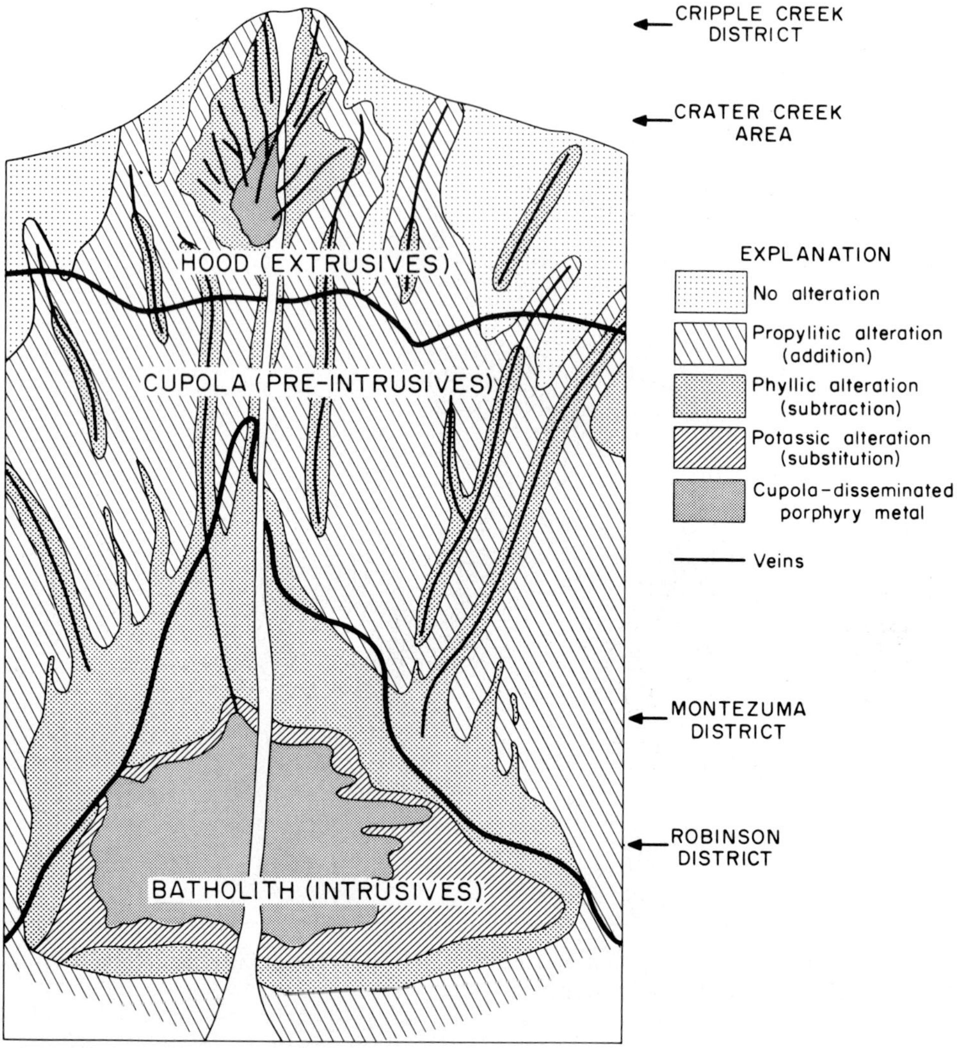

Fig. 12. Approximate erosional levels of the districts studied compared in an imaginary volcano (or porphyry-metal-systems model).

Because of events unique to their genesis, the Coeur d'Alene deposits have not been typified; peculiarities in the dispersion patterns of Te and other elements likewise await correlation with the dispersion patterns of other districts.

Certain problems also exist in the interpretation of the Te data from the other four districts. Te may be present in significant concentrations much further from the Cripple Creek vein deposits than the rather high analytical threshold indicated at the time of the study. The same may be said for the Robinson district. In the Montezuma and Coeur d'Alene districts, the Te dispersion patterns are essentially coextensive with the sampled area; if samples had been taken further away from the deposits, a more complete picture of Te dispersion would have resulted, largely because of increased analytical sensitivity (Watterson, 1976).

Because of sampling, analytical and environmental differences it may not be possible to make strictly valid comparisons between Te dispersion patterns in these districts.

Although a complete assessment of the usefulness of Te as a pathfinder in the search for ore deposits awaits further investigation, we feel that Te is potentially useful as a guide to mineralization. On the basis of these studies, we recommend that Te be considered for inclusion in future geochemical exploration programmes.

REFERENCES

Czamanske, G.K. and Hall, W.E., 1975. The Ag-Bi-Pb-Sb-S-Se-Te mineralogy of the Darwin lead-silver-zinc deposit, southern California. Econ. Geol., 70: 1092—1110.

Goldschmidt, V.M., 1958. Geochemistry (edited by Alex Muir). Oxford University Press, London, 730 pp.

Gott, G.B. and Botbol, J.M., 1973. Zoning of major and minor metals in the Coeur d'Alene mining district, Idaho, U.S.A. In: M.J. Jones (Editor), Geochemical Exploration 1972. Institution of Mining and Metallurgy, London, pp. 1—12.

Gott, G.B. and Cathrall, J.B., 1977. Geochemical exploration studies in the Coeur d'Alene district, Idaho. U.S. Geol. Surv., Open-File Rep., 77-216: 195 pp.

Gott, G.B. and McCarthy, J.H., 1966. Distribution of gold, silver, tellurium and mercury in the Ely mining district, Nevada. U.S. Geol. Surv., Circ., 535: 5 pp.

Gott, G.B., McCarthy, J.H., Jr., VanSickle, G.H. and McHugh, J.B., 1969. Distribution of gold and other metals in the Cripple Creek district, Colorado. U.S. Geol. Surv., Prof. Paper, 625-A: 17 pp.

Griffitts, W.R. and Nakagawa, H.M., 1960. Variations in base-metal contents of monzonitic intrusives. In: Short Papers in the Geological Sciences. U.S. Geol. Surv., Prof. Paper, 400-B: 93—95.

Hobbs, S.W., Griggs, A.B., Wallace, R.E. and Campbell, A.B., 1965. Geology of the Coeur d'Alene district, Shoshone county, Idaho. U.S. Geol. Surv., Prof. Paper, 478: 139 pp.

James, L.P., 1976. Zoned alteration in limestone at porphyry copper deposits, Ely, Nevada. Econ. Geol., 71: 488—512.

Kelly, W.C. and Goddard, E.N., 1969. Telluride ores of Boulder County, Colorado. Geol. Soc. Am., Mem., 109: 237 pp.

Lipman, P.W., 1975. Evolution of the Platoro caldera complex and related volcanic rocks, southeastern San Juan Mountains, Colorado. U.S. Geol. Surv., Prof. Paper, 852: 128 pp.

Lovering, T.G., Lakin, H.W. and McCarthy, J.H., 1966. Tellurium and mercury in jasperoid samples. U.S. Geol. Surv., Prof. Paper, 550-B: 138—141.

Neuerburg, G.J., Botinelly, T. and Watterson, J.R., 1974. Molybdenite in the Montezuma district of central Colorado. U.S. Geol. Surv., Circ., 704: 21 pp.

Oldfield, J.E., Allaway, W.H., Laitinen, H.A., Lakin, H.W. and Muth, O.H., 1974. Tellurium. In: Geochemistry and the Environment, 1. The Relation of Selected Trace Elements to Health and Disease. National Academy of Sciences, Washington, D.C., pp. 64—67.

Steven, T.A. and Ratté, J.C., 1960. Geology and ore deposits of the Summitville district, San Juan Mountains, Colorado. U.S. Geol. Surv., Prof. Paper, 343: 70 pp.

U.S. Geological Survey, 1971. Geological Survey Research 1971. U.S. Geol. Surv., Prof. Paper., 750-A: 11.

Vlasov, K.A. (Editor), 1966. Tellurium. In: Geochemistry and Mineralogy of Rare Elements and Genetic Types of Their Deposits, 1. Geochemistry of Rare Elements (translated from Russian by Z. Lerman). Israel Program for Scientific Translations, Jerusalem, pp. 581—606.

Wallace, R.E. and Hosterman, J.W., 1956. Reconnaissance geology of western Mineral County, Montana. U.S. Geol. Surv., Bull., 1027-M: 575—612.

Watterson, J.R., 1976. Determination of tellurium and gold in rocks to 1 part per billion. U.S. Geol. Surv., Open-File Rep., 76-531: 3pp.

Watterson, J.R. and Neuerburg, G.J., 1975. Analysis for tellurium in rocks to 5 parts per billion. U.S. Geol. Surv., J. Res., 3: 191—195.

Journal of Geochemical Exploration, 8 (1977) 49—71

GEOCHEMICAL PROSPECTING FOR VOLCANOGENIC SULPHIDE DEPOSITS IN THE EASTERN BLACK SEA ORE PROVINCE, TURKEY

M.N. CAGATAY and D.R. BOYLE*

Applied Geochemistry Research Group, Imperial College, London (Great Britain)
Geological Survey of Canada, Ottawa, Ont. (Canada)

(Revised version accepted March 1, 1977)

ABSTRACT

Cagatay, M.N. and Boyle, D.R., 1977. Geochemical prospecting for volcanogenic sulphide deposits in the Eastern Black Sea ore province, Turkey. J. Geochem. Explor., 8: 49—71.

Three related types of sulphide mineralization in the Eastern Black Sea volcanic province (Eastern Pontids) of Turkey were studied with a view towards developing geochemical techniques both for their detection and differentiation: (1) Zn-Cu-Pb massive sulphide deposits; (2) Cu-pyrite stockwork deposits; and (3) pyritic impregnation deposits.

Bedrock surveys utilizing samples of the host dacites and overlying basic rocks were effective in outlining various types of primary haloes associated with the sulphide deposits. The most indicative elements are F, Pb, Zn, Cd, Cu, Ag, As and Ba; the dispersion of Hg is restricted, and the element is not a good indicator.

Quantitative evaluation of the clay (kaolinite plus montmorillonite plus illite-sericite)/feldspar ratio by X-ray methods is useful in detecting the presence and intensity of hydrothermal activity related to sulphide mineralization and also for estimating proximity to mineralization.

On a regional basis, the mineral potential of exposed stockwork and impregnation zones can be assessed by their elemental content. Barren impregnation pyritic deposits generally report only anomalous Fe and Cu contents. Cu-pyrite stockworks with no associated massive sulphides are enriched in Fe, As and Cu, whereas those accompanying massive sulphides are marked by enrichments of some or all of Fe, As, F, Cu, Pb, Zn, Ag, Cd and Mo.

INTRODUCTION

Massive, stockwork and impregnation sulphide deposits genetically related to calc-alkaline volcanics of Upper Cretaceous age occur in the Eastern Black Sea metallogenic province of Turkey. Since Medieval time the deposits have

---

*Formerly, Applied Geochemistry Research Group, Imperial College, London, Great Britain.

Paper authorised for publication by the Director General of M.T.A., Ankara, Turkey.

produced Cu and have recently been shown to have a potential as major sources of Zn, Pb, Ag, Au, baryte and gypsum. In many respects the deposits are similar to the "Kuroko" ores of the "Green Tuff" regions of Japan.

Mineral exploration in the province is hindered by high topographic relief, dense vegetation and limited infrastructure. Sulphide deposits found to date are those that outcrop or sub-outcrop and occur in accessible districts. Large areas remain to be explored for concealed mineral deposits, and many areas containing pyritic stockworks and impregnation zones in dacitic rocks require assessment to determine if they are indicative of associated massive sulphide deposits.

The deposits described lie within an area characterized by high topographic relief, a mean annual temperature of 14°C and an average rainfall up to 2 m or more.

The Eastern Black Sea volcanogenic province (Fig. 1) forms the eastern

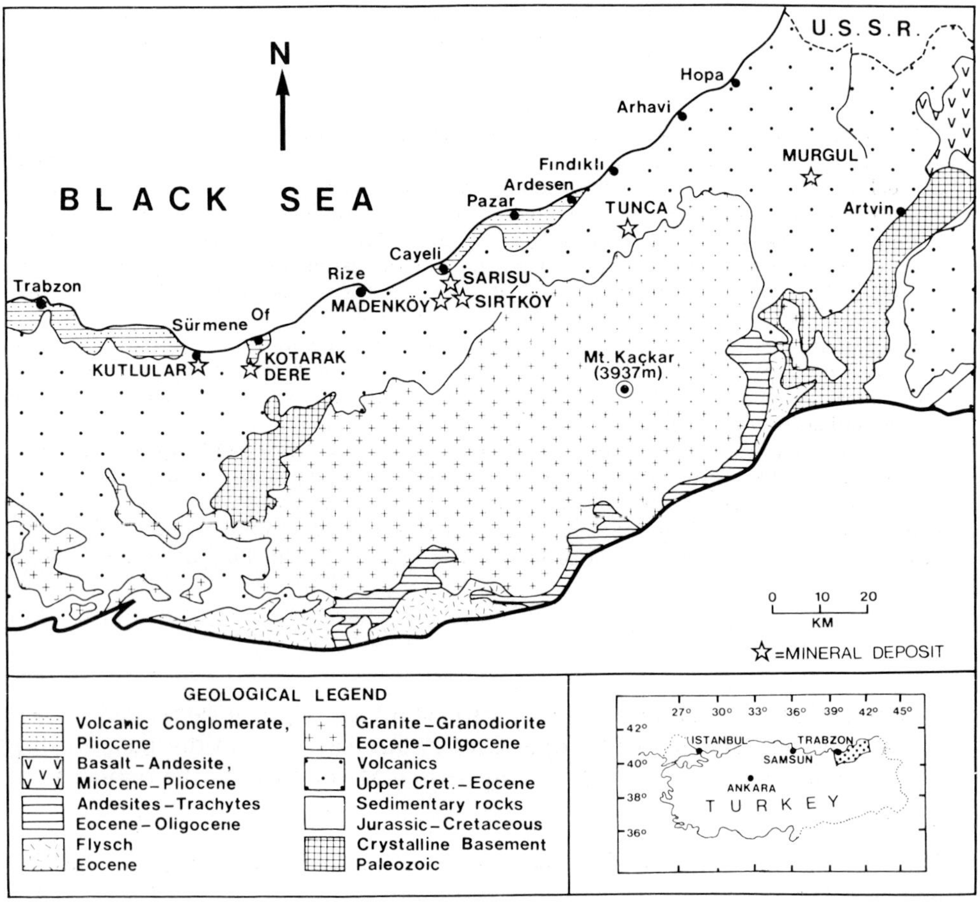

Fig. 1. Geology of eastern part of Eastern Black Sea volcanogenic province of Turkey.

TABLE I

Table of formations, Eastern Black Sea volcanogenic province, Turkey

Age		Series			Lithology
Quaternary					alluvium
Pliocene					terrace deposits
Oligo-Miocene (granites, granodiorites, diorites)		Young Basic Series			basalts, tuffites, marls, limestones
Alpine					
Eocene		Upper Basic Series	(1000 m)	-flysch-	andesitic, tuff breccia, basalts, limestone, shales conglomerates, marls
		unconformity			
Cretaceous	Upper	Dacitic Series			dacites, andesites
	Lower	Lower Basic Series	(4000 m)	-flysch-	andesites, basalts, tuffs, agglomerates, shales, conglomerates, sandstone, marls
Jurassic (peridotites, gabbros, diorites)	Upper				limestones, conglomerates
		unconformity			
	Middle	(transgression, erosion, uplift)			
Cimmerian					
	Lower				sandstones, conglomerates, marls spilites, basic tuffs
		unconformity			
Triassic					limestones, (often absent)
Permo-Carboniferous (granites, diorites)			(2000 m)		arkose, schists, quartzites, limestones
Hercynian		unconformity			
Pre-Carboniferous (undifferentiated)			(700—800 m)		mica schists, quartzites, greywackes, pyllites, graphitic schists marbles, gneisses, granites
Caledonian					

part of the Pontids; and comprises formations involved in the Caledonian, Hercynian, Cimmerian and Alpine orogenies (Table I). The volcanism which commenced during lower Jurassic time is marked by the Upper Cretaceous Dacitic Series which is capped by the Upper and Young Basic Series of Upper Cretaceous and Oligo-Miocene ages respectively. All of the above types of mineralization in the area are associated with the Dacitic Series.

The Eastern Pontids have a germano-type (horst and graben) tectonic style with faulting predominating (Kraeff, 1963a,b; Kronberg, 1970). Extensive block-faulting along NE-SW and NW-SE preferred directions has played a major role in exposing the Dacitic Series in places, a feature which provides windows of dacitic rocks ranging from single to relatively large discontinuous exposures (e.g. Cayeli area) suitable for detailed rock geochemical surveys.

Mineralization

Four types of sulphide mineralization associated with the Dacitic Series are recognized including: (1) massive, (2) stockwork, (3) pyritic impregnation, and (4) vein. The vein-type deposits comprising Cu-Pb-Zn sulphides in quartz and carbonate gangue, are small and non-economic, and are not included in the study. The other deposit types are not independent of one another since all of the massive sulphide deposits in the area have associated stockwork and pyritic impregnation zones. Cupriferous pyrite stockwork deposits seemingly mark environments in which massive sulphides have not formed (e.g. Murgul, Sirtköy) or, alternatively, the massive sulphide component has been elevated by faulting and subsequently eroded (e.g. Kotarak Dere). Pyritic impregnation zones in dacites may be present in the absence of massive sulphides or stockwork mineralization. In some places (e.g. Madenköy), only part of the stockwork and/or impregnation zone of an associated massive sulphide body is exposed. These zones have geological characteristics similar to the barren pyritic impregnation occurrences, and there is thus an obvious need for exploration parameters capable of differentiating between productive and non-productive mineralized zones in dacites.

A descriptive summary of the various types of sulphide deposits studied has been compiled in Table II. More comprehensive descriptions are given by Cagatay (1976). A brief review of all types of sulphide bodies in the volcanogenic province is given by Vujanovic (1974), and descriptions of some of the other sulphide bodies not studied by the writers are given by Kraeff (1963b), Tugal (1969), Leitch (1975) and Hattie (1977).

BEDROCK GEOCHEMICAL SURVEYS

Bedrock geochemical surveys in the vicinity of a number of sulphide deposits in the Eastern Black Sea volcanic province were mounted with the intention of developing the following diagnostic parameters: (1) detection of primary haloes and alteration zones in the hanging wall and footwall forma-

TABLE II

Volganogenic sulphide mineralization, Eastern Black Sea province, Turkey

Deposit (type)	Reserves(tonnes) and grade (%)	Massive ore	Footwall zone	Hanging wall zone	Remarks
Madenköy (massive)	27m; 4%Zn,3%Cu (Ag+Au)	concealed; pyrite sphalerite, chalcopyrite, minor galena, gypsum, barite	chalcopyrite-pyrite stockwork and pyritic impregnation zones in dacitic tuffs	haematitized pumice tuffs overlain by basalts and andesitic tuff breccias	undisturbed; part of stockwork-impregnation zone exposed
Kutlular (massive)	1.5m; 2.2%Cu, 0.4m; 1.5%Zn	concealed; pyrite, chalcopyrite, sphalerite, minor galena, barite	same as Madenköy	mineralized dacitic tuff breccia overlain by basalts	undisturbed; stockwork and impregnation zones exposed; post-ore hydrothermal activity
Tunca (massive)	0.3m; 1.5%Cu, 1.0%Zn	exposed; pyrite, chalcopyrite, sphalerite, galena	same as Madenköy	haematitized porphyritic dacites overlain by basalts	stockwork, impregnation, and massive ore zones exposed by steeply dipping fault zone
Kotarak Dere (stockwork)	—	small remnant of massive ore	chalcopyrite-pyrite stockwork (minor sphalerite) and pyritic impregnation zones in dacites	basalts	most of stockwork and impregnation zones exposed
Murgul* (stockwork)	47m; 1.4%Cu	not formed	chalcopyrite-pyrite-stockwork with minor sphalerite, galena, and barite in dacitic tuffs; peripheral gypsum	haematitized dacites and tuffs overlain by basalts	extensive mining activity since medieval time
Sirtköy (stockwork)	—	not formed	pyrite and minor chalcopyrite stockwork (0.2% Cu) in porphyritic dacite	basalts	basalts are locally pyritized
Sarisu (pyrite impregnation)	—	not formed	pyrite impregnation zone in dacites	basalts and andesitic tuff breccias	no associated stockwork

*This deposit was not studied in detail.

tions of blind orebodies; (2) determination of criteria (i.e. directional vectors) for indicating the position of concealed massive sulphide deposits and thus for the siting of exploratory drill holes; and (3) criteria for discriminating barren stockwork and impregnation zones from their more productive counterparts associated with concealed massive sulphides.

Bedrock geochemical surveys in this region are aided by the presence of a single Dacitic Series in which the known sulphide orebodies occur. This feature obviates the need for detailed stratigraphic control which is often required in volcanic areas characterized by numerous productive and barren calc-alkaline cycles. Chemical exhalite zones (Fe-Mn carbonates, cherts, jasperoids, iron formations, etc.), such as those in Precambrian and some younger rocks, are not a feature of the Eastern Black Sea volcanogenic province; exhalite formation representing the proximal and distal environments about the massive sulphide orebodies appears to be limited to the formation of haematitized pumice tuffs (or other haematitized volcanics) blanketing the mineralized dacites. Where present, as in the Madenköy, Tunca and Murgul areas, they constitute an ideal horizon for the detection of primary haloes.

Sampling and analysis

The bedrock surveys were conducted by taking composite chip samples of outcrop exposures. Despite the fact that the area has a high topographic relief, heavy vegetation and soil cover severely limit the amount of outcrop available, and careful representative sampling of accessible outcrop is required. Every attempt was made to sample material relatively free of the effects of weathering.

Composite chip samples were crushed to minus 10-mesh using a jaw crusher. A representative split of the crushed sample was then ground in a Tema mill to minus 200-mesh. Samples were analyzed for Cu, Pb, Zn and Cd by atomic absorption spectrophotometry after a hot HNO_3 attack. The precision of the method for Cu, Pb and Zn is better than ±20% at the 90% confidence level over the concentrations observed. For Cd, the precision (95% confidence level) is ±30% for values greater than 1 ppm. As and Mo were analyzed colorimetrically after a bisulphate fusion; the precision for both methods is ±30% for concentrations greater than 1 ppm. Total fluorine was determined by specific ion electrode after an NaOH fusion; precision is ±10% for concentrations greater than 10 ppm. Fe, Mn, Ba, Sr, V, Co, Ni, Cr and Ti were determined by direct reading spectrography using a DC arc source (precision, ±30%).

Hydrothermal alteration was quantified by means of a fast and simple X--ray diffraction method. A Philips X-ray diffractometer was used operating at 34 kV and 24 mA with a sensitivity of 400 cps. Special care was taken to produce the same close packing and smooth flat surface for all samples. The rock samples were scanned from 4° to 40° at 2°/minute using Co-K_α radiation.The feldspar, kaolinite, montmorillonite, and illite-sericite peak intensities were measured at *d*-spacings of 3.18–3.21, 7.10, 14.7 and 10.0 Å respectively. The peak heights (intensities) recorded under the above specified conditions are related semi-quantitatively to the mineral concentrations.

Data processing

All of the elemental distributions in the various rock types were best approximated by a $\log_{10}$ normal distribution. For each element the geometric mean was calculated for the various mineralized and unmineralized rock types. As an estimate of the dispersion of the data about the mean, the mean plus one standard deviation of the $\log_{10}$ transformed data was also calculated.

Class intervals for distribution plots of As, Pb, Zn and F in the rocks of the Madenköy area correspond to the 20th, 50th, 80th and 95th percentiles of the $\log_{10}$ transformed data for dacitic rocks of the Eastern Black Sea region as a whole. Distribution patterns in the dacites and overlying pumice tuffs (these rocks have a very similar composition to dacites) can, therefore, be interpreted directly; those in rocks of the Upper Basic Series must be interpreted in the light of the mean concentrations for the elements compiled in Table III.

Reconnaissance bedrock surveys

During routine geological mapping of the Cayeli area, composite chip samples were taken from available bedrock exposures. This area contains examples of all three types of sulphide mineralization and is characterized by an extensive cover of basalts and andesitic tuff breccias of the Upper Basic Series in which there are three dacitic windows produced by extensive block faulting in NE-SW and NW-SE directions at Madenköy, Sirtköy and Sarisu. In the Madenköy and Sirtköy areas, the dacites are discontinuous with small outcrops occurring within a thin cover of basalts and haematitized pumice tuffs (violet tuffs), and in the eastern half of the area the absence of the andesitic-tuff breccia member of the Upper Basic Series exposes smaller, unmineralized outcrops of dacites.

For reconnaissance surveys, As (Fig. 2), F (Fig. 3), Pb and Zn effectively detect and distinguish the three types of mineralization, the first two elements being sensitive indicators of mineralizing processes (pyritization) and hydrothermal activity, respectively. In the Madenköy area, both elements exhibit good vector properties for locating mineralization, and over the blind Madenköy massive sulphide orebody the primary halos of F and As have an area of 200 m × 400 m, extending over both the dacite and basic cover rocks. In the Sirtköy area, F shows a slight enrichment in parts of the exposured stockwork zone; the overlying basic rocks exhibit no enrichment. Arsenic is, however, anomalous in both the mineralized dacites and overlying basic rocks. At Sarisu, F is slightly enriched in the hydrothermally altered

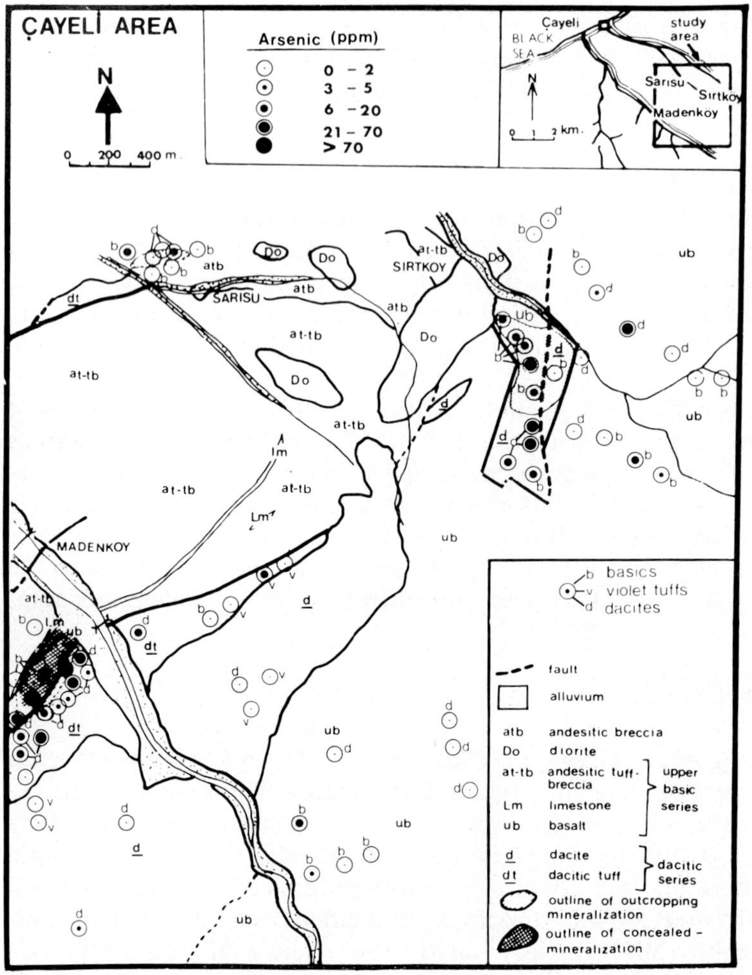

Fig. 2. As content of bedrocks, Cayeli area.

dacites as compared with unmineralized dacites. Arsenic exhibits no enrichment in the mineralized dacites and the element would, therefore, appear to be an active mineralizer only during processes of pyritization associated with stockwork and massive sulphide mineralization.

Pb and Zn are also good indicators, both elements clearly showing stronger anomalies in the vicinity of massive sulphides (Madenköy) as compared to the stockwork (Sirtköy) and disseminated pyrite (Sarisu) occurrences (Cagatay, 1976,see also Table IV). Cu is not a good indicator of the type of mineralization, although it is anomalous in the mineralized dacites as a whole.

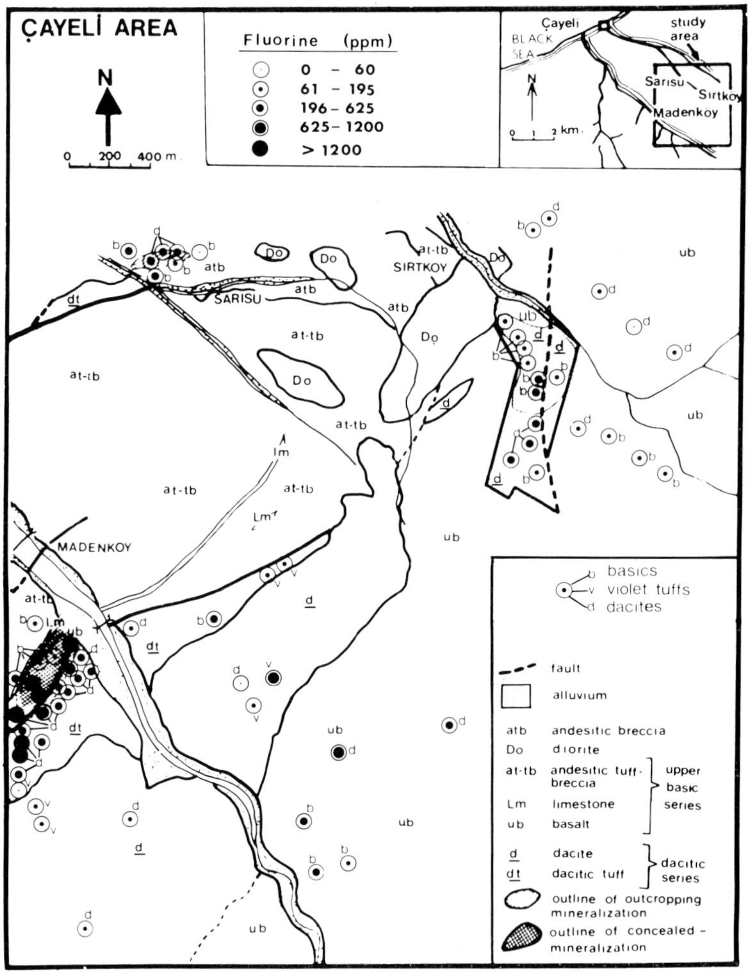

Fig. 3. F content of bedrocks, Cayeli area.

Primary haloes in dacitic host rocks

Dacitic rocks hosting massive sulphide orebodies generally contain a stockwork zone surrounded by more extensive pyritic impregnation and argillitic alteration zones. In the vicinity of suspected ore, the dispersion patterns of certain indicator elements in the dacitic host rocks may be particularly useful in indicating both the presence and position of massive sulphide ore. Initial exploration drilling will, of course, depend to a large extent on such detailed bedrock studies as those for As, F, Zn and Pb in the vicinity of the Madenköy deposit (Figs. 4, 5, 6 and 7).

The exposed section of dacitic rocks associated with the Madenköy deposit shows distinct signs of hydrothermal alteration (Fig. 9) and pyritic

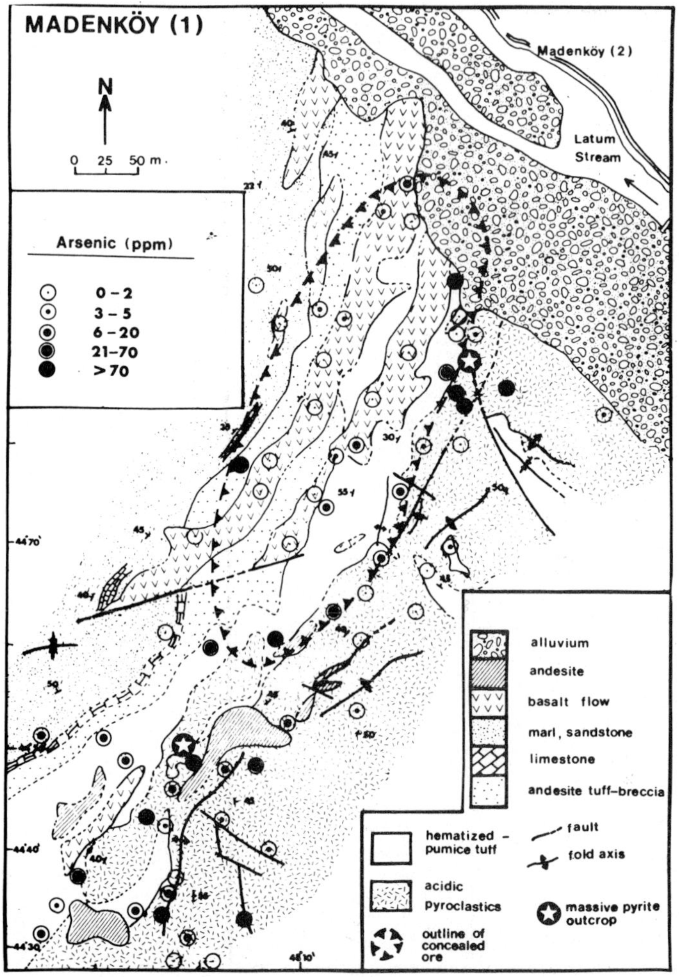

Fig. 4. As content of bedrocks, Madenköy deposit.

impregnation. Only a small part of the stockwork zone is exposed. As, F and, especially, Pb and Zn form anomalous haloes in the dacitic tuffs along the entire length of the dacite-pumice tuff contact. Although the average F and As contents (466 and 7 ppm, respectively, see Table IV) of the Madenköy dacites are noticably greater than those for unmineralized dacites (201 and 3 ppm, respectively) their distributions in the footwall rocks are only marked by a few sporadic anomalies; Cu displays a similar pattern to As. On the other hand, Pb and Zn exhibit almost continuous anomalies in the footwall zone and form a lateral halo extending 300 m to the south of the concealed zone. On a plan view the anomalous zones for As, Pb, Zn and F appear to extend 100—200 m away from the dacite-pumice tuff contact, but

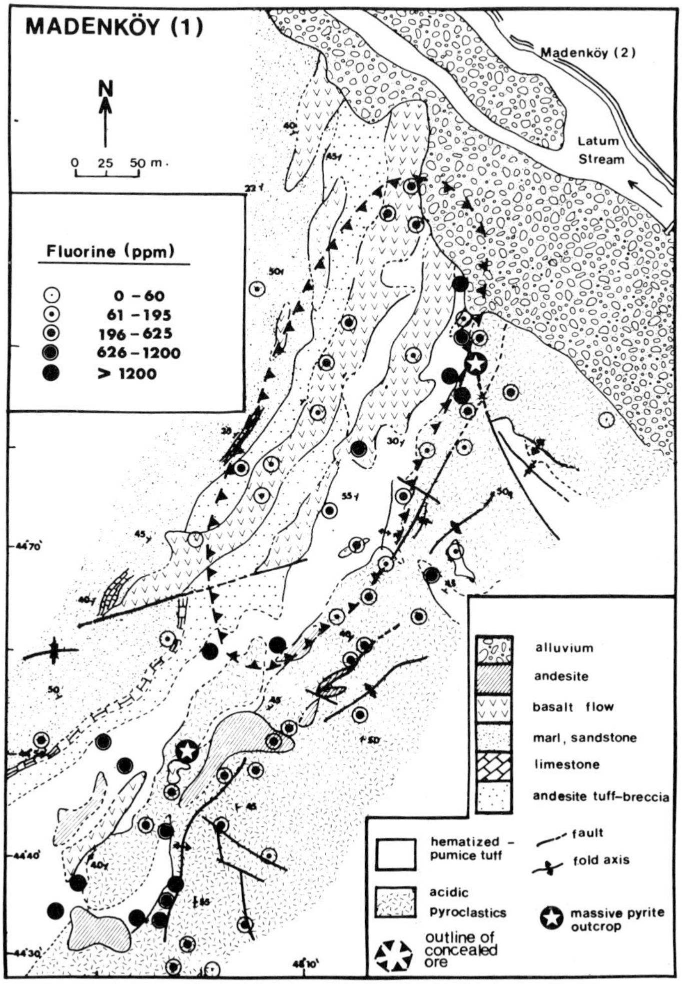

Fig. 5. F content of bedrocks, Madenköy deposit.

since the ore zone dips 50° to the northwest, these haloes actually represent a much narrower horizontal zone. Alternatively, the true lateral extension of these haloes out of the plane of the map, towards the southeast of the massive ore zone, may be as great as the lateral haloes extending along the southwest hinge line. These studies indicate that As, F, Pb, Zn and Cu may be used as vectors to indicate the location of massive sulphides during bedrock surveys of the dacitic formations.

Primary haloes in cover rocks

Exposures of dacitic rocks in the Eastern Black Sea region are scattered

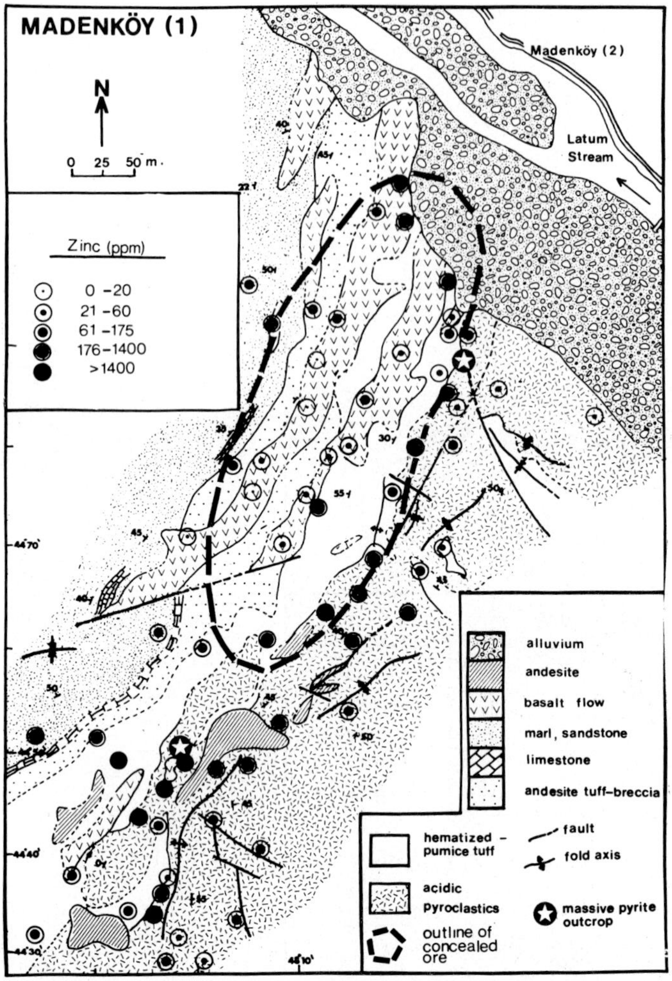

Fig. 6. Zn content of bedrocks, Madenköy deposit.

and there is much geological evidence suggesting that massive sulphide deposits are preserved (e.g. Madenköy, Kutlular) where sufficient cover of Upper Basic rock exists. Consequently, detection of these deposits is possible only if they have an anomalous elemental expression (halo) in the overlying Upper Basic Series. As shown by the aforementioned reconnaissance survey of the Cayeli area, a number of elements show such an expression (Figs. 2 and 3). More detailed sampling and analysis were, therefore, carried out in the Madenköy and Sirtköy areas to determine the exact nature of the primary haloes in the hanging wall (cover) rocks, the results being shown in Table III and Figs. 4, 5, 6 and 7.

In the Madenköy area, pumice tuffs (violet tuffs) lie directly over the massive sulphide mineralization. These rocks are similar to the haematitized

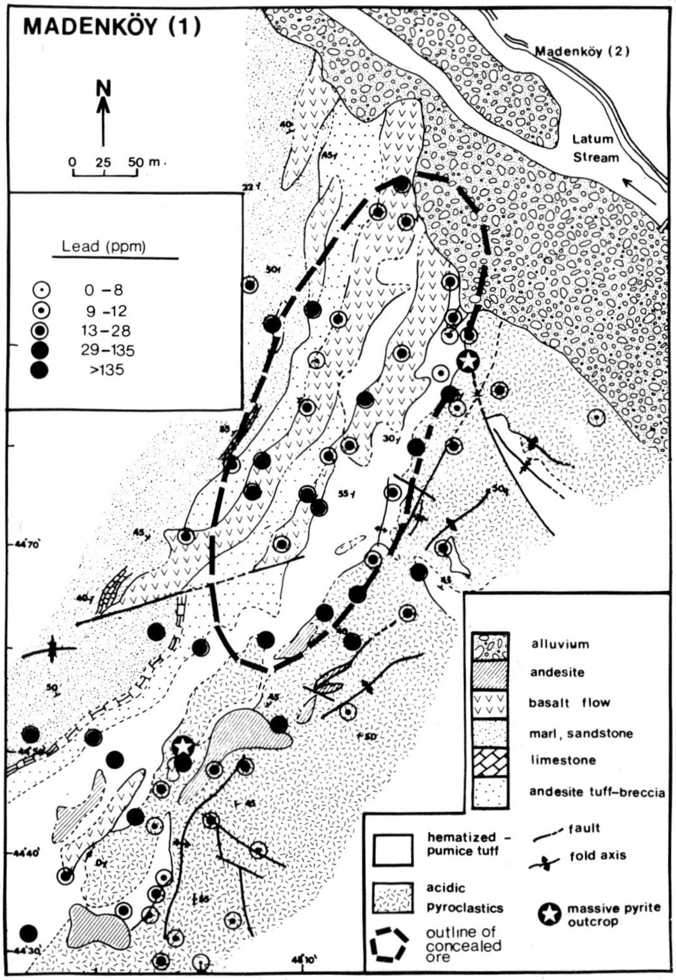

Fig. 7. Pb content of bedrocks, Madenköy deposit.

tuffaceous rocks overlying the Tunca and Murgul deposits. In terms of chemical composition, they are similar to dacites; the Fe content is represented almost entirely by haematitization. Compared to the Upper Basics, they have much lower Mn, V, Co, Ni, Cr and Ti contents and should, therefore, be considered as a transition phase between the Dacitic and Basic Series. On a visual basis, pumice tuffs from background and mineralized areas cannot be differentiated. Background pumice tuffs have similar Fe contents (mean 2.3%) to those covering mineralization (mean 2.8%). However, the tuffs covering the Madenköy deposit are considerable enriched in As, Cu, Pb, Zn, Mn, Ag, Ba and, particularly, F (Table III); for most of these elements their means exceed the means for background tuffs by more than one standard deviation. For the Madenköy tuffs, As, F, Zn, and Pb (Figs. 4, 5, 6 and 7) form anomalous

TABLE III

Composition of cover rocks associated with volcanogenic sulphide mineralization in the Cayeli area, Eastern Black Sea region. Values are given for antilog of the mean and mean + one standard deviation (in italics) of log_{10} transformed data (all values are in ppm except where otherwise noted)

	Background Upper Basics (11)		Background Pumice Tuffs (9)		Madenköy Upper Basics (19)		Madenköy Pumice Tuffs (31)		Sirtköy Upper Basics (9)	
Fe(%)	6.7	*(8.5)*	2.3	*(3.7)*	4.1	*(5.2)*	2.8	*(4.5)*	9.6	*(4.8)*
Mn	1850	*(3640)*	457	*(1007)*	1415	*(2300)*	730	*(2650)*	2400	*(5860)*
As	2	*(5)*	2	*(4)*	3	*(6)*	5	*(21)*	7	*(19)*
Cu	27	*(132)*	8	*(21)*	77	*(181)*	19	*(110)*	171	*(703)*
Pb	23	*(39)*	9	*(14)*	32	*(53)*	21	*(70)*	18	*(32)*
Zn	90	*(175)*	60	*(95)*	100	*(345)*	123	*(330)*	106	*(255)*
Ag	< 0.1	*(—)*	< 0.1	*(—)*	0.3	*(1.3)*	0.7	*(2.5)*	< 0.1	*(—)*
Cd	2	*(3)*	1	*(2)*	2	*(3)*	2	*(3)*	2	*(3)*
Mo	1	*(2)*	1	*(2)*	2	*(4)*	2	*(3)*	2	*(4)*
Ba	64	*(146)*	20	*(39)*	90	*(214)*	86	*(257)*	146	*(430)*
Sr	65	*(197)*	61	*(145)*	127	*(272)*	34	*(79)*	20	*(62)*
V	238	*(367)*	15	*(20)*	145	*(345)*	23	*(53)*	316	*(435)*
Co	21	*(42)*	33	*(49)*	19	*(35)*	4	*(12)*	17	*(66)*
Ni	12	*(28)*	10	*(35)*	36	*(105)*	13	*(73)*	14	*(37)*
Cr	15	*(38)*	17	*(51)*	108	*(415)*	17	*(41)*	26	*(59)*
Ti	5566	*(8340)*	703	*(1470)*	2870	*(4305)*	790	*(2000)*	5956	*(7762)*
F	140	*(237)*	105	*(160)*	256	*(482)*	450	*(1365)*	115	*(191)*

Notes: (1) numbers in brackets below each rock type represent the number of outcrops from which composite chip samples were taken; (2) Upper Basic rocks represent basalts, spilites and basaltic flows haloes along the full length of the dacite-tuff contact zone and extend more than 200 m southwest of the ore zone. When present, this formation constitutes an ideal horizon for sampling during detailed bedrock surveys.

Compared to their background counterparts, the Upper Basics (mainly basalts) overlying the Madenköy deposit are strongly enriched in F, Pb, and Ag (Table III), and these elements are, therefore, good indicators of underlying mineralization. As, Cu, Zn, Ba and Sr are only slightly enhanced in the Madenköy Basics, but as a group they may also aid interpretation. The Upper Basics covering the Sirtköy Cu-pyrite stockwork deposit have increased concentrations of As and Cu only, thus reflecting the type of mineralization over which they lie. There is a distinct diagnostic difference between the composition of basic rocks overlying massive sulphide ore (high F, Pb, and Ag) and those overlying Cu-pyrite stockwork mineralization (high As and Cu). These parameters can, therefore, be used to indicate the type of ore underlying anomalous zones in the Upper Basics.

Studies of the dispersion of Hg above and below the Madenköy orebody have produced disappointing results (Cagatay, 1976). Although the ore itself contains 1000—2000 ppb Hg, the overlying rocks (including the pumice tuffs) rarely have values greater than 40 ppb. In fact, the cut-off in primary dispersion into the pumice tuffs amounts to less than 0.30 m. In the footwall rocks, Hg shows a slight enrichment in the stockwork zone, but this dies out quickly in the impregnation zone. Hg is obviously not a good indicator in this province.

To date, studies in the Eastern Black Sea region of the primary dispersion of elements into hanging wall formations of massive sulphide deposits indicate that primary haloes in cover rocks are generally characterized by sporadic highs and are thus discontinuous in nature. During reconnaissance bedrock surveys, even single exposures of basic rocks displaying a higher than normal content of indicator elements such as F, Pb, Ag, As and Cu should form the foci for more detailed studies in which local population statistics similar to those generated in Table III can be compared with background data to ascertain the representivity of the reconnaissance anomaly in terms of possible blind ore.

Alteration as a guide to ore

In the Eastern Black Sea volcanogenic province, the sulphide deposits are associated with extensive zones of argillitic alteration mainly in the dacitic rocks but also in places in the hanging wall rocks. Kaolinite, sericite, montmorillonite and quartz are the most characteristic alteration minerals; chlorite is minor. In this study the clay (kaolinite plus montmorillonite plus illite-sericite)/feldspar ratio was found to be the best index for assessing the degree of hydrothermal alteration associated with mineralization. Based on the frequency distribution of this ratio, intervals of 0—0.2, 0.3—0.7, 0.8—6.5 and >6.5 were selected for purposes of pattern recognition. Interference

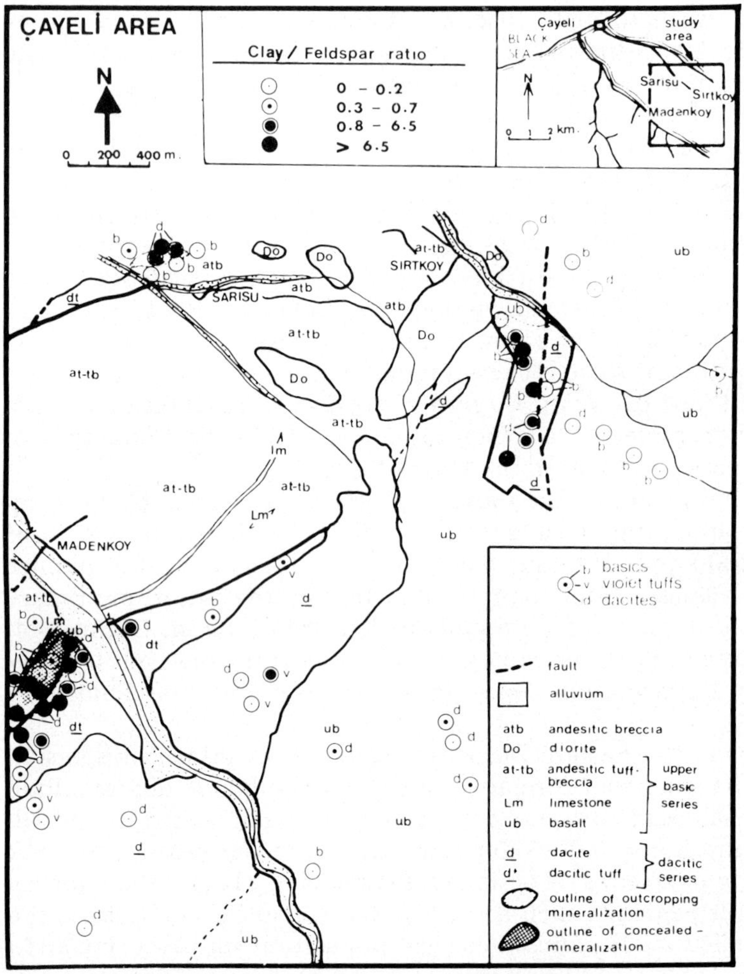

Fig. 8. Clay/feldspar ratio of bedrocks, Cayeli area.

from clay minerals formed as a result of weathering was kept to a minimum by selecting the freshest rock materials available.

The results of a reconnaissance survey of the Cayeli area and a detailed study of the hydrothermal alteration at the Madenköy deposit are presented in Figs. 8 and 9. For the reconnaissance survey, areas of strong alteration, as indicated by the clay/feldspar ratio, correspond to Madenköy, Sirtköy and Sarisu mineralizations. There is little difference in the intensity of alteration in these areas, and the ratio is, therefore, indicative of hydrothermal alteration only and not of deposit type. In the area of the Sarisu disseminated pyrite deposit, the dacites are highly altered but the overlying basic rocks are fresh. In the Sirtköy area, both the dacitic porphyry and the overlying basalts are strongly altered. Outside of the mineralized zone at Sirtköy, the

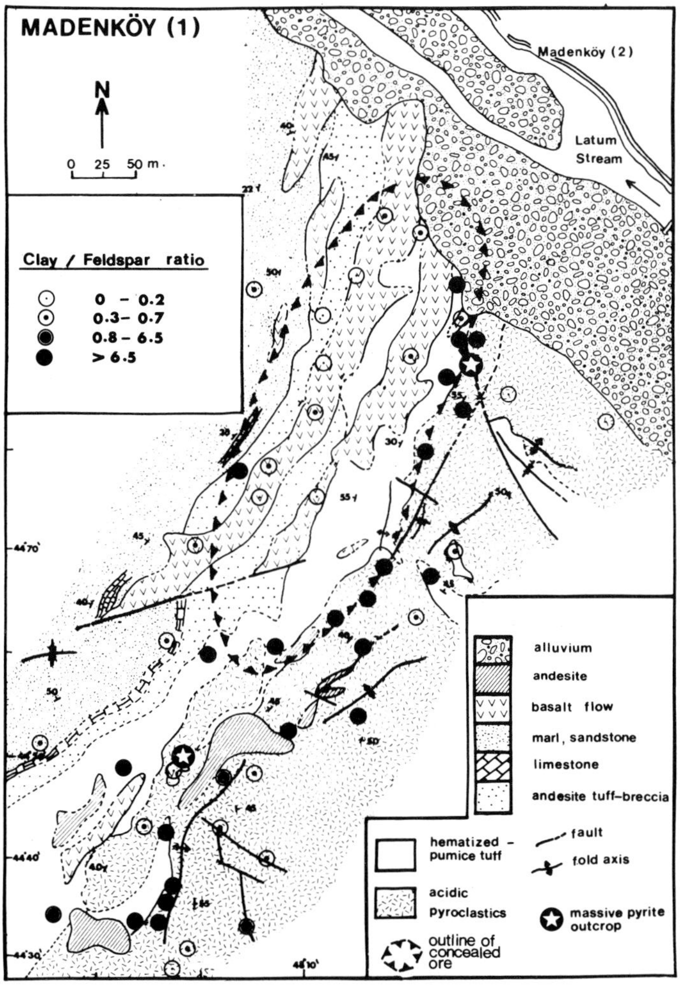

Fig. 9. Clay/feldspar ratio of bedrocks, Madenköy deposit.

dacites and basic rocks display no signs of alteration. In the large but discontinuous exposure of dacites in the Madenköy area, the clay/feldspar ratio outlines an alteration halo over the concealed massive sulphide deposit approximately 200 m in width and 500 m in lateral extent. Compared to the F and As haloes over this deposit, the clay/feldspar halo is more extensive.

In the immediate vicinity of the Madenköy deposit (Fig. 9), an extensive alteration zone extends along the entire pumice tuff-dacite contact zone. Above the ore zone, only the pumice tuffs display an anomalous alteration pattern; the basic rocks exhibit few signs of alteration and the clay/feldspar halo in the hanging wall rocks is, therefore, not as pronounced as the primary elemental haloes for F, Pb and Ag. Intensive alteration in both the dacites and pumice tuffs continues for about 300 m southwest of the ore zone; this trend is also evident for the base metals, F and As. The alteration halo in the dacites is much more uniform than the primary haloes for the indicator elements.

The relative proportions of the clay alteration products may be used to estimate the magnitudes of hydrothermal and mineralizing processes. Sericitization and kaolinization become more pronounced as the stockwork and massive sulphide ore is approached, a feature that permits their use as directional vectors during trenching and drilling operations.

Regional assessment of mineralized stockwork and impregnation zones

Mineralized zones in dacites are widespread in the Eastern Black Sea volcanogenic province; some are associated with sulphide orebodies, others are not. Another feature of the province is that some of the massive sulphide deposits are completely concealed beneath a cover of Upper Basics, their only signature being an exposed portion of the mineralized dacitic footwall rocks. During this study, an attempt was made to develop geochemical criteria for differentiating the indicator qualities of the various mineralized zones. This was done by bedrock sampling in stockwork-impregnation zones for which a priori information on mineral potential was known (i.e., Madenköy, Kutlular, Tunca, Kotarak Dere, Sirtköy and Sarisu deposits).

In many cases, the exposures of and demarcation between the mineralized zones is not good, and a decision as regards mineral potential is possible only if care is taken to sample the same type of material at each outcrop. On this account, it was decided to sample only altered rocks visibly free of sulphides mainly because this type of material comprises the largest percentage of the available rock exposures. Since it is often difficult, with limited exposures, to determine whether the stockwork or impregnation zones (or transitions between the two) are being sampled, it was decided on logistical grounds to classify exposures merely as mineralized dacites, noting any observations that might aid interpretation. For the Madenköy deposit, Cu was the only element that exhibited a marked difference in concentration between stockwork and impregnation zones. Fortunately this element is not a strong indicator of the deposit type.

The average compositions (and mean + one standard deviation) of the mineralized dacites associated with massive sulphide, Cu-pyrite stockwork, and barren pyritic impregnation deposits are given in Table IV. The compositions of the mineralized dacites as a whole (excepting those at Sarisu) and the unmineralized dacites from the Cayeli area are also given.

Compared to their background counterparts, rocks taken from the mineralized stockwork-impregnation zones (whole population) are significantly enriched in Fe, As, Cu, Pb, Zn and F and to a lesser extent in Ag, Cd, Mo and Ba. Mn, Sr, Co, Ni and Ti are not noticeably enriched in these mineralized zones, indicating their lack of participation in mineralizing processes.

TABLE IV

Composition of dacitic rocks associated with volcanogenic sulphide mineralization in the Eastern Black Sea region. Values are given for antilog of the mean and mean + one standard deviation (in italics) of $\log_{10}$ transformed data (all values are in ppm except where otherwise noted)

	Unmineralized dacites* (19)		Mineralized dacites** (83)		Massive sulphide deposits						Stockwork deposits				Barren	
					Madenköy (36)		Kutlular (10)		Tunca (7)		Kotarak Dere (14)		Sirtköy (16)		Sarisu (10)	
Fe(%)	1.2	*(4.4)*	2.5	*(6.0)*	1.8	*(3.7)*	2.0	*(3.6)*	2.9	*(8.9)*	3.0	*(6.4)*	5.2	*(6.8)*	5.3	*(7.4)*
Mn	303	*(1080)*	400	*(1125)*	362	*(860)*	208	*(365)*	279	*(720)*	537	*(1472)*	675	*(2167)*	248	*(700)*
As	3	*(9)*	6	*(26)*	7	*(22)*	21	*(97)*	40	*(156)*	3	*(9)*	7	*(28)*	2	*(4)*
Cu	13	*(42)*	36	*(215)*	40	*(295)*	32	*(131)*	120	*(845)*	38	*(260)*	43	*(272)*	26	*(79)*
Pb	14	*(27)*	21	*(63)*	25	*(83)*	29	*(126)*	70	*(367)*	17	*(32)*	13	*(23)*	16	*(27)*
Zn	72	*(142)*	95	*(475)*	184	*(627)*	11	*(27)*	505	*(4850)*	148	*(895)*	74	*(170)*	20	*(63)*
Ag	< 0.1	*(–)*	0.2	*(1.0)*	0.2	*(1.2)*	0.5	*(2.8)*	1.3	*(7.5)*	0.3	*(1.1)*	0.1	*(0.3)*	0.1	*(–)*
Cd	1	*(2)*	2	*(6)*	2	*(8)*	1	*(2)*	5	*(29)*	2	*(8)*	2	*(2)*	2	*(3)*
Mo	1	*(3)*	2	*(6)*	2	*(4)*	10	*(30)*	2	*(9)*	3	*(7)*	2	*(4)*	2	*(4)*
Ba	84	*(270)*	113	*(437)*	90	*(356)*	226	*(525)*	816	*(2770)*	208	*(757)*	90	*(323)*	50	*(84)*
Sr	27	*(60)*	25	*(82)*	23	*(75)*	9	*(13)*	33	*(142)*	11	*(25)*	40	*(104)*	73	*(260)*
V	17	*(36)*	28	*(94)*	11	*(22)*	24	*(38)*	29	*(36)*	37	*(70)*	87	*(337)*	305	*(577)*
Co	3	*(7)*	3	*(9)*	2	*(5)*	2	*(3)*	2	*(4)*	4	*(10)*	12	*(35)*	7	*(35)*
Ni	12	*(35)*	13	*(36)*	13	*(29)*	22	*(48)*	24	*(47)*	24	*(37)*	7	*(17)*	9	*(29)*
Cr	20	*(48)*	21	*(54)*	20	*(40)*	32	*(62)*	38	*(59)*	33	*(61)*	8	*(22)*	41	*(145)*
Ti	914	*(2430)*	1075	*(3515)*	454	*(1336)*	955	*(1810)*	1680	*(1995)*	1660	*(3148)*	2570	*(7870)*	5148	*(8055)*
F	201	*(432)*	306	*(740)*	466	*(1100)*	154	*(250)*	750	*(1406)*	360	*(711)*	173	*(587)*	284	*(385)*

*From Cayeli area.

**Includes dacites from all deposits except Sarisu

Note: Numbers in brackets for the various areas represent the number of outcrops from which composite chip samples were taken.

The results for the individual deposits indicate that no single element can be adequately used as a discriminator. For example, due probably to a much higher level of hydrothermal activity at Kutlular, some of which post-dates the main stage of mineralization, the more readily leached elements such as Zn, Sr and F exhibit a considerable depletion in the mineralized dacites (it should be pointed out that part of the Kutlular massive ore contains about 1.5% Zn). Furthermore, Mo, which is only slightly anomalous in other deposits, displays a very high mean concentration in the Kutlular dacites; to a lesser extent the same is also true for Ag and Ba. A further problem affecting the interpretation of single-element data relates to the proximity of the sampled zone to the orebody. For the Tunca dacites, the relatively high concentrations observed for all of the indicator elements is related to the fact that a fault bisects the ore zone thus permitting the exposure of the stockwork-impregnation zone immediately below the ore.

Since each of the massive sulphide deposits have different characteristics with respect to tenor or ore, degree of hydrothermal activity and degree of exposure, an adequate assessment of mineralized dacitic exposures can only be made by examining, on a relative basis, a more comprehensive selection of trace elements. Thus, the dacites associated with massive sulphide ores form a clearly anomalous group having abnormally high concentrations of most, and in some cases (Madenköy, Tunca) all, of the base metal indicators (Cu, Pb, Zn, Ag) as well as As and F. Compared to the other types of mineralization outlined in Table IV, Pb is the only element which is consistently higher in the footwall rocks of massive sulphide deposits; greater weight can, therefore, be given to the observed levels for this element.

Although the Kotarak Dere deposit has been classed as a stockwork deposit, there is considerable evidence to suggest that it represents the exposed root of an eroded massive sulphide deposit. A small remnant of massive pyritic sulphide mineralization occurs in the middle of the exposure and disseminations of sphalerite (unlike Sirtköy) have been found in the stockwork zone. Geochemically, the Kotarak Dere dacites are similar to the footwall rocks of the Madenköy deposit; only As and Pb exhibit lower levels in these rocks. Although it is probable that the massive sulphide orebody has been eroded, it is possible that massive ore is present below the covered flanks. Present drilling has been confined to assessing the mineral potential of the stockwork zone.

The Sirtköy stockwork has a chemistry distinctly different from that related to massive sulphide mineralization. Of the ore indicators, only Fe, As and Cu are enriched in this deposit. The other base metals, together with F, have values similar to those for unmineralized dacites. The basic elements Mn, V, Co and Ti are also high in these rocks but Ni and Cr exhibit no particular enrichment. It is suggested that these elements were incorporated as the result of contact-metasomatic processes between intrusive dacites and Upper Basics. They may, therefore, be used to indicate stockwork deposits having more of an intrusive nature rather than a submarine exhalative one.

Finally, the Sarisu impregnation deposit is characterized by high levels of Fe and Cu only. Geochemically, deposits of this type can be differentiated from more productive mineralized zones by their lack of high As, F and base metals (excepting Cu).

Depending on the degree of exposure, a minimum of ten to fifteen samples representing the areal extent of the mineralized zone should be taken for a regional assessment. Where only single mineralized outcrops are exposed great care should be taken to sample material free of sulphides. This method of visual cut-off in sampling was found to give the most meaningful results in the Eastern Black Sea province. Exposures which have higher than normal contents of most of the base metals (particularly Pb), as well as As and F, should be given a high priority for detailed surveying.

CONCLUSIONS AND GUIDELINES FOR EXPLORATION

Bedrock surveys utilizing samples of the host dacites and rocks of the Upper Basic Series have proven effective in outlining various types of primary haloes associated with the sulphide deposits. For reconnaissance surveys, As, F, Pb and Zn are of most value for detecting and differentiating between the three types of mineralization. F and As are sensitive indicators of hydrothermal activity and mineralizing processes (pyritization) respectively, and these elements consequently possess good vector properties for pointing out mineralization on a reconaissance level. On a detailed basis, it has been shown that the elements F, As, Pb, Zn and Cu have characteristic primary haloes with high anomaly contrasts in the dacitic host rocks around mineralization. The concentrations of the indicator elements increase significantly in the direction of massive sulphide mineralization and can, therefore, be used as directional vectors during drilling programmes. In the overlying haematitized pumice tuffs and rocks of the Upper Basic Series, significant primary haloes marked by enrichments of some or all of As, F, Cu, Pb, Zn, Cd, Ag and Ba are associated with the various types of underlying sulphide deposits.

Quantitative evaluation of the clay minerals/feldspar ratio is useful in detecting the presence and intensity of hydrothermal activity related to sulphide mineralization. The relative proportions of the alteration products (clay minerals), as well as the alteration ratio, can be used as directional vectors during drilling programmes.

Diagnostic parameters have been formulated for assessing the mineral potential of exposed stockwork and impregnation zones on a regional basis. Barren pyritic impregnation deposits as at Sarisu have only anomalous Fe and Cu contents, whereas Cu-pyrite stockwork deposits such as those at Sirtköy, not associated with massive sulphide ore, are enriched in Fe, As and Cu. Stockwork-impregnation zones associated with massive sulphides are significantly enriched in most but not necessarily all of the following: Fe, As, F, Cu, Pb, Zn, Ag, Cd and Mo.

The present studies demonstrate a strong relationship between fluorine and calc-alkaline submarine volcanism leading to the formation of massive sulphide and Cu-pyrite stockwork deposits. In view of the fact that, in the Eastern Black Sea province, F migrates greater distances into hanging wall rocks and shows a larger anomaly contrast than any of the other elements investigated, it would be worthwhile to study its behaviour in other volcanic environments.

Guidelines for exploration in the Eastern Black Sea province of Turkey are dependent upon a sound knowledge of the chemistry and alteration characteristics of both the dacitic host rocks and basic cover rocks, genetic and structural features of the various types of deposits, and adequate orientation data on the concentration levels and dispersion patterns exhibited by the various indicator elements in the primary and secondary environments. As in many other areas, the efficacy of bedrock surveys is severely affected by limited exposure; for this reason the present studies have also been directed towards determining whether primary haloes in the basic rocks have a signature in the overlying soils. This would permit both soils and rocks to be sampled on a regular grid pattern (metallometric surveying), thus increasing the effectiveness of both reconnaissance and detailed surveys. Preliminary results on this aspect of the research in the Black Sea region are encouraging.

ACKNOWLEDGEMENTS

The present work forms a part of a more extensive Ph. D. research programme by M.N. Cagatay with the Applied Geochemistry Research Group, Imperial College, London, England. The authors would like to thank Professor S. Alpan, General Director, Maden Tetkik ve Arama Enstitusu (M.T.A., Ankara, Turkey) for making funds available to carry out this research and for authorizing publication. The generous assistance of the M.T.A. geologists in the Black Sea region is much appreciated. The analytical staff of the Applied Goechemistry Research Group are thanked for their help in carrying out some of the analytical work. Assistance with computional programming by Mr. S. Earle is gratefully acknowledged.

REFERENCES

Cagatay, M.N., 1976. Development of geochemical exploration techniques for massive sulphide ore deposits, Eastern Black Sea region, Turkey. Ph.D. Thesis, University of London, London.

Hattie, D.W., 1977. The geology and mineralogy of the Lahanos and Kizilkaya deposits. Ph. D. Thesis, University of Western Ontario, London, Ont. (in preparation).

Kraeff, A., 1963a. A contribution to the geology of the region between Sirya and Ardanuc. Maden Tetkik Arama Enst. Bull., 60: 37—44.

Kraeff, A., 1963b. Geology and mineral deposits of the Hopa-Murgul region (western part of the province of Artvin, N.E. Turkey) Maden Tetkik Arama Enst. Bull., 60: 45—60.

Kronberg, P., 1970. The tectonics of the eastern Black Sea mountains, N.E. Turkey. Maden Tetkik Arama Enst. Bull., 74: 57—65 (in Turk).
Leitch, C.H.B., 1975. Lateral distribution of trace metals in two Turkish volcanogenic sulphide deposits. M. Sc. Thesis, University of London, London, 212 pp.
Nebioglu, T.Y., 1975. Geological map of Madenköy. Maden Tetkik Arama Enst. Map, No. 31134.
Tugal, T., 1969. The pyritic sulphide deposits of the Lahanos Mine area, Eastern Black Sea region, Turkey. Ph. D. Thesis, University of Durham, Durham (unpublished).
Vujanovic, V., 1974. The basic mineralogic, paragenetic and genetic characteristics of the sulphide deposits exposed in the Eastern Black Sea coastal region (Turkey). Maden Tetkik Arama Enst. Bull., 82: 21—36.

Journal of Geochemical Exploration, 8(1977) 73—83

ANOMALOUS TRACE ELEMENTS IN PYRITE IN THE VICINITY OF MINERALIZED ZONES AT WOODLAWN, N.S.W., AUSTRALIA

W.R. RYALL

Division of Mineralogy, Commonwealth Scientific and Industrial Research Organization, North Ryde, N.S.W. (Australia)

(Revised version accepted February 10, 1977)

ABSTRACT

Ryall, W.R., 1977. Anomalous trace elements in pyrite in the vicinity of mineralized zones at Woodlawn, N.S.W., Australia. J. Geochem. Explor. 8: 73—83.

Pyrite from both chloritized and silicified alteration zones around the Woodlawn, N.S.W., Australia, volcanogenic massive sulphide deposit is characterized by anomalous trace element levels which reflect the presence of base metal inclusions.

Ag, As,Bi, Cd, Co, Cu, Mn, Ni, Pb, Sb, Sn and Zn contents of 76 pyrite samples from 48 drill cores have been determined by a semi-quantitative emission spectroscopic method and the samples classified by a cluster analysis technique which defined four groups. Group 1 represents pyrite from the zone of chloritization in the footwall, and members of this group have been recognized at distances greater than 0.5 km from mineralized zones. The group is characterized by high Pb, Ag, Zn and Cu contents and a mean Co/Ni ratio of 0.8. Inclusions, peripheral replacement by base metal sulphides, and rounded crystal edges are commonly observed features.

Group 2 includes pyrite generally found as fine-grained aggregates in cross-cutting, chlorite-rich bands in footwall units. Pyrite of this group contains lower base metal values than members of group 1. Group 3 members, from the zone of silicification around the ore deposit, have the highest Zn, Cu, Cd, Sn and Bi contents, high As, Pb, Sb and Ag, and low Ni contents, with a high Co/Ni ratio (mean 21.0). Group 4 pyrite is from rocks equivalent to the hanging wall remote from mineralization and has low base metal, Ag and Ni, but high As and Bi contents, with a high Co/Ni ratio (mean 7.9).

Affinity is shown between groups 1 and 2, and between groups 3 and 4 on the basis of Co, Ni, As and Bi contents and Co/Ni ratios, but enrichment of group 1 and 3 members in the ore-forming elements shows them to have been associated with mineralizing solutions.

The distribution and mineralogical nature of inclusions in pyrite from alteration zones encountered in volcaniclastic piles has potential to indicate proximity to mineralization and whether the pile itself is mineralized.

INTRODUCTION

Pyrite is commonly encountered in the search for economic base metal deposits and, being the most common sulphide in the deposits themselves, has held a special fascination for the exploration geologist. However, little success has been achieved in interpreting the exploration significance of its areal distribution, trace element contents or included phases. The trace element content, particularly the Co/Ni ratio and the Se content, has been widely used to indicate the environment of deposition (Fleischer, 1955) and it is commonly found that pyrite of sedimentary origin has a Co/Ni ratio of less than unity, whereas pyrite of igneous origin has a Co/Ni ratio of greater than unity. Many exceptions have been recorded, however, and Loftus-Hills and Solomon (1967) have shown that the Co/Ni ratio alone is not diagnostic, and that concentration levels must also be considered.

Pyrite from mineralized zones has frequently been reported to have appreciable base metal contents, but these have been shown to be due, at least in part, to included phases (Hegeman, 1941; Fleischer, 1955; Sutherland, 1967).

The exploration potential of determining trace element contents of pyrite and interpreting the significance of included phases have been rarely exploited because of a poor understanding of the nature and genetic significance of pyrite in host rocks, the difficulty of dealing with frequent local variability of trace element levels and problems associated with obtaining samples in sufficient number and amount.

Sinclair (1967), Dawson and Sinclair (1974) and Sinclair et al. (1976) have demonstrated that the trace element content of pyrite, after appropriate statistical treatment, reflects mineralization in "porphyry-type" deposits in British Columbia. However, the utilization of pyrite as an indicator of base metal mineralization in volcaniclastic terrains has not been widespread. In such areas, further complications are posed by development of pyrite in the sedimentary environment and by frequent rapid changes in facies, each with its characteristic pyrite population. Pyrite *is* used as an important exploration guide in the Green Tuff region of Japan, where its presence, either as disseminations in the country rocks or "pyritized zones", is considered an indicator of proximity to Kuroko mineralization (Ohtagaki et al., 1974; Tanimura et al., 1974).

In volcanogenic terrains less well known than those in Japan, the widespread occurrence of disseminated pyrite requires that some discriminatory information be sought. Branov et al. (1972) showed that pyrite from alteration aureoles of massive sulphide deposits in the Ural Mountains contains inclusions of chalcopyrite, sphalerite and galena, although the dimensions of the aureoles are not stated. Smirnov et al. (1974) have also shown that disseminated pyrite, for an unstated distance both below and above the ore in the Tisinsk massive sulphide deposit, is enriched in Ag, Bi, Cu, Pb and Zn relative to pyrite remote from mineralization.

As part of a study of the trace element content of sulphide phases in Silurian volcanic units of the Lachlan Orogenic Belt in eastern Australia, pyrite from the main ore lens and surrounding alteration zones of the Woodlawn deposit has been examined to provide data on trace element variability and the nature of included phases.

DISTRIBUTION OF PYRITE AT WOODLAWN

The Woodlawn Zn-Pb-Cu massive sulphide deposit, located some 200 km southwest of Sydney, N.S.W., Australia (Fig. 1), is situated on the flank of a folded Silurian volcaniclastic pile (Fig. 2) and has been ascribed a submarine volcanogenic origin (Malone et al., 1975). Pyrite, the principal sulphide phase, making up 50—75 vol.% of the ore zone, is also widely disseminated in the country rocks, becoming more abundant marginal to the ore lenses.

Within the main ore zone, pyrite has a bimodal size distribution. Ayres et al. (1976) have distinguished a coarse-grained group, consisting largely of euhedral crystals from 0.1 mm to greater than 0.5 mm across with fractures filled by other sulphide phases, and a second, finer-grained group with subhedral to euhedral crystals less than 0.1 mm across, which are generally clustered into stringers.

Away from the ore zones, pyrite occurs most commonly in rocks showing a high degree of both chlorite and silica-sericite alteration. Where chlorite is intensely developed, the rocks are termed chlorite schists and these grade laterally and vertically into chloritized tuffs which are strongly developed in the footwall, and extend southwards over 0.5 km from the main ore zone. Relict porphyritic textures, visible in some rocks, have led Malone et al. (1975) to suggest that these units represent altered volcaniclastic rocks.

Pyrite in the chlorite schists ranges from about 50 vol.% close to the ore zone to less than 1 vol.% remote from it and occurs as disseminated crystals and stringer aggregates, which in places are accompanied by chalcopyrite and

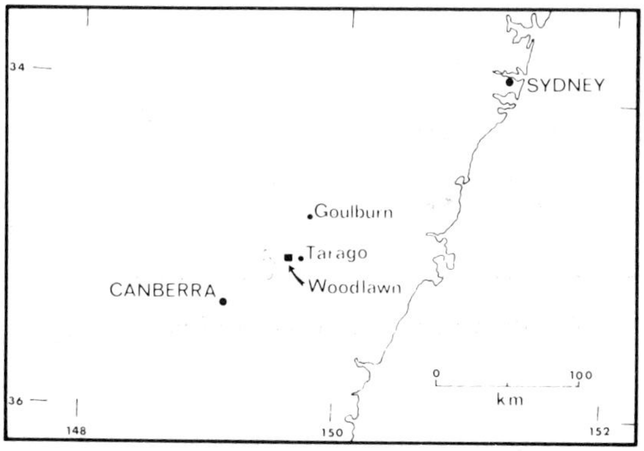

Fig. 1. Location map, Woodlawn, N.S.W., Australia.

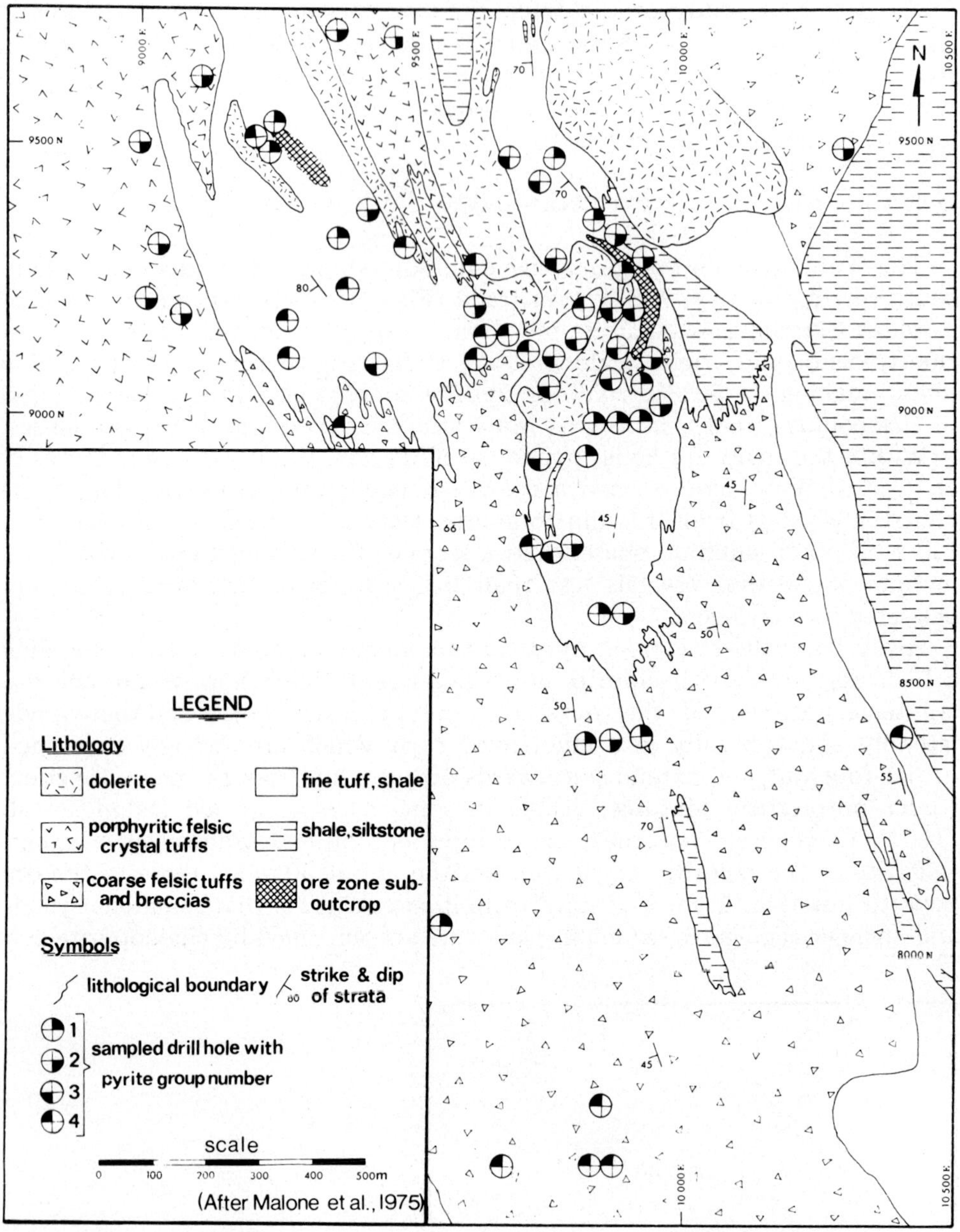

Fig. 2. Generalized geological map of the Woodlawn area showing location of sampled drill holes and pyrite classification.

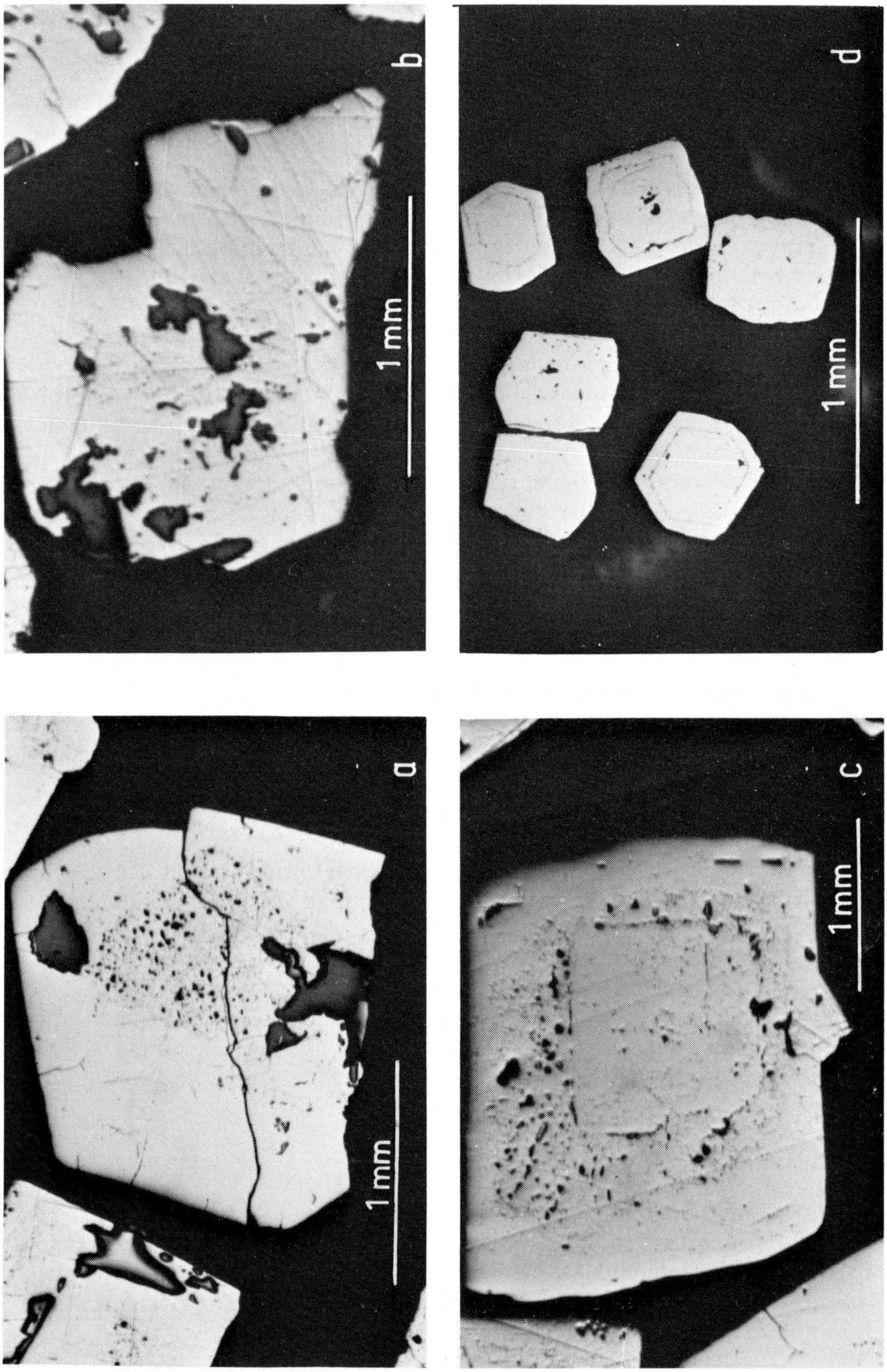

Fig. 3. (a) Pyrite from chlorite schist zone showing rounded edges, base metal inclusions, and peripheral replacement by sphalerite and galena. (b) Embayed pyrite from chlorite schist zone with included and peripheral galena and sphalerite. (c) Pyrite showing zoned inclusions of galena and chalcopyrite. From chlorite schist zone. (d) Euhedral pyrite from silicified zone showing zoned sphalerite and chalcopyrite inclusions.

sphalerite. Pyrite in some footwall rocks and disseminated pyrite in some hanging wall units are accompanied by quartz-filled pressure shadows indicating pre-tectonic formation: grain size ranges from 0.2 mm to 25 mm, with the mode around 1 mm. Most crystals are cubes, although larger crystals near the ore zones have rounded edges (Fig. 3a, b, c). Inclusions of galena, sphalerite and, to a lesser extent, chalcopyrite, each generally from 5 to 50 μm across, are commonly present and define zones oblique to crystal edges. Peripheral replacement by sphalerite, galena and, less commonly, chalcopyrite is widespread close to ore zones and in mineralized stringer zones. Crystals etched with 1:1 HNO_3 commonly reveal complex concentric and colloform growth patterns.

An irregularly shaped zone of silicification, with accompanying minor sericite and chlorite, extends from below the main ore zone to 50 m above it, and over 0.5 km southwards from the deposit. Pyrite in this zone is typically present as fine-grained disseminated cubes and stringers ranging from 0.1 mm to 0.5 mm across. Inclusions are generally of sphalerite or chalcopyrite and are commonly concentrated near present crystal surfaces (Fig. 3d). Peripheral replacement by galena and sphalerite has been observed only in bands of siliceous tuff ore. Etched crystals commonly show zonal structures similar to those described above.

SAMPLE PREPARATION AND ANALYTICAL PROCEDURES

Seventy-six pyrite concentrates from 48 drill cores were selected to provide data both close to the main ore zone, and from regions removed from known mineralization. Samples were prepared by crushing in an agate mortar, followed by heavy liquid and magnetic separation. Samples used for analysis were hand-picked under the binocular microscope and a representative portion examined in polished section.

Samples were ground to minus 75 μm in an agate mortar and Ag, As, Bi, Cd, Co, Cu, Mn, Ni, Pb, Sb, Sn and Zn determined by a semi-quantitative emission spectrographic method using 15 of sample mixed with 15 mg of graphite. Cr, Ga, Ge, In, Mo, V and W were sought, but were mostly below detection limits.

The analytical precision for each element, determined from eight duplicate samples using the method of Garrett (1969), is shown in Table I. Precision levels are mostly consistent with emission spectrographic results but, where high, may be due to sample inhomogeneity.

CLASSIFICATION

In order to measure the ability of the trace element contents of pyrite to indicate the proximity to mineralization at Woodlawn, it is necessary to compare the trace element contents of pyrite from rocks displaying chloritic and siliceous alteration with the trace element contents of pyrite from un-

TABLE I

Detection limits for emission spectrographic method with mean values, and analytical precision based on eight duplicate samples

Element	Detection limit (ppm)	Arithmetic mean (ppm)	Precision at 95% confidence level (%)
Ag	0.1	10	45
As	50	1040	26
Bi	1	63	15
Cd*	3		
Co	5	130	37
Cu	1	812	46
Mn	10	37	40
Ni	5	266	27
Pb	1	307	18
Sb*	30		
Sn	1	3	47
Zn	20	575	31

* Below detection limit for most samples.

altered hanging wall and footwall rocks remote from mineralization.

Although pyrite from rocks close to the ore zone, which clearly display siliceous or chloritic alteration, was generally easily classified, pyrite in rocks remote from the ore zone was difficult to classify due to the presence of overlapping hydrothermal alteration zones and the presence of chlorite of metamorphic origin. Classification of the pyrites was effected by an *R*-mode cluster analysis procedure using program MULCLASS (Lance and Williams, 1967) with logarithmically transformed analytical data and a similarity measure based on a Euclidean distance standard, defined by Burr (1967). Initial sorting was by a group average strategy (Sokal and Michener, 1958) and a more intense sorting was achieved using a function which minimized incremental sums of squares.

RESULTS

Four groups of pyrite were defined as shown in Fig. 4. The mean element concentrations for each group are given in Table II. The large standard deviations mainly reflect the variability of inclusion development and reduce the usefulness of mean values in characterizing the populations. A similar problem confronted Dawson and Sinclair (1974).

Group 1A pyrite is located in chlorite schist alteration zones in tuffs, both around and south of the main ore zone. Other group 1A pyrite is associated with chlorite schists around a mineralized lens located about 0.7 km northwest of the main deposit. Pyrite assigned to this subgroup has high Pb, Ag, Zn and Cu levels and a mean Co/Ni ratio of 0.8.

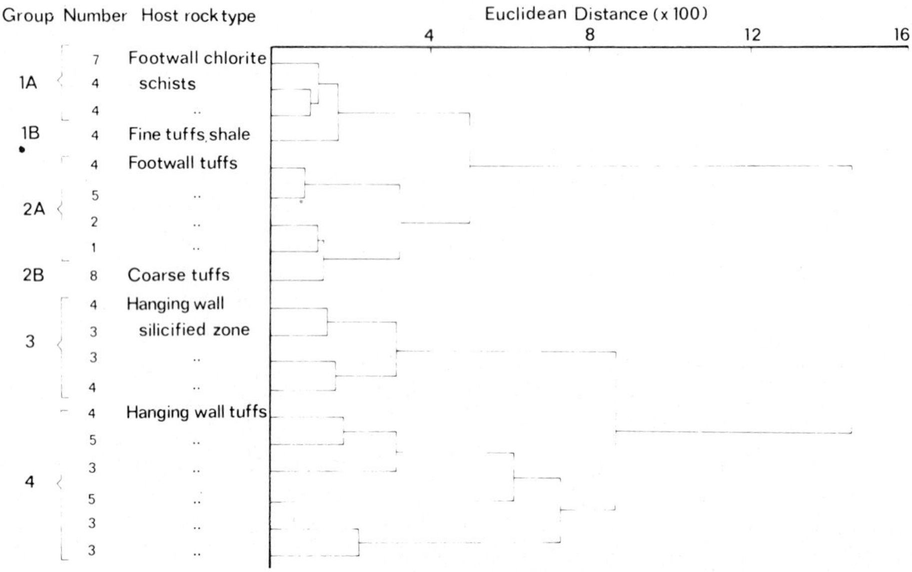

Fig. 4. *R*-mode cluster analysis dendrogram for 76 pyrites

TABLE II

Mean element concentrations for groups defined by cluster analysis (standard deviations are shown in parenthesis)

Element	Group 1A (ppm)	Group 1B (ppm)	Group 2A (ppm)	Group 2B (ppm)	Group 3 (ppm)	Group 4 (ppm)
Ag	13(7)	12(5)	4(3)	3(2)	19(8)	3(2)
As	163(82)	1100(135)	276(150)	508(367)	425(216)	494(207)
Bi	6(5)	14(6)	2(2)	55(19)	118(34)	29(11)
Cd*	3	<1	2	< 1	5	2
Co	119(65)	105(47)	316(185)	148(76)	84(21)	95(38)
Cu	389(250)	315(209)	289(190)	183(111)	820(403)	181(56)
Mn	24(12)	25(13)	42(18)	37(15)	28(9)	11(8)
Ni	156(73)	188(74)	290(95)	122(99)	4(2)	12(8)
Pb	858(465)	290(94)	236(127)	123(116)	363(201)	144(75)
Sb	26(8)	27(7)	31(12)	43(14)	36(20)	23(18)
Sn*	1	<1	< 1	< 1	9(8)	1
Zn	710(452)	216(93)	75(40)	95(43)	1220(502)	187(46)
Co/Ni	0.8	0.6	1.1	1.2	21.0	7.9
Number of members	15	4	12	8	14	23

* Most samples below detection limits.

Pyrite from fine-grained volcaniclastic-sedimentary units located distant from mineralization, and showing only minor chlorite development, has been distinguished as group 1B. The Co/Ni ratio and Co, Ni, Cu, Mn and Sb contents are similar to group 1A, but it contains less Pb and Zn, and greater As and Bi than group 1A.

Group 2 pyrite has generally been found as fine-grained aggregates and stringers in cross-cutting, chlorite-rich bands in coarser-grained footwall units at moderate distances from mineralization. Pyrite from coarse-grained volcaniclastic units remote from mineralization, which has a similar Co/Ni ratio and Ag, Cu, Sb and Zn contents, less Co, Ni and Pb, and greater As, Bi and Mn contents, has also been classified with this group.

Group 3 pyrite occurs as disseminated, fine-grained cubes or stringer aggregates in silicified-sericitized tuffs that show minor chloritization. Pyrite from this group has the highest Zn, Cu, Cd, Sn and Bi, high As, Pb, Sb and Ag, and low Ni contents with a large Co/Ni ratio (mean 21.0).

Group 4 contains pyrite from hanging wall rocks remote from mineralization which have low base metal, Ag and Ni, and high As and Bi levels and moderately high Co/Ni ratios (mean 7.9).

Affinity between groups 1 and 2 is indicated by the similarity of the Co/Ni ratio, and the Co, Ni, As and Bi contents. However, the Zn, Pb, Cu and Ag enrichment of group 1 pyrite reflects the presence of base metal inclusions, and peripheral replacement shows them to have been associated with base-metal-bearing solutions.

Groups 3 and 4, which contain pyrite from hanging wall units, show mutual similarities in their Co/Ni ratios, and Co and Ni contents which imply a common origin. The increased levels of Cu, Pb, Zn and Ag in group 3 members reflect the presence of inclusions in pyrite from the silicified zone.

The location of widespread silicification in hanging wall units indicates post-ore development, and group 3 pyrites may have formed either by modification of existing pyrite by hydrothermal solutions carrying base metals or by precipitation from these solutions themselves.

CONCLUSIONS

Although pyrite developed in the vicinity of the Woodlawn volcanogenic massive sulphide deposit can probably be ascribed to more than one formational process, the distribution of the pyrite and the nature of included and peripheral replacement phases can be interpreted in terms of a genetic model which reflects part of the history of the volcaniclastic pile. The model proposed for the development of pyrite at Woodlawn is that the diagenetic pyrite developed in footwall tuffs (now classified in group 2) has been modified near the ore zone by base-metal-bearing solutions, to give pyrite which is peripherally replaced by base metals and is now classified in group 1. The large size of some crystals, and the greater volume of pyrite present in chlorite-rich footwall tuffs, suggests that some of the pyrite was deposited within the channel-ways.

It is also proposed that pyrite of group 4, which is present in hanging wall units remote from mineralization, may have been modified by post-ore, hydrothermal fluids to give rise to pyrite now belonging to group 3.

The trace element contents of pyrite from within zones of silica or chlorite alteration developed around volcanogenic massive sulphide deposits can be referred to some of the events related to the mineralizing processes. A proper interpretation of the nature and extent of base metal inclusions in the pyrite has potential to indicate proximity to mineralization.

At Woodlawn, pyrite from the various lithological units remote from mineralization has a distinctive Co/Ni ratio. The ratio remains similar in alteration zones associated with mineralization in which pyrite is enriched in base metal sulphide inclusions.

Examination of the nature of inclusions and the trace element chemistry of pyrite from volcaniclastic piles, especially from commonly encountered zones of chloritization or silicification, potentially provides an additional method of discriminating barren from mineralized terrains.

ACKNOWLEDGEMENTS

The ready assistance of the staff of Jododex Australia Pty. Ltd., in particular Messrs. E.J. Malone and T. Nicholas, is gratefully acknowledged. Discussions with my colleague, John Wilmshurst, are much appreciated.

REFERENCES

Ayres, D.E., Lambert, I.B. and Peterson, M.D., 1976. Geochemical and mineralogical setting of the Woodlawn Zn-Pb-Cu orebody (abstract). 25th Int. Geol. Congr., Sydney, N.S.W., Sect. 4A.

Branov, E.N., Zasukhin, G.N., Karpukhina, V.S., Loginova, L.A. and Bukharova, V.A. 1972. Occurrence of Cu, Zn, Pb and other elements in pyrite from aureoles of pyrite deposits. Geochem. Int., 9: 834—843.

Burr, E.J., 1967. Cluster sorting with mixed character types, I. Standardization of character values. Aust. Comput. J., 1: 97—99.

Dawson, K.M. and Sinclair, A.J., 1974. Factor analysis of minor element data for pyrites, Endako molybdenum mine, British Columbia, Canada. Econ. Geol., 69: 404—411.

Fleischer, M., 1955. Minor elements in some sulfide minerals. In: A.M. Bateman (Editor), Economic Geology Fiftieth Anniversary Volume 1905—1955, Part II. The Economic Geology Publishing Co., Urbana, Ill., pp. 970—1024.

Garrett, R.G., 1969. The determination of sampling and analytical errors in exploration geochemistry. Econ. Geol., 64: 568—569.

Hegeman, F., 1941. Die isomorphen Beziehungen von Mn, Zn, Co, Ni und Cu zu Pyrit und Magnetkies. Z. Kristallogr., 103: 168—177.

Lance, G.N. and Williams, W.T., 1967. Mixed data classificatory programs, I. Agglomerative systems. Aust. Comput. J., 1: 15—20.

Loftus-Hills, G. and Solomon, M., 1967. Cobalt, nickel and selenium in sulphides as indicators of ore genesis. Miner. Deposita, 2: 228—242.

Malone, E.J., Olgers, F., Cucchi, F.G., Nicholas, T. and McKay, W.J., 1975. Woodlawn copper-lead-zinc deposit. In: C.L. Knight (Editor), Economic Geology of Australia and Papua New Guinea, 1. Metals. Australasian Institute of Mining and Metallurgy, Melbourne, Vic. pp. 701—710.

Ohtagaki, T., Takahashi, H. and Obara, K., 1974. Geology of the Hanawa Mine, Akita Prefecture. Min. Geol. (Japan), Spec. Issue, 6: 157—168.
Sinclair, A.J., 1967. Trend surface analysis of minor elements in sulfides of the Slocan Mining camp, British Columbia, Canada. Econ. Geol., 62: 1095—1101.
Sinclair, A.J., Fletcher, W.K., Price, B.J., Bentzen, A. and Wong, S.S., 1976. Minor elements in pyrites from some porphyry-type deposits, British Columbia (abstract). AIME Annu. Meet. Las Vegas, Nev.
Smirnov, V.I., Eremin, N.I., Kelch, V.E., Okrugin, V.M. and Sergeeva, N.E., 1974. The investigation of trace elements in the sulphides of pyrite deposits with the aid of the LMA1 laser microspectral analyzer of VEB Carl Zeiss Jena. Jena Rev., 4: 240—244.
Sokal, R.R. and Michener, C.D., 1958. A statistical method for evaluating systematic relationships. Sci. Bull. Kansas Univ., 38: 1409—1438.
Sutherland, J.K., 1967. The chemistry of some New Brunswick pyrites. Can. Mineral., 9: 71—84.
Tanimura, S., Shimoda, T. and Sawaguchi, T., 1974. On the Fukazawa ore bodies, Akita Prefecture. Min. Geol. (Japan), Spec. Issue, 6: 147—155.

Journal of Geochemical Exploration, 8 (1977) 85—103 85

APPLICATION OF LEAD ISOTOPES AND TRACE ELEMENTS TO MAPPING BLACK SHALES AROUND A BASE METAL SULPHIDE DEPOSIT

BRIAN L. GULSON*

Division of Mineralogy, Commonwealth Scientific and Industrial Research Organization, North Ryde, N.S.W. (Australia)

(Revised version received February 23, 1977)

ABSTRACT

Gulson, B.L., 1977. Application of lead isotopes and trace elements to mapping black shales around a base metal sulphide deposit. J. Geochem. Explor., 8: 85—103.

Pb isotopic analyses have been used in southeastern New South Wales, Australia, to distinguish Ordovician black shales, which have no associated mineralization, from Silurian black shales in which mineralization is known to occur. The more radiogenic nature of the Ordovician Pb, as shown by analysis of the sulphide, whole rock, acid leach or residue, reflects a higher U/Pb environment compared with the Silurian which is due to the absence of volcanism in the Ordovician.

V concentrations and Co/Ni ratios in sulphides may prove to be an important indicator to distinguish Silurian and Ordovician black shales but lithology, mineralogy, whole rock major and trace elements, and C and S isotopes give ambiguous answers. Co/Ni ratios in the Silurian black shale sulphides average ~ 0.26 compared with 0.04 in the Ordovician sulphides.

On the basis of Pb isotope and V concentration data, an unconformity between Middle-Upper Ordovician and Middle-Upper Silurian rocks and overturning of the strata is proposed for a drill hole previously mapped on palaeontological evidence as entirely Ordovician.

No correlation between organic C and U concentration was observed in the whole rocks.

An approximate negative correlation exists between $^{208}Pb/^{206}Pb$ and V and Ni contents; with higher Ni and V concentrations the sulphides have lower $^{208}Pb/^{206}Pb$ ratios but higher $^{206}Pb/^{204}Pb$ ratios. Shales distant from mineralization appear to have been wholly influenced by marine processes in comparison with samples close to the mineralization which are affected by volcanic conditions.

INTRODUCTION

In southeastern N.S.W., Australia, Pb-Zn-Cu mineralization such as at Woodlawn and Captains Flat is restricted to Middle-Upper Silurian volcanics

---

* Also: Visiting Fellow, Research School of Earth Sciences, Australian National University, Canberra, A.C.T., Australia.

which are intimately associated with black and dark grey shales. It is these volcano-sedimentary sequences which are the "target" for exploration. The Middle-Upper Ordovician sedimentary sequence, which has no known associated mineralization but also contains black shales, unconformably underlies the Silurian rocks and because of the similarity of the black shales in hand specimen, difficulties have been encountered in distinguishing between them, particularly during core logging. A paucity of fossils (Felton and Sherwin, 1974) compounds the problems of mapping.

Attempts to distinguish the shales other than by lithological means have proved inconclusive. McKay (1973) first tried major and trace element whole rock data on a limited number of samples and later, Petersen and Lambert (in preparation) have carried out a more extensive chemical survey on rocks which yields no greater discrimination. X-ray diffraction data (this study) are ambiguous, the mineralogy for both types being quartz (dominant), chlorite, muscovite and traces of feldspar, sulphide and calcite. Similar S and C isotopic values (see later section) are obtained for the Silurian and Ordovician rocks.

During the course of a detailed Pb isotopic investigation around Woodlawn, initial analyses of one Ordovician and one Silurian pyrite showed marked differences in Pb isotopic ratios. Consequently an extended study was undertaken in this area but with *no prior* knowledge of the stratigraphy of the samples.

This study presents additional data following on the earlier preliminary work (Gulson, 1976), and includes data for a proposed unconformity in a drill hole logged solely as Ordovician. It also presents data for localities in central western N.S.W. chosen as a check on the regional extent of this Pb isotopic distinction between the Ordovician and Silurian. A summary of trace element data on the sulphides is also given.

Pb isotopic investigations make use of variations in the daughter isotopes ^{206}Pb, ^{207}Pb and ^{208}Pb produced by radioactive decay from their respective parents ^{238}U, ^{235}U and ^{232}Th. The variations are usually expressed as a ratio of the particular daughter isotope to the orphan isotope ^{204}Pb (i.e. ^{204}Pb has no known parent).

GEOLOGICAL SETTING OF THE WOODLAWN DEPOSIT

The Woodlawn polymetallic sulphide deposit occurs approximately 200 km southwest of Sydney (Fig. 1). It is located in a sequence of acid volcanics and sediments of Middle to Upper Silurian age which form part of the Lower Palaeozoic succession in the Lachlan Fold Belt. The geological setting is similar to that found at Captains Flat, some 70 km south of Woodlawn.

In the immediate vicinity of Woodlawn, sedimentary rocks are poorly represented in the lower part of the Silurian succession and a sequence (called the Woodlawn volcanics) of acidic tuffs, rhyolites, ignimbrite agglomerates and acid volcanoclastics is developed (Felton, 1974). The immediate

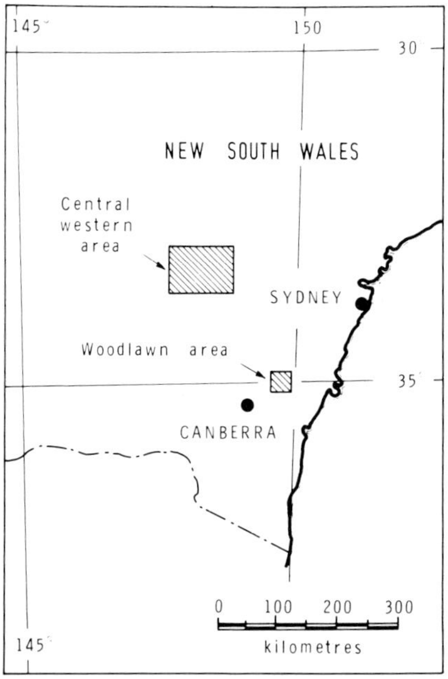

Fig. 1. Locality map of the Woodlawn area and central western N.S.W., Australia, from where samples were selected.

host rocks to the massive sulphides consist of rhyolitic tuffs and flows, black shales, chloritic schists and rare chert-like rocks.

The Silurian succession is unconformably underlain by Middle to Upper Ordovician sediments and is overlain by the Currawang Basalt.

The whole sequence has been regionally metamorphosed to produce highly deformed lower greenschist facies rocks and is intruded by Upper Silurian dolerite and Devonian granite (Malone et al., 1975).

The main massive sulphide lens is a polymetallic sulphide body consisting of pyrite as the main constituent and variable amounts of sphalerite, galena and chalcopyrite. Silicate gangue minerals are chlorite, quartz, talc, sericite and feldspar. Stringers of sulphides are commonly found throughout the volcanic succession.

EXPERIMENTAL PROCEDURES

Sulphides. Pyrite and pyrrhotite were separated from the shales and volcanics by crushing, heavy liquid separation using tetrabromoethane and methylene iodine, sieving and then magnetic separation of the minus 100, plus 200-mesh fraction. The samples (usually > 90—95% pure) were washed in 6*N* HCl for 15 minutes ($\sim$ 2 -3 minutes for pyrrhotite), rinsed with water and boiled in water for about 15 minutes (all reagents were prepared using sub-boiling distillation methods in silica glass or Teflon stills).

Pyrite was dissolved in $7N$ HNO_3 /$6N$ HCl (3:1) with a few drops of bromine water, whilst pyrrhotite was dissolved in $6N$ HCl. For the shales, an insoluble residue of carbonaceous material was usually present which amounted to less than 3% by weight of the sample. The solutions were ion exchanged on a 2-cm^3 anion column in bromide form and further cleaned on a 0.5-cm^3 anion column in chloride form.

Rocks. Minus 200-mesh samples, prepared under clean conditions, were "decomposed" in Teflon bombs using HF-HNO_3 but, in all cases, even with nitric acid/sodium chlorate (E. Kiss, personal communication, 1975) and more HF treatment, some residue of carbonaceous matter, quartz and chloritic material was always present.

Acid leaches were effected with $7N$ HNO_3 /$6N$ HCl (5:2) to dissolve the ore component, mainly pyrite. The residue was then dissolved in HF as for rocks. The ion exchange procedure for the rock analyses was the same as for the sulphides.

Mass spectrometry. Samples were analyzed on the high ion beam mass spectrometer at the Research School of Earth Sciences, Australian National University. The "within-run" percent standard deviation (1σ) of a set of 12—20 ratios ranged from 0.06 to 0.7%, with the average usually 0.1—0.2%. Runs with standard deviations greater than 0.3% are indicated in Tables I and II. Duplicate dissolution data are given in the tables.

All analyses were performed with a mixed $^{207}Pb/^{204}Pb$—^{235}U double spike to control mass fractionation.

DISCUSSION OF RESULTS

Lead isotopes from the Woodlawn area

Analytical data for sulphides and whole rocks are given in Tables I and II and plotted in Figs. 2, 3 and 4. Most of the data in Table IA are from Gulson (1976). Minor differences in values for 6 samples, particularly for $^{207}Pb/^{204}Pb$ ratios, in Table IA and Gulson (1976) are due to recalculation of the fractionation correction. For clarity, only the field of data for the Silurian sulphides is outlined. The preponderance of Silurian over Ordovician data is purely a reflection of drilling for "target" beds and this limits the availability of Ordovician samples, particularly away from mineralization.

The variability and radiogenic nature, not only of the Ordovician sulphides, but also the rocks, compared with the Silurian black shale data are the obvious features of the results:

	$^{206}Pb/^{204}Pb$	$^{207}Pb/^{204}Pb$	$^{208}Pb/^{204}Pb$
Ordovician	20.4—38.6	15.8—16.7	38.8—43.6
Silurian	17.9—19.3	15.5—15.7	37.7—40.7

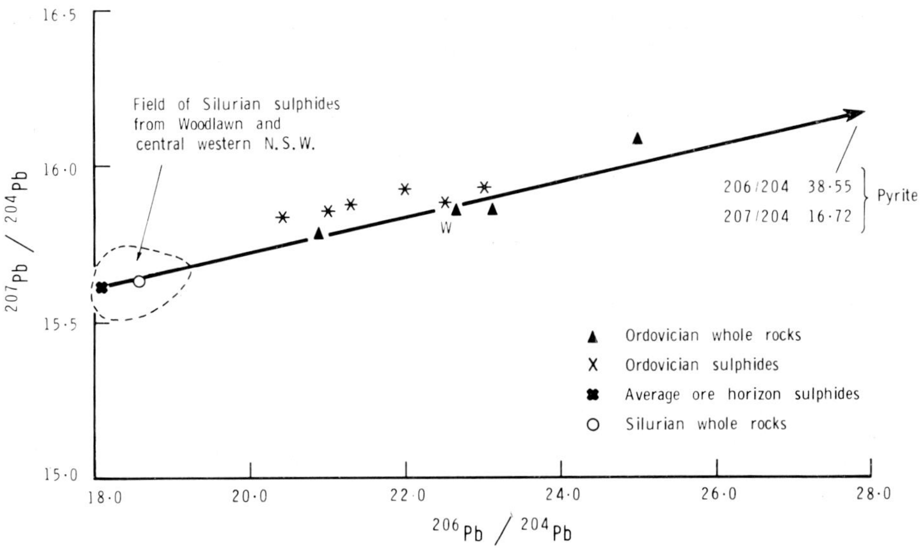

Fig. 2. $^{207}Pb/^{204}Pb$ vs. $^{206}Pb/^{204}Pb$ plot for black shales from the Woodlawn and central western N.S.W. ares.

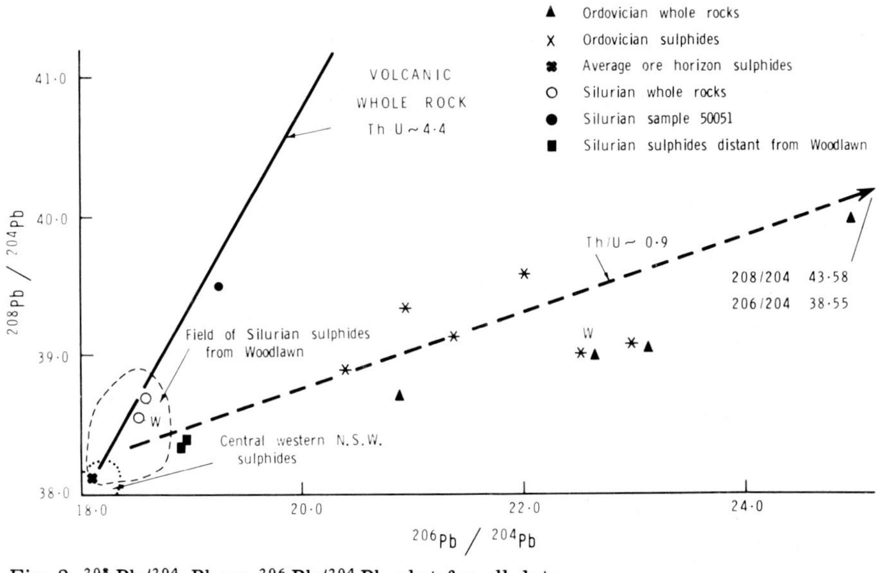

Fig. 3. $^{208}Pb/^{204}Pb$ vs. $^{206}Pb/^{204}Pb$ plot for all data.

When subdivided, mainly on the basis of $^{206}Pb/^{204}Pb$ ratios, into possible Silurian and Ordovician samples, the data agreed in every case with the fossil and stratigraphic controls.

The isotopic values for the Silurian samples close to mineralization (< 2 km) are similar to those for Pb in sulphides from the ore horizon (pyrite, galena, chalcopyrite, pyrite) and pyrite from the immediate ore host volcanics (the average value of the ore Pb is given by a solid cross in Figs. 2, 3 and 4).

Similar isotopic ratios and U and Pb concentrations were obtained for coexisting pyrite and pyrrhotite. In most cases, the transformation of pyrite to pyrrhotite could be traced to the proximity of the shales to intrusive dolerites.

The variability in isotopic ratios and elemental concentrations over short distances in either Silurian or Ordovician sulphides is marked, e.g. in diamond drill hole (DDH) CB-1, W-43, WE-1. This variability not only applies to Pb isotopes and trace elements (Tables I and III) but also to S and C isotopes and organic C contents (Table IV).

U and Pb concentrations in sulphides from the Ordovician and Silurian overlap and, of course, are highly variable. Although the data are limited, particularly for the Ordovician, the Ordovician sulphides do appear to have, on the whole, higher U and lower Pb concentrations than the Silurian sulphides.

On the $^{207}Pb/^{204}Pb$ vs. $^{206}Pb/^{204}Pb$ plot in Fig. 2, the solid line is that for the whole rock Silurian acid volcanics from around Woodlawn (Gulson, 1977a). Most of the high-quality data for the black shales lie to within experimental error of this line. Its slope is equivalent to an apparent age of about 350 m.y. compared with a stratigraphic age of about 430 m.y.; however, in this age range small variations in slope result in disproportionately

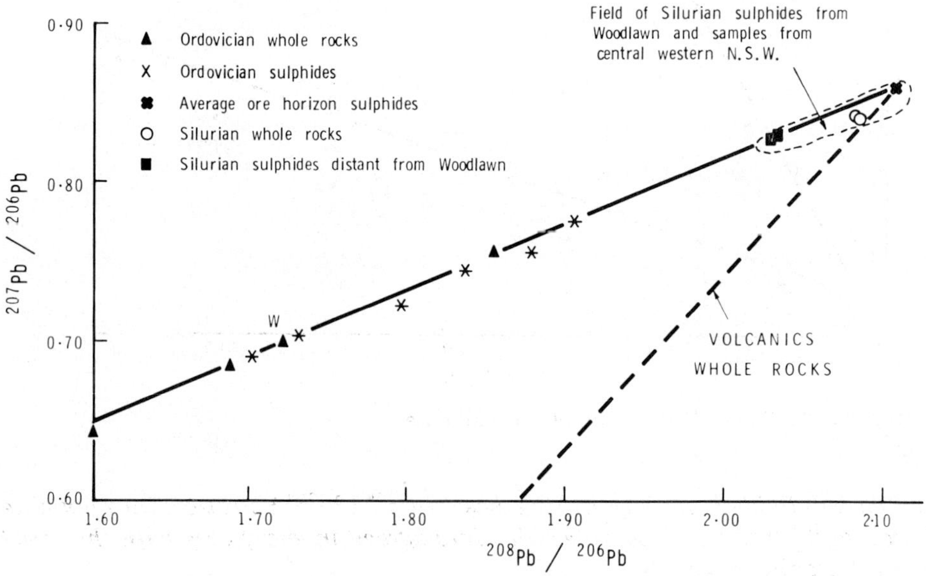

Fig. 4. $^{207}Pb/^{206}Pb$ vs. $^{208}Pb/^{206}Pb$. Symbols as in previous figures. At the left-hand extremity of the solid line lies the extremely radiogenic Ordovician pyrite.

TABLE 1A

Pb isotopic ratios and U and Pb analyses of pyrite (py) and pyrrhotite (po) from shales close (< 2 km) to Woodlawn

Sample No.		Drill hole No.	Depth (m)	Pb (ppm)	U (ppm)	$^{208}Pb/^{206}Pb$	$^{207}Pb/^{206}Pb$	$^{206}Pb/^{204}Pb$	$^{207}Pb/^{204}Pb$	$^{208}Pb/^{204}Pb$	$^{238}U/^{204}Pb$
Silurian											
39289*	py	W-1	62.2	383	1.3	2.1025	0.8607	17.93	15.43	37.70	0.22
39291-1-A	py**	W-20	183.0	68.7	3.2	2.1076	0.8557	18.34	15.69	38.66	2.95
-1-B	py			68.4	—	2.1062	0.8457	18.46	15.62	38.89	2.98
-2	py			76.5	1.4	2.1051	0.8556	18.28	15.64	38.49	1.2
39291	po			89.2	1.5	2.1070	0.8590	18.21	15.64	38.37	1.04
39292	py**	W-30	33.5	84.2	2.5	2.1140	0.8485	18.41	15.62	38.92	1.91
46184	py	W-30	75.3	33.3	3.5	2.0385	0.8330	18.74	15.61	38.20	6.72
39293	py	W-52	132.9	136	0.9	2.0943	0.8526	18.32	15.62	38.38	0.42
39296	py	W-280	15.2	1350	4.4	2.1053	0.8607	18.16	15.63	38.23	0.21
39295	py	W-54	470.3	—	—	2.1062	0.8619	18.18	15.67	38.29	—
46183	py**	W-6	243.4	446	0.3	2.1061	0.8617	18.10	15.59	38.11	0.04
46186*	py**	W-283	46.1	733	0.7	2.1121	0.8638	18.19	15.71	38.41	0.06
46187	py	W-43	195	26.4	1.3	2.0669	0.8458	18.51	15.65	38.25	3.16
49018	py	W-43	185.6	51.1	1.9	2.0196	0.8245	19.00	15.67	38.38	2.39
50051	py	W-65	272.5	12.7	0.15	2.1148	0.8131	19.27	15.67	40.74	0.79
Ordovician											
39290*	py**	W-43	183.8	19.2	20.6	1.7316	0.7051	22.53	15.89	39.02	72.7
	po			29.2	21.5	1.7002	0.6918	22.99	15.91	39.09	50.8
WL 15-1	py	W-E1	~78	—	—	1.9062	0.7760	20.42	15.84	38.90	
-2	py			—	—	1.8363	0.7454	21.31	15.88	39.13	
-2	(acid wash)**	(HCl)		—	—	2.0001	0.8175	19.40	15.86	38.79	
-3	py**		78.2	74.0	8.6	1.1307	0.4338	38.55	16.72	43.58	9.88
49016	py	W-E1	70.1	35.4	2.5	1.7967	0.7232	22.03	15.93	39.59	4.72
46185*	py**	W-35	82.3	36.3	4.1	1.8787	0.7568	20.95	15.86	39.36	8.88

-A, -B: new load from same beaker; -1, -2: new dissolution.

* Fossil locality.

** Poor-quality data, 206/204 = 0.3—0.5% (1σ).

TABLE 1B

Analyses of pyrite (py) and pyrrhotite (po) from Silurian shales distant (> 10 km) from Woodlawn

Sample No.		Drill hole No.	Depth (m)	Pb (ppm)	U (ppm)	$^{208}Pb/^{206}Pb$	$^{207}Pb/^{206}Pb$	$^{206}Pb/^{204}Pb$	$^{207}Pb/^{204}Pb$	$^{208}Pb/^{204}Pb$	$^{238}U/^{204}Pb$
46188*-1	py	CB-1	72.1	471	1.6	2.1074	0.8610	18.08	15.57	38.10	0.23
-2	py			462	1.6	2.1071	0.8613	18.14	15.62	38.22	0.22
46189*-1	py	CB-1	86.2	163	8.9	2.0687	0.8442	18.52	15.63	38.31	3.50
-2	py			150	4.8	2.0683	0.8457	18.50	15.65	38.26	2.04
46190*-1	py**	CB-1	87.8	240	0.7	2.0906	0.8557	18.19	15.57	38.04	0.18
-2	py			236	0.7	2.0922	0.8562	18.26	15.63	38.20	0.18
46191*-1	py	CB-1	91.1	419	6.7	2.0767	0.8493	18.42	15.64	38.25	1.01
-2	py			389	7.3	2.0825	0.8513	18.42	15.68	38.35	1.20
50050	py	CB-2	188.0	200	1.7	2.0753	0.8490	18.45	15.66	38.29	0.50
46192	po**	MO-1	378.9	45.1	2.2	2.0306	0.8322	18.71	15.57	37.99	3.08
50044	py	LH-1	125.0	18.4	1.3	2.0275	0.8275	18.93	15.67	38.38	4.70
50045	py	LH-2	130.8	64.4	0.04	2.1120	0.8584	18.21	15.63	38.46	0.06
50046	py	CV-1	284	43.5	0.7	2.0998	0.8527	18.31	15.62	38.46	1.70
50047	py	CV-1	287	93.2	3.1	2.0323	0.8307	18.80	15.61	38.20	2.1
50048	py	CV-1	299	114.4	2.2	2.0281	0.8287	18.90	15.66	38.34	1.3
50049	py	CV-2	115.4	101.1	8.8	2.1077	0.8415	18.59	15.64	39.17	5.59

-1, -2: new dissolution.

* Fossil locality.

** Poor-quality data, 206/204 = 0.3–0.5% (1σ).

TABLE 1C

Pyrite (py) and oxidized (ox. py) pyrite analyses from surface samples from central western N.S.W.

Sample No.		Location	Pb (ppm)	U (ppm)	$^{208}Pb/^{206}Pb$	$^{207}Pb/^{206}Pb$	$^{206}Pb/^{204}Pb$	$^{207}Pb/^{204}Pb$	$^{208}Pb/^{204}Pb$	Age
48317*	py	Keenans Bridge, shale	22.5	0.07	2.0950	0.8548	18.25	15.60	38.23	Upper Ordovician
48318*	py	Keenans Bridge, sandstone	191	0.2	2.0800	0.8488	18.32	15.55	38.11	Upper Ordovician
	py and ox. py	Keenans Bridge, sandstone	165	0.3	2.0822	0.8489	18.32	15.55	38.15	Upper Ordovician
48323*	py**	Forbes Pistol Range, marly shale	258	0.3	2.1013	0.8580	18.22	15.63	38.28	Lower Silurian
	py and ox. py	Forbes Pistol Range, marly shale	122	0.6	2.1016	0.8565	18.20	15.59	38.25	Lower Silurian
48324*	py**	Quarry W. Forbes, marl	4.2	0.10	2.0805	0.8454	18.28	15.45	38.03	Ordovician-Silurian
48329*	py**	Endeavour Prospect (DDH 1 38.2 m), black shale	49.3	1.3	2.0043	0.8241	18.93	15.60	37.93	?

* Fossil locality.
** Poor-quality data, 206/204 = 0.3—0.5% (1 σ).

TABLE II

Data for whole rock black shales around Woodlawn

Sample No.	Locality	Pb (ppm)	U (ppm)	$^{208}Pb/^{206}Pb$	$^{207}Pb/^{206}Pb$	$^{206}Pb/^{204}Pb$	$^{207}Pb/^{204}Pb$	$^{208}Pb/^{204}Pb$	$^{238}U/^{204}Pb$
Silurian									
39289 — rock	W-1 62.2 m	29.2	—	2.0839	0.8419	18.57	15.63	38.69	
39295 — rock	W-54 479.7 m	37.5	3.1	2.0825	0.8435	18.51	15.61	38.55	5.4
				(2.0783	0.8429	18.48	15.58	38.41)	
Ordovician									
49016 — rock	WE-1 70.1 m	9.8	10.3	1.7219	0.7001	22.65	15.86	39.00	71.0
52010 — rock	W-43 181.8 m	6.8	8.7	1.6890	0.6858	23.13	15.86	39.06	87.8
52011 — rock	W-43 183.1 m	3.0	3.7	1.6012	0.6439	24.97	16.08	39.98	209.2
50053 — rock	Woolaroo-Woodlawn boundary	7.8	4.5	1.8549	0.7560	20.88	15.78	38.72	38.0

Data given in brackets are run on the AVCO mass spectrometer and are uncorrected for fractionation using the double spike.

TABLE III

Summary of emission spectrographic analyses of sulphides from black shales in the Woodlawn area (in ppm)

	Ag	As	Co	Cu	Mo	Ni	Pb	V	Zn	Bi	Co/Ni
Silurian close to mineralization (< 2 km)											
Average (10)	23	1230	75	160	3	340	290	n10	800	1	0.29
Range	n1—200	n200—6000	10—225	10—600	n0.3—10	60—900	13—1350	—	30—2000	n1—6	0.05—5
Silurian distant (> 10 km) from mineralization											
Average (12)	50	1150	100	400	8	660	240	20	2500	20	0.23
Range	n1—250	n200—8000	15—400	40—1000	n0.3—15	60—2500	30—1000	n10—80	100—>1%	n1—200	0.06—6
Ordovician											
Average (5)	10	500	120	640	26	4300	40	100	1100	8	0.04
Range	n1—30	n200—2000	40—250	25—1500	6—200	575—8000	19—60	60—150	n30—3000	n1—20	0.01—0.07
Insoluble residue (1)	n1	n200	10	4	n3	6	20	100	—	—	1.6
Insoluble residue (4)	n1	n200	n10	1	4	6	n20	250	—	—	—

n1 = not detected at 1 ppm.
Elements underlined indicate significant differences from Silurian sulphides close to mineralization.

Analyst: N. Morgan.

large variations in age, so the apparent age deficiency cannot be regarded as significant.

The scatter of the data is greater, particularly for the Ordovician samples, on a $^{208}Pb/^{204}Pb$ vs. $^{206}Pb/^{204}Pb$ plot (Fig. 3, cf. Fig. 2). However, a number of important features, besides the more radiogenic nature of the Ordovician Pb, may be noted:

(1) The low $^{208}Pb/^{204}Pb$ ratio of the Ordovician shales which is equivalent to a Th/U ratio of 0.4—0.9 compared with about 4.4 for the Silurian volcanics.

(2) The acid leach-rock-residue data for the Silurian shales lie closer to the volcanic line than to the Ordovician shale line. A single, comparatively radiogenic pyrite (50051) also lies close to the volcanic line and is similar to radiogenic sulphides separated from the volcanics (Gulson, in preparation).

(3) Some sulphide data for Silurian black shales (e.g. 50044, 50048, 46192) distant (> 10 km) from the Woodlawn mineralization are comparatively radiogenic and plot on the Ordovician black shale line. The distinction can be more clearly seen in Fig. 4, a plot of $^{207}Pb/^{206}Pb$ vs. $^{208}Pb/^{206}Pb$, the advantage of which is to overcome problems of measurement of the small ^{204}Pb isotope.

Correction of the Silurian sulphides for Pb derived by radioactive decay from U since about 430 m.y. results in less scatter than the uncorrected data, but this correction may be meaningless because of changes in the U/Pb ratio due to post-depositional processes or even during mineral separation where the ratio of sulphide to carbonaceous material may be important (Peltola, 1968). Furthermore, the enormous scatter on a $^{238}U/^{204}Pb$ vs. $^{206}Pb/^{204}Pb$ plot indicates either recent movement or partitioning during mineral separation of U.

Lead isotopes indicative of an unconformity and overturned sequence at Woodlawn

The first analysis of pyrite and pyrrhotite from DDH W-43 at 183.8 m (39290; Table IA, Ordovician) exhibited Ordovician-type ratios (and also Ni and V contents) which agreed with the presence of Ordovician graptolites found in the upper part of the hole. A new sample at 195 m (46187; Table IA, Silurian) contained only pyrite and the isotopic data and trace elements are more typical of Silurian shales. As the possibility existed that this sample had been mixed with another during either sampling or processing, a further sample was analyzed from 185.6 m (49018; Table IA, Silurian). Although it was more radiogenic than the pyrite from 195 m, the pyrite (no pyrrhotite) also gave isotopic and trace element data of the Silurian type. At this stage, it became apparent that there was possibly an unconformity and overturned succession in the hole, so a further two (whole rock) samples higher in the core were analyzed. Both of these, from about 182 (52010; Table II) and 183 m (52011; Table II), gave values typical of those found in other

Ordovician samples, supporting the inferred unconformity and overturning. The paucity of fossils in this area and difficulty of finding them in drill core makes palaeontological confirmation remote.

Whole rock and acid-leaching experiments

Because of the time involved in separation of the sulphides and their absence in some cases due to oxidation, whole rock samples were investigated as a possibly simpler method of distinguishing the Ordovician and Silurian black shales. The data, given in Table II, confirm the radiogenic nature of the Ordovician samples and thus the use of whole rock samples.

The data for the acid-leaching experiments (Gulson, 1977b) on both Silurian and Ordovician whole rocks show that the acid leaches and residues are consistently different as in the sulphides and whole rocks.

Carbon and sulphur isotopes

C and S isotopic analyses were performed by J. Smith and M. Burns in a search for alternative and less expensive methods, besides the elusive fossils, to substantiate the Pb isotope data. The limited isotopic data, given in Table IV, for the four Silurian samples from the one drill hole, showed slight variation in $\delta^{13}C$ values (—27.8 to —29.0‰) but large differences in $\delta^{34}S$ values (—0.1 to +13.5‰).

TABLE IV

C and S isotopic data and organic C and U concentrations for rocks and sulphides from selected samples

Sample No.	Locality	Rocks			Sulphides
		$\delta^{13}C$(‰)	org. C (%)	U (ppm)	$\delta^{34}S$(‰)
Silurian					
46188	CB-1 72.1 m	—28.4	0.68	4.5	— 0.1
46189	CB-1 86.2 m	—29.0	1.26	8.2	+ 12.1
46190	CB-1 87.8 m	—28.3	0.81	13.6	+ 13.5
46191	CB-1 91.1 m	—27.8	0.78	11.1	+ 5.3
Baryte	Clare Vale	—	—	—	+ 30.3
Co-existing pyrite	Clare Vale	—	—	—	+ 11.5
Ordovician					
51482	WE-1 70.1 m	—29.7	0.31	15.1	+ 12.9
46185	W-35 82.3 m	—28.6	0.72	8.2	+ 4.4
39290 (po)	W-43 183.8 m	—	—	—	+ 1.6

Analysts: M. Burns and J. Smith.

There is no significant difference between the Ordovician and Silurian isotopic C values and the S isotope variations within samples of the same age are so large that interpretation is difficult. Relatively uniform values of $\delta^{34}S$ of about $+8^0/_{00}$ have been obtained for sulphides from the Woodlawn ore horizon and host volcanics (Ayres, Smith and Burns, in preparation).

Even though the few C and S isotope values do not differentiate the Ordovician and Silurian shales unequivocally, the data may possibly be used to indicate their environment of deposition. The Silurian (and Ordovician) samples are distant from the mineralization (> 10 km) and so the higher values of $\delta^{34}S$ in 46189 and 46190 may be attributed to Silurian marine influence; the S isotopic value of seawater at this time is given by the baryte ($+30.3^0/_{00}$). If sulphides from Silurian samples close to mineralization were analyzed, these would probably have $\delta^{34}S$ values closer to those of the sulphides from the volcanics, reflecting a greater volcanic than marine influence.

No trends in organic C and U concentration were discernible in the Woodlawn data (Table IV) in contrast to results for other shales obtained by Wampler and Kulp (1964) and Baturin (1973).

Trace element concentrations in sulphides

A summary of semi-quantitative data on certain trace elements in sulphides is given in Table III. Only V and possibly Ni, Mo and Pb seem to distinguish between the Ordovician and Silurian shales close to mineralization (< 2 km).

Samples from central western N.S.W.

Results for pyrite and oxidized sulphide separated from shales, sandstone, and marls of Ordovician and Silurian age in central western N.S.W. (Fig. 1) are given in Table IC. In contrast to the Woodlawn area, differences in Pb isotopic ratios were not observed and this precludes the use of Pb isotopes as a mapping tool in central western N.S.W. The similarity in Pb isotopes may be attributed to the extensive volcanic activity with limited associated mineralization (mainly Cu) which occurred during both the Ordovician and Silurian in the central west (Sherwin, 1973) with resulting low U/Pb ratios. The data, except for the low-precision analysis 48324, lie within the field of the Woodlawn Silurian black shale data.

Trace element contents of the sulphides from central western N.S.W., in keeping with the isotopic data, show no differences between Ordovician and Silurian but they are too few in number to be useful.

Co/Ni ratios

Loftus-Hills and Solomon (1967) suggested that pyrite formed by sedimentary or diagenetic process had Co/Ni < 1 in contrast to those of volcanic origin unaccompanied by Pb and Zn minerals, which had Co/Ni > 1. At Woodlawn, Co/Ni ratios of sulphides from Silurian black shales vary from 0.29 (range 0.05—5) close to mineralization (< 2 km), to 0.23 (range 0.06—6) distant from mineralization. However, the Ordovician black shale sulphides have an average Co/Ni ratio of 0.04 (range 0.01—0.07) reflecting the larger contributions of Ni in these samples. These results are in contrast with the ratio of > 3 for pyrite from volcanoclastics in the Woodlawn area (Gulson, in preparation) and substantiate the syngenetic as against volcanogenic nature of the sulphides in the black shales.

Possible relationship of nickel and vanadium and lead isotopic ratios

Because of the occasional high V and Ni values in Silurian sulphides distant from mineralization and in the Ordovician shales, these data were plotted against $^{208}Pb/^{206}Pb$ (Fig. 5). Although few in number, the data indicate a negative correlation which is well defined for Ni but less so for V; that is, as $^{208}Pb/^{206}Pb$ decreases (or $^{206}Pb/^{204}Pb$ increases) there is an increase in Ni and V. The poor correlation for V may be partly one of sampling — a different split is used for the Ni and V determinations and the sulphide used in the Pb analysis is initially washed with 6*N* HCl. Furthermore, the emission analysis is for the whole sample (i.e. sulphide plus insoluble impurity) whereas the isotopic analysis is only of the acid-soluble sulphide.

Goldschmidt (1954) suggested that the V content in marine shales is higher than in corresponding fresh-water deposits whereas Degens et al. (1957) maintained that this is only valid for the carbon fractions of the deposits and for a limited areal extent. When considered with the other data and the marine environment in which the Ordovician shales formed, the iso-

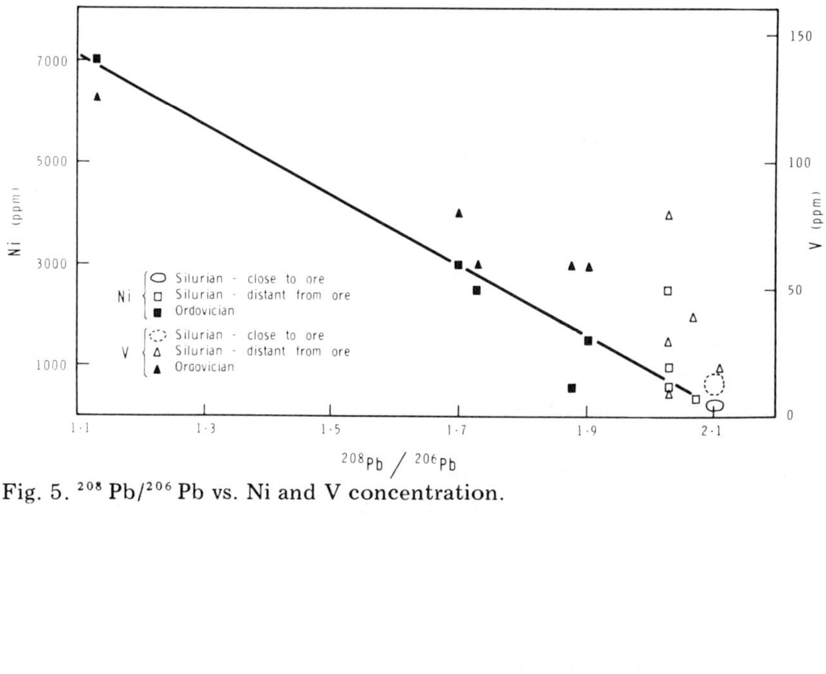

Fig. 5. $^{208}Pb/^{206}Pb$ vs. Ni and V concentration.

topic and V and Ni information suggests that the Silurian shales distant from mineralization were influenced largely by marine conditions, in contrast to the dominance of volcanism near the mineralization and volcanic centre(s), which overrides the influence of the marine conditions.

Black shales and metallogenesis

The major aim of this work was to investigate the use of Pb isotopes in the Woodlawn and other areas in distinguishing Ordovician from Silurian carbonaceous shales, because of the economic potential for base metal mineralization in Silurian rocks. However, because of the ubiquitous association of carbonaceous shale with most mineralization, whether it be Cu-Pb-Zn, Au, or U, it is worthwhile briefly considering certain aspects of the problems of black shales, sulphides and metals (particularly U) and how they relate to the present study.

Wampler and Kulp (1964) carried out an extensive investigation of pyrite from black shales of varying ages on a world-wide basis. They suggested that the Pb in sedimentary pyrite was representative of the Pb in the environment in which it formed, provided that the isotopic composition was not changed by radioactive decay or by recrystallization of the sulphides long after deposition. They suggested that U occurs in surficial positions and much of it could be removed by thorough leaching with hot HCl. The surficial radiogenic Pb derived from U may be lost from the sample or it may remain on or near the surface of the pyrite. They further suggested that there may be a correlation between the amount of carbonaceous material and U observed.

In contrast, the data given in Table IV indicate that little or no correlation exists between organic C and U concentration for the Woodlawn area. Pyrites from the Woodlawn volcanics and chloritic "schists" show a homogeneous distribution of U throughout the grains (Gulson, unpublished data). U distribution studies on pyrite from black shales from a number of Australian localities are in progress.

Saxby (1976) suggests that the "rank" of carbonaceous matter in the black shales is critical to any discussion of bonding to metals. For low-rank organic material, metals can be attached to the carbonaceous matter through carboxyl groups. At higher grades, such as at Woodlawn where partial graphitization has occurred, Saxby considers it unlikely that metal ions (e.g. V, Ni, U) bonded to carboxyl or other groups would survive. However, metals could be retained in the sediment if they reacted with some sulphur (or oxygen), forming a sulphide (or oxide), which remained in close association with high-rank organic matter.

Peltola (1968) has described a protective coating of organic C surrounding pentlandite and uraninite in the Outokumpu black slates. Such a coating around sulphides or oxides would inhibit metal ion extraction that might occur in nature or during laboratory dissolution of the pyrite for analysis. This protection hypothesis is important for trace element and isotopic analyses because of the possibility that some of the carbonaceous material might be dissolved during sample dissolution. Many sulphide concentrates analyzed in this study and other studies in the literature on black shales (e.g. Wampler and Kulp, 1964) contained varying amounts of carbonaceous material, some of which occurs as inclusions in the sulphide. Because of the affinity of U for the carbonaceous matter, it was of some concern that some of the C plus U (and its radiogenic Pb) could be dissolved during the dissolution stage and that this may account for the radiogenic Pb in the Ordovician samples. Although such concern was discounted by the whole rock analyses, the possibility was further studied. As the residue after the acid treatment was almost impossible to get into solution for U analysis by isotope dilution, an approximation of the amount of organic C dissolved, if any, could be gauged from the V and Ni concentrations. V is concentrated with organic material and Ni in the sulphide fractions (Peltola, 1968; Kholodov and Gavrilov, 1974). Partitioning of V into the hydrocarbon phase is well known in the petroleum industry. The analyses of the insoluble residues (Table III) for sample 39292 and a composite of 46188-91 show complete partitioning of V into the residues and about 1% of the Ni. In view of the negligible amounts of V and high concentrations of Ni in the Silurian black shale sulphides, these data suggest that little carbonaceous material (and thus U) is lost during the dissolution stage.

As part of his trace element investigation of the black schists and inherent sulphides of the Outokumpu area (Finland), Peltola (1968) analyzed whole rock, sulphide and carbon fractions for 5 samples. U averaged 10.5 ppm in the rocks and 34 ppm in the carbon fraction and the ratio of U in carbonaceous matter to U in sulphide was about 2:1. He suggests the U occurs as uraninite grains enveloped by a thick carbonaceous shell, and because of the shell, U and, partly, radiogenic Pb are thus enriched in the carbon fraction during mineral separation. His observation of discrete uraninite grains would appear to differ from the observations of Wampler and Kulp (1964). If the uraninite was coated with a thick protective carbon shell, it is highly unlikely that much of the U would be extracted with only an HCl wash as indicated by Wampler and Kulp (1964).

The statement by Peltola (1960) that Th has not been found in the Outokumpu black schists would indicate that they had low Th/U ratios. The Pb isotopic data for the Woodlawn Ordovician rocks and sulphides suggest they have low Th/U ratios of 0.4—0.9. Petersen and Lambert (in preparation) have found that Ordovician black shales usually have low Th/U ratios of < 1 and certainly < 2. In contrast, Silurian black shales near mineralization usually have Th/U ratios > 3 and often up to 10 or 20 whereas those distant from mineralization have Th/U ratios < 1. The very high Th/U ratios and often high concentrations of Th (> 30 ppm) in black shales near mineralization compared with relatively normal values distant from it indicate substantial mobility of Th in this environment. In contrast, the Th/U ratios of $\sim$ 3—5 in many of the tuffs, tuffaceous shales and shales are "normal". As mentioned previously, various data suggest that the Ordovician and Silurian black shales distant from mineralization are dominantly marine in environment and this is further supported by the low Th/U ratios in the shales. Perhaps low Th/U ratios were typical of seawater during the Ordovician and Silurian periods as they are at the present time. The low $^{208}Pb/^{206}Pb$ ratios for the Woodlawn shales compare with similar ratios found in Black Sea sediments (Cooper et al., 1974) but contrast with higher values found in the Baltic Sea and extreme ratios from the North American lakes and Hudson Bay (Chow and Johnstone, 1963, 1964; Hart and Tilton, 1966). The North American sediments have average $^{208}Pb/^{204}Pb$ and $^{206}Pb/^{204}Pb$ values which plot close to the upper limit of the Woodlawn acid volcanic line.

CONCLUSION

Pb isotopic studies of either whole rocks or sulphides appear to give unequivocal answers as to the Ordovician or Silurian age of black shales in the fossil-poor Woodlawn area. The V concentration and Co/Ni ratios of sulphides also appear to be possible means of characterization but further studies are necessary. Although Pb isotopic analyses are expensive, so is drilling. Unravelling of the stratigraphy, structures and thus ore reserves and future drilling programmes may be assisted in this case by sophisticated techniques.

ACKNOWLEDGEMENTS

The author is grateful to Dr. W. Compston who generously provided access to the mass spectrometer at the Australian National University which he and Dr. Clement developed and maintained. The "on-line" data processing system was designed and developed by Dr. P.A. Arriens. I would like to thank S. Mudie for technical assistance, N. Morgan for emission spectrographic measurements on the sulphides, J. Smith and M. Burns for the carbon and sulphur isotope data, and G.F. Taylor and his group for the X-ray diffraction traces. The staff of Jododex Aust. Pty. Ltd. generously provided samples and useful information and many fruitful discussions were held with Messrs. E. Malone, T. Nicholas and W. McKay.

REFERENCES

Ayres, D.E., Burns, M.S. and Smith, J.W., in preparation. A sulphur isotope study of the Woodlawn massive sulphide orebody (for submission to J. Geol. Soc. Aust.).

Baturin, G.N., 1973. Uranium in the modern sedimentary cycle. Geochem. Int., 9: 1031—1041.

Chow, T.J. and Johnstone, M.S., 1964. Lead isotopes in the sediments of three Canadian Baltic Sea. Trans. Am. Geophys. Union, 44: 890—891 (abstract).

Chow, T.J. and Johnstone, M.S., 1964. Lead isotopes in the sediements of three Canadian shield lakes. Trans. Am. Geophys. Union, 45: 110—111 (abstract).

Cooper, J.A., Dasch, E.J. and Kaye, M., 1974. Isotopic and elemental geochemistry of Black Sea sediments. Am. Assoc. Pet. Geol., Mem., 20: 554—565.

Degens, E.T., Williams, E.G. and Keith, M.L., 1957. Environmental studies of carboniferous sediments, I. Geochemical criteria for differentiating marine and fresh-water shales. Bull. Am. Assoc. Pet. Geol., 41: 2427—2455.
Felton, E.A., 1974. Stratigraphic revisions in the Tarago-Woodlawn-Mount Fairy area. Q. Notes, Geol. Surv. N.S.W., 17: 7—12.
Felton, E.A. and Sherwin, L., 1974. A new Ordovician graptolite locality east of Tarago, New South Wales. Q. Notes, Geol. Surv. N.S.W., 17: 1—2.
Goldschmidt, V.M., 1954. Geochemistry. Oxford University Press, Oxford, 730 pp.
Gulson, B.L., 1976. Exploration and mapping around a base metal sulphide deposit using trace lead isotopes. Miner. Deposita, 11: 1—5.
Gulson, B.L., 1977a. Isotopic and chemical studies on crustal effects in the genesis of the Woodlawn Pb-Zn-Cu deposit. Contrib. Mineral. Petrol. (in press.)
Gulson, B.L., 1977b. Lead isotope results of acid leaching experiments on acid volcanics and black shales in an ore environment (submitted to Chem. Geol.)
Gulson, B.L., in preparation. A lead isotopic study of the Woodlawn Pb-Zn-Cu deposit (for submission to J. Geol. Soc. Aust.).
Hart, S.R. and Tilton, G.R., 1966. The isotope geochemistry of strontium and lead in Lake Superior sediments and water. In: J.S. Steinhart and T.J. Smith (Editors), The Earth Beneath the Continents. Am. Geophys. Union, Geophys. Monogr., 10: 127—137.
Loftus-Hills, G. and Solomon, M., 1967. Cobalt, nickel and selenium in sulphides as indicators of ore genesis. Miner. Deposita, 2: 228—242.
Kholodov, V.N. and Gavrilov, Yu.O., 1974. Distribution patterns of trace elements in the Chokrak-Karagan rocks of the Yaryk-Su river (eastern Ciscaucasia). Lithol. Miner. Resour. (USSR)., 6: 726—739.
Malone, E., Olgers, F., Cucchi, F.G., Nicholas, T. and McKay, W., 1975. Woodlawn copper-lead-zinc deposit. In: C.L. Knight (Editor), Economic Geology of Australia and Papua-New Guinea, 1. Metals. Australasian Institute of Mining and Metallurgy, Melbourne, Vic., pp. 701—710.
McKay, W., 1973. Major-trace elements of Woodlawn black shales. Jododex Aust. Pty. Ltd., Intern. Rep., 12 pp (unpublished).
Peltola, E., 1960. On the black schists in the Outokumpu region in Eastern Finland. Bull. Comm. Geol. Finl., 192.
Peltola, E., 1968. On some geochemical features in the black schists of the Outokumpu area, Finland. Bull. Geol. Soc. Finl., 40: 39—50.
Petersen, M.D. and Lambert, I.B., in preparation. Implications of petrographic mineralogical and geochemical investigations of rocks around the Woodlawn copper-lead-zinc orebody, southeastern New South Wales (for submission to J. Geol. Soc. Aust.).
Saxby, J.D., 1976. The significance of organic matter in ore genesis. In: K.H. Wolf (Editor), Handbook of Stratabound and Stratiform Ore Deposits. Part 1: Principles and General Studies, 2. Geochemical Studies. Elsevier, Amsterdam, pp. 111—133.
Sherwin, L., 1973. Stratigraphy of the Forbes-Bogan Gate district. Rec., Geol. Surv. N.S.W., 15: 47—101.
Wampler, J.M. and Kulp, J.L., 1964. An isotopic study of lead in sedimentary pyrite. Geochim. Cosmochim. Acta, 28: 1419—1458.

Journal of Geochemical Exploration, 8 (1977) 105—115
© Elsevier Scientific Publishing Company, Amsterdam — Printed in The Netherlands

USE OF PRIMARY DISPERSION FOR EXPLORATION OF MISSISSIPPI VALLEY-TYPE DEPOSITS

H.L. BARNES and N.G. LAVERY*

Ore Deposits Research Section, Pennsylvania State University, University Park, Pa. 16802 (U.S.A.)

(Revised version accepted February 10, 1977)

ABSTRACT

Barnes, H.L. and Lavery, N.G., 1977. Use of primary dispersion for exploration of Mississippi Valley-type deposits. J. Geochem. Explor., 8: 105—115.

In unmineralized, carbonate rocks of the Wisconsin district, the Zn content is less than 4.7 ppm and varies with the clay content. Therefore, by subtracting an amount proportional to the clay content from the analysis of unweathered mineralized samples, the remainder is the amount of Zn added hydrothermally. These adjusted analyses reveal that the lateral extent of primary hydrothermal dispersion is a function of the width of the adjacent sulphide vein and may extend to at least 53 m from a major orebody. The adjusted concentration and its gradient are indicators of the distance to, and size of, veins in this district.

The utility of this relationship for exploration depends upon optimizing several factors to maximize the probability of detecting any orebody in the area searched at minimum cost. To detect orebodies with a minimum width of 1 m, the optimum sampling interval was found to be 9 m with a threshold concentration of 5 ppm Zn (adjusted value). This combination achieves a 95% probability of detecting any orebody over 1 m thick within the area sampled. There is a 27% probability that a consecutive pair of samples might be anomalous due to sampling within the narrow dispersion of two random fractures or minor veins. A second cycle of samples taken midway between each pair of anomalous samples eliminate such false anomalies.

Where samples can be collected across the trend of orebodies, either through potentially transecting drill core or along apparently barren drifts, the cost of testing an apparently barren 300-m section is roughly equivalent to that of 12 m of diamond drilling. This procedure complements a drilling programme by greatly increasing the probability of detecting orebodies at minimum cost for those areas where the initial, background metal concentration can be reliably established either from its constancy in the host rock or by a normalizing procedure such as that used here.

INTRODUCTION

In principle, primary dispersion near epigenetic orebodies should be useful

*Present address: Minerals Department, Exxon Company, Salt Lake City, Utah, U.S.A.

for evaluating the potential of new ground within a mining district. For this application, the measurement of primary dispersion requires that some means be found of distinguishing between concentrations of metals or other trace elements added by mineralization, and those initially present in the host rocks. It also depends upon having accurate analyses that can resolve within a few percent of background concentrations. This dispersion becomes more useful in mineral exploration if the primary dispersion halo is broad. Consequently, the more mobile and abundant elements are most useful for this purpose, provided of course that samples can be found that are essentially unaffected by later, secondary dispersion.

The geochemical characteristics of zinc distribution in the Illinois-Wisconsin zinc-lead district have been found to meet, at least partially, the above requirements for use in exploration (Lavery and Barnes, 1971). We shall present here a statistical evaluation of the optimum sampling scheme and threshold level for exploration in this district as well as the probabilities for success and overall cost-effectiveness.

CHARACTERISTICS OF PRIMARY DISPERSION

In this Mississippi Valley-type district, the ores occur in three Middle Ordovician carbonate formations, the Galena, Decorah, and Platteville Formations (Fig. 1). Unmineralized samples show that here, as in other normal carbonate sedimentary rocks (Wedepohl, 1972), the initial Zn content prior to mineralization is dominantly in the clay fraction. The regression line for Zn content as a function of acid-insoluble residue (Lavery and Barnes, 1971, fig. 3) is:

$$\text{Zn (ppm)} = 0.931 + 3.067 \log (\% \text{ insoluble residue})$$

for 14 unmineralized samples with 1.7—4.7 ppm Zn; the standard error of estimated is 0.4 ppm. Consequently, by measuring the proportion of insoluble residue in unweathered, mineralized samples, the initial Zn content can be inferred and then subtracted from the total content as determined analytically, to find the concentration added hydrothermally.

These adjusted contents, attributable to mineralization, indicate that the lateral extent of primary hydrothermal dispersion is a function of the width of the adjacent sulphide vein (see Fig. 2). Extending north from the North Gensler orebody, the primary Zn dispersion curve of Fig. 2 is detectable in visually barren carbonate samples to about 53 m from the ore/host rock contact. The three curves of Fig. 2 show that the slope and curvature of the dispersion gradient, if taken roughly perpendicular to the vein, are directly related to the distance to, and size of, the adjacent mineralization. Therefore, the length and regularity of the dispersion apron clearly merits quantitative evaluation for exploration.

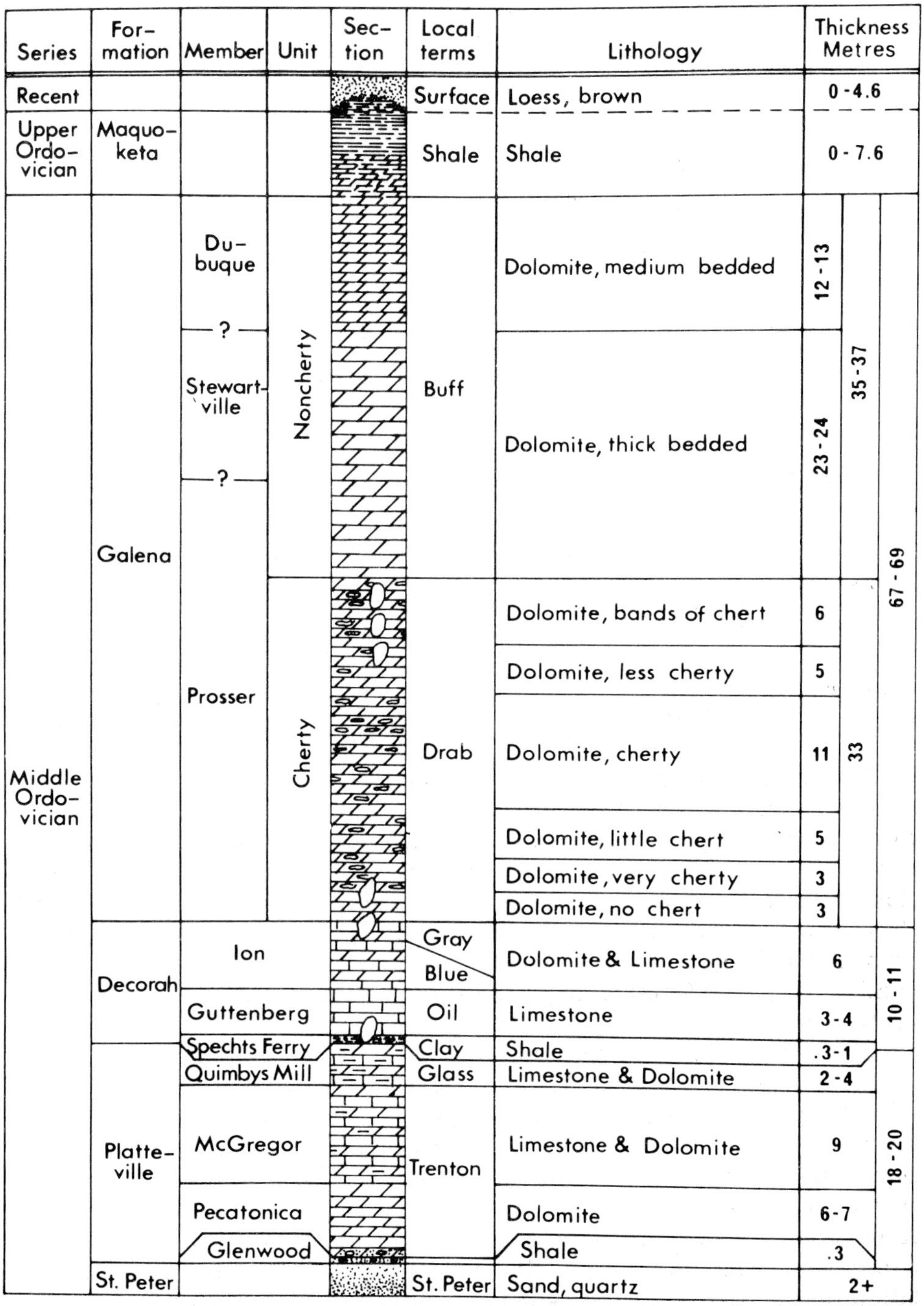

Fig. 1. The strongly mineralized section of the stratigraphy in the Wisconsin zinc-lead district. (Modified from Heyl et al., 1959.)

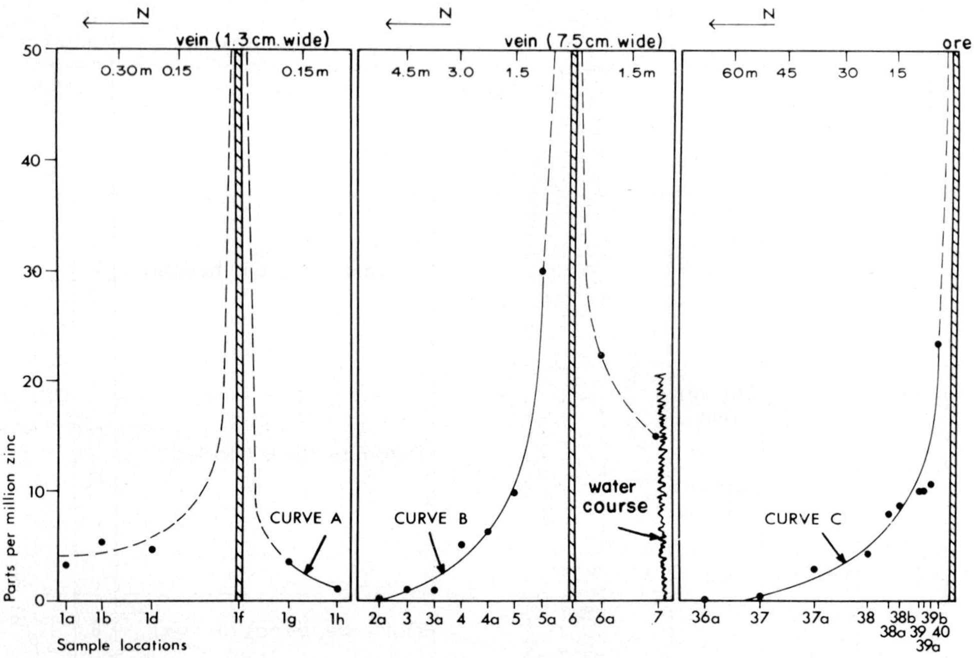

Fig. 2. Primary Zn dispersion adjacent to three veins of differing thicknesses along the Sedgewick-North Gensler drift. The analytical uncertainty is within the diameter of each point. (Modified from Lavery and Barnes, 1971, fig. 6.)

The extent of dispersion can be directly associated with vein width for assumed threshold levels as shown by the empirical curves in Fig. 3. Although we have derived an analytical differential equation describing the overall dispersion process (Lavery and Barnes, 1971, equation 11), its evaluation involves an intractable complementary error function and reveals only that the solution is an exponential function. Consequently, Fig. 3 has been derived simply by reading from the dispersion curves for veins of specific widths in Fig. 2, corresponding values for the length of dispersion and concentration of Zn.

Fig. 3 shows that, as expected, the lower the threshold level for samples to be classified as anomalous, the longer the resolvable dispersion apron. For example, at a threshold of 5 ppm Zn (adjusted), a 1-m vein causes anomalous dispersion to 8.2 m, but at 15 ppm, it only extends 3.0 m. This figure can also be used to predict distance to, and size of, nearby mineralization where only a portion of a dispersion curve is known. An example illustrates this procedure. If along a regular dispersion curve, the distance between the 5 ppm and 10 ppm Zn concentrations is 6.4 m, then by interpolation on the figure, the adjacent mineralization is about 3.0 m thick and its contact is 7.2 m from the 10-ppm sample in the direction toward increasing concentrations.

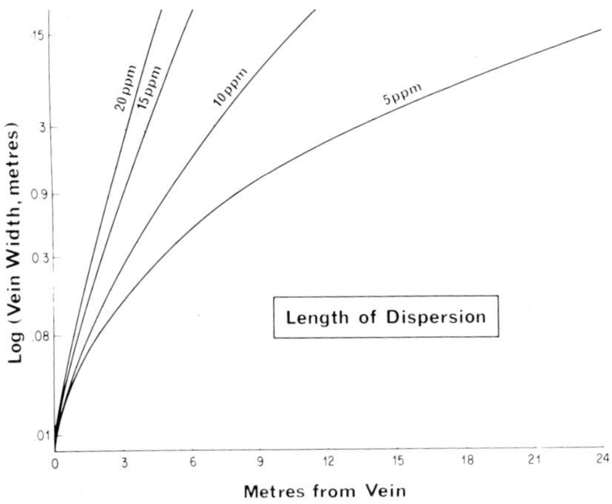

Fig. 3. Correlation of log (vein width) with the length of dispersion out to four adjusted Zn contents.

SAMPLING TACTICS

The cost-effectiveness of using dispersion curves in exploration can be maximized by having sampling intervals as wide as possible while still minimizing the risk of missing dispersion related to important orebodies. A single anomalous sample is insufficient to distinguish between a minor and an important anomaly but two adjacent anomalous samples are adequate. A single sample may frequently be anomalous due to the influence of an adjacent minor vein but only adjacent anomalous samples provide an indication of the width of an anomaly. Consequently, our objective for exploration is to be certain that at least one pair of samples will be taken within the dispersion apron of any vein thick enough to be an orebody. A means of evaluating probabilities for pairs of samples in the search for veins is illustrated by Fig. 4. At the sampling interval, s, any sample falling in the uncross-hatched zone will have an anomalous mate lying within a dispersion apron of length, d; however, samples from the cross-hatched zone would not have a mate in the anomalous region and the importance of the anomaly would be undetected. In general terms, the ratio of the total of the combined lengths of the two unshaded zones (as fixed by the chosen value of s) to the length of the apron, d, equals the fractional probability of detecting by two samples the anomaly associated with the vein.

Starting with a pair of adjacent samples, the probabilities of detecting an anomaly can be calculated from the above approach using values interpolated from Fig. 3 for various sizes of dispersion aprons associated with different vein widths and threshold levels. In this district, a vein 1 m thick is worth mining but one only 0.3 m thick probably is not, so that the discrimination level may best be fixed to favour the discovery of the thicker vein. The results of this probability calculation are given in Table I for threshold levels of

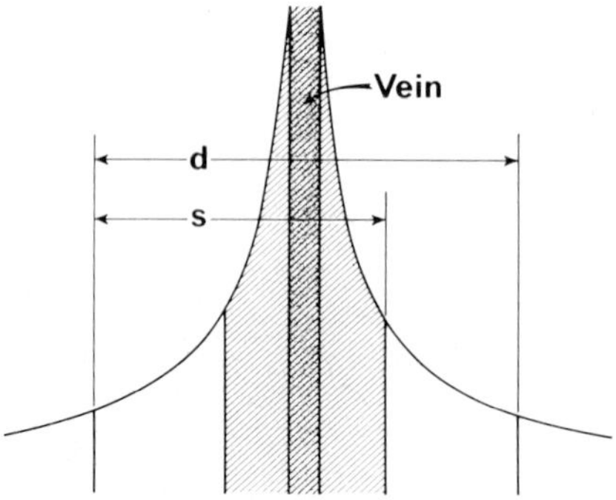

Fig. 4. Dispersion adjacent to a vein schematically showing metal content vertically versus distance laterally. The width of dispersion to a selected threshold level is given by *d* and the sampling interval by *s*. See text for discussion of the cross-hatched zone.

TABLE I

Probabilities of detecting dispersion aprons using adjusted Zn analyses

Sample interval (m)	Vein width (m)	Probability (%) at selected Zn threshold values			
		5 ppm	10 ppm	15 ppm	20 ppm
6	0.3	89	26	0	0
	1.0	100	86	26	0
7.5	0.3	61	0	0	0
	1.0	100	57	0	0
9	0.3	33	0	0	0
	1.0	95	29	0	0
12	0.3	0	0	0	0
	1.0	60	0	0	0

5 ppm Zn and above. At a sampling interval of 6 m there is an 89% probability that a pair of adjacent samples would fall within the dispersion apron of an 0.3-m vein if threshold were at 5 ppm. Clearly, this is too high a probability for detecting non-economic mineralization. In addition, the 100% probability of detecting the 1.0-m vein associated with this combination of variables indicates that either a higher threshold and/or a longer sampling interval would be preferred both to improve discrimination between ore and mineralization and to reduce sampling frequency and cost.

There is little advantage to using threshold levels above 5 ppm as shown by the results in Table I. If threshold is set at 10 ppm, then sampling intervals of no more than 6 m provide adequate discrimination between ore and non-ore vein widths, together with high probabilities of finding 3-m-wide veins. Higher threshold values have probabilities that are too low for reliable detection of ore. Apparently, a better choice is a lower threshold level and a longer sampling interval. Optimum for useful discrimination at least cost appears to be the combination of a 9-m sampling interval and the 5-ppm threshold because there is a 95% probability of detecting the anomaly of a valuable 1.0-m vein but only a 33% probability for the unimportant 0.3-m vein. Sampling intervals wider than 9 m have significant probabilities of not revealing veins of ore width.

False alarms

Near large orebodies, the associated long dispersion gradients generally have superimposed on them some smaller dispersion aprons from veins of varying sizes (see Fig. 5). The additive effect of the major and minor dispersion aprons is to make the samples more obviously anomalous. However, such minor veins also occur commonly in otherwise barren rocks. Consequently, it is possible that a pair of consecutive samples may fall on dispersion aprons of two separate minor veins rather than on a single, dominant gradient. The probability of such false alarms occurring can also be evaluated from the available adjusted analyses.

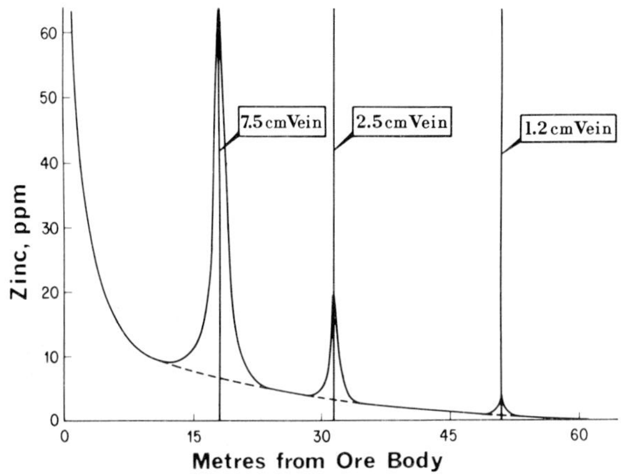

Fig. 5. Schematic diagram of typical multiple dispersion aprons near a major orebody. Superimposed on the dispersion from the orebody at the left are minor dispersion aprons from small veins.

The Zn content of a population of 101 random samples collected along 1.04 km of drifts is shown in Fig. 6. It is evident that many samples (47%) contain less than 5 ppm Zn and are not anomalous; therefore, false alarms caused by neighbouring samples lying on separate dispersion curves should not happen often enough to be a major problem. From data on 71 pairs of neighbouring samples, probabilities of both samples exceeding chosen threshold levels have been compiled and are shown in Table II. Excluded from

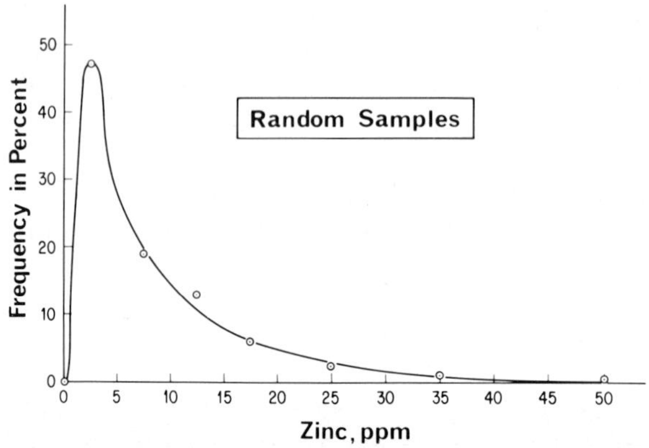

Fig. 6. Zn content added by mineralization in a population of 101 samples of carbonate host rocks. The points represent the midpoints of the histogram intervals beginning from the origin.

TABLE II

Probabilities for adjacent samples both exceeding selected threshold levels due to the influence of dispersion from random, separate veins

Threshold level (ppm)	False alarm probability (%)	Observed frequency
5	27	1 in 4
10	17	1 in 6
15	6	1 in 18
20	1	1 in 71

this set of samples were any pairs found to be on a single dispersion curve, so that this population represents only those pairs subject to the influence of random veins and their dispersion aprons. At the threshold of 5 ppm, there is a 27% probability that any single pair would be anomalous although not close to ore, but this probability drops to 17% if the threshold is 10 ppm.

False alarms can easily be evaluated by taking a second sample midway between those of any anomalous pair. The second set of analyses, besides verifying the continuity of anomalies between the two initial samples, also provides a third point on the dispersion-gradient curve so that its curvature can be estimated. As we have shown before (Lavery and Barnes, 1971), the curvature of the dispersion gradient can be used as an indicator of the magnitude of associated mineralization, with low curvatures being related to larger orebodies.

Sampling requirements

Two combinations of threshold levels and sampling intervals provide adequate distinction between anomalies associated with minor mineralization and those near orebodies (Table I): a 10-ppm threshold and a 6-m interval, or a 5-ppm threshold and 9-m interval. The probabilities for needing intermediate samples to evaluate false alarms differ significantly between the two threshold levels and are 17% and 27%, respectively (Table II). The number of samples needed to explore a typical 300-m interval is given on Table III. The optimum combination, for minimum cost and maximum reliability in detecting anomalies near ore-width veins but not lower-grade mineralization, is clearly a 9-m interval with a 5-ppm threshold level.

TABLE III

The number of samples required to explore for primary Zn dispersion 300 m of host rocks if no orebodies are found

Threshold (ppm Zn):	5	10
Sampling interval (m):	9	6
Samples in first cycle	33	50
Samples in second cycle	9	9
Total samples per 300 m	42	59

SUMMARY AND CONCLUSIONS

Evaluation of primary dispersion should be useful in various situations. Where veins occur near, but do not cross, drifts or drill holes, then their dispersion haloes may be detected by routine sampling. In drill holes where the intersected vein was either locally thin or too friable to be collected as core, then dispersion can provide supplementary evidence for its existence and value. Where several closely spaced drill holes intersect a bed, then the metal content may be contoured as a guide to where further holes, if any, should be drilled. The most important use of primary dispersion is probably to check the ends of blind drifts and drill holes for proximity to ore and to determine whether further testing is warranted.

Primary Zn dispersion can be used effectively in this district in the above situations to evaluate barren carbonates for proximity to ore. Samples analyzed for Zn content may be from surface or subsurface openings or from drill core, so long as they are unaffected by weathering and secondary dispersion. About 42 samples per 300 m of rock should be sufficient to locate, by at least two neighbouring anomalous samples, any ore-related dispersion. The optimum sample spacing during the first of two cycles is near 9 m, ideally along one bed in a direction perpendicular to the trend of veins. Check samples must then be collected in a second cycle from midway between each pair of anomalous samples from the first cycle. Samples are considered anomalous optimally only if above a threshold of 5 ppm adjusted Zn content. The adjusted Zn content is that added by mineralization; it is simply the total Zn content less that initially present as indicated by the content of clay-rich, acid-insoluble residue. Besides the utility of this approach in simply identifying anomalies, the curvature in the dispersion gradient also is indicative of proximity to ore.

The low costs involved in checking for primary dispersion favour its routine use. In quantity, suitably accurate trace analyses cost the equivalent of about 0.3 m of exploratory diamond drilling. Because major orebodies may generate dispersion haloes as long as 53 m, then samples taken from barren drifts or cores across the trend of veins are useful to check for unexposed mineable mineralization within a few tens of metres of the sample line. This method is especially valuable at the end of drifts or drill holes because it provides a basis for deciding whether further drilling or drifting is warranted. By three analyses costing the equivalent of 1 m of drilling, a distance of up to 9 m can be tested for a 1-m vein and up to 53 m for a major orebody, at a cost-benefit ratio compared to drilling of between 0.11 and 0.02.

Primary dispersion is useful in any district only if it is possible to differentiate between metal contents added by mineralization and those originally present. In carbonate rocks, adjusting analyses on the basis of insoluble residue content may be effective as used here for this district. Alternatively, some host rocks are sufficiently constant in original content that it may be simply subtracted from the total present to find the increment added by ore solutions. For example, shales are frequently sufficiently homogeneous to be treated in this manner and preliminary testing in the Parral District, Chihuahua, has shown that the dispersion gradient from veins in the black shale is detectable to at least 200 m (Barnes, 1959). However, the problem of establishing the initial background metal content may become unwieldly and primary dispersion of little use in exploration in mining districts where several types of host rocks are commonly mineralized.

ACKNOWLEDGEMENTS

Our colleagues, A.W. Rose and J.C. Griffiths pointed out several flaws in the original manuscript for which we were grateful. This research was supported by the Earth Sciences Section of the National Science Foundation under grants NSF-G-21997, NSF-GA-477 and NSF-GP-2597, and also by the Ore Deposits Research Section of the Pennsylvania State University.

REFERENCES

Barnes, H.L., 1959. The effect of metamorphism on metal distribution near base metal deposits. Econ. Geol., 54: 919—943.

Heyl, A.V., Jr., Agnew, A.F., Lyons, E.J. and Behre, C.H., Jr., 1959. The geology of the Upper Mississippi Valley zinc-lead district. U.S. Geol. Surv., Prof. Paper, 309: 310 pp.

Lavery, N.G. and Barnes, H.L., 1971. Zinc dispersion in the Wisconsin zinc-lead district. Econ. Geol., 66: 226—242.

Wedepohl, K.H., 1972. Zinc: abundance in common sediments and sedimetary rocks. In: K.H. Wedepohl (Editor), Handbook of Geochemistry. Springer-Verlag, Berlin, Section 30-K, pp. 1—13.

Journal of Geochemical Exploration, 8 (1977) 117--137 117

SULPHUR ISOTOPE AND TRACE METAL COMPOSITION OF STRATIFORM SULPHIDES AS AN ORE GUIDE IN THE CANADIAN SHIELD

P.K. SECCOMBE

Department of Geology, University of Newcastle, Newcastle, N.S.W. (Australia)

(Revised version accepted February 4, 1977)

ABSTRACT

Seccombe, P.K., 1977. Sulphur isotope and trace metal composition of stratiform sulphides as an ore guide in the Canadian Shield. J. Geochem. Explor., 8: 117—137.

Sulphur isotope and metal data are used to discriminate between ore, sub-economic and barren mineralization in a region of the Canadian Shield containing a number of stratiform base metal sulphide deposits.

A statistical study indicates that ore sulphides from the Zn-Cu-Ag massive sulphide orebodies of South Bay Mines, northwestern Ontario, Canada which exhibit a narrow (2‰) range and proximity to zero permil, are similar isotopically to sub-economic sulphides in comparable host rocks. Sulphides from other deposits in the region demonstrate a broad (greater than 6‰) spread in isotopic values and averages generally removed from, and commonly lighter than, zero permil.

High contents of the trace metals Ag, Sn, Cd and, to a lesser extent, hight In, and high but variable Co/Ni ratios characterize both the ore association and several of the sub-economic deposits. The best representation of ore affinity is obtained from Cu-Pb-Zn, Pb-Zn and Sn-Zn plots. Fields delineated by plots of Pb-δ^{34}S, Sn-δ^{34}S and Cd-δ^{34}S developed from sulphide compositional data best define an ore association. Plots of host rock silica content and average δ^{34}S values for individual deposits discriminate between ore, sub-economic deposits in geological settings similar to that of the South Bay Mine, other sub-economic deposits and zones of barren mineralization.

Although directed specifically towards strata-bound base metal ores in a volcanic environment, the criteria used in this study to identify an ore association should be equally applicable as a genetic and possible exploration guide to other classes of ores in differing environments.

INTRODUCTION

Recent geochemical and isotopic studies of base metal ore deposits in volcanic environments have added much to our knowledge of this important ore type. Many studies relate either to individual deposits or are directed towards the correlation of distribution patterns of genetically similar deposits. However, a detailed geochemical comparison has not been made previously between an economic ore deposit of this type and bodies of low-grade or barren mineralization existing within similar geological environments. Determination of the parameters that distinguish "ore" from "oc-

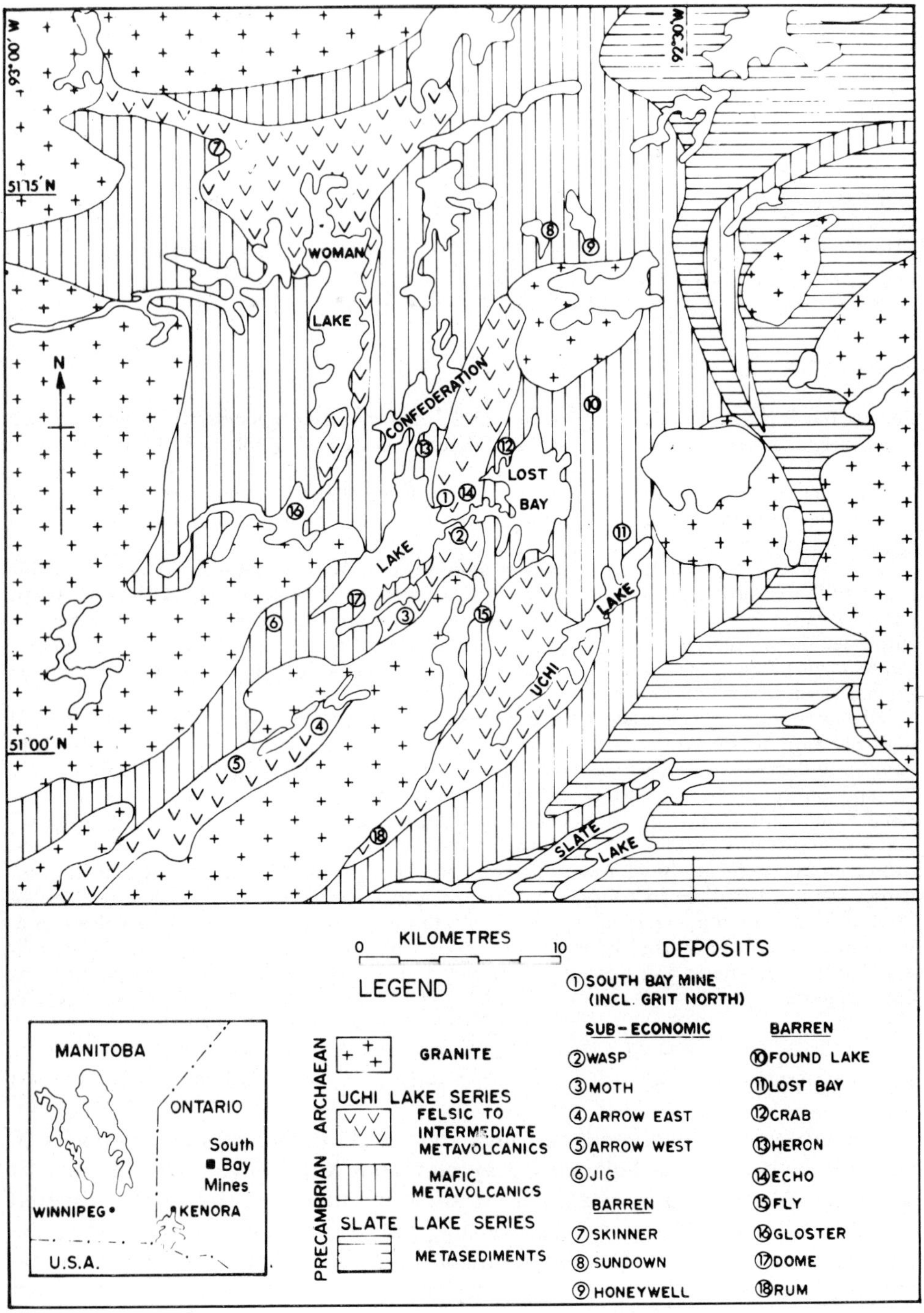

Fig. 1. Location and general geology of the southern part of the Birch-Uchi greenstone belt, northwestern Ontario, Canada. (After Ontario Department of Mines Map 2175, 1968.)

currence" in a particular geological environment can provide useful information for developing a theory of ore formation and has important economic implications in the search for further ore deposits.

With this above approach in mind, a study was made to determine the distribution of trace metals and sulphur isotopes in sulphides from deposits throughout a major portion of the Birch-Uchi volcano-sedimentary belt of northwestern Ontario (Fig. 1). The Birch-Uchi belt is a typical Archaean greenstone belt located in the Superior structural province of the Canadian Shield. The objective was to compare barren and sub-economic deposits of the belt with the Zn-Cu-Ag orebodies of South Bay Mines Limited at Confederation Lake (the South Bay or Uchi deposit). The South Bay deposit was discovered by Selco Mining Corporation in 1968.

The generalized geology of the south and central sections of the Birch-Uchi greenstone belt, the regional location and the location of the South Bay deposit and the seventeen areas of sub-economic and barren mineralization selected for study are indicated in Fig. 1.

REGIONAL GEOLOGY

The Birch-Uchi greenstone belt has the regional structure of a broad syncline and is surrounded by extensive granitic terrain. The belt trends meridionally and is approximately 80 km in length and averages 30 km in width. Within the greenstone belt, a group of metasedimentary rocks termed the Slate Lake Series by Bateman (1939) forms the basement. These are conformably overlain by the predominantly volcanic Uchi Series; both rock series have been intruded subsequently by stocks and dykes which range in composition from gabbro to granite.

The South Bay Mines deposit and the seventeen other mineralized areas included in this study are all contained within the Uchi Series.

The Uchi Series is the dominant unit in the area and comprises basic to acid volcanic rocks and associated sedimentary rocks, together with derived schists. Goodwin (1967) found that the volcanic and sedimentary components of this group are distributed in two successive cycles, in which each cycle displays a progression from older basic volcanics to younger acid rocks. Goodwin calculated an average stratigraphic thickness of 10,000 m for the present Uchi Series and estimated a thickness of 13,000 m for the original volcanic-sedimentary pile.

The mafic metavolcanic rocks comprise lava flows, tuffs, breccias and associated sills and dykes ranging in composition from basalt to andesite.

The major rock types in the felsic to intermediate volcanic units are fragmental rocks of rhyodacite, dacite and andesite composition. Rhyolites, although widely distributed throughout the volcanic rock series, are notably subordinate.

Small, semi-conformable bodies of quartz feldspar porphyry, ranging in size from 100 m in lateral extent to 4000 m by 800 m, are enclosed within

TABLE I

Geology of the deposits

Area	Mineralization			Host rocks
	form	mineralogy*	maximum metal content	
Economic deposits				
South Bay Mine	Stratiform massive sulphide lenses to 140 m in length and 15 m in width localized within felsic metavolcanic rocks marginal to the folded contact of a body of quartz feldspar porphyry	py, sp, cp (gn), (cst), <asp, <po <mt	15% Zn, 2.4% Cu 107 g/t Ag	Altered rhyolite tuffs, flows, lapilli tuff, quartz feldspar porphyry
Sub-economic deposits				
Arrow	Massive sulphide lenses to 2 m in width, stringers, veins and disseminations within metavolcanic rocks and marginal to a felsic porphyritic pyroclastic unit. Shear zone mineralization in the Arrow East area	po, py, mt (sp), (cp), <gn, <asp	5.5% Zn, 1.2% Pb, 0.5% Cu, 28 g/t Ag	Felsic to intermediate metavolcanic rocks, derived schists. Felsic porphyry
Wasp (including Wasp South prospect)	Massive sulphide lenses to 2 m in width, stringers and disseminations in felsic metavolcanic rocks marginal to quartz feldspar porphyry 1100 m and 1700 m southeast of the South Bay orebodies	po, py, (cp), (sp)	6.3% Zn, 0.5% Cu, 24 g/t Ag, 0.9 g/t Au over a 1-m drill intersection	Sheared and altered volcanic rocks of the South Bay Mine rhyolite series. Rhyolite flows, tuff, lapilli tuff, agglomerate and breccia; argillite. Quartz feldspar porphyry
Moth	Sulphide stringers and disseminations in metavolcanic rocks within 60 m of felsic intrusions	po, py (cp), (sp)	2% Cu, 0.7% Zn, 9 g/t Ag over over 2 m	Fragmental felsic to intermediate volcanic rocks, derived schist
Jig	Massive sulphide zones and stringers within metavolcanic rocks, metasediments and schists and at the contact of intermediate to mafic metavolcanic units and gabbro intrusions	po, py, mt, cp, (sp), <gn	5.8% Zn, 3.7% Cu 24 g/t Ag, 0.6 g/t Au over 1 m	Felsic to intermediate tuffs, metasediments, derived schists; gabbro

Barren deposits				
Grit North	Sulphide veins, stringers and disseminations. Contact mineralization 200—300 m north of the South Bay orebodies	po, py, <sp, <cp	1.8% Zn, 0.3% Cu 9 g/t Ag over 0.6 m	Tuffs and flows of the mine rhyolite series. Quartz feldspar porphyry
Echo	Sulphide veins and stringers to 0.2 m in width. Contact mineralization 900—1500 m northeast of the South Bay orebodies	po, (py), <sp, <cp	1% Zn, 0.1% Cu, 3 g/t Ag	Felsic to intermediate tuffs, agglomerate and flows; minor argillite. Quartz feldspar porphyry
Crab	Sulphide bands, stringers and disseminations	po, (py), <sp, <cp	0.5% Zn, 0.1% Cu, 1.8 g/t Ag over 0.6 m	Basalt and andesite, commonly amygdaloidal. Minor dacite and rhyolite tuff, lapilli tuff, flows
Dome	Sulphide stringers and bands	po, py	No assays, very restricted mineralization	Andesite, dacite tuffs
Fly	Sulphide stringers, bands and disseminations	po, py, <cp	0.3% Cu, 0.04% Zn, 1.8 g/t Ag, 2.1 g/t Au over 0.6 m	Andesite, commonly amygdaloidal. Andesite and dacite tuffs, diorite, minor metasediments
Found Lake	Sulphide disseminations; shear zone mineralization with quartz and magnetite	po, py, <sp, <cp	1.9% Zn, 0.05% Cu, 1.8 g/t Ag over 0.6 m	Felsic to intermediate tuffs, lapilli tuff, gabbro
Gloster	Massive sulphide lenses 1—3 m wide, stringers and disseminations	py, po, <cp	No assays	Intermediate to felsic metavolcanic rocks, shale, argillite, limestone, iron formation
Heron	Sulphide stringers, subconformable to shearing	py, po, <cp, <sp	No assays, very restricted mineralization	Felsic to intermediate tuffs marginal to mafic metavolcanic rocks
Honeywell	Massive sulphide bands; shear zone mineralization	po, (py), <cp, <sp	No assays minor mineralization	Intermediate to felsic metavolcanic flows and pyroclastics, felsic dykes, diorite

TABLE I (continued)

Area	Mineralization			Host rocks
	form	mineralogy*	maximum metal content	
Lost Bay	Sulphide disseminations and bands; shear zone mineralization	po, py	No assays	Intermediate metavolcanic tuffs and agglomerate, minor metasediments
Rum	Sulphide stringers and bands associated with quartz veining	py, po	No assays, very restricted mineralization	Biotite and sericite schists, mafic to intermediate metavolcanic rocks
Skinner	Disseminated pyrite. Stringers and bands of pyrrhotite and pyrite to 0.6 m in width, conformable to lithology	py, po, <cp, <sp	0.03% Cu, 0.03% Zn, 1.5 g/t Ag over 1 m	Graphitic argillite, argillaceous tuffite; felsic to mafic metavolcanic rocks
Sundown	Sulphide stringers and bands	po, py, <cp	0.6% Cu, 0.3% Zn, 1.5 g/t Ag over 1 m	Mafic to intermediate metavolcanic rocks, commonly spherulitic

*Abbreviations and symbols: po = pyrrhotite, py = pyrite, sp = sphalerite, cp = chalcopyrite, gn = galena, asp = arsenopyrite, mt = magnetite, cst = cassiterite, qz = quartz, ( ) = minor, < = trace dacite and rhyolite volcanics along an 8-km zone close to the central axis of the greenstone belt. The South Bay deposit is localized in rhyolite tuffs along the contact of the largest porphyry body. Quartz feldspar porphyry is characteristically noted as a wallrock to Archaean stratiform base metal mineralization in the Canadian Shield (Hopwood, 1976), although its actual character is somewhat enigmatic. Pollock et al. (1972) suggest that the South Bay porphyry is, in part, an extrusion. Chemical similarities with the rhyolite volcanic rocks (Seccombe, 1973) and the recognition of intrusive features such as transgressive contacts, fine-grained marginal zones and rare wallrock inclusions suggest a shallow-seated origin, co-magmatic with the acid lavas.

The sedimentary rocks, consisting mainly of tuff, greywacke, arkose, impure quartzite, conglomerate, breccia and minor iron formation and limestone are found to be associated more commonly with the upper, predominantly acid to intermediate phases of the volcanic cycles.

Metamorphism in the area is generally limited to the low-rank greenschist facies.

GEOLOGY OF THE DEPOSITS

The geology and mineralogy of the South Bay deposit and the seventeen areas of sub-economic and barren mineralization selected for the comparative study are summarized in Table I. A brief summary of the geology of the South Bay deposit is given below. For a more detailed description of this deposit and the other mineralized areas, reference should be made to Pollock et al. (1972) and Seccombe (1973).

Table I indicates that many of the deposits form in felsic metavolcanic rocks and associated sediments. Several, including the South Bay ores, the sub-economic deposits Wasp and Arrow and the barren deposits Grit North and Echo, are localized within fragmental felsic metavolcanic host rocks marginal to bodies of felsic porphyry.

The distinction between sub-economic and barren mineralization is based primarily on grade; the sub-economic deposits yield metal values as high as 4% Cu, 6% Zn, 1% Pb, 30 g/t Ag and 0.1 g/t Au, whereas the barren zones rarely exceed 0.5% Cu, 0.5% Zn and 3 g/t Ag. The sub-economic deposits possess grades comparable with the South Bay Mine but lack sufficient tonnage of contained metal to prove a viable deposit. Mineralization in the barren deposits is generally of limited extent and principally comprises disseminated to massive pyrrhotite and pyrite.

The South Bay orebodies form a number of closely related, steeply dipping massive sulphide lenses localized at the contact of an embayment of rhyolite tuff penetrating quartz feldspar porphyry. The rhyolite tuff is the uppermost unit of a steepy dipping volcanic sequence, the younger of the two sequences of upward-grading, mafic to felsic volcanism recognized in the Uchi area.

The ore zones are massive and essentially lacking in enclosed gangue or disseminated mineralization and generally show sharp contacts with the enclosing rocks. Pyrite is the dominant sulphide, and is accompanied by abundant sphalerite and lesser amounts of chalcopyrite.

The metals are considered to be volcanogenic. In their present configuration, the ores and sub-economic sulphides located in felsic volcanic rocks are suggested to be either in-situ volcanic exhalative deposits (cf. Pollock et al., 1972; Coomer and Schwarcz, 1974) or epigenetic deposits of hydrothermal origin (Seccombe, 1973; Seccombe and Clark, in preparation). Barren deposits are considered to be relatively in-situ syngenetic deposits (Seccombe, 1973).

METAL DISTRIBUTION IN SULPHIDES

The average trace metal values for 31 sulphide samples separated from drill core, compiled in Table II, represent the South Bay deposit and major mineralized areas. Individual analyses which form a basis for this table are plotted as metal variation diagrams in Fig. 2.

TABLE II

Mean trace metal content of sulphides — Birch-Uchi greenstone belt

Deposit	Sulphide*	Metal content (ppm)**							Number of samples
		Ag	Sn	In	Cd	Co	Ni	Co/Ni	
Economic deposits									
South Bay	sp	150	1710	74	3000	n.d.	7	—	3
Mine	cp	205	620	118	101	1850	8	231	2
	py	121	659	466	48	79	—	>79	4
Sub-economic deposits									
Arrow	po	5	7	28	—	50	63	0.79	5
	py	—	11	61	—	216	—	>216	1
	sp	20	590	206	1700	n.d.	8	—	1
Wasp	po	8	26	85	—	16	159	0.10	2
Jig	po	10	1	22	—	120	430	0.28	4
Barren deposits									
Crab	po	—	1	7	—	84	74	1.14	2
Sundown	po	5	—	42	—	64	144	0.44	2
	py	—	—	62	—	1510	26	58.1	1
Miscellaneous areas	po	1	6	37	—	174	121	1.44	4

*Sulphide abbreviations: sp = sphalerite, cp = chalcopyrite, py = pyrite, po = pyrrhotite.
**Detection limits: Cd, 10 ppm; Ag, Sn and Ni, 1 ppm.
n.d. = not determined.

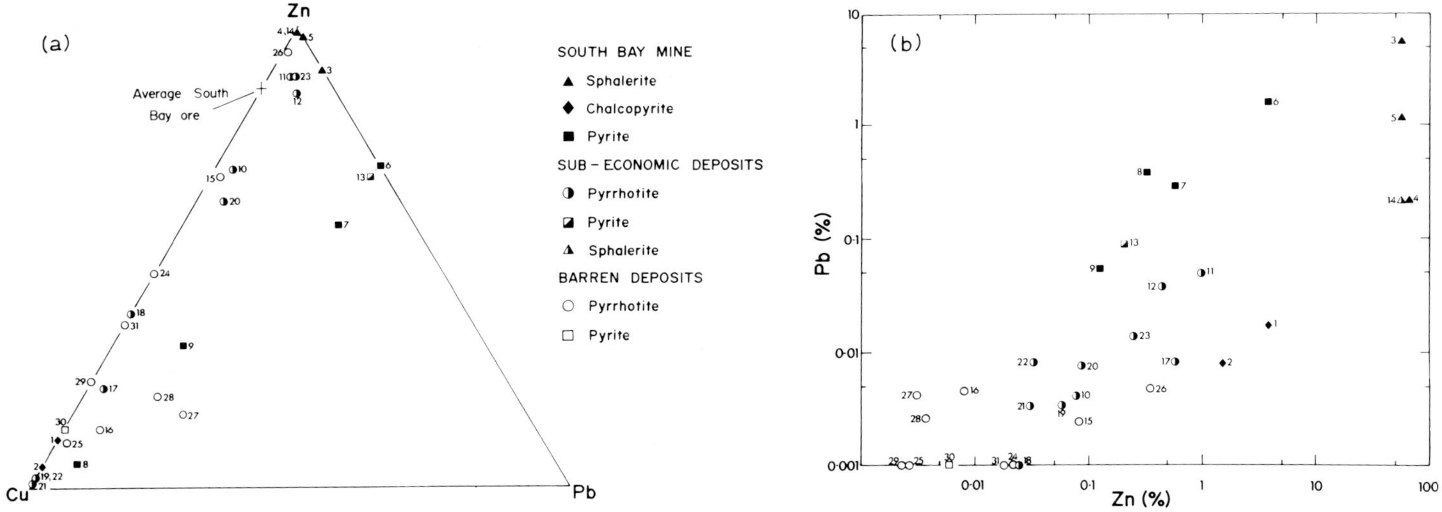

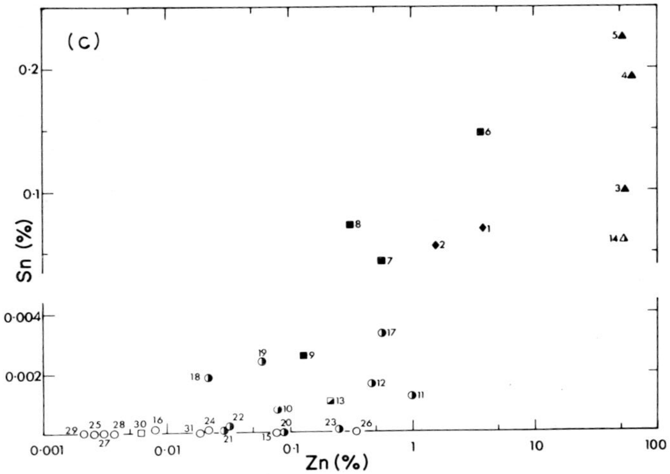

Fig. 2. Metal variation diagrams. (a) Cu-Zn-Pb diagram. Note extensive scatter for barren pyrrhotite and pyrite (open symbols) close to the Cu-Zn tie line and the Cu vertex. Pyrite and pyrrhotite from the South Bay ore and Arrow deposit are displaced towards the Zn-Pb tie line. Pyrite samples demonstrate constant Zn/Pb ratios. (b) Pb-Zn diagram. The Arrow sub-economic sulphides are grouped with South Bay sulphides at high Pb and Zn values. Other sub-economic and barren deposits show suppressed metal values. (c) Sn-Zn diagram. Note separation of field occupied by South Bay, Arrow, Moth and Wasp sulphides at moderate to high Sn and Zn values from the Jig deposit and all barren deposits.

Key to numbering — *Economic deposits*: samples 1—9 = South Bay Mine. *Sub-economic deposits*: 10—16 = Arrow (10-14 = Arrow ore zone); 17, 18 = Wasp; 19 = Moth; 20—23 = Jig. *Barren deposits*: 24, 25 = Crab, 26 = Found Lake; 27 = Gloster; 28 = Skinner; 29—31 = Sundown.

The data of Table II show that the ore sulphides are characterized by high Ag, Sn, Cd and, to a lesser extent, by high In, and high but variable Co/Ni ratios. The sub-economic deposits, particularly Arrow and Wasp, show moderately high Ag, Sn and In values. Sphalerite from the Arrow area, apart from suppressed Ag content, is chemically very similar to the ore sphalerite. Barren deposits lack ore affinity, although pyrite from the Sundown prospect is substantially enriched in Co.

A study of the geochemical affinities of the sulphides was undertaken using a number of metal variation diagrams. Those plots which were found to be most useful in illustrating ore affinity are reproduced in Fig. 2.

The total spread in Cu-Zn-Pb ratios for the sulphides is similar to that produced by Wilson and Anderson (1959) for Archaean ore deposits. Pyrrhotite and pyrite from barren mineralization, in accord with a relatively low Pb content, show extensive scatter close to the Cu-Zn tie line and the Cu vertex. Pyrite and pyrrhotite from the South Bay ores and the sub-economic prospect Arrow are displaced towards the Zn-Pb tie line. Constant Pb/Zn ratios are implied by the linear trend demonstrated by pyrite from these areas. The apparent ore association of pyrite from the Arrow deposit (sample 13) is evident. The sample co-exists with sphalerite (sample 14) in a lithological and structural setting comparable with that of the South Bay ores.

The Pb-Zn diagram (Fig. 2) serves to distinguish ore and sub-economic sulphides from barren sulphides. The interdependence of Pb and Zn is again illustrated, whereby ore pyrite and chalcopyrite show a substantially linear trend. Pyrrhotite from the Arrow and Jig deposits exhibit a similar, though less distinct trend. As in the Cu-Zn-Pb diagram, pyrite sample 13 from the Arrow area is co-linear with the ore pyrite.

The high Sn content of the ores and the somewhat linear relationship of Sn to Zn, for low to moderate Zn content, is illustrated in the Sn-Zn diagram. The diagram is effective in outlining fields occupied by the ore sulphides, the sub-economic sulphides of ore affinity (Arrow, Wasp and Moth), sub-economic sulphides lacking in ore affinity (the Jig deposit) and barren mineralization.

The work of Bridge (1972) indicates that Sn is present in the South Bay ores as cassiterite and is most commonly associated with sphalerite. The high Sn content of sphalerite (Table II) and the recognition of a sympathetic Sn-Zn relationship are in line with Bridge's conclusions.

SULPHUR ISOTOPE DISTRIBUTION

Sulphur isotope ratios were determined for approximately 260 sulphide samples, obtained principally from drill core, representing the ores and the seventeen other mineralized areas selected for the investigation. Isotope ratio measurements were made on a Varian Mat GD 150 mass spectrometer. $\delta^{34}S$ values are referred to Canon Diablo troilite sulphur with the accepted

$^{32}S/^{34}S$ value of 22.220. An accuracy of $\pm 0.2^0/_{00}$ is indicated for the method. A description of sampling, SO_2 extraction and measurement techniques, and a full listing of results is provided by Seccombe (1973).

Sulphur isotope distribution patterns for the ores and other mineralized areas studied are graphed in Fig. 3. Statistical data for the more important groupings are given in Table III.

The South Bay ore sulphides show a narrow (approximately $2^0/_{00}$) distribution centred close to the meteoritic mean. Ore sulphides skewed towards more positive $\delta^{34}S$ values ("heavy" sulphur) are considered to have been involved in equilibration reactions with "heavy" wallrock pyrite

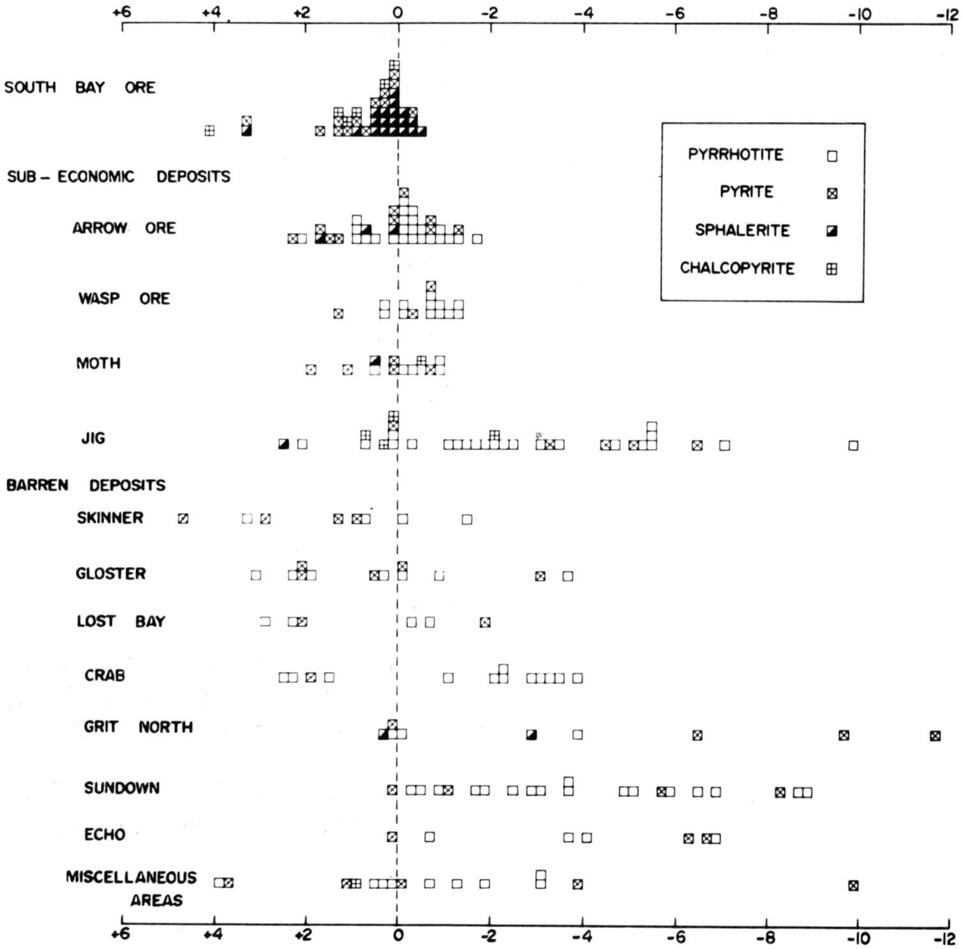

Fig. 3. Sulphur isotope distribution in sulphides from the Birch-Uchi greenstone belt. Note narrow range in isotopic values for sulphides from the South Bay Mine and sub-economic deposits Arrow, Wasp and Moth. All barren deposits and the sub-economic deposit Jig possess disperse and commonly negative $\delta^{34}S$ values.

TABLE III

Sulphur isotope results — Birch-Uchi greenstone belt

Deposit and statistical group	Number of samples	Mean $\delta^{34}S$ (‰)	Standard deviation (‰)
Economic deposits			
South Bay Mine			
total	53	+0.57	2.10
ore	39	+0.68	1.02
wallrock	14	+0.37	3.02
Sub-economic deposits			
Arrow			
total	49	−0.23	1.64
ore zone	37	+0.09	1.01
Wasp			
total	19	−0.97	2.29
ore zone	15	−0.47	0.73
Moth	13	+0.03	0.85
Jig	32	−2.51	2.88
Barren deposits			
Skinner	8	+1.57	2.06
Fly	4	+0.64	0.50
Gloster	12	+0.37	2.14
Lost Bay	5	+0.28	1.92
Found Lake	3	−1.27	1.83
Crab	13	−1.82	2.27
Grit North	9	−3.82	4.57
Sundown	21	−4.00	2.87
Echo	7	−4.05	2.87

which is known to have a distribution range of approximately 12‰ (Seccombe, 1973).

The sub-economic deposits Arrow, Wasp and Moth, located in lithologies similar to the South Bay deposit, show ranges for their ore zones of 4.1, 2.6 and 3.0‰ respectively and narrow distributions comparable with that for the South Bay ores.

The Jig deposit and all barren deposits yield disperse (greater than 6‰) distributions and averages generally removed from, and commonly lighter than, zero permil.

Statistical analysis of sulphur isotope data

Sulphur isotope data were grouped into several categories for statistical processing — total distribution for each deposit, pyrite distribution, pyrrhotite distribution, South Bay ores and Arrow area sulphides. Means and

TABLE IV

Statistical comparison of sulphur isotope distributions for the major deposits. Columns represent those deposits statistically similar (95% confidence level) to the area compared. Statistics based on the application of Student's *t*-test to mean deposit values

	Comparison										
	South Bay ore	Arrow ore	Wasp ore	Moth	Skinner	Gloster	Crab	Jig	Grit North	Echo	Sundown
Statistically similar deposits		South Bay ore		South Bay ore	South Bay ore	South Bay ore					
	Arrow ore		Arrow ore	Arrow ore	Arrow ore	Arrow ore					
		Wasp ore		Wasp ore		Wasp ore					
	Moth	Moth	Moth			Moth					
	Skinner	Skinner				Skinner					
	Gloster	Gloster	Gloster	Gloster	Gloster						
								Crab	Crab	Crab	
							Jig		Jig	Jig	Jig
							Grit North	Grit North		Grit North	Grit North
							Echo	Echo	Echo		Echo
								Sundown	Sundown	Sundown	

TABLE V

Statistical comparison of sulphur isotope values for pyrite from the major pyrite-bearing deposits. Columns represent those deposits statistically similar (95% confidence level) to the area compared. Statistics based on the application of Student's *t*-test to mean values for pyrite from individual deposits

	Comparison								
	South Bay ore	Arrow ore	Wasp ore	Moth	Skinner	Gloster	Sundown	Jig	Grit North
Statistically similar deposits		South Bay ore	South Bay ore	South Bay ore	South Bay ore	South Bay ore			
	Arrow ore		Arrow ore	Arrow ore		Arrow ore			
	Wasp ore	Wasp ore		Wasp ore		Wasp ore			
	Moth	Moth	Moth		Moth	Moth			
	Skinner			Skinner		Skinner			
	Gloster	Gloster	Gloster	Gloster	Gloster		Gloster		
						Sundown		Sundown	Sundown
							Jig		Jig
							Grit North	Grit North	

TABLE VI

Statistical comparison of sulphur isotope values for pyrrhotite from the major pyrrhotite-bearing deposits. Columns represent those deposits statistically similar (95% confidence level) to the area compared. Statistics based on the application of Student's t-test to mean values for pyrrhotite from individual deposits

	Comparison							
	Arrow ore	Wasp ore	Moth	Gloster	Crab	Jig	Echo	Sundown
Statistically similar deposits		Arrow ore	Arrow ore	Arrow ore	Arrow ore			
	Wasp ore		Wasp ore	Wasp ore	Wasp ore			
	Moth	Moth		Moth	Moth			
	Gloster	Gloster	Gloster		Gloster			
	Crab	Crab	Crab	Crab		Crab	Crab	
					Jig		Jig	Jig
					Echo	Echo		Echo
						Sundown	Sundown	

standard deviations were calculated for each significant statistical group in the above categories and as the distributions approached Gaussian, Student's t-statistics was employed to compute a comparison of group means.

Tables IV, V and VI are graphical representations of the results of such comparisons for total sulphides, pyrite and pyrrhotite respectively. For each comparison, the tables list only those areas that are calculated as statistically similar. As the recognition of similarities amongst isotopic distributions is of importance to this study, the 95% confidence level was chosen as the base for comparison.

In general, the South Bay ore is statistically similar in terms of total sulphide (Table IV) and pyrite distributions (Table V) to the sub-economic deposits Arrow, Wasp and Moth. In addition, the ores and sub-economic sulphides show similarities to the barren deposits Skinner and Gloster which are also located in felsic volcanic rocks. The Jig deposit and the barren sulphides of Crab, Grit North, Echo and Sundown show close isotopic similarities to each other and are quite distinct from the ore assemblages, the only exception being pyrrhotite from the Crab area (Table VI) which is statistically similar to the sub-economic deposits Arrow, Wasp and Moth.

SULPHUR ISOTOPE–TRACE METAL ASSOCIATIONS

A study of the relationship between isotopic composition and trace metal content was undertaken with a view to identifying a possible ore association, using the same group of 31 sulphide samples from the Uchi area described previously. Fig. 4 illustrates plots of $\delta^{34}S$ values for these sulphides against the elements Pb, Sn and Cd.

The ore association in the Pb-$\delta^{34}S$ diagram is characterized by meteoritic to "heavy" sulphur and Pb values greater than 0.01%. Samples 11–14 from the Arrow area showing ore affinity represent a sub-economic ore horizon at the contact of dacite tuff and a porphyritic felsic volcanic unit — the analogous situation to the stratigraphic horizon at the South Bay Mine. Sample 15 from the Arrow area, representing barren pyrrhotite within dacite tuff distant from the mineralized contact, is distinct. Pyrrhotite from the mineralized porphyry contact in the Wasp deposit (sample 17) also shows a strong ore affinity. All other sulphides demonstrate suppressed Pb values or a skewness towards light sulphur and suggest a possible genetically influenced relationship between Pb content and isotopic ratio, demonstrated by a general reduction in Pb content in those samples possessing lighter sulphur isotope ratios.

Fields delineated by the Sn-$\delta^{34}S$ diagram allow an excellent distinction to be made between the ore association of the sub-economic deposits Arrow, Wasp and Moth and all other deposits. Sn contents for these deposits, however, are generally at least an order of magnitude lower than the South Bay ores.

Cd, apart from two samples from the Arrow area, is conspicuously absent in the barren sulphide group and hence is present in significant quantities only in those sulphides whose isotopic composition is heavier than, or approximates the meteoritic mean.

ASSOCIATION BETWEEN SULPHUR ISOTOPIC COMPOSITION AND HOST ROCK SILICA CONTENT

Previous studies by Shima et al. (1963) and Smitheringale and Jensen (1963) point to an apparent enrichment of the ^{34}S isotope in the later products of magmatic fractional crystallization.

The relationship between average host rock silica values for the deposits and area $\delta^{34}S$ averages are shown in Fig. 5. Silica values have been obtained for the host rocks of each area from the full chemical analyses reported by Seccombe (1973). These data show that most of the rocks are relatively unaltered and hence effects such as local silicification will not affect the interpretation. Two major fields are outlined in Fig. 5. Field *I* includes the South Bay sulphides and the geologically similar deposits Wasp, Grit North, Echo and the mineralized contact zones of Arrow East and West. Field *II* encompasses the barren prospects that show no ore affinity. Field *I* is further

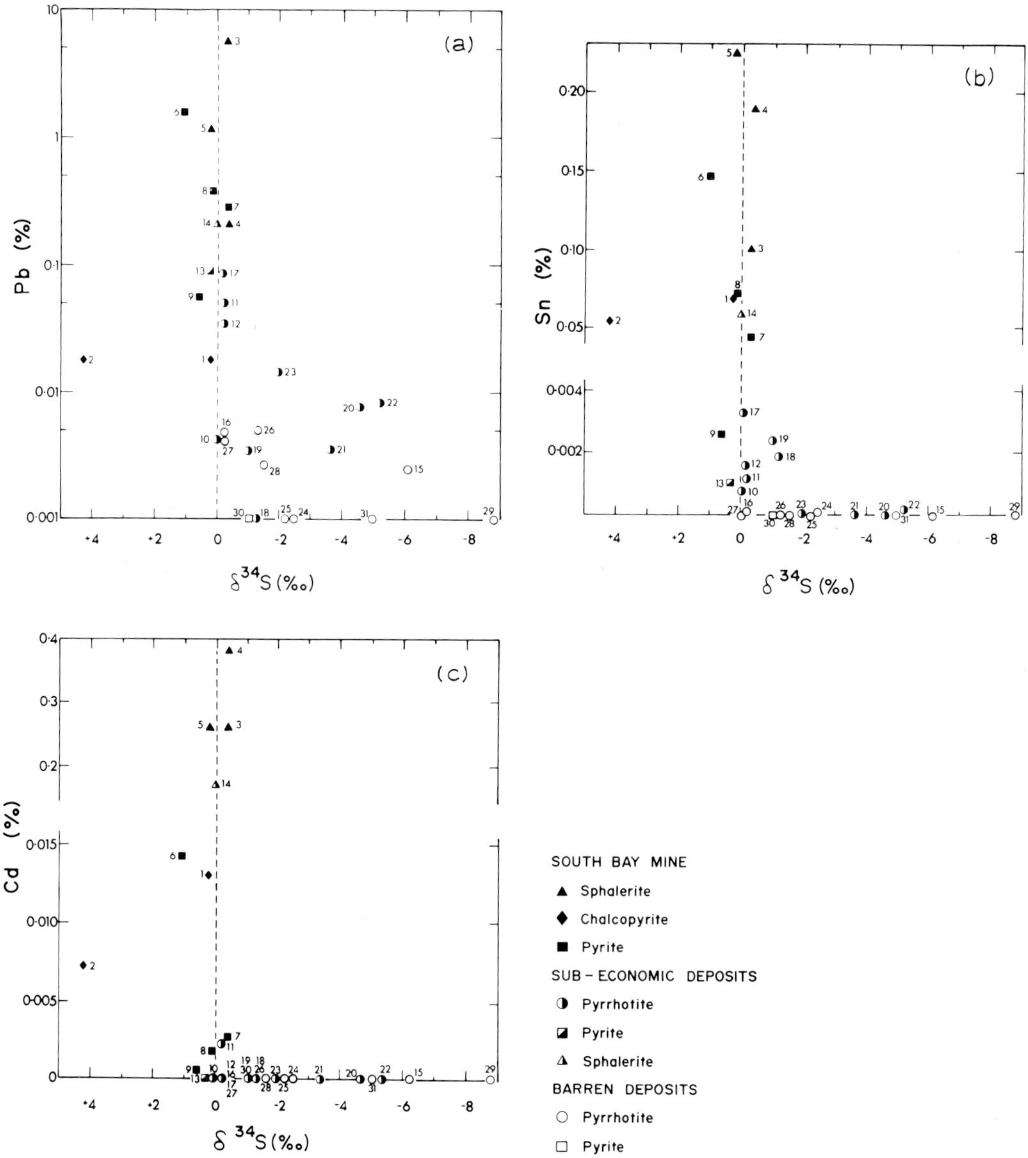

Fig. 4. Variation diagrams — the relationship between $\delta^{34}S$ values and sulphide trace metal abundances. (a) Pb-$\delta^{34}S$ diagram. Note ore association characterized by meteoritic to "heavy" sulphur and Pb values greater than 0.01%. (b) Sn-$\delta^{34}S$ diagram. The field occupied by South Bay ores and the sub-economic deposits Moth and Wasp is distinct from all other deposits. (c) Cd-$\delta^{34}S$ diagram. Note Cd content only in South Bay ores and sulphides of the sub-economic Arrow deposit.

Key to numbering — *Economic deposits:* samples 1—9 = South Bay Mine. *Sub-economic deposits*: 10—16 = Arrow (10—14 = Arrow ore zone); 17, 18 = Wasp; 19 = Moth; 20—23 = Jig. *Barren deposits*: 24, 25 = Crab; 26 = Found Lake; 27 = Gloster; 28 = Skinner; 29—31 = Sundown.

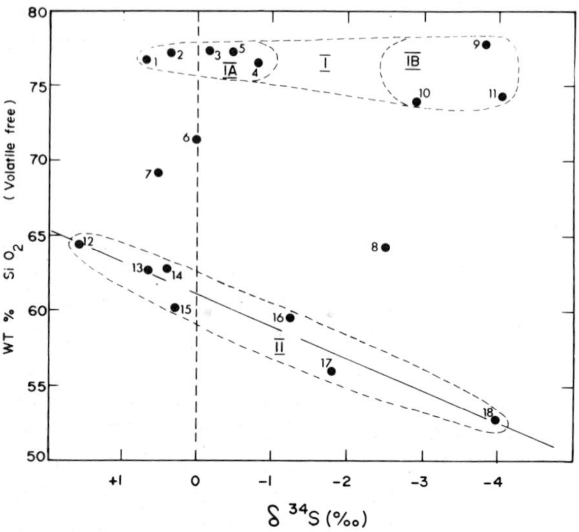

Fig. 5. The relationship between mean $\delta^{34}S$ values for the deposits and host rock silica content. Silica values represent area averages calculated from the data of Seccombe (1973).

Key to numbering — *Economic deposits*: sample 1 = South Bay Mine ore; 2 = South Bay Mine wallrock. *Sub-economic deposits*: 3 = Arrow East; 4 = Arrow West; 5 = Wasp ore; 6 = Moth; 7 = Arrow East shear zone; 8 = Jig. *Barren deposits*: 9 = Grit North; 10 = Wasp South; 11 = Echo; 12 = Skinner; 13 = Fly; 14 = Gloster; 15 = Lost Bay; 16 = Found Lake; 17 = Crab; 18 = Sundown. Field *I* = ore environment (*IA*: ores and sub-economic deposits, *IB*: barren deposits); field *II* = barren deposits lacking in ore affinity.

subdivided into fields *IA* , ores and sub-economic sulphides and *IB*, barren prospects. The sub-economic deposit Moth and the mineralized shear zone of Arrow East which are in somewhat less felsic lithologies and different structural settings from the ores are represented by points 6 and 7 respectively and together with the Jig deposit (8) appear midway between fields *I* and *II*.

Area $\delta^{34}S$ averages from the barren prospects of field *II* indicate a definite trend towards lighter sulphur concomitant with a decrease in host rock silica content. The results of a least-squares fit analysis favours a linear function relating $\delta^{34}S$ to silica content by the equation:

$$y = 61.03 + 2.123x \qquad (1)$$

where y= silica content and $x = \delta^{34}S$. This function is plotted on the diagram. As the isotopic composition relates so closely to the degree of magmatic fractionation, a genetic link is implied between the sulphide deposits involved and their host rocks. Thus, it is considered that the sulphides from these areas represent syngenetic and for the most part, volcanogenic accumulations that are essentially in situ.

Sulphides from the three sub-economic areas differing from the ore environment (samples 6, 7 and 8) show a localization in rocks of a consistently higher silica content than those of the barren deposits. This effect may represent the in-situ accumulation of volcanic exhalative sulphides of variable sulphur isotopic composition within these host rocks. Alternatively, remobilization of syngenetic sulphides from deposits in more mafic rocks, equivalent to the group of barren deposits, may have resulted in epigenetic deposition in more felsic volcanic rocks from relatively homogeneous ore fluids. The shear zone mineralization (sample 7) undoubtedly represents an epigenetic situation.

The dispersion towards light sulphur for the field *IB* sulphides is likely to represent a fractionation process operating in the volcanic environment or epigenetic deposition from a relatively barren hydrothermal fluid depleted both in ^{34}S content and ore metals.

The variation in isotopic composition of sulphides from the mineralized felsic volcanic association (field *IA*) independent of the relatively constant rock composition, contrasting with the well-defined dependence of $\delta^{34}S$ composition on host rock differentiation described previously, implies that the ores and sub-economic sulphides or ore affinity are epigenetic. The sulphur and metal content of these deposits is suggested to have been derived from pre-existing syngenetic volcanic exhalative deposits in volcanic host rocks of dacite-andesite composition (host rocks equivalent to the silica content at $\delta^{34}S = 0$ for equation (1)). Host rocks of this composition have Cu-Zn-Pb ratios consistent with the average South Bay ore value indicated in Fig. 2 (Sandell and Goldich, 1943).

EXPLORATION GUIDELINES AND SUMMARY

Irrespective of a knowledge of the precise nature of more emplacement, the study establishes an ore association with anomalously high values of Ag, Pb, Sn, In, Cd and Co/Ni ratios and $\delta^{34}S$ values close to the meteoritic mean. Homogeneous $\delta^{34}S$ values for the South Bay and sub-economic sulphide ore zones, together with a consistent trace element assemblage and mineralogy, strongly suggest derivation from a hydrothermal metal-bearing fluid; minor variations in isotopic and trace element content amongst the deposits reflect slight fluctuations in composition of the source fluid either spatially or with time. The association of sulphide mineralization with the volcanic rocks of the Uchi Series and the recognition by Davenport and Nichol (1973) of anomalous concentrations of Zn in unmineralized volcanic rocks of the Confederation Lake volcanic cycle indicate that the source of the metals is ultimately the volcanic rocks. Metal associations, sulphur isotope distribution and potential sources of the ores thus become important exploration guides for deposits of this type.

The following points which emerge from the study should be considered in the search for stratiform pyritic Zn-Cu-Ag sulphide ore deposits.

(1) Routine trace metal and $\delta^{34}S$ determinations carried out during an exploration programme on relatively few sulphide samples from mineralized

prospects are likely to assist in distinguishing those prospects containing sulphides of an economic association from barren mineralization.

(2) The trace metals Ag, Sn and Cd and, to a lesser extent, high In and high but variable Co/Ni ratios characterize both the ore association and subeconomic sulphides contained within similar acid volcanic rock types. The best representation of ore affinity in the comparison of the trace metal content of sulphides is obtained from Cu-Zn-Pb, Pb-Zn and Sn-Zn plots.

(3) Ore sulphides are identified by a small spread in $\delta^{34}S$ values (less than 4‰) and proximity to zero permil. Ore associations are defined best by anomalously high metal values and meteoritic sulphur values on Pb-$\delta^{34}S$, Sn-$\delta^{34}S$ and Cd-$\delta^{34}S$ diagrams.

(4) Plots of mean $\delta^{34}S$ values for deposits against mean host rock silica content enables a distinction to be made between ores and barren sulphides and indicates syngenetic or epigenetic character of the sulphides.

Reported data on isotopic and trace metal abundances for volcanogenic Archaean massive sulphide deposits elsewhere in the Canadian Shield such as in the Noranda and Matagami areas (Roscoe, 1965; Ryznar et al., 1967; Coomer and Schwarcz, 1974) and the Flin Flon district (Sangameshwar, 1972) suggest general similarities to the Uchi data. Hence it is suggested that the criteria developed for the Uchi area should apply equally well to other regions of the Canadian Shield. By re-establishing the parameters, the technique should be applicable as a genetic and possible exploration guide to other classes of ores in differing environments.

ACKNOWLEDGEMENTS

The results presented in this paper represent part of a Ph. D research programme undertaken at the University of Manitoba and supervised by Professors G.S. Clark and H.D.B. Wilson.

The cooperation and assistance extended by officials of Selco Mining Corporation and South Bay Mines Limited, in particular G.D. Pollock, J.S. Auston and J.C. Gibson, is gratefully acknowledged.

REFERENCES

Bateman, J.D., 1939. Geology and gold deposits of the Uchi-Slate Lakes area. Ont. Dep. Mines, 48 (8): 1—43.

Bridge, D.A., 1972. The geology and geochemistry of cassiterite as a minor constituent of the zinc-copper massive sulphide deposit at South Bay Mines, Confederation Lake, Ontario. M.Sc. Thesis, University of Manitoba, Winnipeg, Manit. (unpublished).

Coomer, P.G. and Schwarcz, H.P., 1974. Sulfur isotopic study of a massive, volcanogenic Archaean sulfide ore deposit. Geol. Soc. Am. Annu. Meet., Abstr. Progr., pp. 693—694.

Davenport, P.H. and Nichol, I., 1973. Bedrock geochemistry as a guide to areas of base-metal potential in volcano-sedimentary belts of the Canadian Shield. In: M.J. Jones (Editor), Geochemical Exploration 1972. Institution of Mining and Metallurgy, London, pp. 45—57.

Goodwin, A.M., 1967. Volcanic studies in the Birch-Uchi Lakes area of Ontario. Ont., Dep. Mines, Misc. Paper, 6: 96 pp.
Hopwood, T.P., 1976. "Quartz-eye"-bearing porphyroidal rocks and volcanogenic massive sulfide deposits. Econ. Geol., 71: 589—612.
Pollock, G.D., Sinclair, I.G.L., Warburton, A.F. and Wierzbicki, V., 1972. The Uchi orebody — a massive sulphide deposit in an Archean siliceous volcanic environment. 24th Int. Geol. Congr., Sect. 4, pp. 299—308.
Roscoe, S.M., 1965. Geochemical and isotopic studies, Noranda and Mattagami areas. Can. Inst. Min. Metall., Trans., 68: 279—285.
Ryznar, G., Campbell, F.A. and Krouse, H.R., 1967. Sulfur isotopes and the origin of the Quemont orebody. Econ. Geol., 62: 664—678.
Sandell, E.B. and Goldich, S.S., 1943. The rarer metallic constituents of some American igneous rocks. J. Geol., 51: 99—115, 167—189.
Sangameshwar, S.R.R., 1972. Trace element and sulphur isotope geochemistry of sulphide deposits from the Flin Flon and Snow Lake areas of Saskatchewan and Manitoba. Ph.D. Thesis, University of Saskatchewan, Saskatoon, Sask. (unpublished).
Seccombe, P.K., 1973. Sulphur isotope and trace element geochemistry of sulphide mineralisation in the Birch-Uchi greenstone belt, northwestern Ontario. Ph.D. Thesis, University of Manitoba, Winnipeg, Manit. (unpublished).
Seccombe, P.K. and Clark, G.S., in preparation. A sulfur isotope and geochemical study of the South Bay Mine, northwestern Ontario.
Shima, M., Gross, W.H. and Thode, H.G., 1963. Sulfur isotope abundances in basic sills, differentiated granites, and meteorites. J. Geophys. Res., 68: 2835—2847.
Smitheringale, W.G. and Jensen, M.L., 1963. Sulfur isotopic composition of the Triassic igneous rocks of eastern United States. Geochim. Cosmochim. Acta, 27: 1183—1207.
Wilson, H.D.B. and Anderson, D.T., 1959. The composition of Canadian sulphide ore deposits. Can. Inst. Min. Metall. Trans., 62: 327—339.

Journal of Geochemical Exploration, 8 (1977) 139—151

PRIMARY DISPERSION PATTERNS ASSOCIATED WITH THE DETOUR ZINC-COPPER-SILVER DEPOSIT AT LAC BROUILLAN, PROVINCE OF QUEBEC, CANADA

I.G.L. SINCLAIR

Selco Mining Corporation Limited, Toronto, Ont. (Canada)

(Revised version accepted February 4, 1977)

ABSTRACT

Sinclair, I.G.L., 1977. Primary dispersion patterns associated with the Detour zinc-copper-silver deposit at Lac Brouillan, Province of Quebec, Canada. J. Geochem. Explor., 8: 139—151.

The Detour Zn-Cu-Ag deposit is located within a series of acid volcanic rocks in the Archaean Abitibi orogenic belt. The mineralization occurs in two distinct zones some 400 m apart. Results presented in this paper are related to the eastern of the two zones.

A study of the distribution of K_2O, MgO and Hg in the host rocks was conducted to examine relationships between mineralization and primary dispersion patterns.

Results show the existence of two steeply dipping, linear K_2O-deficient zones, the intersection of which coincides with the maximum concentration of Zn-Cu sulphides. There is a central, MgO-low zone, coinciding with the principal concentration of sulphides, which is flanked by irregular MgO-rich zones.

The distribution of Hg and Ag appear to be controlled by volcanic stratigraphy, highest values generally being found in the more permeable tuffs and breccias in the upper part of the succession.

The dispersion patterns, together with the available geological evidence, suggest that mineralization took place during a period of hydrothermal activity following a cycle of explosive acid volcanism.

INTRODUCTION

The Detour Zn-Cu-Ag deposit is located some 100 km west of Matagami in northwestern Quebec within a series of acid volcanic rocks in the Archaean Abitibi greenstone belt (Fig. 1). Several other important volcanogenic, base metal deposits occur in the Abitibi Belt, which extends from eastern Ontario into central Quebec and includes the Timmins, Noranda, Val d'Or, Chibougamau and Matagami mining districts (Douglas, 1968).

The deposit was discovered in 1974 as a result of an airborne electromagnetic survey followed by ground geophysical surveys and diamond drilling (Reed, 1976). The exploration programme, which led to the discovery, was a joint venture by Selco Mining Corporation Limited and Pickands Mather and Company.

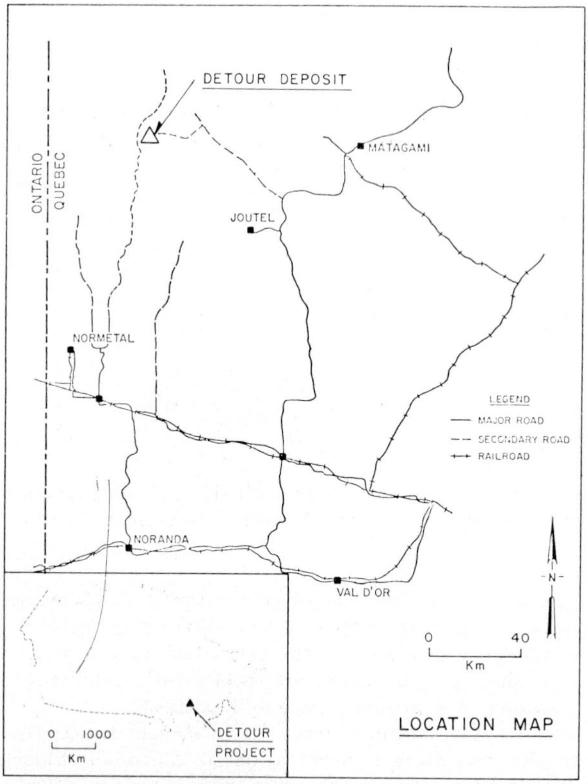

Fig. 1. Location of the Detour deposit relative to other base metal mining camps in northwestern Quebec, Canada.

In the vicinity of the deposit, heavy mantling of the bedrock by glacial till has resulted in a complete absence of outcrops; thus, information concerning the nature and distribution of sulphide zones and their relationship to the host rocks is confined to that available from drill core.

The mineralization occurs in two distinct zones some 400 m apart. The eastern of these, the A-1 zone, was the first to be discovered and to be drilled in detail. In this zone, mineralization consists of fracture fillings and disseminations of Zn-Cu-Ag sulphides with associated pyrite and quartz gangue. Concentrations of economic tenor occur mainly above the −150-m level in an irregularly shaped zone up to 500 m in lateral extent, and grade in all directions, laterally and to depth, into sub-economic mineralization. Steep fracturing and possible block faulting are commonly observed in the host rocks of this zone but they do not appear to have been subjected to shearing or folding to any important extent.

In the western B zone, the host rocks have been severely sheared and the sulphide minerals appear to have been extensively re-mobilized. Most of the economic mineralization is now concentrated along a zone of intense shearing which trends in a northeasterly direction and is some 700 m in length.

During the geological examination of the drill core, irregular zones of sericitization and chloritization were observed in the host rocks but it was difficult to determine any systematic relationship between this alteration and the principal concentrations of sulphides. The study reported here, which is concerned only with the rocks of the A-1 zone, was undertaken to identify chemical alteration patterns related to mineralization. It was considered that recognition of such alteration patterns might be an aid in the location of additional mineralized zones and would be an important contribution to the establishment of a genetic model for the deposit.

GEOLOGICAL SETTING

In northwestern Quebec, the rocks of the Abitibi Belt consist of sequences of volcanic and sedimentary rocks intruded by elongate granitic bodies. The Detour deposit lies near the western extremity of one on these sequences known as the Chibougamau-Matagami zone.

The sulphide mineralization occurs within a section of dominantly pyroclastic, intermediate-to-acid volcanic rocks which exhibit textural features indicative of their having been deposited in a predominantly sub-aerial environment.

A massive, acid pyroclastic unit, known as the Welded Acid Tuff, is the dominant host rock of the deposit. This rock consists of quartz grains, dark chloritic fiamme and rounded to angular lapilli in a cryptocrystalline siliceous matrix. The quartz grains are angular to rounded; they range in size from 0.1 mm to 2.0 mm and constitute about 5 vol.% of the rock. The fiamme, which typically make up 5—10 vol.% of the rock, range from 1 mm to 5 cm in length and generally show some degree of parallel elongation. The lapilli, which consist of angular to rounded fragments of acid volcanic rocks up to 10 mm in diameter, make up about 5 vol.% of the rock.

The texture of this rock closely resembles that described as being typical of welded tuffs by many workers (Ross and Smith, 1961; Cook, 1966; Fitch, 1967; Lambert, 1974) and indicates that this rock was deposited in a sub-aerial environment during a period of explosive acid volcanism.

The thickness of the Welded Acid Tuff is variable up to a maximum of about 100 m (Fig. 2). The unit grades into dacitic tuffs at its base but its upper contact is generally abrupt and has an undulating surface. There is no evidence that this unit has been folded; therefore, this undulating surface may have been developed during an hiatus in the deposition of pyroclastic material, since welded tuffs characteristically have flat upper surfaces (Marshall, 1935; Ross and Smith, 1961).

The Welded Acid Tuff is overlain by an irregularly intergrading sequence of rhyolitic and dacitic tuffs, volcanic breccias and agglomerates. The breccias and agglomerates contain many fragments of the Welded Acid Tuff and many of the tuff bands contain abundant lithophysae. The presence of

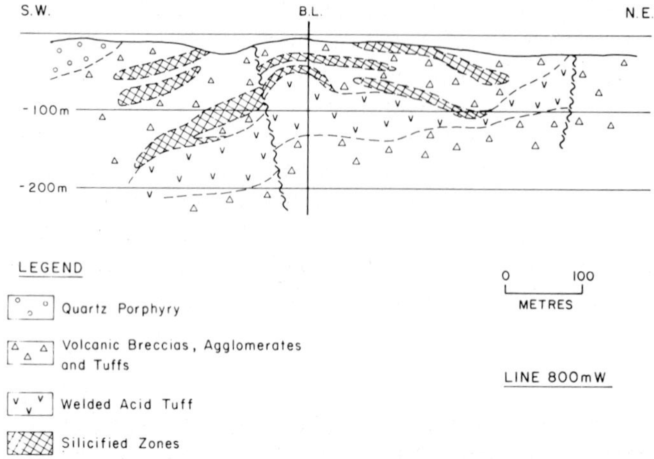

Fig. 2. Geological section across the central part of the A-1 zone.

lithophysae is considered to be indicative of explosive sub-aerial volcanism (Lilljequist and Svenson, 1974). The sill-like quartz porphyry body, portrayed on the geological section (Fig. 2), has been shown by petrographic studies (Hull, 1976) to have originated as an ash-flow deposit.

Within the pyroclastic rocks above the Welded Acid Tuff, there are some bands of massive, stratiform pyrite which appear to pre-date, and to be genetically unrelated to, the Zn-Cu-Ag mineralization. The presence of this stratiform pyrite suggests that there probably were occasional periods of submergence during the evolution of the sequence.

Silicification, sericitization and chloritization are the modes of alteration commonly observed in the host rocks.

The silicification is manifested, firstly, as quartz veining associated with pyrite and Zn-Cu sulphides in fractures; and, secondly, as stratiform silicified zones in the pyroclastic rocks above the Welded Acid Tuff. In these silicified zones, remnant textural features of the pyroclastic rocks can still be recognized within a matrix of cherty siliceous material.

Chloritic and sericitic alteration are ubiquitous throughout the host rocks but the intensities of both these types of alteration is extremely variable and it has not been possible to plot their occurrence systematically by visual examination of the drill core.

The host rocks of the A-1 zone are dissected by an extensive plexus of steep fractures and possible block faults. A granite intrusion occurs to the east of the deposit. Intrusion of this body has not imposed extensive alteration on the volcanic rocks nor has it any apparent genetic relationship to the mineralization.

MINERALIZATION

The principal ore minerals, chalcopyrite and sphalerite, occur mainly within the system of steep fractures which penetrate the host rocks where they are associated with pyrite, quartz and subordinate amounts of carbonate. Stringers and disseminations of the ore minerals also occur in the silicified zones within the pyroclastic rocks above the Welded Acid Tuff.

The sulphide minerals occur as fine to medium-grained, interlocking aggregates. The relative proportions of the sulphide species are locally variable but, in general, sphalerite occurs in important amounts throughout the deposit while the most important concentrations of chalcopyrite occur in the central part of the mineralized zone.

Ag minerals cannot be seen in drill core but argentite and native Ag have been observed during the microscopic examination of ore concentrates.

Preliminary calculations indicate that the A-1 zone contains some 32×10^6 tonnes of ore averaging 2.30% Zn, 0.39% Cu and 36 g/t Ag.

METHOD OF STUDY

The study involved a comparison of the dispersion of K_2O, MgO and Hg with the distribution of Zn, Cu and Ag. K_2O and MgO were selected for analysis because of the frequent presence of sericitic and chloritic alteration in the host rocks. Hg was selected because of the known association of the element with volcanogenic base metal deposits (Friedrich and Hawkes, 1968; Jonasson and Sangster, 1974).

All of the core obtained during the drilling of the deposit was divided into 3-m lengths which were split and assayed for Zn, Cu, and Ag in Selco's Matagami laboratory. Core samples were pulverized and ground to minus 100-mesh and the elements determined by atomic absorption after digestion in a 1:3 mixture of concentrated HNO_3 and concentrated HCl. The results of these determinations were stored in a computerized filing system. This system is programmed to enable the analytical data to be retrieved either in the form of a plot of element distribution on a particular level or section, or in the form of weighted averages across prescribed vertical intervals.

Samples for K_2O, MgO and Hg determinations were obtained by selecting from the set of 3-m core samples already prepared for Zn, Cu and Ag assays, those occuring at the −50-m, −100-m, −150-m and −200-m levels. A few samples from deeper levels were also taken in the central part of the deposit where some drill holes penetrated to −300 m. Thus, while Zn, Cu and Ag values are available for all of the core drilled, K_2O, MgO and Hg values are available only for 3-m samples at the specified levels. The rationale for this difference in sampling density for the two sets of data is that the sampling programme for Zn, Cu and Ag was designed to evaluate the actual amounts of these elements present in the mineralized zone whereas the programme for K_2O, MgO and Hg was designed to identify alteration patterns on certain levels and sections.

K_2O, MgO and Hg determinations were made by Bondar-Clegg and Company in their Ottawa laboratory. K_2O and MgO were determined by X-ray fluorescence on samples which were first ground to minus 200-mesh and pelletized. Hg was determined by the cold-vapour technique using a Perkin-Elmer-Coleman Hg analyzer.

RESULTS

Potassium oxide

The distribution of K_2O in the host rocks of the A-1 zone at the —100-m level is illustrated in the accompanying plan (Fig. 3); also shown are outlines of those parts of the zone in which the combined, weighted mean values of Zn and Cu between the —100-m level and the bedrock surface exceed 2%. Thus, a comparison is made between alteration patterns at the —100-m level and the combined Cu and Zn contents of that part of the orebody above it. It is useful to compare the results in this manner, since the intensity and distribution of alteration at a given level may be related to the passage upwards of mineralizing fluids which have deposited their load of ore-forming elements at higher levels in the host rocks. The plan shows the presence of two linear K_2O-deficient zones, the intersection of which more or less coincides with the maximum concentration of combined Zn and Cu in the orebody. At the —50-m level, these K_2O-deficient zones are broader

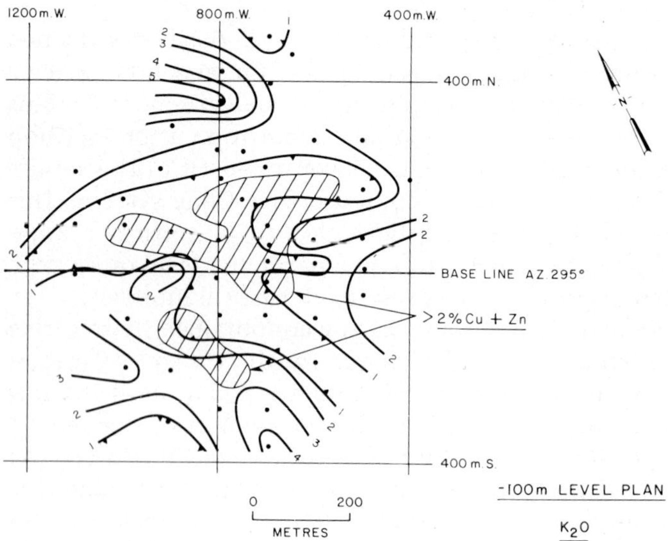

Fig. 3. Comparison of K_2O dispersion at the —100-m level with the Cu+Zn distribution above that level. Contour values are weight percent K_2O in 3-m core samples at points shown. Areas in which the weighted mean Cu+Zn content of all rocks above the —100-m level exceeds 2% are indicated.

and more irregular in outline while, at deeper levels, they pass into narrow, pipe-like features.

Distribution of K_2O, within a vertical section through the central part of the orebody (Fig. 4), shows that the K_2O-deficient zones have steeply dipping margins which transgress lithological boundaries. Comparison of the distribution of K_2O with that of Zn (Fig. 4) indicates that concentrations of greater than 1% Zn generally occur within zones containing less than 1% K_2O.

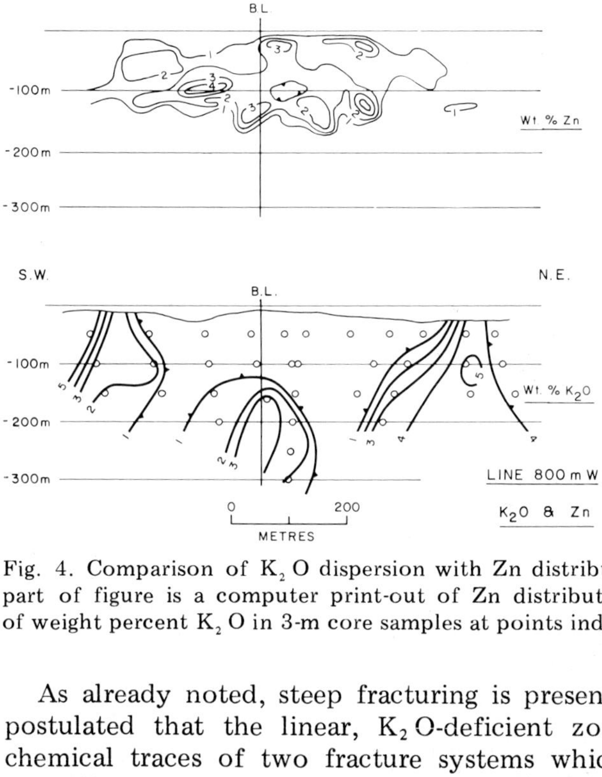

Fig. 4. Comparison of K_2O dispersion with Zn distribution in a vertical section. Upper part of figure is a computer print-out of Zn distribution. Lower part shows contours of weight percent K_2O in 3-m core samples at points indicated.

As already noted, steep fracturing is present in the host rocks and it is postulated that the linear, K_2O-deficient zones may represent the geochemical traces of two fracture systems which acted as channelways for ascending, hydrothermal fluids. The distribution of maximum Cu+Zn values suggests that the intersection of the fracture systems provided optimum conditions for mineral deposition.

In contrast to this depletion of K_2O adjacent to the sulphide zones, Descarreaux (1973) recorded zones enriched in K_2O adjacent to massive sulphide bodies in the Abitibi Belt and Boyle (1974) stated that, in the case of epigenetic, hydrothermal deposits, the K_2O content of the host rocks "invariably increased uniformly towards the ore". However, Goodfellow (1974) has noted a depletion in K_2O in the altered zone underlying the Brunswick No. 12 massive sulphide deposit.

Magnesium oxide

MgO displays a distribution pattern somewhat less clearly defined than that of K_2O; however, the principal concentrations of Zn+Cu tend to occur within a central zone containing less than 5% MgO which is flanked by irregularly outlined, MgO-rich areas (Fig. 5). This general relationship between MgO distribution and mineralization also prevails on the —50-m level but, at levels below —100-m, the MgO distribution is irregular and no systematic relationship with sulphide distribution can be discerned.

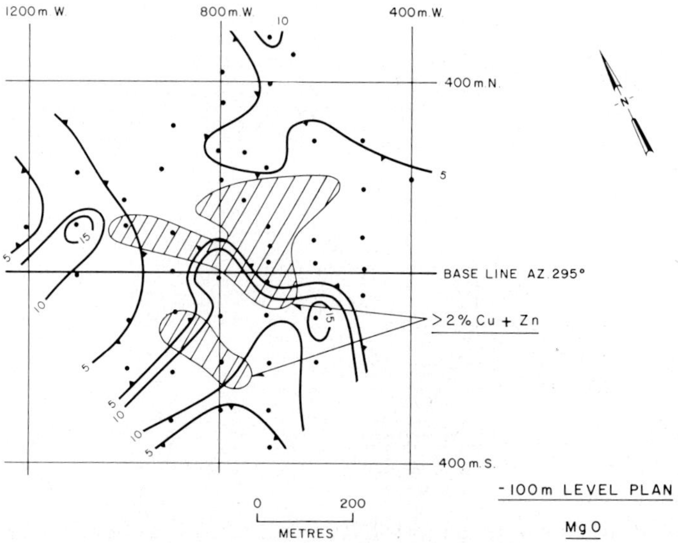

Fig. 5. Comparison of MgO dispersion at the —100-m level with the Cu and Zn distribution above that level. Contour values are weight percent MgO in 3-m core samples at points shown. Areas in which the weighted mean Cu+Zn content of all rocks above the —100-m level exceeds 2% are indicated.

In addition to this general relationship between the total Zn+Cu content of the rocks and the MgO distribution, a vertical section (Fig. 6) shows that concentrations of Cu greater than 1% are generally confined to zones containing less than 5% MgO.

Mineralogical studies by Hull (1976) indicate that the high tenor of MgO in the zones flanking the mineralization is related both to the presence of Mg carbonate in fractures and the presence of chlorite in the matrices of the rocks.

Mercury

Hg values range from 10 to 895 ppb. Concentrations greater than 100 ppb are, for the most part, found in the pyroclastic rocks above the Welded Acid

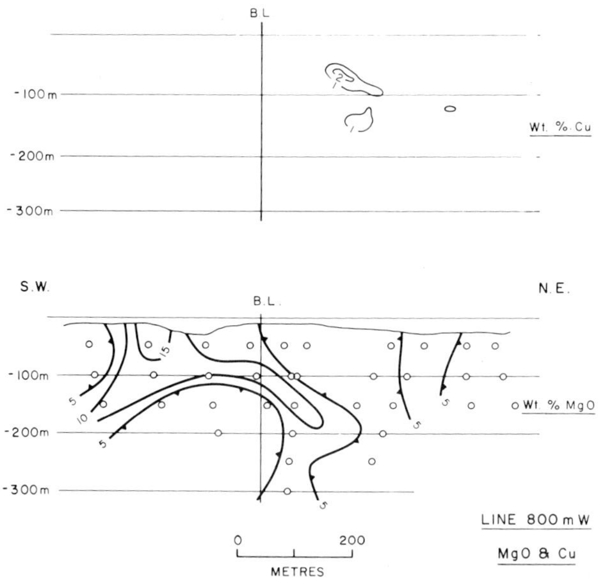

Fig. 6. Comparison of MgO dispersion with Cu distribution in a vertical section. Upper part of the diagram is a computer plot of Cu distribution. Lower part of the diagram shows contouring of weight percent MgO in 3-m core samples at the points shown.

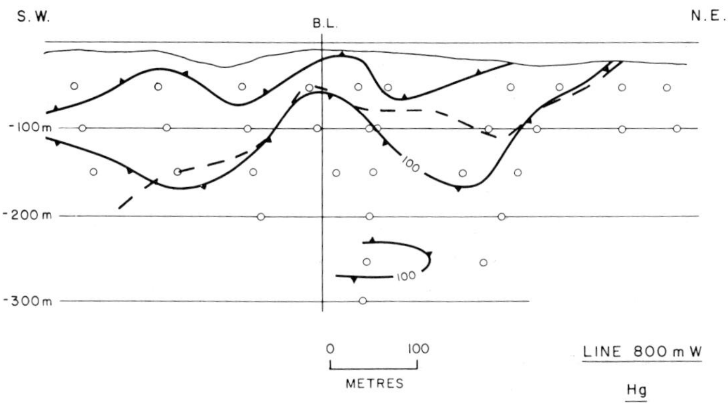

Fig. 7. Hg dispersion within a vertical section. Contours enclose values greater than 100 ppb. Dashed line indicates upper surface of the Welded Acid Tuff.

Tuff (Fig. 7). This is interpreted as indicating that Hg migrated upwards through steep fractures and, upon reaching the upper levels of the pile, migrated laterally through, and was deposited within, the relatively permeable, upper pyroclastic rocks. Ag exhibits strong positive correlation with Hg (Fig. 8), suggesting that its distribution may have been controlled in a similar manner.

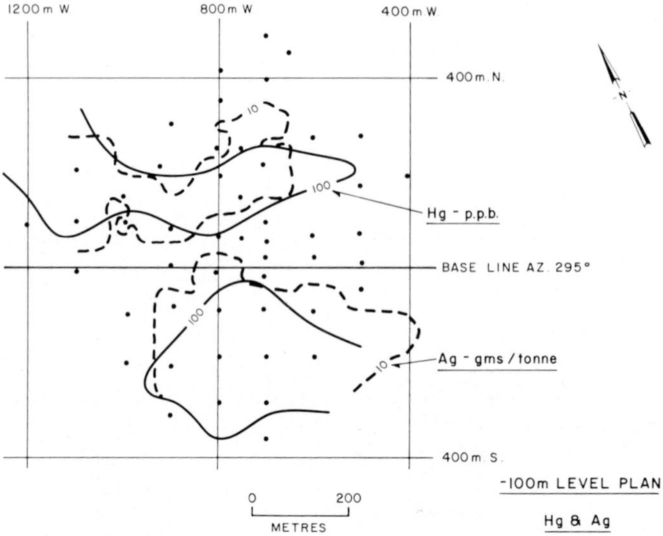

Fig. 8. Distribution of Hg and Ag in 3-m core samples at the points indicated on the −100-m level. Solid lines enclose areas containing values greater than 100 ppb Hg. Dashed lines enclose areas containing values greater than 10.0 g/t Ag.

DISCUSSION

Most of the Zn-Cu deposits in the Abitibi Belt and elsewhere in the Canadian Archaean are associated with acid volcanic rocks and are thought to have been deposited by submarine volcanic processes. The sulphide minerals generally occur in massive lenticular bodies which are apparently conformable with the host rocks and which are often underlain by zones of disseminated sulphides (Sangster, 1972; Spence 1975). Many workers have pointed out the important similarities between these deposits and the Tertiary Kuroko deposits of Japan, the submarine origin of which has been firmly established (Horikoshi, 1969).

In contrast, textural features noted in the host rocks of the Detour deposit suggest that much of the volcanic sequence may be of sub-aerial origin and massive, stratiform Zn-Cu sulphide accumulations have not been found. Mineralization, for the most part, occurs in fractures and silicified replacement zones.

The Detour deposit, therefore, may be more closely related to deposits in areas in which mineralization has occurred within fracture systems in volcanic rocks in a subsurface environment, such as the Creede District of Colorado (Steven and Eaton, 1975), than to those of the Kuroko type.

In the Creede District, the host rocks are predominantly sub-aerial, pyroclastic rocks, the products of a series of Tertiary caldera subsidences. Mineralization took place during the final hydrothermal phase of this period of volcanism with sulphides being deposited from ascending fluids in steep fractures and faults.

Mineral deposition at Detour appears, therefore, to have taken place in a geological environment somewhat different from that prevailing during the deposition of the massive sulphide accumulations. However, chloritic and sericitic alteration, of the type observed at Detour, are also found in the rocks associated with the massive deposits (Sangster, 1972). In such deposits, where massive sulphides have been precipitated at the rock-seawater interface, the alteration is found mainly in rocks stratigraphically beneath the ore. At Detour, however, the sulphides have been emplaced within the rock mass and rocks surrounding, as well as underlying, them have suffered alteration. This difference is of considerable importance in the interpretation of geochemical results and in planning further underground exploration.

The potential importance of pre-ore fracture patterns in controlling the distribution of epigenetic mineralization and hydrothermal alteration is well known (Kraeff, 1963; Goossens and Hollister, 1973; Steven and Eaton, 1975). Results from Detour suggest that zones where transverse fracture systems intersect were favourable loci for the deposition of Zn and Cu and that it may be possible to detect such fracture systems by identifying the geochemical effects of the hydrothermal fluids which passed through them.

In addition to the general tendency for the principal concentrations of Zn+Cu to occur in zones relatively deficient in K_2O and MgO more specific relationships between element distributions and the dispersion of K_2O and MgO have been noted (Figs. 4 and 6). This suggests that Zn and Cu may have been introduced into the deposit during two distinct phases of hydrothermal activity, with expulsion of K_2O simultaneous with the introduction of Zn and expulsion of MgO simultaneous with the introduction of Cu. Evidence is not available to indicate the order of occurrence of these two postulated mineralizing events.

Maximum Hg values are found, for the most part, in pyroclastic rocks above the Welded Acid Tuff, a distribution which fits a model of deposition from ascending fluids in a near-surface volcanic environment. At the McDermitt Mine in Nevada, which is located in a Miocene caldera structure, Hg is considered to have ascended through steep fractures in a welded tuff and to have diffused laterally into more permeable overlying rocks in which economic concentrations are now found (Roper, 1976).

CONCLUSIONS

Results presented here demonstrate that geochemical techniques can be used to outline zones of hydrothermal alteration surrounding the ore zones at Detour. This may provide a useful tool in the search for further ore zones in that area and in similar geological environments elsewhere.

The geochemical results are a useful supplement to available geological information in considering a possible genetic model for the deposit. It is considered that the host rocks, which are predominantly sub-aerial pyroclastics, were laid down during a caldera-related cycle of explosive acid volcanism. Steep fracturing and prolonged terminal hydrothermal activity are known to be typical features of such volcanic cycles (Smith and Bailey, 1968). The steep fracturing observed at Detour and the steeply dipping zones of alteration indicated by the geochemical data accord with this model.

Hydrothermal fluids ascending through the fracture system are believed to have been responsible for the alteration of the host rocks and the deposition of the Zn-Cu sulphides. Deposition of Hg and Ag in the upper part of the succession is believed to have been effected by fluids which ascended through the fracture system and migrated laterally into the pyroclastic rocks above the Welded Acid Tuff.

ACKNOWLEDGEMENTS

The author is grateful to Selco Mining Corporation Limited and Pickands Mather and Company for permission to publish this paper. He also wishes to thank his colleagues in the Exploration and Mining Divisions of Selco for their co-operation and assistance during the course of the study.

Mr. John Auston and Mr. David Hutton reviewed the manuscript; their helpful comments and constructive criticism were greatly appreciated.

REFERENCES

Boyle, R.W., 1974. The use of major elemental ratios in detailed geochemical prospecting utilizing primary halos. J. Geochem. Explor., 3: 345—369.

Cook, E.F. (Editor), 1966. Tufflavas and Ignimbrites: A Survey of Soviet Studies. Elsevier, New York, N.Y., 212 pp.

Descarreaux, J., 1973. A petrochemical study of the Abitibi volcanic belt and its bearing on the occurrences of massive sulphide ores. Can. Min. Metall. Bull., 66 (730): 61—69.

Douglas, R.J.W. (Editor), 1968. Geology and Economic Minerals in Canada. Economic Geology Report No. 1. Geological Survey of Canada, Ottawa, Ont., 838 pp.

Fitch, F.J., 1967. Ignimbrite volcanism in North Wales. Bull. Volcanol., 30: 199—219.

Friedrich, G.H. and Hawkes, H.E., 1968. Mercury dispersion halos as ore guides for massive sulphide deposits, West Shasta District, California. Miner. Deposita, 2: 77—88.

Goodfellow, W.D., 1974. Major and minor element halos in volcanic rocks at Brunswick No. 12 sulphide deposit, N.B., Canada. In: I.L. Elliott and W.K. Fletcher (Editors), Geochemical Exploration 1974. Elsevier, Amsterdam, pp. 279—295.

Goossens, P.J. and Hollister, V.F., 1973. Structural control and hydrothermal alteration pattern of Chaucha porphyry copper, Ecuador. Miner. Deposita, 8: 321—331.

Horikoshi, E., 1969. Volcanic activity related to the formation of the Kuroko type deposits, in the Kosaka District, Japan. Miner. Deposita, 4: 321—345.

Hull, D.N., 1976. A petrologic-geochronologic study of host rocks and wall rocks at the Detour Prospect, a copper-zinc-silver discovery in Brouillan Township, Abitibi—West County, Quebec. M.Sc. Thesis, Wright State University, Dayton, Ohio.

Jonasson, I.R. and Sangster, D.F., 1974. Variations in the mercury content of sphalerite from some Canadian sulphide deposits. In: I.L. Elliott and W.K. Fletcher (Editors), Geochemical Exploration 1974. Elsevier, Amsterdam, pp. 313—331.

Kraeff, A., 1963. Geology and mineral deposits of the Hopa-Murgul region, western part of the Province of Artvin, N.W. Turkey. Bull. Min. Res. Expl. Inst. Turkey (Foreign Edition), 60: 45—60.
Lambert, M.B., 1974. The Bennet Lake cauldron subsidence complex, British Columbia and Yukon Territory. Geol. Surv. Can. Bull., 227: 213 pp.
Lilljequist, R. and Svenson, S.A., 1974. Exceptionally well preserved Precambrian ignimbrites and basic lavas in North Sweden. Geol. Fören. Stockholm Förh. 96: 221—229.
Marshall, P., 1935. Acid rocks of the Taupo-Rotorua volcanic district. Trans. R. Soc. New Zealand, 64: 323—366.
Reed, L.E., 1976. The discovery of the Brouillan zinc-copper-silver deposits, northwestern Quebec. Paper presented at the 46th Annu. Meet., Soc. Explor. Geophys., Houston, Texas.
Roper, M.W., 1976. Hot springs mercury deposition at McDermitt Mine, Humboldt County, Nevada. AIME Annu. Meet., Las Vegas, Nev., Preprint No. 76-S-28.
Ross, C.S. and Smith, R.L., 1961. Ash-flow tuffs: their origin, geologic relations and identification. U.S. Geol. Surv., Prof. Paper, 366: 81 pp.
Sangster, D.F., 1972. Precambrian volcanogenic massive sulphide deposits in Canada. Geol. Surv. Can., Paper, 72-22: 44 pp.
Smith, R.L. and Bailey, R.A., 1968. Resurgent cauldrons. In: R.R. Coats, R.L. Hay and C.A. Anderson (Editors), Studies in Volcanology. Geol. Soc. Am., Mem., 116: 613—662.
Spence, C.D., 1975. Volcanogenic features of the Vauze deposit, Noranda, Quebec. Econ. Geol., 70: 102—113.
Steven, T.A. and Eaton, G.P., 1975. Environment of ore deposition in the Creede Mining District, San Juan Mountains, Colorado I. Geologic, hydrologic and geophysical setting Econ. Geol., 70: 1023—1037.

Journal of Geochemical Exploration, 8 (1977) 153—168

GEOCHEMISTRY OF THE MAMMOTH COPPER DEPOSIT' NORTHWEST QUEENSLAND, AUSTRALIA

K.M. SCOTT and G.F. TAYLOR

Division of Mineralogy, Commonwealth Scientific and Industrial Research Organization, North Ryde, N.S.W. (Australia)

(Revised version accepted February 4, 1977)

ABSTRACT

Scott, K.M. and Taylor, G.F., 1977. Geochemistry of the Mammoth copper deposit, northwest Queensland, Australia. J. Geochem. Explor., 8: 153—168.

The stratabound Mammoth Cu deposit, situated north of Mount Isa, northwest Queensland, Australia, consists of a number of orebodies in the Proterozoic (Carpentarian) Myally Subgroup. The sulphides are predominantly chalcocite/bornite, chalcopyrite and pyrite.

Only No. 1 and No. 2 orebodies are being mined and the majority of the data presented is obtained from core samples from them. Each sample has been analyzed for 10 major elements, 27 trace elements, C, S and H_2O, and the major mineralogical components have been determined by X-ray diffraction.

Immediately above the middle levels of No. 2 orebody is a bleached zone with depleted Fe, Mn and Mg. A similar bleached zone is also present above the No. 3 orebody but there is no conspicuous bleaching in the upper levels of No. 2 orebody or associated with No. 1.

Significant geochemical variations between the hanging and the footwalls have been detected. K, Na and Ba are more abundant in the hanging walls while Co, Ni, Fe, Mn, H_2O and, in the case of No. 2 orebody, Mg are concentrated in the footwalls. Trace elements associated with the sulphides are Ag, As, Bi, Co, Mo, Ni, Pb, Sb and Tl. Of these Tl and As are concentrated in pyrite, whilst Ag and Bi are associated with Cu minerals.

Deep weathering has caused oxidation of the ore in the upper levels of the mine and some chalcocite in No. 1 orebody may be the result of supergene enrichment. Gossans and oxidized ore were removed during open-cut operations. However, at the nearby Esperanza/Pluto prospect where similar mineralization is known to occur, anomalous concentrations of As, Sb, Mo and Pb associated with Cu in the ironstone capping and oxidized zone appear to be significant for future exploration.

INTRODUCTION

Cu mineralization is known throughout the Middle Proterozoic (Carpentarian) sediments of northwest Queensland and has been mined since the first discovery at Cloncurry in 1867. By far the greatest known concentration occurs within the Urquhart Shales at Mount Isa. The only other major copper mine is that at Mammoth, 115 km to the north of Mount Isa. Other Cu deposits have been worked spasmodically during the last century and exploration for further Cu mineralization continues. Much of this exploration is concentrated in the Western Basin (the less metamorphosed area extending north of Mount Isa) and is becoming increasingly oriented towards finding blind orebodies.

This paper presents preliminary data on the geochemistry of the ore of the Mammoth deposit and its enclosing host rocks and relates this to the geochemistry of associated ironstone capping.

LOCATION

The Mammoth deposit is one of a number of known Cu deposits occurring along the Mount Gordon Fault zone in northwest Queensland (Fig. 1). Detailed descriptions of the deposit and regional geology have been given by Moore (1974), Mitchell and Moore (1975) and Holyland et al. (1976).

Known mineralization at Mammoth is confined to a number of discrete, closely related, lenticular orebodies which occur within the steeply west-dipping sandstones, arkoses, quartzites and siltstones of the Alsace Quartzite, of the basal unit of the Myally Subgroup (Derrick et al., 1976; Holyland et al., 1976). The orebodies underlie an erosional unconformity which

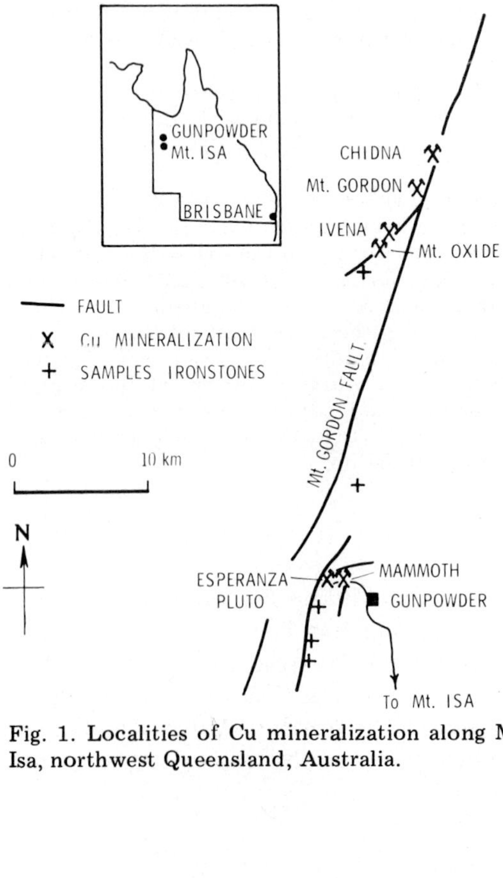

Fig. 1. Localities of Cu mineralization along Mount Gordon Fault zone, north of Mount Isa, northwest Queensland, Australia.

is marked by a basal conglomerate in the overlying Gunpowder Creek Formation (Fig. 2). They are adjacent to two major faults, the Mammoth and Portal, and their relationship to these faults is shown in Fig. 3.

Sandstone is the most common host to ore which occurs in veins up to several centimetres thick. Within each orebody, thicker veins enclose brecciated arenaceous rock. Disseminated sulphides may also occur within the

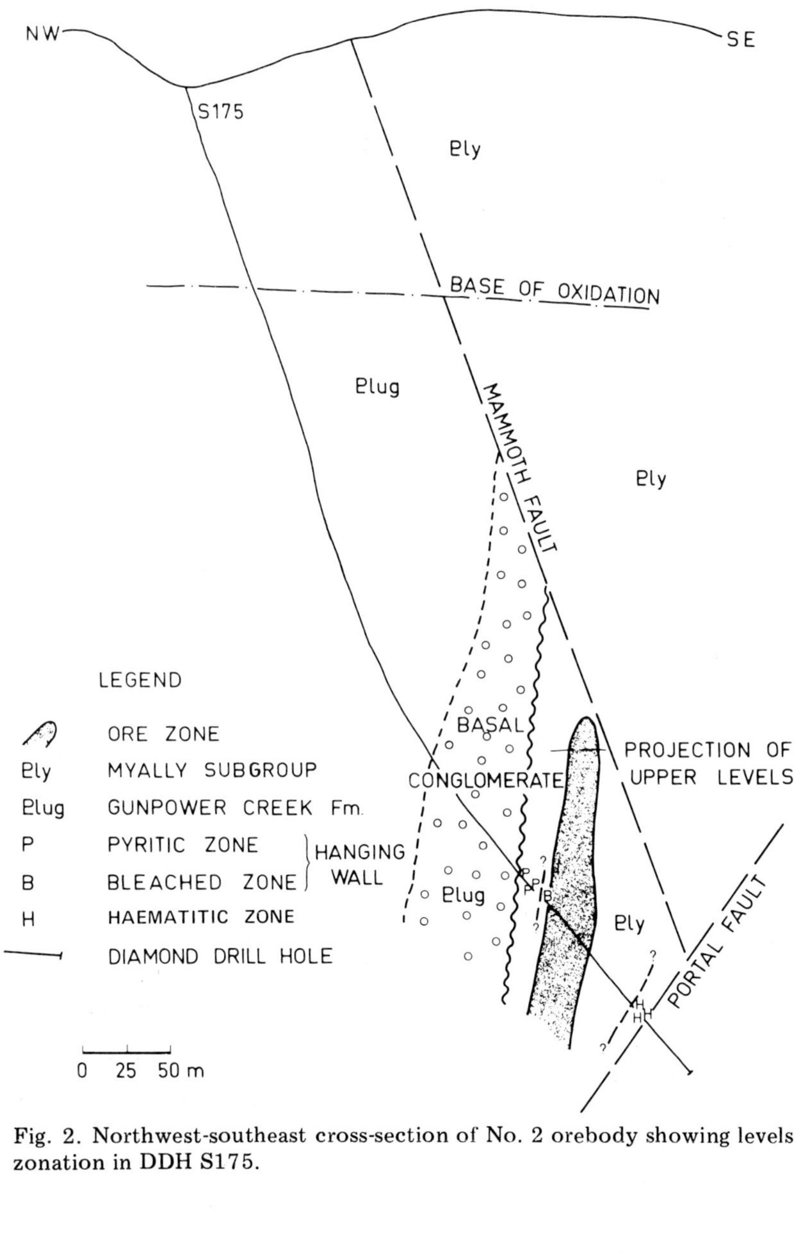

Fig. 2. Northwest-southeast cross-section of No. 2 orebody showing levels considered and zonation in DDH S175.

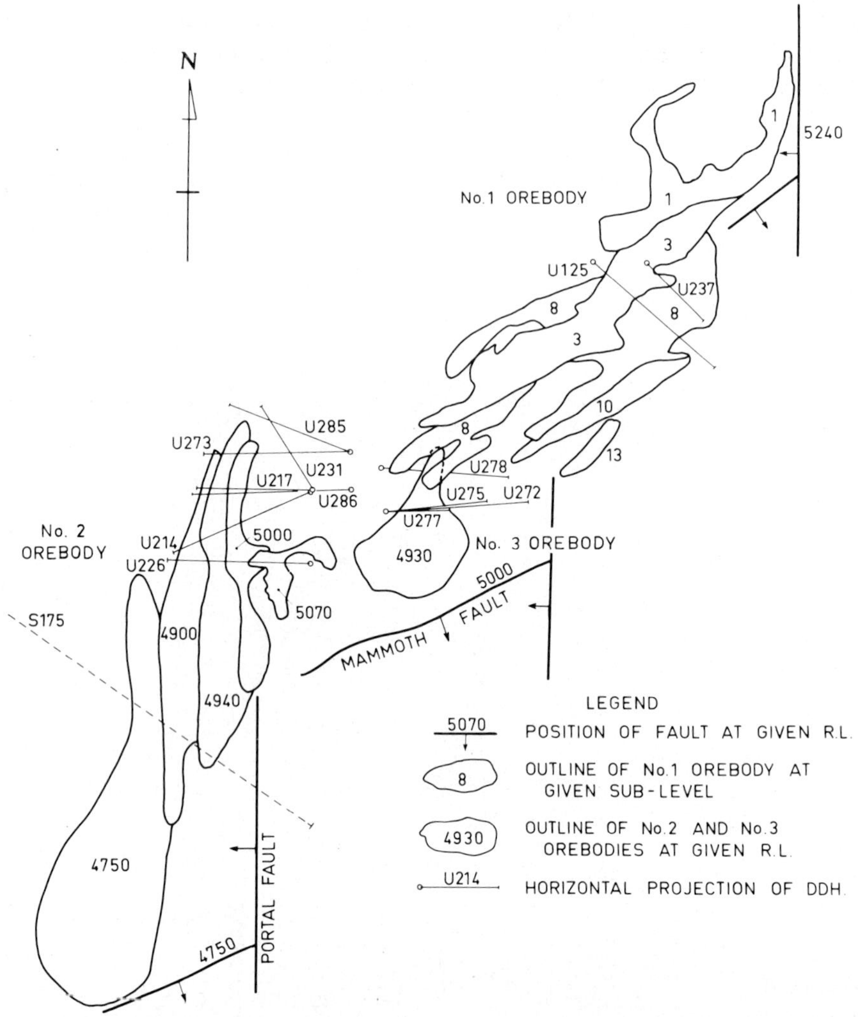

Fig. 3. Composite plan showing relation of Mammoth orebodies to Mammoth Fault and Portal Fault. All cores used in this study projected onto plan.

more silty component of the orebodies. The sulphides are predominantly chalcocite, chalcopyrite and pyrite in No. 1, and chalcocite, bornite and pyrite in No. 2 orebody. There is an upward stratigraphic sequence from chalcocite to chalcocite/bornite to pyrite in No. 2 orebody.

In the No.1 orebody, samples were taken from the cores, U125* and U237, together with hand specimens from the 8 sub-level. For the upper

---

* Prefixes U and S refer to underground and surface cores respectively.

levels of the No. 2 orebody, six cores, U214, U217, U226, U231, U285 and U286 were sampled. The holes U217 and U226 intersect ore, whilst the others pass through the lateral equivalent of the ore horizon. Hand specimens from 5030 and 5000 sub-levels were also used to determine the mineralogy. A large number of samples were also taken from S175, which passed through the middle levels of No. 2 orebody and the Portal Fault (Fig. 2). Samples from cores U272, U275, U277 and U278, from the upper levels of the No. 3 orebody were also examined. The positions of all these cores are projected onto a composite plan of the mine (Fig. 3).

TECHNIQUES

Samples approximately 5 cm in length were selected from the cores, with all lithological and physical changes being represented. Overall, the sampling interval never exceeded 10 m and in many instances was less than 1 m.

A portion of each sample was ground to minus 75 μm in a manganese-steel Siebtechnik ring mill. Individual sulphide minerals were drilled from some samples to determine their chemistry. Prior to analysis, the major mineralogical components of each sample were determined by an X-ray diffraction technique using graphite crystal-monochromated copper radiation, scanning over the range 4—75° 2θ at ½° 2θ /minute.

For most of the samples, major elements, As, Ba, Rb, Sr, Y and Zr were determined by X-ray fluorescence spectrometry (XRF). In those samples in which the copper sulphide content exceeded 5%, the major elements were determined by atomic absorption spectrometry (AAS) after a sodium peroxide sinter (Corbett et al., 1974). Na, Li, Cu, Pb and Zn were all determined by AAS; other trace elements by optical emission spectroscopy (ES); S and C using a Leco analyser (Corbett and Mizon, 1976); and H_2O by a gravimetric technique. As no standards with comparable matrix were available, trace element data for separated sulphides and samples with a high sulphide content are generally only semi-quantitative.

RESULTS

As it is the purpose of this paper to identify geochemical trends, the data presented in Table I are average values for the various zones associated with each orebody. Because the dip of No. 1 orebody is the reverse of that for the other two orebodies, the terms hanging wall and footwall are used in a stratigraphic sense only. These values have been calculated from the data for individual samples in all cores which pass through a particular zone. Thus for No. 1 orebody, the hanging wall and footwall data are from U125, while for the ore zone, both U125 and U237 values were averaged. Because the ore is closely associated with sandstones and quartzites, values for siltstones, which are a minor component of the Alsace Quartzite, have been excluded. No results are presented for the overlying Gunpowder Creek Formation.

TABLE I

Average compositions of sandstones from zones about the Mammoth orebodies (major components in wt.%, trace elements in ppm)

	No. 1 orebody			No. 2 orebody —upper levels						No. 2 orebody — middle levels			
	hanging wall	ore zone	footwall	ore stratigraphy			equivalent stratigraphy			pyritic hanging wall	bleached hanging wall	ore zone	footwall
				hanging wall	ore zone	footwall	hanging wall	ore equivalent	footwall				
Number of samples	6	23	3	4	5	7	7	8	14	5	2	3	3
SiO_2	74.0	56.5	87.4	74.2	62.7	67.6	66.4	72.5	72.2	69.8	80.9	74.3	66.5
Al_2O_3	11.1	3.49	2.83	10.6	11.3	11.3	11.1	9.50	10.0	8.75	8.26	9.52	11.2
FeO	2.25	13.5	5.26	3.39	8.29	8.01	6.18	8.19	5.94	4.63	2.64	4.15	6.71
MgO	0.32	0.07	0.09	1.33	0.67	3.24	4.24	2.69	3.37	3.22	2.41	4.18	7.68
CaO	0.03	0.04	0.03	0.92	0.05	0.96	0.82	0.21	0.32	2.99	0.31	0.26	0.52
Na_2O	0.07	0.01	0.01	0.06	0.04	0.05	0.05	0.03	0.02	0.04	<0.01	0.01	0.01
K_2O	7.89	1.40	0.38	5.15	3.15	1.57	4.15	1.16	1.06	3.40	0.76	0.55	0.13
TiO_2	0.38	0.13	0.04	0.35	0.39	0.41	0.36	0.26	0.31	0.24	0.19	0.32	0.34
P_2O_5	0.04	<0.01	<0.01	0.53	0.04	0.70	0.20	0.08	0.11	2.21	0.04	0.08	0.29
MnO	0.02	0.10	0.06	0.04	0.09	0.06	0.08	0.11	0.06	0.10	0.05	0.14	0.14
CO_2	1.33	0.58	0.48	0.15	0.46	0.18	0.75	0.25	0.27	0.96	0.38	0.57	0.05
S	1.25	12.7	0.06	0.01	3.25	0.01	0.01	0.01	0.01	1.40	0.37	0.88	<0.01
H_2O	0.94	1.28	1.70	2.47	2.94	4.37	3.79	4.82	4.36	2.53	4.03	4.44	5.68
Ag	2	9	5	<1	16	2	1	2	5	<1	<1	3	<1
As	260	5200	58	48	670	220	29	87	64	330	85	43	<20
B	63	25	25	33	84	39	23	22	26	18	25	25	17
Ba	920	180	73	510	190	150	350	110	150	1600	25	20	28
Bi	16	140	3	1	140	7	3	15	3	8	2	26	<1
Co	62	2600	180	70	90	180	110	300	410	58	50	80	220
Cr	54	17	20	40	52	34	30	27	35	29	22	18	32
Cu	6300	87000	1700	71	45000	68	72	51	64	5600	2300	16000	92
Ga	13	3	3	6	6	9	9	9	7	11	8	14	13
Li	6	4	21	24	19	36	29	40	37	21	33	45	43
Mo	1	25	<0.6	0.7	1	1	0.6	1	0.7	<0.6	<0.6	<0.6	<0.6
Ni	23	1200	220	33	73	140	59	180	230	31	19	33	53
Pb	20	96	<20	<20	26	21	45	41	200	24	<20	<20	<20
Rb	n.d.	n.d.	n.d.	65	76	41	89	32	35	n.d.	n.d.	n.d.	n.d.
Sb	30	35	30	40	42	30	30	30	30	25	<25	<25	<25
Sr	180	29	15	31	38	39	41	35	40	170	15	48	14
Tl	2	28	<0.6	<0.6	11	<0.6	<0.6	<0.6	<0.6	8	0.6	<0.6	<0.6
V	77	33	47	55	120	69	71	65	74	51	30	60	48
Y	10	10	30	23	12	32	28	27	33	56	22	18	10
Zn	24	12	48	23	29	42	26	32	25	62	27	42	67
Zr	200	89	90	130	120	180	200	180	150	130	130	110	120

n.d. = not determined.

No. 1 orebody

Quartz is abundant throughout the orebody and its surrounding host rocks. Both microcline and sericite appear to be concentrated in the hanging wall but are diminished in the ore zone and are almost absent in the footwall. Although traces of chlorite occur throughout the whole sequence, iron-rich chlorite (and haematite) is concentrated in the narrow footwall zone adjacent to the Mammoth Fault.

Pyrite and chalcopyrite are abundant in the ore zone and pyrite extends for approximately 20 m into the hanging wall. In the lower levels of the orebody, intersected by U237, chalcocite and bornite are more abundant and cobaltite occurs in the top 5 m of the ore zone.

Mineralogical and lithological variations are reflected in the elemental concentrations, some of which are shown in Fig. 4 and represented as average compositions in Table I. The steady decline in the Al, K and Na contents from the hanging wall to the footwall, shown in Fig. 4, reflects the reduced feldspar content toward the footwall. Some Al occurs within the small amount of chlorite, immediately adjacent to the Mammoth Fault. Ca values are uniformly low.

Mg content is highly variable but is concentrated in the hanging wall where it occurs within magnesium-rich chlorite. The ore zone is particularly iron-rich, mainly reflecting the contribution from the sulphides. In the footwall, where Fe is also concentrated, there are no sulphides and both haematite and iron-rich chlorite are abundant. Mn appears to correlate with Fe in each zone. Both pyrite and haematite are present in the hanging wall but the Fe content is lower than in the other zones. Ti, P and carbonate are slightly enriched in the feldspathic sandstones of the hanging wall and are depleted in the footwall quartzite.

There are a number of elements other than Cu which appear to be directly related to the sulphides as shown by the average composition of the ore zone in Table I. As, Bi, Co, Mo, Ni, Pb, Tl and, to a lesser extent, Ag and Sb are associated with the sulphides. Of these, As and Tl are closely associated with pyrite (Table II). Co and Ni are correlated with Cu within the ore zone (Fig. 4) and with S within the hanging wall and thus appear to be associated with both pyrite and copper sulphides. In the footwall, where no sulphides are present, concentrations of Co and Ni occur in haematite and iron-rich chlorite.

Ag, Bi, Pb and possibly Sb are associated with Cu and probably occur within minerals of the tetrahedrite-tennantite series, which have previously been identified in the ore (G.P. Moore, personal communication, 1975). Mo, which is concentrated in the ore zone, cannot readily be correlated with a particular sulphide mineral. The distribution of Zn is not presently understood. Analysis of individual sulphides from hand specimens serves to emphasize these associations.

B, Ba, Ga and Sr are all enriched in the hanging wall and depleted in the

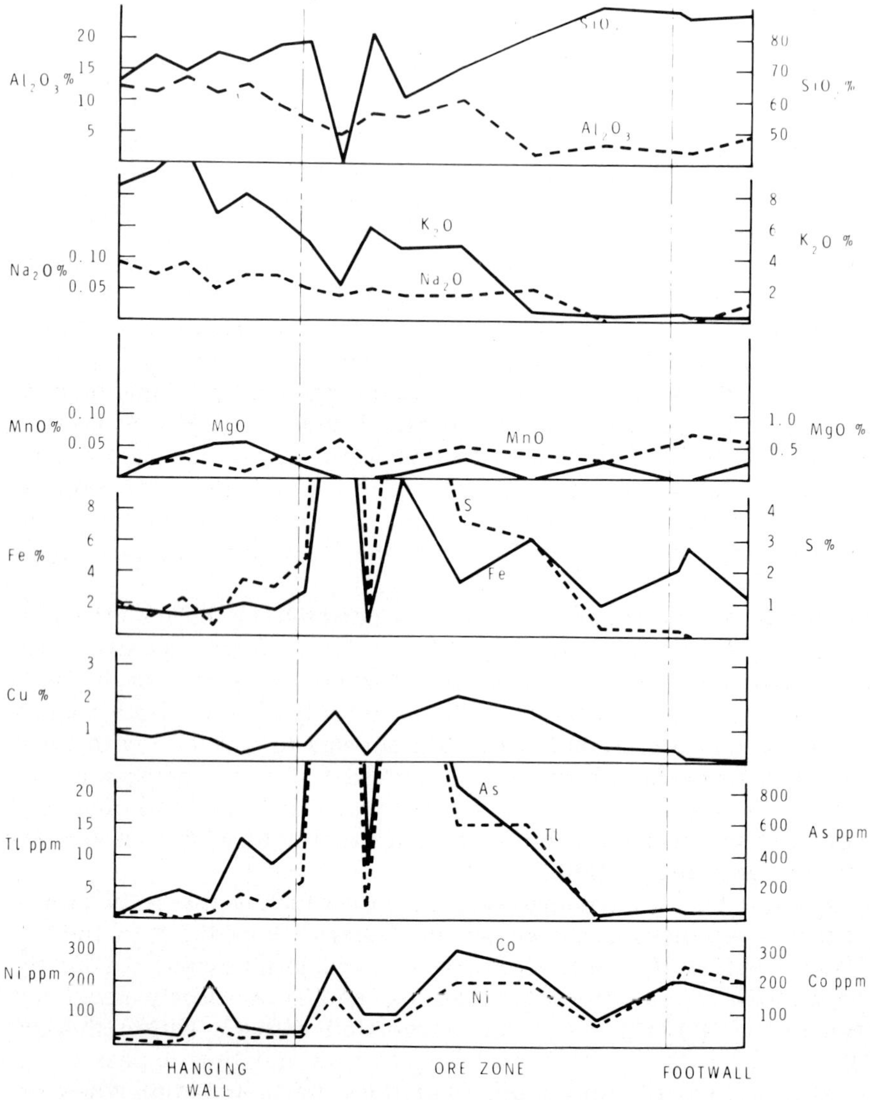

Fig. 4. Compositional trends for Mammoth No. 1 orebody (DDH U125).

footwall. Ba and Sr substitute for K in microcline, while B and Ga are associated with the aluminosilicates. Cr, V and Zr are also concentrated in the hanging wall while Li and Y appear to occur in the footwall. The distribution of these elements is probably related to the lithological change from a sandstone to a quartzite.

TABLE II

As and Tl contents of pyrite from No. 1 and No. 2 orebodies

	Type		
	vein (orebody No. 1)	vein (orebody No. 2)	disseminated (orebody No. 2)
Number of samples	3	3	2
As (ppm)	12,000	13,000	4000
Tl (ppm)	260	240	10

No. 2 orebody

Upper levels. The mineralogy is similar to that described above for the No. 1 orebody except that chlorite is more abundant, particularly in the ore and for 15 m into the footwall zone, where it is iron-rich. In the hanging wall, chlorite is magnesium-rich. Sulphide mineralogy is characterized by chalcocite/bornite, pyrite and minor chalcopyrite. Pyrite increases in abundance towards the hanging wall.

Because the ore is slightly transgressive, an apatite band which occurs in the host rocks of the No. 2 orebody may occur in the hanging wall or in the footwall. Variations in phosphate content in Table I are therefore not significant. The occurrence of this apatite band also affects the average values of Ca, Sr and Y which are therefore not considered further.

Na, K, Ba and Rb are concentrated in the hanging wall (Table I) where they occur mainly in microcline and they are depleted with increasing depth (Fig. 5). K/Rb ratios are lowest in the ore zone and highest in the hanging wall.

In the footwall, Fe, Mn and Mg contents are approximately double those of the hanging wall (Table I). Fe and Mn are also concentrated in the ore zone but Mg is significantly depleted. All three elements are concentrated in the lateral equivalent of the ore zone but the distinction between hanging wall and footwall is not developed.

The H_2O content is significantly higher in the footwall and ore zone than in the hanging wall, which together with the trends in Fe, Mn and Mg, reflects the development of chlorite. In the laterally equivalent sections, there are only small differences in the H_2O contents of the various zones (Table I). The lowest value occurs in the hanging wall and probably reflects a smaller chlorite content. For a distance of approximately 15 m, the chlorite in

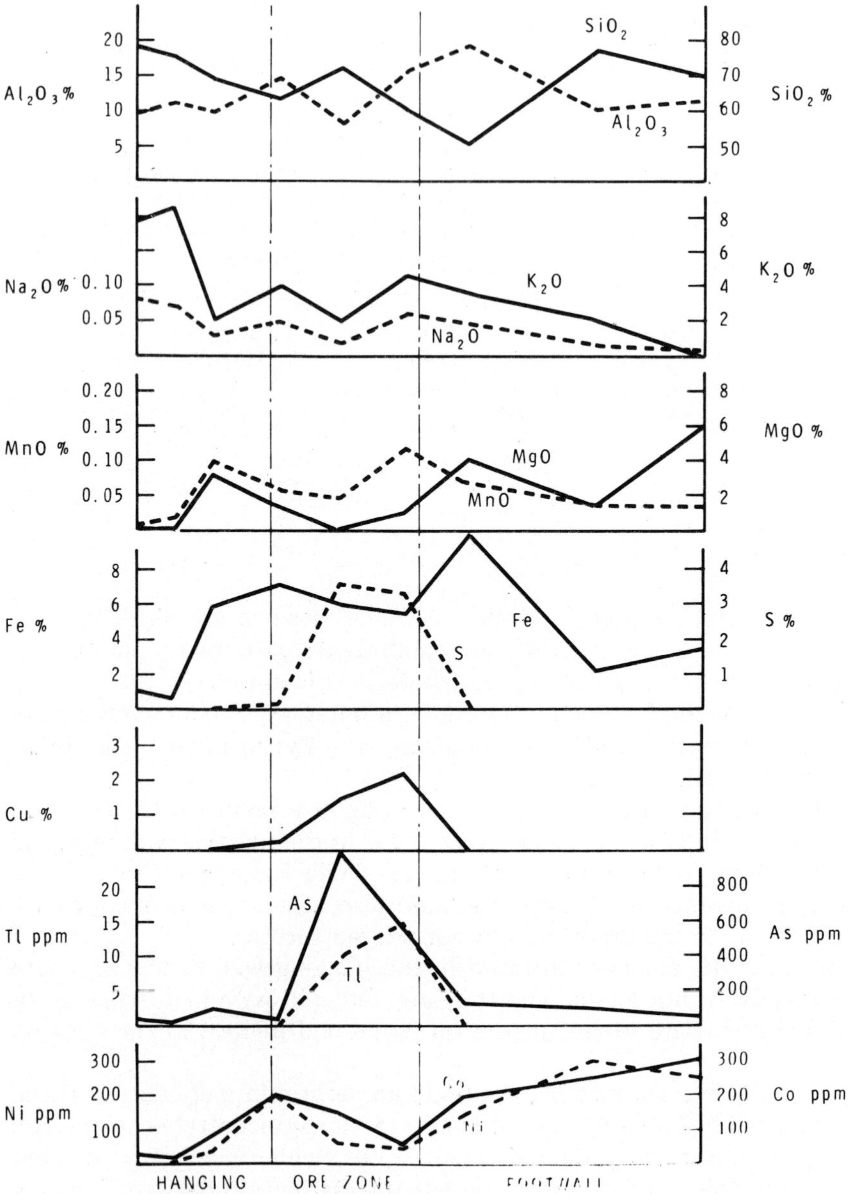

Fig. 5. Compositional trends for upper levels, Mammoth No. 2 orebody (DDH U226).

the zone laterally equivalent to ore is iron-rich. Thus, an envelope of iron-rich chlorite extends 15 m laterally and vertically beneath the ore zone.

Trace elements associated with the sulphides in the upper levels of the ore zone are Ag, As, Bi, Co, Mo, Ni, Sb and Tl (Table I). Co and Ni are also concentrated in the footwall (Fig. 5) where they occur in either traces of sulphides or, more likely, within chlorite and haematite. Significantly higher values of As, Bi and Mo can also occur in the footwall. Zn is depleted in the ore zone relative to either the footwall or hanging wall.

As, Bi, Mo, Sb and Zn occur in the lateral equivalent of the ore zone (Table I). Again, they probably occur with the haematite and chlorite but may be present in minute traces of tetrahedrite-tennantite and chalcopyrite. Co and Ni occur stratigraphically beneath the lateral equivalent of the ore zone. Electron microprobe analysis has shown that they occur within the chlorite. Ag, Pb and Sb also occur in this zone which may indicate the presence of minute quantities of sulpho-salts and galena.

Middle levels. A section approximately 200 m below the levels described above has also been examined in S175. At this level, considerably more bands of siltstone are present than in either No. 1 orebody or the upper levels of No. 2 orebody. Within these silty bands, chlorite and sericite are extensively developed and disseminated sulphides may be present. The envelope of iron-rich chlorite is, however, not present. The apatite band occurs in the hanging wall, although isolated higher phosphate values were obtained close to the Portal Fault. A striking feature of the lower levels is the well-developed bleached appearance of rocks in the ore zone and for a distance of up to 10 m into the hanging wall. Within this bleached zone of the hanging wall, there is less Cu and S than in either the ore zone or overlying pyritic hanging wall (Table I).

The vertical stratigraphic sequence from chalcocite to chalcocite/bornite/pyrite within the ore zone and, finally, into a pyritic hanging wall, is best developed in these levels of No. 2 orebody. Trace chalcopyrite occurs mainly along shears, especially in the hanging wall, and for at least 30 m into the footwall.

Pyrite and haematite are almost mutually exclusive. Pyrite occurs in the section from the top of the Alsace Quartzite to about the middle of the ore zone. Haematite in this section is present in trace amounts only but is more abundant elsewhere, especially immediately adjacent to the Portal Fault (Fig. 2).

Significant trends within the pyritic hanging wall are consistently high values for Na, K, S, As, Ba and Tl. The Na, K and Ba occur in the abundant microcline although minor K is present in muscovite and some Ba occurs as barite. S, As and Tl reflect the dominance of pyrite in this zone. Fe, Mn, Mg and H_2O are depleted in the bleached zone (Table I) reflecting the absence of chlorite.

The highest values of Fe, Mn, Mg, H_2O, Co, Cr, Ni and Zn occur in the footwall zone. All of these elements, other than Zn, are probably associated with chlorite as in the upper levels of No. 2 orebody. The mode of occurrence of Zn is again puzzling but as in the other sections (considered above), it is less abundant in the ore zone than in the foot or hanging walls. It is interesting to note, however, that both Pb and Zn contents are higher in siltstones than in the sandstone at this level.

Within the ore zone high values of Ag, Bi and Cu are associated with the chalcocite/bornite (Table I) and Li and B with the clay fraction.

No. 3 orebody

The chemical associations and mineralogy of the No. 3 orebody are similar to those found in the other two orebodies. Thus, there are close associations of Fe, Mg, Mn, H_2O, Co and Ni, and of K, Na, Ba, Rb and Sr. Unlike the other orebodies, however, the siderophile elements are concentrated in the hanging wall of No. 3 orebody which also has the greatest depletion of alkalis. The hanging wall of No. 3 is, however, within the footwall of No. 2 and there is some suggestion that the two may be limbs of the same orebody. The top of the No. 3 orebody shows some development of a bleached zone similar to that in the middle levels of No. 2 orebody. Such a bleached zone is best developed in the lateral equivalent of the steeply dipping ore horizon.

Ag, As, Bi, Cu, Tl and S are associated with the ore zone. As in the other orebodies, As and Tl occur with the pyrite while Ag and Bi are associated with the copper sulphides.

Secondary dispersion

Deep surface weathering is common throughout the whole of northwest Queensland and in some places may be as deep as 150 m (Guy, 1975). Along the Mount Gordon Fault zone the depth of weathering varies greatly, being in excess of 300 m at Esperanza/Pluto (G.P. Moore, personal communication, 1975), while at Mammoth it was probably restricted to within 20 m of the surface (Moore, 1974). Below this level occurrences of carbonate and oxides are restricted to fracture surfaces.

Early reports indicate that two gossans cropped out over the Mammoth deposit but were removed by early gougers. Most of the oxidized mineralization has been removed during open-cut mining operations so that no samples were available for analysis. Other mineralization along the Mount Gordon Fault zone, particularly that at the nearby Esperanza/Pluto prospect, is similar to that at Mammoth. The oxidized zones of this and other mineralization remain largely intact so that samples have been collected from them as well as from many other ferruginous outcrops in the area. The results of analyses of samples from some of these areas are presented in Table III.

However, as previously described by Glasson (1973) and Taylor (1973), there are many ferruginous outcrops in the Western Basin which carry anomalous concentrations of Cu but are not related to mineralization. Many of these "false" or pseudo-gossans occur along the Mount Gordon Fault zone and several have been analyzed (see Table III).

TABLE III

Average elemental compositions (in ppm) of some ferruginous outcrop (ironstones) from the Mount Gordon Fault zone

	"True" gossans			"False" gossans		
	Esperanza/Pluto		Ivena gossan	Esperanza ironstone	Fault ironstone	Manitoba ironstone
	gossan	decline				
Number of samples	5	3	3	6	2	1
Fe_2O_3 (%)	50.8	40.6	62.0	18.1	74.9	5.89
Ag	<1	14	7	<1	<1	<1
As	2000*	4540	2800	150	340	30
Bi	4	11	140	<1	<1	<1
Co	10	100	100	30	50	<10
Cu	1130*	1470	6400	700	1400	50
Mo	4	24	11	1	1	1
Ni	5	70	25	23	25	40
Pb	110*	480	260	37	8	6
Sb	350	770	2170	130	<30	<30
Tl	<0.6	<0.6	<0.6	<0.6	<0.6	<0.6

*13 samples.

DISCUSSION

Marked similarities in the mineralogy and chemistry exist in each of the orebodies and the surrounding host rocks considered in this investigation. These are:

(1) Microcline, sericite and magnesian-chlorite occur in the hanging wall zones.

(2) Iron-chlorite is developed within the ore zone, its lateral equivalent and for some distance into the footwall in some regions about the orebodies.

(3) Pyrite is more abundant towards the top of the ore zones and, except for the upper levels of No. 2 orebody, extends for at least 20 m into the hanging walls, whilst haematite is abundant in the footwalls, especially beside the faults.

(4) Sulphides are essentially pyrite, chalcocite/bornite and chalcopyrite.

(5) Na, K, Ba and Sr are concentrated in the hanging walls and generally depleted in the footwalls.

(6) Fe is concentrated in the ore zones as sulphides and the footwalls as haematite and iron-chlorite with less in the hanging walls as pyrite.

(7) There is a variable distribution of Mg.

(8) As and Tl are associated with pyrite.

(9) Ag, Bi, Co, Mo, Ni, Pb and Sb are associated with the ore.

(10) Zn is generally depleted in the ore zones, and is more abundant in the footwalls than the hanging walls.

(11) Co and Ni are associated with haematite or chlorite in footwall zones.

The dispersion haloes about the orebodies (as listed above) appear to be consistent with the currently favoured genetic model (Holyland et al., 1976) in which the bulk of the mineralization is presumed to have entered the consolidated rocks along the Mammoth and Portal Faults.

Dispersion haloes of the No. 1 and No. 2 orebodies are largely confined to the footwall zones, although the lateral equivalents of the ore zone do contain significant anomalies for at least 15 m from the mineralization. However, there is a relative concentration of K in the hanging wall. Marked K metasomatism has also been noted in the host rocks at Mount Isa (Croxford, 1964; van den Heuvel, 1969) whilst a sill adjacent tot the Mammoth orebodies has been almost entirely altered to K-feldspar and sericite, with an average K_2O value of 12%.

In addition to the massive vein sulphide mineralization, disseminated sulphides are also present. As is shown in Table II, the vein pyrite is much richer in As and Tl than is the disseminated pyrite. This feature is considered to be primary as the vein pyrite is much later than the disseminated and is related to the vein Cu mineralization.

Thus, it would appear that the recognition of K enrichment and the As and Tl contents of pyrite might be suitable primary dispersion haloes for use in geochemical exploration. However, of greater significance for regional exploration is the secondary dispersion of elements associated with the sulphide ores.

A comparison of the chemistry of samples of oxidized mineralization (Table III) with the chemistry of the ore zone at Mammoth (Table I) shows that some of the elements associated with sulphides also occur in the ferruginous outcrop. In particular, Fe, Cu and As and, to a lesser extent, Mo, Pb and Sb are retained while Ag, Bi, Co, Ni and Tl are depleted or absent. All elements other than Tl occur in the subsurface oxidized zone of the Esperanza/Pluto mineralization penetrated by the decline which encountered highly oxidized ore containing abundant cuprite.

Significantly, although Cu is concentrated in "false" or pseudo-gossans (Table III), other elements commonly associated with mineralization are absent, depleted or only very slightly concentrated. Of these, As appears to be the most reliable pathfinder element for Mammoth-type Cu mineralization since it is concentrated in the ore zone and, during oxidation, is retained in both surface outcrop and the oxidized zone. Boyle and Jonasson (1973) have indicated that As is relatively mobile in the highly acid conditions of the supergene zone and may migrate for considerable distances. As the solutions become less acidic, As becomes less mobile and coprecipitates with hydrous iron oxides.

Tl, which is concentrated in the pyrite associated with the Mammoth mineralization (Table II), is not found in either surface outcrop or in the subsurface oxidized zone (Table III). Thus its high mobility in the oxidized zone limits its use in geochemical exploration.

CONCLUSIONS

Primary dispersion haloes have been found associated with three orebodies at the Mammoth Cu deposit in northwest Queensland. It would appear that K enrichment of the host rocks and the As and Tl contents of pyrite may be useful geochemical indicators. Other dispersion haloes have been found to be limited to within 20 m of the orebodies.

As surface weathering has resulted in the oxidation of the upper levels of the Mammoth and other mineralization along the Mount Gordon Fault zone, the dispersion of the elements associated with mineralization is of particular interest. At the surface, Cu, As and, to a lesser extent, Mo, Pb and Sb are concentrated within ferruginous outcrop. Tl is completely lost. These elements, especially As, may be used to distinguish between "true" and "false" Cu anomalies in ironstones. This is of particular importance to the exploration geologist, as interpretation of their textural features is difficult and often impossible.

Table III would suggest that once a target zone has been so delineated, analysis of material from the subsurface oxidized zone for Cu, Ag, As, Bi, Co, Mo, Ni, Pb and Sb appears warranted.

ACKNOWLEDGEMENTS

We wish to express our appreciation to Gunpowder Copper Limited who have encouraged this work and provided samples and access to company data. In particular we wish to thank G.P. Moore and other members of the geological staff for many fruitful discussions.

Mr. D.G. Reynolds, Exploration Manager, C.G.F.A., who initiated the study, has made many helpful suggestions during discussion of this work. His contribution is gratefully acknowledged.

The preparation and analysis of these samples was done by J.A. Corbett, S.C. Goadby, A.R. Horne, W.V. Hutchings, K.M. Kineally, A. Martinez, K.J. Mizon, N.C. Morgan, and K.W. Riley at the CSIRO Minerals Research Laboratories, North Ryde. To these colleagues we extend our special thanks.

REFERENCES

Boyle, R.W. and Jonasson, I.R., 1973. The geochemistry of arsenic and its use as an indicator element in geochemical prospecting. J. Geochem. Explor., 2: 251—296.

Corbett, J.A. and Mizon, K.J., 1976. The determination of carbon in materials containing high concentrations of sulphur. Chem. Geol., 17: 155—159.

Corbett, J.A., Godbeer, W.C. and Watson, N.C., 1974. The application of sodium peroxide sintering techniques to the analysis of minerals and rocks by atomic absorption spectroscopy. Proc. Australas. Inst. Min. Metall., 250: 51—54.

Croxford, N.J.W., 1964. Origin and significance of volcanic potash-rich rocks from Mount Isa. Trans. Inst. Min. Metall., 74: 33—43.

Derrick, G.M., Wilson, I.H. and Hill, R.M., 1976. Revision of stratigraphic nomenclature in the Precambrian of northwestern Queensland, 11. Haslingden Group. Queensl. Gov. Min. J., 77: 300—306.

Glasson, K.R., 1973. A study of anomalous copper-lead-zinc values in relation to rock types and weathering profiles in an area north of Mount Isa. Australas. Inst. Min. Metall., W. Aust. Conf., pp. 27—40.

Guy, B.B., 1975. Stratiform copper mineralization in the Proterozoic dolomites of northwestern Queensland, Australia. Econ. Geol., 70: 1414—1425.

Holyland, P.J., Moore, G.P., Scott, K.M. and Taylor, G.F., 1976. Geological setting and mineralogy of the Mammoth copper deposit, northwest Queensland. 25th Int. Geol. Congr. Sydney, N.S.W., Sect. 4, Abstract Vol. I: 166-167.

Mitchell, J.W. and Moore, G.P., 1975. Mammoth copper deposit. In: C.L. Knight (Editor), Economic Geology of Australia and Papua New Guinea, 1. Metals. Australasian Institute of Mining and Metallurgy, Melbourne, Vic., pp. 383—389.

Moore, G.P., 1974. Geology and mineralization of the Mammoth copper deposit. In: Recent Technical and Social Advances in the North Australian Minerals Industry. Australas. Inst. Min. Metall., N. Queensl. Reg. Meet., pp. 17—25.

Taylor, G.F., 1973. The geochemistry of siderite in relation to ironstone in the Paradise Creek Formation, northwest Queensland. J. Geochem. Explor., 2: 367—382.

van den Heuvel, H.B., 1969. Sedimentation, stratigraphy and post-depositional changes in the sediments of the upper formations of the Mount Isa Group, northwest Queensland. Ph.D. Thesis, University of Queensland, St. Lucia, Queensl. (unpublished).

Journal of Geochemical Exploration, 8 (1977) 169—188

GEOCHEMICAL INDICATIONS OF CONCEALED COPPER MINERALIZATION IN AN AREA NORTHWEST OF MOUNT ISA, QUEENSLAND, AUSTRALIA

K.F. BAMPTON, A.R. COLLINS, K.R. GLASSON and B.B. GUY

Union Minière Development and Mining Corporation Ltd., Double Bay, N.S.W. (Australia)

(Revised version accepted March 22, 1977)

ABSTRACT

Bampton, K.F., Collins, A.R., Glasson, K.R. and Guy, B.B., 1977. Geochemical indications of concealed copper mineralization in an area northwest of Mount Isa, Queensland, Australia. J. Geochem. Explor., 8: 169--188.

Ferruginous outcrops (ironstones) are a common feature of deeply oxidized dolomitic members of the Middle Proterozoic dolomitic and carbonaceous/siliceous sequence in northwestern Queensland, Australia. The textural and geochemical interpretation of the ironstones is important in regional and detailed exploration for Cu, Ag, Pb and Zn. In the Mount Kelly area, primary Cu mineralization occurs in breccias, veins and disseminations in dolomitic-rich bands. Because of structural complexities, the moderate- to high-grade (>2% Cu) zones do not outcrop. The geochemistry of this primary mineralization is described, no element other than Cu being detected characteristic of the chalcopyrite.

Surface indications of mineralization are Cu anomalous ironstones in the dolomitic units and some secondary Cu mineralization. The ironstones are classified as (1) gossans forming directly from in-situ oxidation of primary mineralization; (2) fault ironstones; (3) "stratigraphic" ironstones which are broadly distributed along particular units but which transgress lithological boundaries in detail.

Whole rock geochemistry of the fault and "stratigraphic" ironstones reveals a significant positive correlation of Cu with Ag, a weak positive correlation with Ni and a strong negative correlation of Cu with Co. There is no statistical or spatial evidence of Cu being related to Mn. Co and As anomalies can reflect the presence of pyrite. The Fe in the ironstones is derived from dolomite and sulphides in the sequence. Cu in the ironstones is derived from three sources, namely: disseminated chalcopyrite in the sequence, Cu in pyrite, and specific higher-grade Cu concentrations. These three sources cannot readily be differentiated. The magnitude of a Cu anomaly is no indication of Cu values in the primary zone.

Exploration should be aimed at locating ironstones and prospective stratigraphic horizons which are anomalous in Cu, followed by subsurface testing, but not necessarily with the subsurface continuation of ironstones as drilling targets. Where the primary dolomitic rocks average over 1500 ppm Cu over significant thickness, the sequence should be considered prospective.

INTRODUCTION

In the Precambrian mineral belt of northwest Queensland and more particularly the Carpentarian (Middle Proterozoic) sediments, ferruginous outcrops are a common feature. They have long been an object of interest as indicators of mineral potential, and thus warrant careful appraisal (Glasson, 1973).

Geophysical methods for the detection of mineralization in the Western Geosyncline have proved to be of limited application for the following reasons: (1) depth of complete oxidation reaches 100 m in some areas, (2) the high carbonaceous content of many of the rocks has restricted the use of electrical and electromagnetic methods to the thick arenaceous units, and (3) the lack of magnetic phases associated with base metal mineralization renders such magnetometer surveys applicable only on the airborne scale and as an aid to regional mapping.

Thus, as the use of geophysical methods is severely restricted, the geochemical approach is essential in the search for non-outcropping mineralization in conjunction with detailed stratigraphic and structural studies.

This paper is concerned with geochemical surface indications of mineralization that occurs in this deeply oxidized terrain at Mt. Kelly (Fig. 1) and more particularly mineralization that for stratigraphic and structural reasons has no in-situ surface expression. The surface geochemical anomalies are generally manifested as "ironstones" many of which have been considered to be of lateritic derivation (cf. De Keyser, 1958; Taylor, 1973) and thus of little value in mineral exploration. However, in this paper it is shown that the geochemistry of the ironstones may be used as an indicator of primary mineralization.

ANALYTICAL METHODS

Analytical work has been carried out by Eastern Copper Mines N.L., Mt. Isa, and Geochemical and Mineralogical Laboratories, Sydney. Table I lists the analytical methods with detection limits for elements determined.

GEOLOGICAL SETTING

Regional. The Mount Kelly area is within the northwest Queensland Province, one of the principal Proterozoic tectonic elements of northern Australia (Plumb and Derrick, 1975). In the vicinity of Mount Isa, this province is divided into Eastern and Western "Geosynclines" separated by a tectonic welt known as the Kalkadoon-Leichhardt Block (Fig. 1). Immediately to the west of the welt is the Mount Isa Trough, which is bounded to the northwest by the Lawn Hill Platform across the Mt. Gordon Fault zone. The Mount Kelly area occurs on the platform adjacent to a north-striking branch of the Mount Gordon Fault and near a basement high.

In the vicinity of Mount Kelly, the stratigraphic sequence of the Lawn

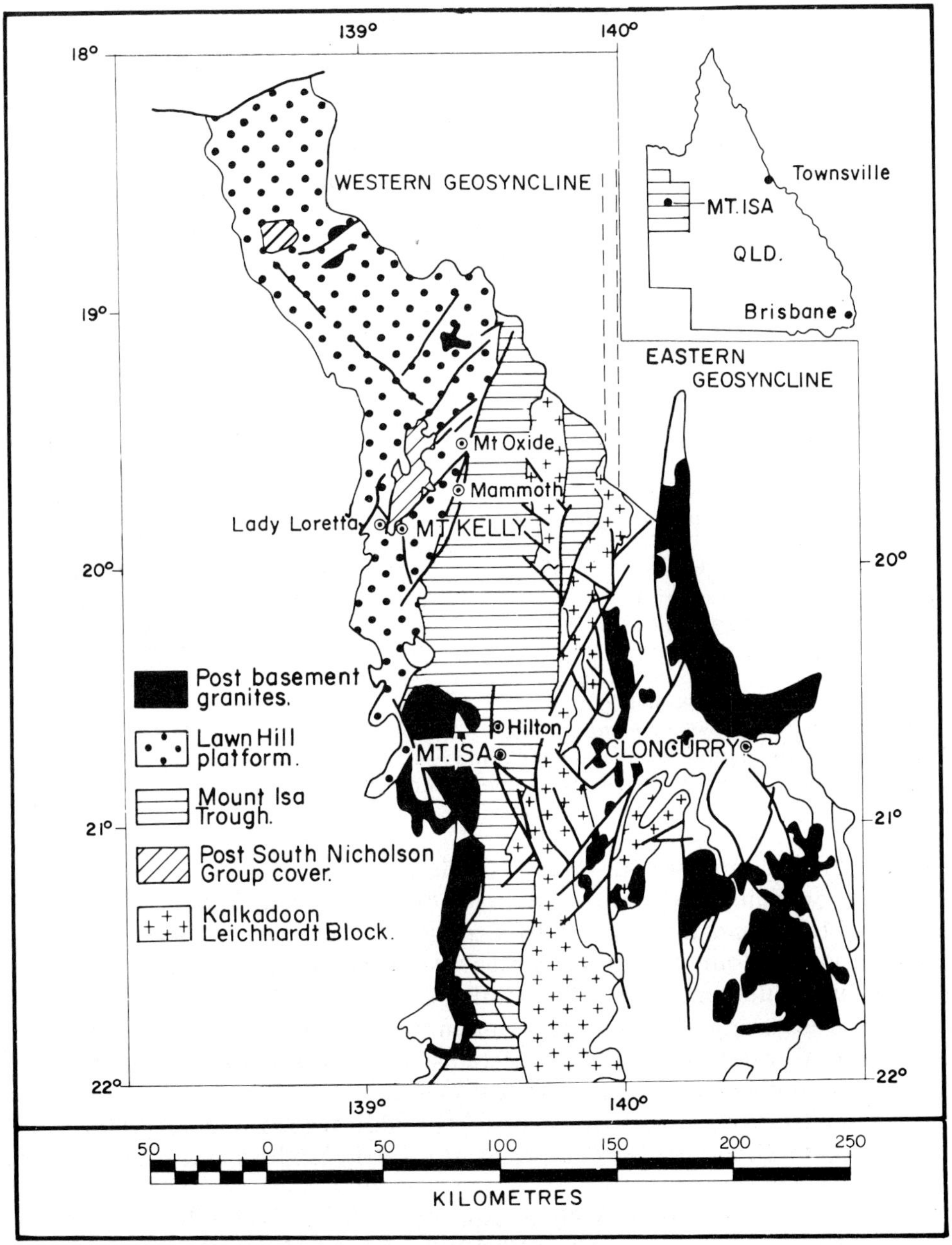

Fig. 1. Locality map showing distribution of Proterozoic rocks and granitic masses in northwest Queensland (modified after Plumb and Derrick, 1975). The main tectonic elements of the Northwest Queensland Province are indicated.

TABLE I

Analytical methods and detection limits

Element	Method (digestion)	Detection limit (ppm, unless otherwise given)	Precision (at level)
Ag	AAS (aqua regia)	0.2	20% (2 ppm)
As	colorimetric[1] (nitric-perchloric)	1	30% (50 ppm)
Bi	AAS (aqua regia)	1	30% (50 ppm)
CaO	AAS (total[2])	0.1%	5%
Cd	AAS (aqua regia)	0.2	20% (50 ppm)
Co	AAS (perchloric)	3	10% (80 ppm)
Cu	AAS (perchloric)	1	10% (30 ppm)
Fe	AAS (total[2])	0.01%	5%
Hg	AAS[3]	0.005	20% (0.5 ppm)
K_2O	AAS (total[2])	0.1%	5%
MgO	AAS (total[2])	0.1%	5%
Mn	AAS (hydrochloric-perchloric)	2	10% (1000 ppm)
Na_2O	AAS (total[2])	0.01%	5%
Ni	AAS (perchloric)	1	10% (100 ppm)
Pb	AAS (perchloric)	2	10% (300 ppm)
V	AAS (perchloric)	5	20% (1000 ppm)
Zn	AAS (perchloric)	1	10% (300 ppm)

[1] Phosphomolybdate colorimetric method.
[2] Peroxide fusion, followed by acid uptake.
[3] Determined after vaporization and distillation.
Acid attacks carried out at temperatures in excess of 180° C

Hill Platform comprises the Myally Beds of the Haslingden Group unconformably overlain by Gunpowder Creek Formation and Paradise Creek Formation (Plumb and Derrick, 1975). Structurally, the area is characterized by open basin and dome style folds which are broken up by E-W and N-S-trending dip-slip faults and later NW-trending strike-slip faults. Cleavage is almost always subordinate to bedding. Metamorphism is low greenschist facies grade. The Gunpowder Creek and Paradise Creek Formations are considered to be time equivalents of the Mount Isa Group which is developed in part of the Mount Isa Trough and which hosts the Mount Isa Cu and Ag-Pb-Zn and Hilton Ag-Pb-Zn orebodies (Carter et al., 1961; Mathias and Clark, 1975).

Mt. Kelly area. The Mount Kelly Prospect is located 100 km NNW of Mount Isa, Queensland, at latitude 19°51′S, longitude 139°08′E (Fig. 1). The prospect lies immediately northwest from the old Mt. Kelly copper workings. These are located at the intersection of important N-S- and E-W-trending faults, where upper Gunpowder Creek Formation and Lower Paradise Creek Formation rocks occur adjacent to an upfaulted block of lower Gunpowder Creek Formation quartzite and Myally Beds. The detailed stratigraphy is summarized in Table II.

TABLE II

Summary of the stratigraphic sequence in the Mount Kelly area

Formation	Symbolic designation of unit	Thickness (m)	Lithology
Paradise	Pcd	50+	undifferentiated dolostones
Creek	Pcq	0—10	dolomitic sandstones
Formation	Pci	150	banded dolomitic siltstones, carbonaceous siltstones
	Pch	7	chert and dolomitic siltstones
Gunpowder	Gcf	7	finely laminated shales
Creek	Gcp	15	massive graphitic siltstones
Formation	Gch	70	banded dolomitic siltstones and shales
	Gcb	300	well-laminated silty-pyritic-carbonaceous shales
	Gcv	30+	andesite
	Gcl	up to 1000	quartzites
	Gcc	up to 30	conglomerates
		Unconformity	
Myally Beds	Mbv	approx 100	basic to intermediate volcanics
	Mbq } Mbp }	approx. 100	interbedded quartzites and purple mudstones

Structure. Folding is of open concentric style on NW-SE-trending axes. There are plunge reversals at about 2-km intervals along strike. In contrast to the simple fold style, faulting is extremely complex. Models of penecontemporaneous faulting have been proposed for the region (Smith, 1969; Mitchell and Moore, 1975) but there is no conclusive evidence at this locality, and all identified faults are interpreted as post folding. Surface and subsurface investigations have revealed that an early phase of low-angle overthrusting and block tilting (mostly E-W to SE-NW in strike) was followed by later steeply dipping strike-slip faulting of which the NW-SE-oriented Mt. Kelly Fault is the most prominent example. A composite cross-section is shown in Fig. 2.

THE PRIMARY ZONE AT MT. KELLY

Distribution of mineralization

Primary mineralization occurs at a number of stratigraphic levels. The

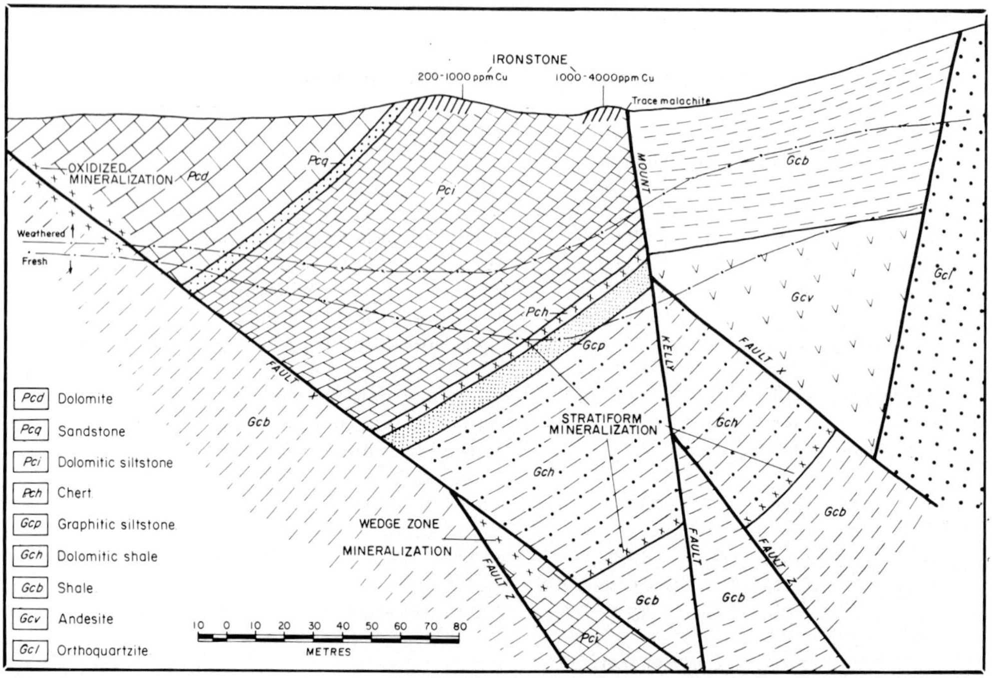

Fig. 2. Composite geological section through the Mount Kelly area showing stratigraphic sequence, folding and complex faulting, zones of mineralization and typical shape of oxidized zone boundaries. Zones of stratiform and wedge zone mineralization are greater than 2% Cu.

major host is the banded dolomitic shale unit Gch, and more particularly its lower part. Secondly, the Pch chert horizon is well mineralized in several drill holes and single drill intersections of greater than 2% Cu have been made in each of the Pci and Pcd dolomitic units. In addition, weak disseminated stratiform or vein mineralization (grade less than 0.5% Cu) is locally present in all units of the sequence from Gcb to Pcd inclusive (see Table II).

The largest zone of primary mineralization discovered to date has a strike length of 500 m, an average width of approximately 4 m and an average grade of 2.5% Cu. Other zones of mineralization have been intersected but their dimensions have not been completely defined.

Mineralogy and texture of mineralization

Chalcopyrite is the only primary Cu mineral present, and is associated with pyrite in a ratio of approximately 2:1. The only other sulphide minerals present are rare pyrrhotite, arsenopyrite and sphalerite.

Most occurrences of greater than 1% Cu mineralization are in breccias. This mineralization, together with lower-grade vein mineralization, is accom-

TABLE III

Mean values for major and minor elements/oxides in some stratigraphic units at Mt. Kelly (ppm, unless otherwise given)

Unit	Number of analysis	Ag	As	Bi	CaO (%)	Cd	Co	Cu	Fe_2O_3 (%)	Hg	K_2O (%)	MgO (%)	Mn	Na_2O	Ni	Pb	V	Zn
Pci	8	1	21	26	11.4	2.4	57	425	4.8	0.020	3.6	7.7	1100	0.03	47	35	16	15
Pch	2	0.5	8	20	5.3	2.0	20	950	2.4	—	4.1	4.3	500	0.04	45	23	20	26
Gcf	2	0.3	28	5	0.4	0.2	28	140	1.8	0.005	—	0.6	55	—	225	22	4	12
Gcp	17	0.5	200	13	0.6	2.0	94	600	3.9	0.005	3.9	1.37	70	0.04	158	34	28	68
Gch	350	1.2	27	28	10.0	1.4	80	1600	4.66*	0.020	3.5	7.0	650	0.05	65	40	12	15
Gcb	15	0.8	33	13	1.2	1.0	42	66	4.50	0.005	4.0	5.8	130	0.02	89	25	77	21
Gcl	1	0.4	65	7	0.1	0.2	16	46	2.9	0.005	—	2.50	44	—	180	20	8	22

— data not available

*Determined on a limited group of samples (15).

For Gch units, hosts to the majority of Cu mineralization, analytical data have been determined on continuous 0.6-m (2 ft) core lengths through this entire unit for three drill holes (designated KB 10, KB 19W, and KB 26) that are distributed over approximately 800 m of strike. For rock units other than Gch, samples assayed are restricted to individual random drill core specimens.

panied by a gangue of dolomite and subordinate quartz. Where mineralization is disseminated, chalcopyrite occurs as individual grains or rimming pyrite in the dolomitic layers of the rock.

Geochemistry

Some data have previously been published (Guy, 1975) on the geochemistry of the host rocks and primary Cu mineralization at Mt. Kelly. These data are summarized in Table III together with additional information on the geochemistry of the main rock units. A total of sixteen elements have been determined in an attempt to categorize the major rock types and the mineralization.

Stratigraphic units

As noted above, mineralization is distributed through several stratigraphic units. The geochemistry of the units is summarized in Table III. The mean Cu values in Gch, Pch and Pci are high relative to other stratigraphic horizons.

Gch unit. The principal rock types in this unit, which hosts the best development of high-grade mineralization, are generally characterized in the analytical data by the following elements:

carbonaceous siliceous units: K_2O; dolomitic units: Bi, CaO, MgO, Mn

In the three drill holes sampled in detail, the geochemistry has revealed that Gch may be subdivided into a "central" carbonaceous-rich unit, between two dolomite-rich units (Fig. 3). Means and standard deviations for some elements in these units are presented in Table IV. Using the average composition of dolomites as $(Ca_{52}, Mg_{39}, Fe_{8.5})(CO_3)_{100}$ (Guy, 1975), a total dolomite content has been calculated from the CaO contents of the analytical data (Fig. 3).

Mineralization

Electron microprobe and bulk sample geochemical data have been collected on coexisting pyrite and chalcopyrite concentrates (Guy, 1975). Elements determined include: As, Ag, Bi, Co, Cr, Cu, Hg, Ni, Pb and Zn. No elements were detected characteristic of chalcopyrite either in the chalcopyrite structure or in included submicroscopic phases within chalcopyrite grains. The available data, however, indicate that As, Co and Pb are concentrated in pyrite, with Co being most abundant — generally 2000 ppm. Cu contents of pyrite have been found to vary markedly but average 500 ppm.

In the bulk rock analyses through Gch (see footnote Table III), Co anomalies coincide with zones of chalcopyrite mineralization while As is more er-

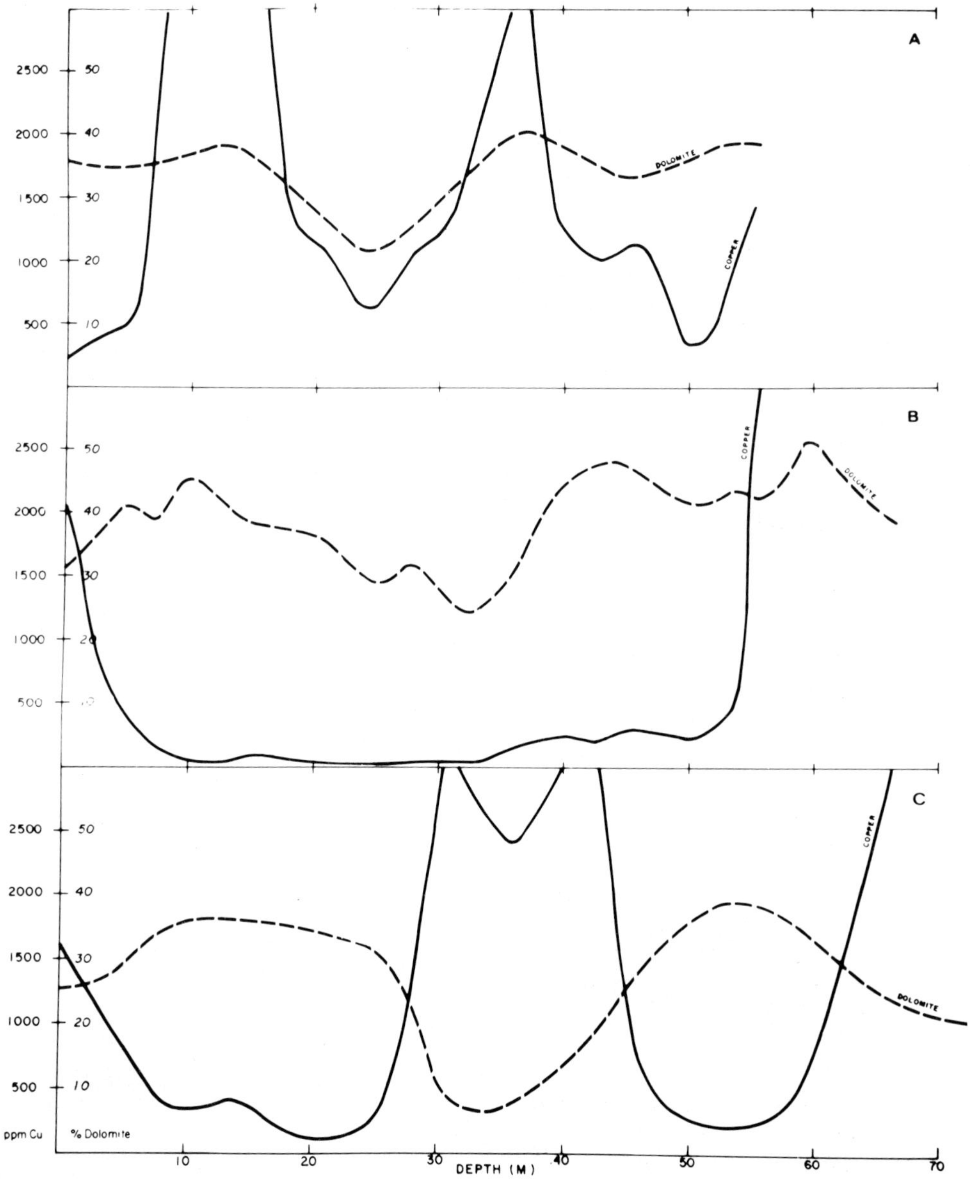

Fig. 3. Curves of best fit to histograms based on 5-m sample intervals showing variation in dolomite (dashed line) and Cu (solid line) contents with position in Gch units in three drill holes: (A) KB 10; (B) KB 19W; (C) KB 26. Zero depth corresponds to Gcp/Gch boundary. Depth scale has been corrected to true thickness of the unit.

TABLE IV

Element values — Gch unit

Element/oxide	KB10				KB 19W				KB 26			
	dolomitic units		carbonaceous unit		dolomitic units		carbonaceous unit		dolomitic unit		carbonaceous unit	
Bi (ppm)	35	(4)	29	(5)	30	(5)	20	(3)	30	(9)	16	(5)
CaO(%)	11.0	(0.7)	7.0	(1.7)	14.0	(2.0)	7.0	(2.0)	12.0	(2.7)	2.7	(3.6)
MgO (%)	7.4	(1.7)	4.9	(1.0)	10.2	(2.0)	5.3	(1.7)	8.8	(1.4)	3.2	(2.4)
Mn (ppm)	750	(175)	587	(110)	950	(150)	600	(150)	500	(80)	200	(190)
K_2O(%)	2.9	(0.6)	4.1	(0.7)	3.0	(0.7)	5.5	(0.7)	3.0	(0.4)	4.6	(0.8)

Unweighted means with standard deviations, in parentheses for some element/oxides in dolomitic and carbonaceous-rich units encountered in drill holes KB 10, 19W and 26. Data collected from continuous bulk rock samples 0.6 m in length through Gch unit.

The subdivision into units is based on total dolomite content as shown in Fig. 3. For KB 10, dolomitic units are greater than 30% dolomite, in KB 19W greater than 35% dolomite and in KB 26 greater than 25% dolomite. This variation in dolomite content reflects lateral variation in the unit.

TABLE V

Means (ppm) and standard deviations of elements in some Mt. Kelly ironstones (25 samples)

Element	Mean	Standard deviation
Ag	2.0	0.5
As	63	53
Co	104	49
Cu	3050	2100
Hg	0.015	0.020
Mn	530	810
Ni	97	76
Pb	32	11
Zn	54	35
S	262	106

ratic but does attain maximum values near to zones of Cu enrichment. Hg is variable throughout the Gch unit and appears to be localized in carbonaceous units (Guy, 1975). This element, however, is higher near some zones of mineralization but rarely is this value greater than one standard deviation outside the mean.

Fig. 3 shows the Cu distribution in drill holes KB 10, 19W and 26. Although disseminated chalcopyrite is visually preferentially localized in carbonate bands and lenses, up to a thickness of 10 cm, there is no direct correlation between carbonate and Cu in samples bulked over 1 m true width. This is shown in KB 26 where copper-rich zones are localized in the broad carbonaceous unit, whereas in KB 10 and KB 19W they are in the broad dolomite-rich units. This is supported by multiple regression correlation analysis which has not revealed any significant correlation of chalcophile elements with the major elements.

THE OXIDIZED ZONE AT MT. KELLY

Surface ironstones

Excluding iron-rich rocks with pisolitic textures developed on the peneplain surface, and thus classified as laterites, the term ironstone is used here to cover all surface and subsurface rocks with a high proportion of haematite and/or goethite. On the basis of spatial distribution and textures, the ironstones may be classified as: (1) gossan; (2) fault ironstones; (3) "stratigraphic" ironstones.

The ironstones are generally restricted to the carbonate-rich lithologies. The siliceous and carbonaceous units contain only weak localized iron oxide accumulation on joint surfaces.

In the Mt. Kelly area, the surface expression of carbonate units is either as bleached kaolinitic outcrops or as ironstones. Ironstone constitutes some 20% of the exposed dolomitic sequence. Surface and subsurface exploration of ironstones in the Mt. Kelly area have not revealed any that have formed directly from the oxidation of the moderate- to high-grade (i.e. >2% Cu) mineralization. In other ironstones (e.g. Hinge Ironstone, Fig. 5), textures indicative of the presence of minor disseminated and vein sulphides occur and these may be classified, in part, as gossans. However, in most ironstones, solution and re-deposition of iron oxides and silica has obliterated rock textures so that most have been placed in the fault or stratigraphic categories. In addition, examination of the leached dolomitic rocks did not reveal any evidence of vein or disseminated sulphides. Fault ironstones may extend 20 m from faults in carbonate units and the major dimension of such ironstones is parallel to the fault regardless of the strike of the beds. In siliceous rocks, major iron oxide development is restricted to the fault zone. "Stratigraphic" ironstones are distributed along particular stratigraphic horizons (e.g. basal Gch), however, dispersion of Fe has resulted in such ironstones transgressing lithological boundaries.

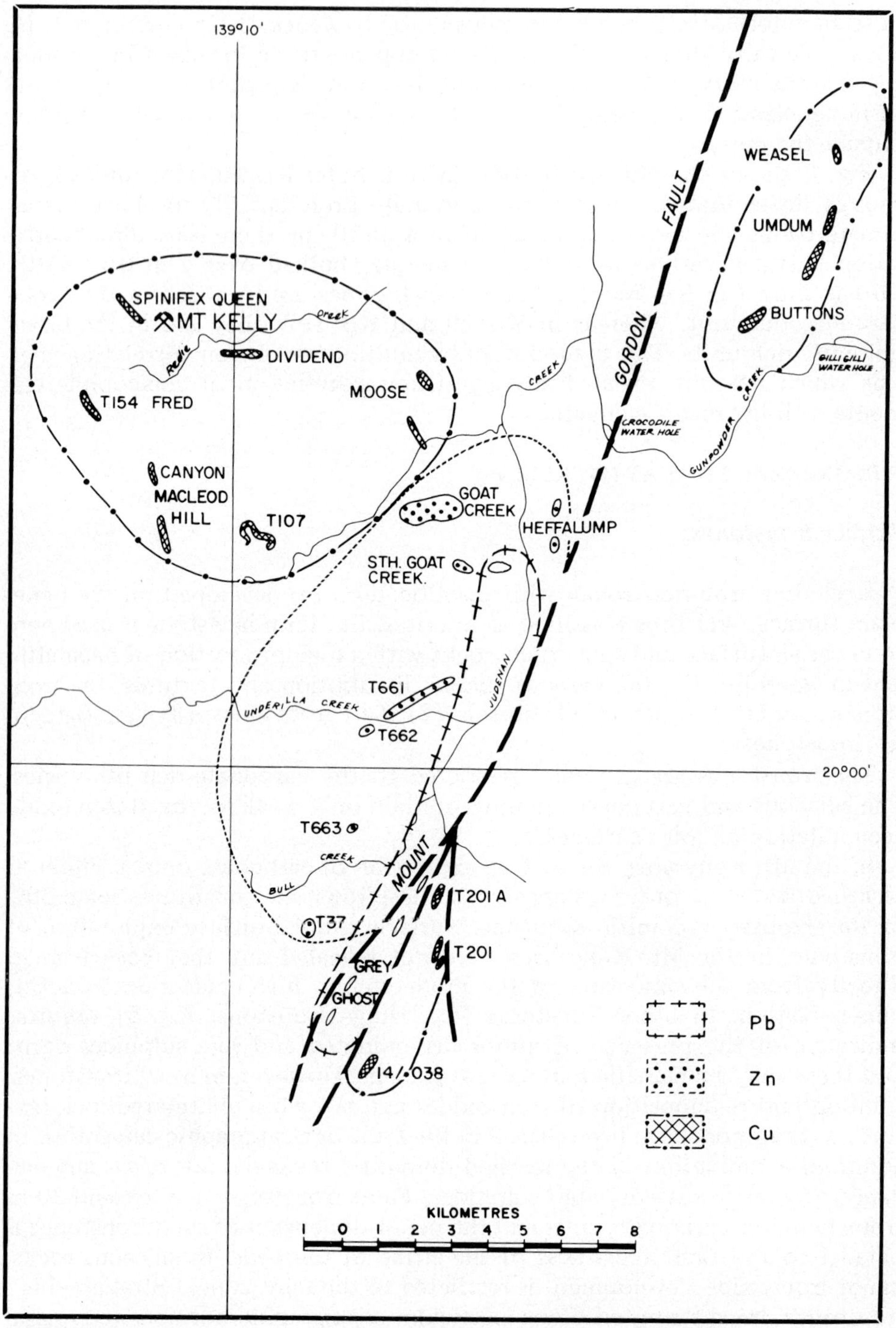

Fig. 4. Regional distribution pattern of ironstones showing anomaly numbers (e.g. T107, 14/038) in the vicinity of Mount Kelly and type of geochemical anomalies.

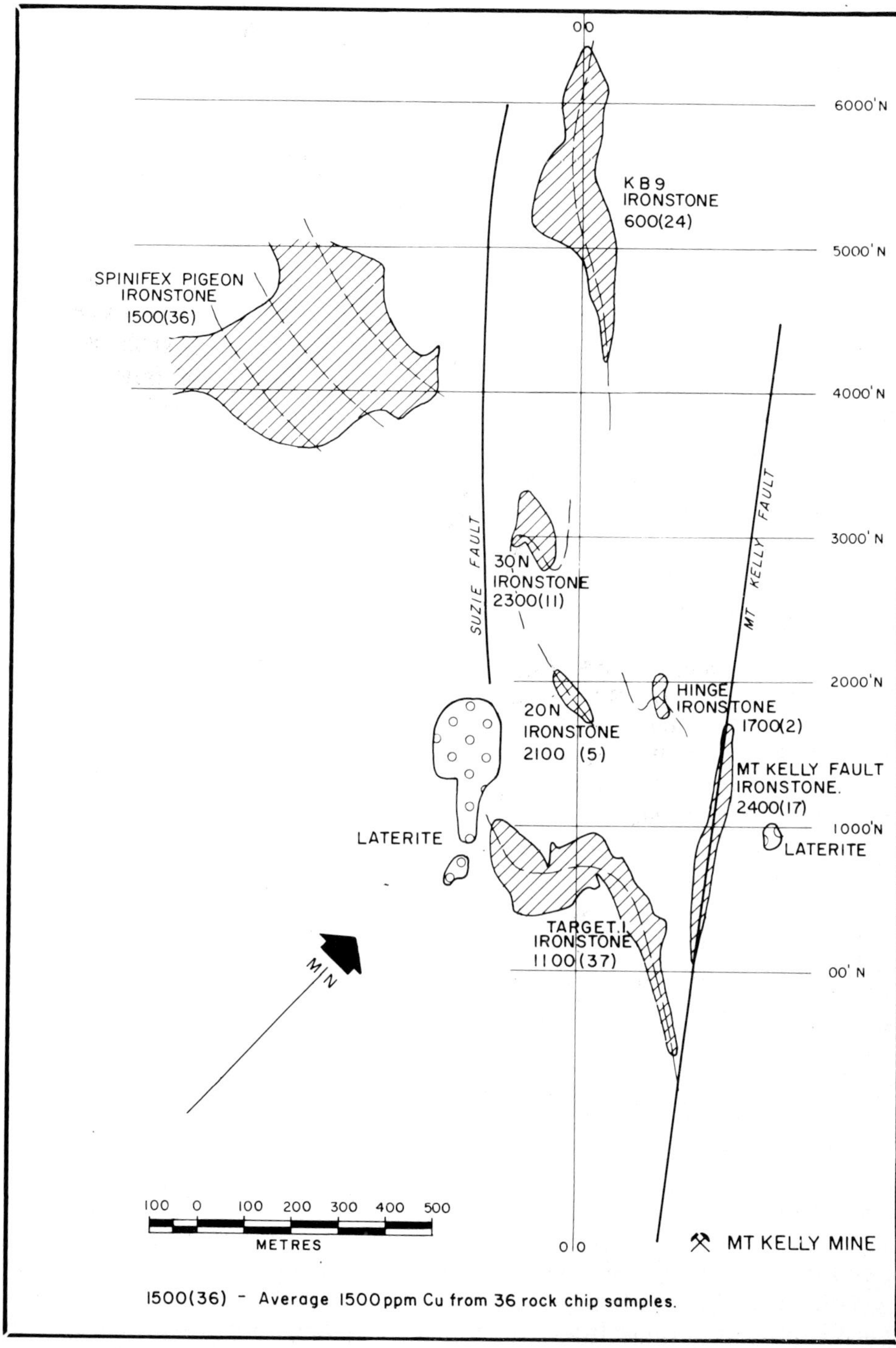

Fig. 5. Distribution of some ironstones at Mount Kelly. Note that the grid coordinates are in feet. Ironstones diagonal hatch, laterite circles.

Fig. 4 shows the distribution of some of these ironstones on a regional scale, and their chemical variation, and Fig. 5 the distribution at Mt. Kelly.

Mineralogy and texture

The mineral assemblages of the ironstones are relatively simple, namely: quartz, goethite and minor mica with or without haematite, manganese oxides and clay minerals. Kaolinite, feldspar and chlorite occur as minor constituents of these rocks (Taylor and Scott, 1976). The rocks display limited evidence of any feature of original textures of the associated sedimentary rocks of the Gunpowder-Paradise sequence but, in ironstones with goethite contents less than 50%, some original grain shapes together with bedding features may be distinguished. In the more iron-rich varieties, fragments of the sedimentary rocks may be observed in a massive goethite base, which may display evidence of concentric growth layers.

Pseudomorphs and voids after pyrite have been observed in some rocks in the Mt. Kelly area, but no definite chalcopyrite textures have been noted. Other voids, not developed after sulphide, are common in the ironstone and may attain dimensions up to 1 cm across.

Geochemistry

Mineral phases. Electron microprobe studies on the ironstones have been carried out in order to determine whether any relation exists on a microscopic scale between anomalous elements (e.g. Cu) and Fe and Mn contents. The following observations were made:

(1) Grains of Fe and Mn-Fe oxides may display banding which in some cases is truncated by adjacent banded grains. In particular, the banding in the Mn-Fe-rich grains is abruptly terminated by grain boundaries of the iron oxides.

(2) Cu is distributed randomly within both Fe and Mn-Fe grains, with no specific concentration observed.

(3) Co, while occurring in significant quantities in Fe and Mn-Fe grains is preferentially localized in the latter.

These observations imply more than one period of iron oxide accumulation, with manganese-rich grains in places pre-dating these iron oxides. There is no evidence to indicate that Cu and Co are localized at the same time with final Cu localization post-dating Co and Mn.

Rocks. Average Cu values for detailed systematic grid sampling of some of the ironstone bodies are shown in Fig. 5. By way of comparison, the average Cu content of non-iron-stained exposed dolomitic sequence is 20 ppm and of laterite 290 ppm. Bulk rock multi-element geochemistry of some 25 ironstone samples from those listed in Table VI have been obtained — means and standard deviations being shown in Table V. From multiple regression analy-

TABLE VI

Classification and origin of Mt. Kelly area ironstones

Name	Classification	Origin of Fe	Origin of Cu
Target 1	stratigraphic	derived from dolomitic rocks	disseminated mineralization in upper Pci and Pcq
Mt. Kelly Fault	fault	exotic from sulphides and dolomite in the sequence	exotic from sulphides at various stratigraphic levels
Hinge	stratigraphic (possibly gossan)	dolomite, pyrite and? chalcopyrite	weak mineralization in lower Gch (cf. Fig. 3B)
20N	stratigraphic	dolomite and sulphide in middle Gch	pyrite and weak disseminated chalcopyrite in Gch
30N	fault	exotic from dolomite and sulphide in sequence	exotic from sulphides at various stratigraphic levels
KB 9	fault/stratigraphic	exotic from dolomite/ pyrite in sequence	pyrite in sequence
Spinifex Pigeon	stratigraphic	dolomite and pyrite in Pcd sequence	pyrite in sequence

sis of these data and the data in Table V the following points can be made:

(1) There is a significant positive correlation of Cu with Ag, a barely significant positive correlation of Cu with Ni, and a very significant negative correlation of Cu with Co.

(2) Apart from ironstones developed over fault zones containing Hg values of the order of 0.05 ppm, there are no significant geochemical differences between the stratigraphic and fault ironstones. This Hg content does not in any way reflect the existence of Cu mineralization.

Subsurface iron accumulations

Complete oxidation extends to an average depth of 60 m and is separated from fresh rock by a transition zone approximately 20 m in thickness. These boundaries vary in rough parallelism with each other but not with the sur-

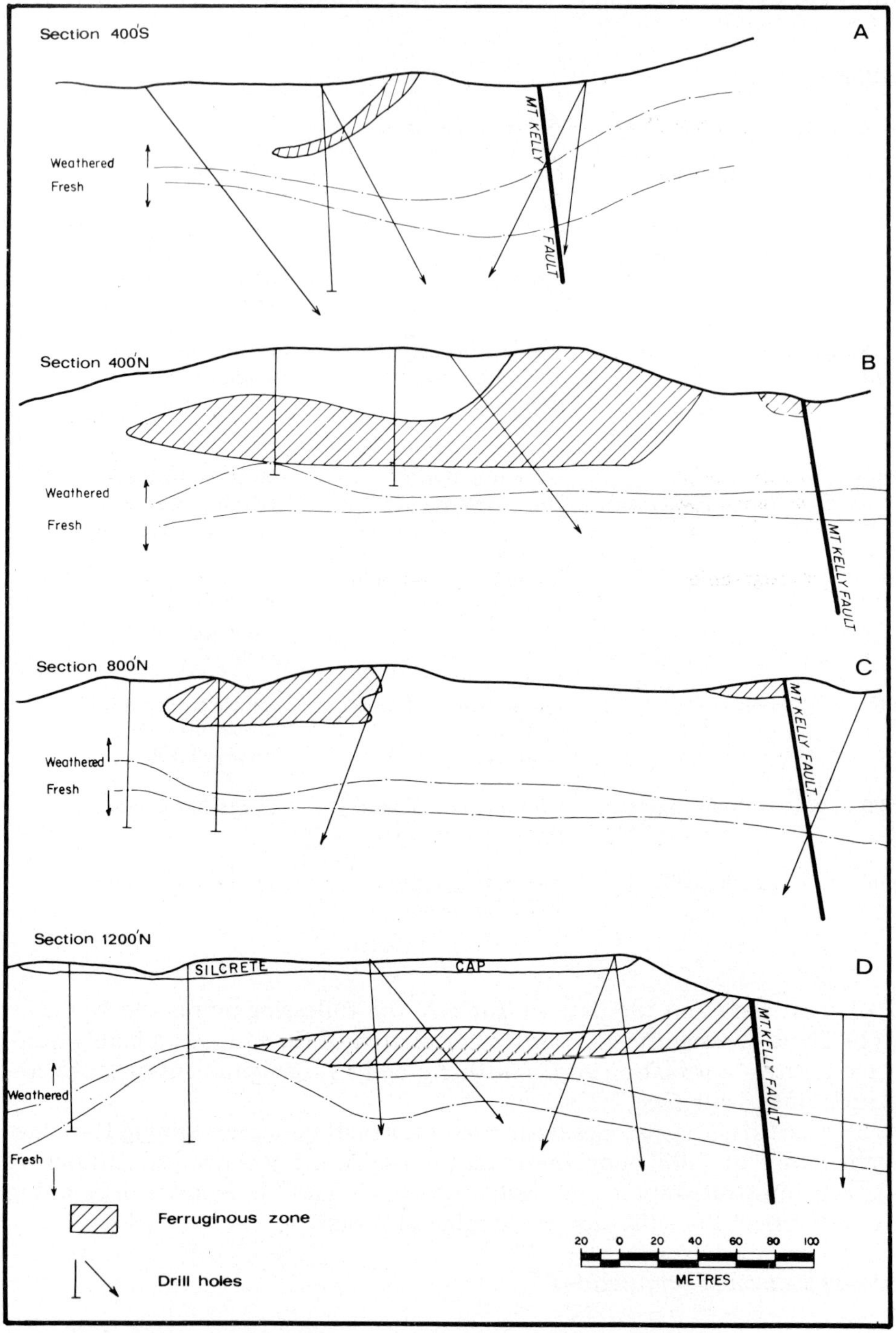

Fig. 6. Interpretation from drill hole data of the subsurface continuation of some of the ironstones in Fig. 5. A. Target 1 ironstones on a fold limb. B. Target 1 ironstone on a fold limb with extensive redistribution of iron in the oxidized zone. C. Target 1 ironstone in a synclinal keel. D. Mount Kelly Fault ironstone extending in a subhorizontal zone away from the fault.

face (see Fig. 6). Ferruginous accumulations are restricted to the zone of complete oxidation but typically cease about 20 m above its base.

The Target 1 ironstone (Fig. 5) is an example of the stratigraphic variety and Fig. 6A shows the subsurface continuation of part of this ironstone on a fold limb (cf, Fig. 2) while in Fig. 6C the ironstone is shown in a synclinal keel. The ironstone along the Mount Kelly Fault is an example of the fault class (Fig. 5). Drill hole data indicate that this ironstone extends away from the fault in a subhorizontal zone (Fig. 6D).

Secondary copper occurrences

In the zone of oxidation, Cu occurs dominantly as malachite followed closely in abundance by chrysocolla with traces of azurite, pseudomalachite, cuprite and tenorite. In the supergene zone chalcocite, bornite, and native Cu have been observed but few data are available. Most occurrences of Cu in the oxidized zone are very limited in extent and either occupy zones of intense fracturing at surface, or occur at local lows in the weathering profile subsurface. They are in general exotic deposits with little or no associated Fe and with the Cu minerals occurring as films on fractures. Mt. Kelly and Mt. Clarke, the two most substantial surface showings, are essentially in-situ occurrences after low-grade primary material but with extensive redistribution by surface agencies. No secondary sulphides have been observed below these surface showings.

DISCUSSION

At Mt. Kelly, the geological structure is such that none of the secondary Cu occurrences or ironstones can be interpreted as direct oxidation products of the greater than 2% Cu mineralization intersected in the primary zone. Zones of weak disseminated and vein mineralization do reach the surface but the textural expression of the Cu mineralization is largely obliterated by oxidation processes.

However, the high- and low-grade mineralizations do have a chemical expression on the surface and in the oxidized zone, namely the Cu anomalous ironstones and secondary Cu occurrences. For example, the stratiform and wedge zone mineralization in Fig. 2 is concealed mineralization giving rise to the oxidized Cu minerals along fault X and the Cu anomalous ironstones along the Mt. Kelly Fault. As noted above, Mt. Kelly and Mt. Clarke are examples of indications of low-grade primary mineralization.

Due to stratigraphic and structural features, it is not possible to study progressive element dispersal that occurs between the primary zone and the surface. The lack of a characteristic trace element association with chalcopyrite and inadequate knowledge of Co, As and Pb distribution between pyrite occurring by itself and in association with chalcopyrite render practical application of trace element geochemical methods difficult (Guy, 1975). Although Ag shows a true correlation with Cu in the ironstones, this correlation is not present for the primary mineralization. Thus, the use of Ag to indicate anomalies derived from a copper-rich source is not warranted. The correlation between Ag and Cu in ironstones warrants further investigation. However, the low levels of Ag in the ironstones are close to the detection limit for methods commonly used in routine analysis of geochemical samples, and thus render this element of dubious value as an exploration guide.

The classification and origin of Cu and Fe in the ironstone bodies at Mt. Kelly (Fig. 5) is summarized in Table VI. Two main points, from this table, are worthy of consideration from a geochemical viewpoint: (1) the origin of Fe in the ironstones; (2) Cu localization in the oxidized zone.

Origin of Fe in ironstones. Available data indicate that most of the stratigraphic units contain moderate amounts of Fe (usually 1.8% Fe^{3+}, Table III) which, in the dolomitic units, is located principally in dolomite itself (cf. Taylor, 1973). Although no data are available on the behaviour of Fe in such an environment as Mount Kelly, any Fe in dolomitic units would oxidize principally in a neutral to weakly alkaline environment near the surface. Most Fe would eventually form haematite or goethite either in situ or after only moderate dispersal (Garrels and Christ, 1965) around permeable zones such as Pcq (Fig. 2). The more siliceous and carbonaceous units would result in a lower pH environment ultimately leading to greater dispersion of Fe. This would be particularly the case for the Gcb unit rocks which contain significant pyrite. Any Fe dispersed from the siliceous/carbonaceous units would tend to deposit in a higher pH environment (e.g. dolomitic units such as Gch).

Fault ironstones are located in areas of high permeability and the Fe in this variety is considered to be exotic. It would be derived from all units through which circulating groundwater passed.

Cu localization in the oxidized zone. The distribution of Cu in the oxidized zone would be subject to similar constraints as those influencing Fe, namely lithological variation and permeable channel ways.

There is no evidence following the exploration programme in the Mt. Kelly area of more than minor quantities of secondary copper sulphides being present. Although such secondary minerals may have existed at some stage, it appears more likely that the primary copper sulphides oxidized directly to non-sulphide phases. Cu^{2+} ions may remain in solution up to a higher value than Fe^{3+} (Garrels and Christ, 1965, fig. 7.24). Secondary Cu occurrences without associated ironstones such as Mt. Kelly and Mt. Clarke may have this origin.

SUMMARY AND CONCLUSIONS

The study of the surface distribution of geochemical anomalies and subsurface mineralization has indicated that:

(1) Geochemical surface anomalies are manifested as "ironstones" and minor secondary Cu mineralization.

(2) The distribution and textures of the ironstones indicates that they are quite distinct from laterites in the region.

(3) The texture of the ironstones indicates that most are not gossans, but developed by precipitation of iron oxides either along fault zones or within dolomitic parts of the sequence.

(4) The "Fe" of the ironstones is derived in part from the dolomites and in part from sulphides in the sequence (Table VI).

(5) Cu localized in the ironstones is apparently derived from three main sources: (a) specific Cu concentrations in stratigraphic units; (b) from disseminated chalcopyrite (less than 0.5% Cu) through stratigraphic units; (c) from Cu within pyrite. Diamond drilling through the basal Gch section in the T107 area 7 km SE of Mt. Kelly (Fig. 4) revealed abundant pyrite and no chalcopyrite in the primary zone under a copper anomalous ironstone (Glasson, 1973). Since the Cu concentration in the surface geochemical anomalies does not directly reflect the Cu values in the primary zone, each of the sources noted above cannot be readily differentiated from each other.

(6) There is no indication that volcanics in the sequence give rise directly to ironstones or geochemical anomalies in the Mt. Kelly area.

(7) Geochemical studies of primary mineralization have not revealed any trace element characteristic of chalcopyrite. However, Co and As are concentrated in pyrite. Surface Co and As anomalies therefore directly reflect the presence of pyrite, but such pyrite may occur as the only sulphide phase or in association with chalcopyrite.

(8) Exploration in the Mt. Kelly area and surrounding district has revealed a number of regions, each of these containing concentrations of ironstones with a distinctive geochemical signature (Fig. 4). The detailed programme on one of these regions, namely Mt. Kelly, has indicated that a combined geochemical-stratigraphic-structural method is necessary for exploration in the region. This method is: (a) airborne search for the location of areas of ironstone concentrations; (b) sampling and mapping of ironstones to determine the geochemistry and the stratigraphic and structural setting of the ironstone; (c) geological mapping of areas with anomalous geochemistry to locate prospective stratigraphic horizons (e.g. Gch); (d) drill testing of prospective units, principally in areas of ironstone concentration, but not necessarily with the subsurface continuation of ironstones as drilling targets; and (e) where the average value for primary dolomitic rocks is in excess of 1500 ppm Cu, detailed subsurface investigations for Cu appear warranted.

ACKNOWLEDGEMENTS

The authors wish to thank the Directors of Union Minière S.A. and Union Minière Development and Mining Corporation Limited for permission to publish this paper. This paper would not have been possible without the efforts of the UNIMIN geological team at Mount Kelly over the past six years. Dr. Graham Taylor, CSIRO, kindly provided unpublished information on his study of ironstones in the district. Mr. C. Grey, CSIRO, kindly assisted with statistical and computing problems.

REFERENCES

Carter, E.K., Brooks, J.H. and Walker, K.R., 1961. The Precambrian mineral belt of north-western Queensland. Bur. Miner. Resour. Aust. Bull., 51: 344 pp.

De Keyser, F., 1958. Geology and mineral deposits in the Paradise Creek area, North-west Queensland. Bur. Miner. Resour. Aust. Rec., 100 (unpublished).

Garrels, R.M. and Christ, C.L., 1965. Solution, Minerals and Equilibria. Harper and Row, New York, N.Y., 450 pp.

Glasson, K.R., 1973. A study of anomalous copper-lead-zinc-values in relation to rock types and weathering profiles in an area north of Mount Isa. Australas. Inst. Min. Metall., W. Aust. Conf., pp. 27—40.

Guy, B.B., 1975. Stratiform copper mineralization in the Proterozoic dolomites of north-western Queensland, Australia. Econ. Geol., 70: 1414—1425.

Mathias, B.V. and Clark, G.J., 1975. Mount Isa copper and silver-lead zinc orebodies — Isa and Hilton Mines. In: C.L. Knight (Editor), Economic Geology of Australia and Papua New Guinea, 1. Metals. Australasian Institute of Mining and Metallurgy, Melbourne, Vic., pp. 351—371.

Mitchell, J.W. and Moore, G.P., 1975. Mammoth copper deposit. In: C.L. Knight (Editor), Economic Geology of Australia and Papua New Guinea. 1. Metals. Australasian Institute of Mining and Metallurgy, Melbourne, Vic., pp. 383—389.

Plumb, K.A. and Derrick, G.M., 1975. Geology of the Proterozoic rocks of the Kimberley to Mount Isa region. In: C.L. Knight (Editor), Economic Geology of Australia and Papua New Guinea, 1. Metals. Australasian Institute of Mining and Metallurgy, Melbourne, Vic., pp. 217—252.

Smith, W.D., 1969. Penecontemporaneous faulting and its likely significance in relation to Mount Isa ore deposition. Spec. Publ. Geol. Soc. Aust., 2: 225—235.

Taylor, G.F., 1973. The geochemistry of siderite in relation to ironstone in the Paradise Creek Formation, northwest Queensland. J. Geochem. Explor., 2: 367—382.

Taylor, G.F. and Scott, K.M., 1976. The geochemistry of ironstones and core samples from the Mount Kelly area, north-west Queensland. C.S.I.R.O. (written communication).

Journal of Geochemical Exploration, 8 (1977) 189—202

THE DISCOVERY OF THE LADY LORETTA ZINC-LEAD-SILVER DEPOSIT, NORTHWEST QUEENSLAND, AUSTRALIA — A GEOCHEMICAL EXPLORATION CASE HISTORY

R. COX and R. CURTIS

Placer Exploration Ltd., G.P.O. Box 4315, Sydney, N.S.W. (Australia)
Triako Mines N.L., York Street, Sydney, N.S.W. (Australia)

(Revised version accepted February 23, 1977)

ABSTRACT

Cox, R. and Curtis, R., 1977. The discovery of the Lady Loretta zinc-lead-silver deposit, northwest Queensland, Australia — a geochemical exploration case history. J. Geochem Explor., 8: 189—202.

Zn-Pb-Ag mineralization at Lady Loretta was indicated by a geochemical soil survey in 1968. It was confirmed in 1969 by diamond drilling beneath the peak of the Pb anomaly, which intersected 7.6 m of 21.2% Pb as anglesite and cerussite in the weathered zone. Subsequent drilling led to the delineation of a major deposit of Mount Isa-type Zn-Pb-Ag stratiform mineralization. By July 1973, drill-indicated mineral reserves totalled 8.69×10^6 tonnes of 18.1% Zn, 6.7% Pb, 110 g/t Ag (using a cut-off grade of 14% combined Zn + Pb), over a strike length of 720 m, a vertical depth of 460 m and a maximum true width of 41 m. The Lady Loretta deposit is the most significant recent discovery of Mount Isa-type Zn-Pb-Ag mineralization in Australia.

Stratiform Zn-Pb-Ag mineralization at Lady Loretta is localized within a sequence of dolomitic-sideritic and carbonaceous pyritic shales forming part of the Paradise Creek Formation (Carpentarian) of the Proterozoic belt of northwest Queensland. Ore grades are mainly confined to a distinct stratigraphic horizon (Ore Horizon) within a basin-shaped structure (Small Syncline). The primary sulphides consist essentially of massive, banded pyrite, sphalerite, galena and minor tetrahedrite, interbedded with laminae of pyritic carbonaceous shale, siltstone, chert, tuff and barite.

Oxidation and leaching extend generally to 100 m below surface and, adjacent to faults and other channelways, to 300 m below surface. There is a highly variable pattern of oxidation and leaching of the ore minerals but the Zn-Pb pyritic shales form the most resistant feature of the stratigraphic sequence, outcropping as a lateritized plateau surface 60 m above the surrounding plains.

Pb and Ag distributions in soils over mineralized and unmineralized sediments accurately reflect bedrock Pb and Ag distribution and delineate lithologic and ore boundaries. Zn distribution in soils reflects the high mobility of this element under the prevailing weathering conditions, with maximum concentration reached downslope from the Pb-Ag anomalies. The Zn-Pb pyritic shales at Lady Loretta are conspicuously low in Cu and the Cu content of soils is of a similarly low order. However, at the nearby Lady Annie copper mine, the distribution of Cu in soils accurately reflects the structually controlled Cu mineralization.

Geochemical surveys using soil and weathered bedrock samples are most effective, rapid and economical techniques for detecting outcropping stratiform Zn-Pb-Ag and structurally controlled Cu mineralization in this deeply weathered environment of northwest Queensland.

INTRODUCTION

The Lady Loretta Zn-Pb-Ag deposit is one of the major recent massive sulphide discoveries in Australia. It is located in northwest Queensland at latitude 19°46′S, longitude 139°03′S, 115 km north-northwest of Mount Isa and 2 km east of the old Lady Annie copper mine. The area consists of an eroded and dissected plateau with generally low relief (Fig. 1a), 300—400 m above sea level. The tropical climate is hot and arid, with a distinct summer wet season related to north and northwest monsoons. Evaporation is high, resulting in a low water table which promotes strong leaching. The thin stony soils support sparse vegetation, mainly spinifex grass (*Triodia* spp.), with isolated clumps of trees and shrubs (*Eucalyptus* and *Acacia* spp.) 3—8 m high.

Mining and exploration history

The Cu mineralization at the nearby Lady Annie copper mine was discovered prior to 1920, though because of economic considerations no attempt at mining was made until 1937. From 1937 to 1959, a number of small syndicates mined an estimated total of 25,000 tonnes of secondary ore averaging 6.43% Cu. In 1964, geological mapping, induced polarization and resistivity surveys, air rotary and diamond drilling were carried out at the Lady Annie Mine by Carpentaria Exploration Company Pty. Ltd. Five diamond drill holes totalling 446 m, and 85 air rotary holes totalling 3439 m outlined 913,000 tonnes of secondary ore averaging 2.33% Cu.

During 1966 and 1967 a Triako-Mitsui Mining & Smelting Co. Ltd. joint venture diamond-drilled 67 holes totalling 5085 m, and 87 air rotary holes totalling 4243 m.

In 1968, Placer Prospecting (Australia) Pty. Ltd. optioned the property and commenced detailed exploration. The programme was directed towards assessing the potential for increasing secondary Cu ore reserves at the Lady Annie Mine and locating primary Cu ore. As an initial step, 5085 m of core from 67 holes drilled during earlier exploration were relogged, in order to realize fully the information available, with the hope that some form of stratigraphic correlation could be established in the intensely drilled secondary orebody. Most of the holes terminated at or near the base of weathering. Concurrently with the relogging, detailed surface and underground mapping, reconnaissance geochemical soil sampling and induced polarization surveys were carried out.

In the regional reconnaissance geochemical survey, soil sampling was

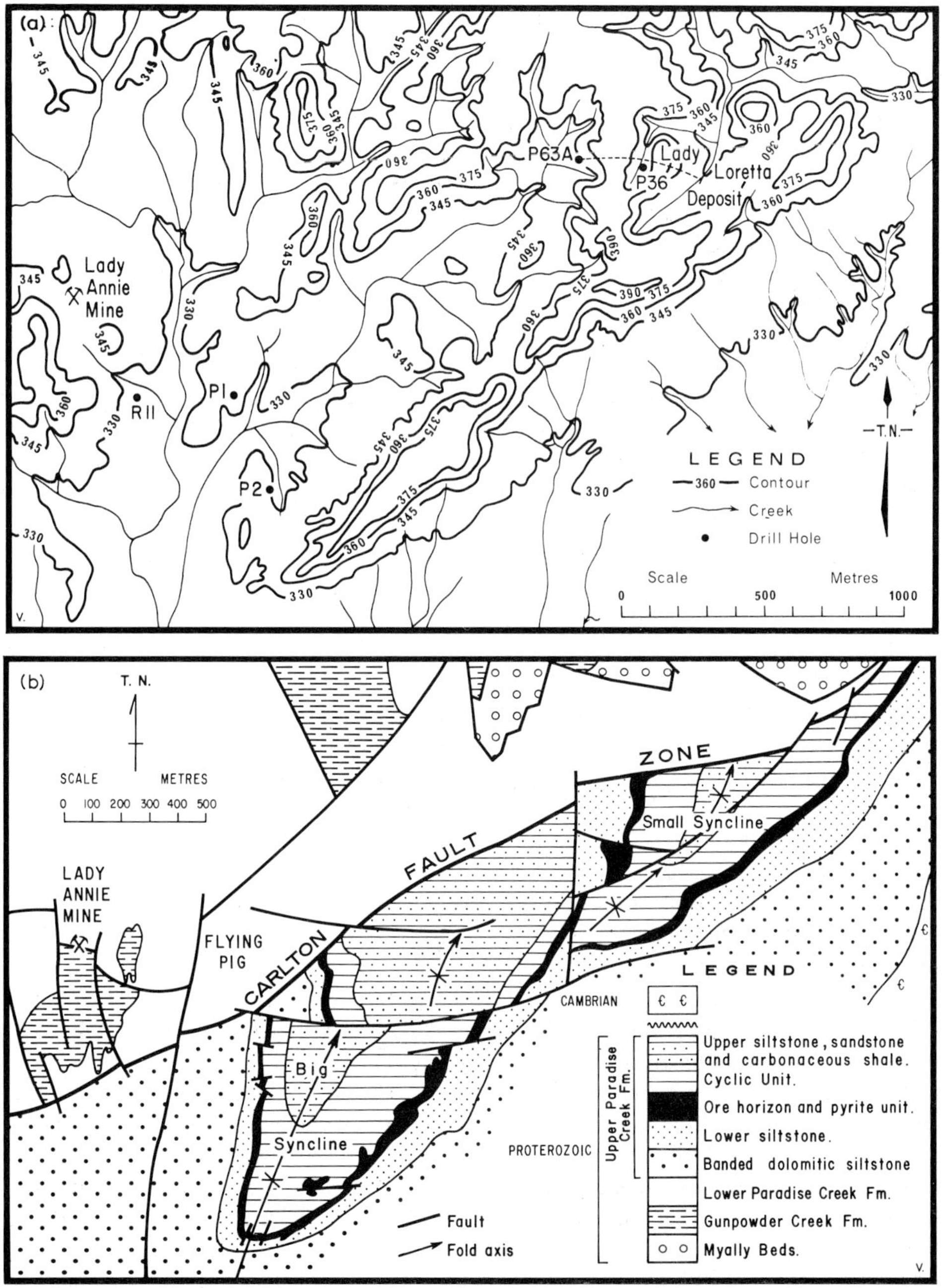

Fig. 1. Lady Loretta—Lady Annie area. (a) Topography plan; (b) geology plan.

undertaken on a 400-ft north-south by 100-ft east-west (122×30.5 m) grid established by tape and compass. The survey was centred on the Lady Annie Mine and an adjacent area of secondary Cu mineralization defined previously by Mitsui, known as the Flying Pig. Initially, the samples were assayed only for Cu but as the relogging of the core progressed, it became evident from the few drill holes with appreciable lengths in the unweathered zone, that the succession of predominantly dolomitic and carbonaceous shales were macroscopically identical to the host beds of the Mount Isa orebodies. Also, where disturbed by faulting or severe folding, the beds were cut by quartz-dolomite veins carrying chalcopyrite. This evidence, together with detailing of haematite shales by surface mapping, suggested strongly that the depositional environment of the Lady Annie succession was identical to that of Mount Isa (Curtis, 1969). Consequently, every tenth sample was assayed for Pb, Ag and Zn to reveal a major Pb anomaly, with associated high Ag and Zn (Fig. 2a). The areas in which Pb exceeds 421 ppm cover approximately 0.5 km^2 and are distributed in three main zones (*A*, *B* and *C* in Fig. 2a). Curtis (1969) noted that the southernmost group (*A*) occurs on the east and west scarp of the prominent ridge east of the Lady Annie Mine (Fig. 1a), and that the northern two anomalies (*B* and *C*) "appear to be controlled by a combination of structure and stratigraphy and are likely to enclose primary mineralization". These anomalies were subsequently found to outline the outcrops of the Lady Loretta Ore Horizon (Fig. 1b).

An induced polarization (I.P.) survey, covering approximately 10 km^2 centred on the Lady Annie Mine, ended before the full geochemical data were available and did not extend to the Small Syncline (Fig. 1b) containing the Lady Loretta deposit. However, I.P. anomalies were established along the eastern limb and part of the western limb of the Big Syncline (Fig. 1b), to the southwest of Lady Loretta (i.e. Small Syncline) and two diamond drill holes were sited to test these anomalies. Diamond drill hole (DDH) P1 (Fig. 1a) cut thinly bedded dolomitic and carbonaceous shales while in DDH P2 beds of fine-grained pyrite were prominent in the succession, establishing a further analogy between Mount Isa and the Lady Annie—Lady Loretta area.

During 1969, exploration drilling for Cu mineralization (Lewis, 1970) at the Lady Annie Mine continued to have highest priority. In December 1969, a diamond drill hole sited on the peak of the Pb anomaly (*B* in Fig. 2a) confirmed the presence of mineralization; hole P36 (Fig. 1a) intersected 7.6 m of 21.2% Pb as anglesite and cerussite in the weathered zone. Subsequent drilling led to the intersection of significant stratiform, massive sulphide, Zn-Pb-Ag mineralization in December 1970; hole P63A (Fig. 1a) intersected 13.6% Zn, 8.9% Pb and 100 g/t Ag over 45 m.

During 1971—1973 exploratory drilling delineated a major deposit of Mount Isa-type mineralization. By July 1973 published drill-indicated mineral reserves totalled 8.69×10^6 tonnes with an average grade of 18.1% Zn, 6.7% Pb and 110 g/t Ag, using a cut-off grade of 14% combined Pb + Zn

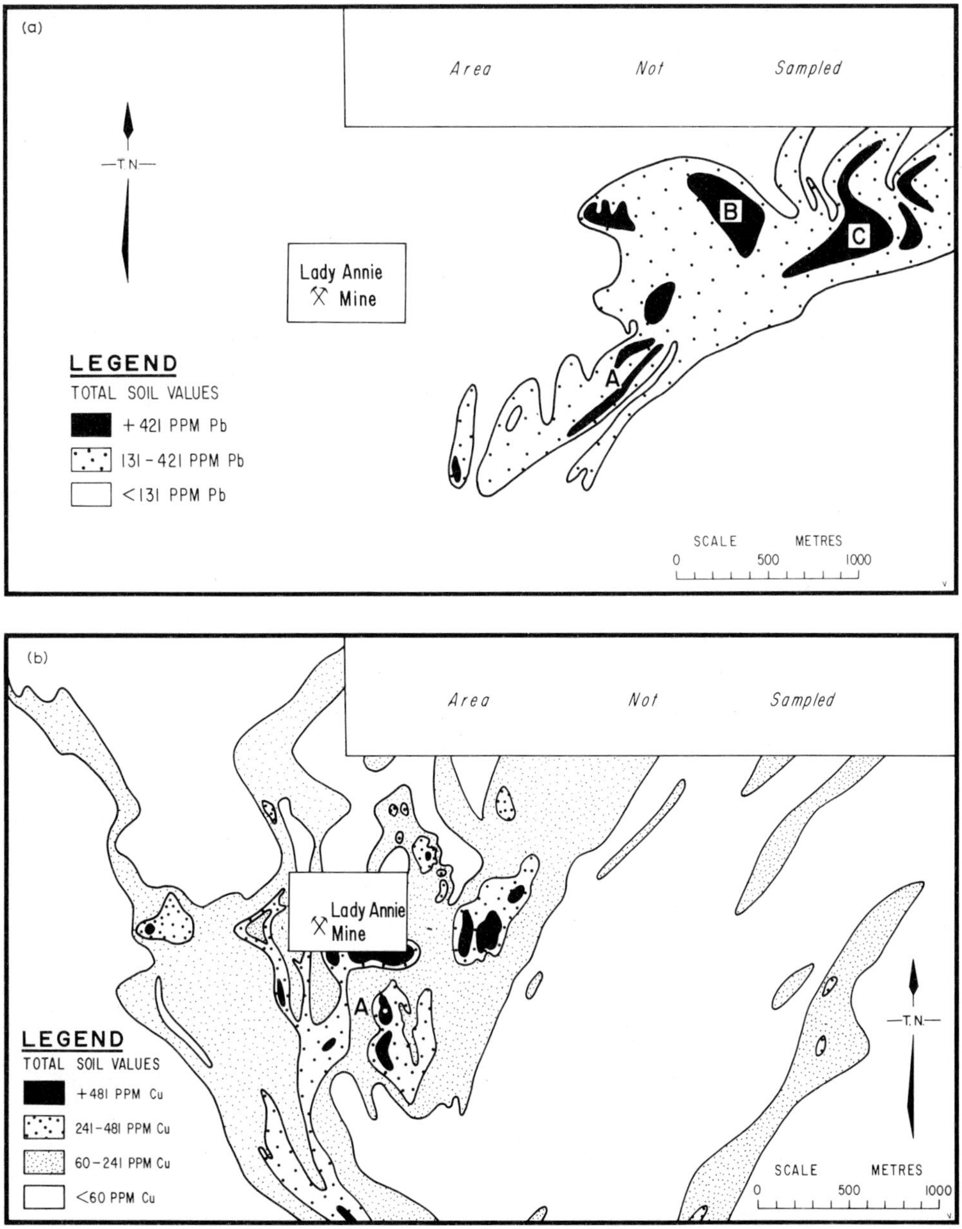

Fig. 2. Regional reconnaissance soil survey plan (1968). (a) Pb; (b) Cu.

(Rennie et al., 1973). These reserves, located in a synclinal structure, have a strike length of 720 m, a vertical depth of 460 m and a maximum true width of 41 m.

In March 1975 Mount Isa Mines Ltd. purchased the assets of Placer Prospecting (Australia) Pty. Ltd. which included an option to earn a 64% interest in the Lady Loretta—Lady Annie property. Triako holds the remaining 36% as a carried interest.

DESCRIPTION OF AREA

Geology

The geology of the Lady Loretta deposit has been discussed in two complementary papers by Alcock and Lee (1974b) and Loudon et al. (1975). The secondary Cu deposits at Lady Annie have been described by Lewis (1975). Only a brief summary of the pertinent geological features of the area are given here, sufficient to facilitate a meaningful appreciation of the results of the geochemical soil surveys.

The Lady Loretta deposit occurs near the top of the Paradise Creek Formation (Carr, 1971; Alcock et al., 1975) within the western basin of the Carpentarian orogenic belt of northwest Queensland (De Keyser, 1958; Carter et al., 1961).

The Paradise Creek Formation here conformably overlies the Gunpowder Creek Formation (which overlies the Myally Beds) and is unconformably overlain by Middle Cambrian sediments. Subdivisions of the Upper Paradise Creek unit at the deposit are shown in Table I.

The detailed geology of the Lady Loretta—Lady Annie area is shown in Fig. 1b. The Lady Loretta mineralization occurs in a major northeast-trending synclinal structure folded and faulted into two domains, referred to as the Big and Small Synclines. The northwestern ends of these synclines are truncated by the sub-vertical Carlton Fault zone. Ore-grade mineralization and published drill-indicated mineral reserves are almost entirely restricted to the Small Syncline, which has a basinal form. The results of a detailed structural analysis are discussed by Lee (1972).

TABLE I

Subdivision of Upper Paradise Creek Formation at Lady Loretta deposit, Queensland (after Loudon et al., 1975)

	Thickness (m)
Upper Siltstone Unit	+ 60
Sandstone Unit	60
Upper Carbonaceous Shale Unit	120
Cyclic Unit of Siltstones and Shales	110
Lower Carbonaceous Shale Unit	5
Ore Horizon	25
Pyritic Unit	35
Lower Siltstone Unit	+ 90

Mineralization

The stratiform, massive sulphide Zn-Pb-Ag mineralization at Lady Loretta reaches ore grade over mineable widths in only one stratigraphic horizon, referred to as the Ore Horizon. The primary sulphide minerals are pyrite, sphalerite and galena with minor chalcopyrite and tetrahedrite. The stratiform sulphides are locally remobilized in tension gashes, fold hinges and zones of structural complexity. Metal values and widths increase down the limbs of the basin structure and reach a maximum on the west limb and in the keel of the Small Syncline. The gangue comprises finely laminated dolomitic siltstone, carbonaceous shale, bedded pyrite and irregularly distributed barite. The mineralization is usually extremely fine-grained, except where it has been locally remobilized.

The underlying Pyrite Unit consists of massive, fine-grained pyrite interbedded with minor carbonaceous shale and siltstone. Intraformational slump structures are often developed. Irregular lenses of sub-economic Zn-Pb-Ag mineralization, barite and chert occur throughout the unit.

The surface expression of the massive sulphide mineralization within both the Ore Horizon and the underlying Pyrite Unit is an elongate ridge (Fig. 1a and b) capped by Tertiary laterite remnants and siliceous, limonitic and haematitic gossan (with lenses of barite and chert). Although the gossanous zone is up to 60 m wide, anomalous rock Pb-Ag values are restricted to the outcrop of the Ore Horizon (which averages 25 m wide); the gossan related to the Pyrite Unit shows generally low Pb-Ag values.

Lateritization

Lateritized plateau surfaces occur at 385—395 m above sea level in the Lady Loretta area (Fig. 1a). They mark remnants of Mesozoic peneplains which underwent deep weathering and lateritization during the Tertiary period. The laterite is commonly preserved as cappings on small mesas and ridges.

Oxidation and leaching

Drilling at Lady Loretta (and Lady Annie) has revealed that all sediments have been oxidized to depths of approximately 100 m. Carbonate is commonly leached to greater depths. Adjacent to major faults and shear zones, however, oxidation has been recorded below 300 m.

Supergene alteration of the ore minerals is highly variable. Galena oxidizes to anglesite with minor cerussite below a few metres from the surface. These oxidation products have been intersected in excess of 100 m below surface at some locations. In a vertical profile within the Ore Horizon, Pb values typically increase from approximately 1% Pb at surface (gossan) to 20% Pb in the oxide zone then decrease to 7% Pb in the primary sulphide zone. Zn is completely leached to below 100 m; no oxidized Zn minerals have been recorded.

Soils

The typical soils of the area are lithosols and solodic soils (Stace et al., 1968). The lithosols are shallow (0—15 cm), sandy and/or loamy soils with a large proportion of weathered rock fragments, characterized by stoniness and a lack of pedological differentiation (other than A_1 due to organic matter accumulation). Lithosols are characteristically developed over areas of strong topographic relief (Fig. 1a), particularly on hill crests and the steeper slopes, where natural erosion has been active enough to maintain a thin soil cover. Certain rocks more resistant to weathering (e.g. Myally Quartzite) give rise to lithosols under gentler topographic conditions.

Downslope in areas of gentler topographic relief, lithosols grade into deeper (5—25 cm) solodic soils. These latter are characterized by a greater profile development; a well-developed, bleached A horizon (typically loose, coarse sand to massive clay loams with low organic content) overlies, with a very abrupt, wavy boundary, a mottled grey-yellow brown B horizon (typically dense, hard, impermeable sandy clay to medium clay), which grades into a strongly alkaline, moderately saline C horizon (typically softer, less dense, with a lower clay content and carbonate nodules, and grey to yellow brown in colour). These characteristic properties are developed by strong leaching, with associated hydrolysis of the clay, in the A and B horizons.

GEOCHEMICAL SURVEYS

Analytical procedures

Soil fraction analysis, by sieving through 12-, 22-, 44- and 80-mesh screens showed that 40—50% (up to a maximum of 80%) of each sample was plus 12-mesh and that the coarser fractions contained proportionately more of the Pb and Ag, usually in rock and gossan fragments; hence routine sieving of soil samples was not carried out (Arthur, 1972).

Each sample was crushed to minus 200-mesh. Analytical procedures consisted of: (1) digesting in perchloric acid at 165° C for 8 hours and analyzing by atomic absorption spectrometry (AAS) for Pb, Zn and Cu, and (2) digesting in nitric acid with Hg ions at 135° C for 2 hours, adding an organic extractant (tri-iso-octyl phosphorothioates) and analyzing by AAS for Ag.

Regional reconnaissance soil survey, 1968

Commencing in April 1968, a regional geochemical soil survey was undertaken over the Lady Annie mineral claims, but excluding the immediate vicinity of the old mine where soils were contaminated due to proximity of dumps. A total of 3093 soil samples were taken on a 122-m × 30.5-m grid. The thin soils, rarely more than 10 cm deep, were sampled immediately below the surface.

The results of this programme (Curtis, 1969) defined anomalous Pb (+421 ppm Pb) and Ag (+2 ppm Ag) values over a northeast-trending strike length of 1 km (Fig. 2a). Anomalous Zn values (+241 ppm Zn) were associated only with the highest Pb-Ag values. Background Cu values (—60 ppm) only were associated with the Pb-Ag-Zn anomaly; anomalous Cu values (+481 ppm) were confined to the general vicinity of the Lady Annie copper mine but were nevertheless successful in delineating new, untested Cu mineralized zones within this area; for example, anomaly *A* (Fig. 2b) was tested by drill hole R11 (Fig. 1a) which intersected 0.64% Cu over 15 m.

This regional geochemical soil survey was successful in defining:

(1) A major Pb-Ag anomaly with associated Zn values, the testing of which led to the discovery of the Lady Loretta stratiform massive sulphide, Zn-Pb-Ag deposit.

(2) Cu anomalies adjacent to the old Lady Annie Mine, the testing of which led to the intersection of secondary Cu mineralization additional to the then known Lady Annie reserves.

To illustrate these geochemical case histories, the regional soil survey Pb and Cu values are shown in Fig. 2.

Detailed soil surveys, 1971—1974

Following the discovery intersection (hole P63A, Fig. 1a) of stratiform, massive sulphide Zn-Pb-Ag mineralization in the Small Syncline in December 1970, a programme of detailed geochemical soil sampling was carried out in conjunction with detailed mapping of the Small Syncline. During 1971, a total of 362 soil samples were collected from an area covering 0.3 km^2 of the Small Syncline; sampling was carried out at 15-m intervals along lines 60 m apart running at 140° True (Graylin et al., 1972).

Similarly, during 1973—1974, a programme of detailed geochemical soil sampling was carried out in conjunction with detailed mapping of the Big Syncline (Alcock and Lee, 1974a). A total of 1000 soil samples were collected from an area covering 1.8 km^2. Samples were taken at 30-m intervals along lines 60 m apart running at 140° True. The lines were extended northwest of the Carlton Fault to include some of the Cu anomalous area not previously sampled in detail.

The samples from both of these detailed soil surveys were analyzed for Pb, Ag, Zn and Cu. The combined results from both of these surveys and the 1968 regional survey are presented as contour plans of Pb (Fig. 3a), Ag (Fig. 3b), Zn (Fig. 4a) and Cu (Fig. 4b) distribution in soil (total sample) covering the Lady Loretta/Lady Annie areas. The contour intervals are based on log

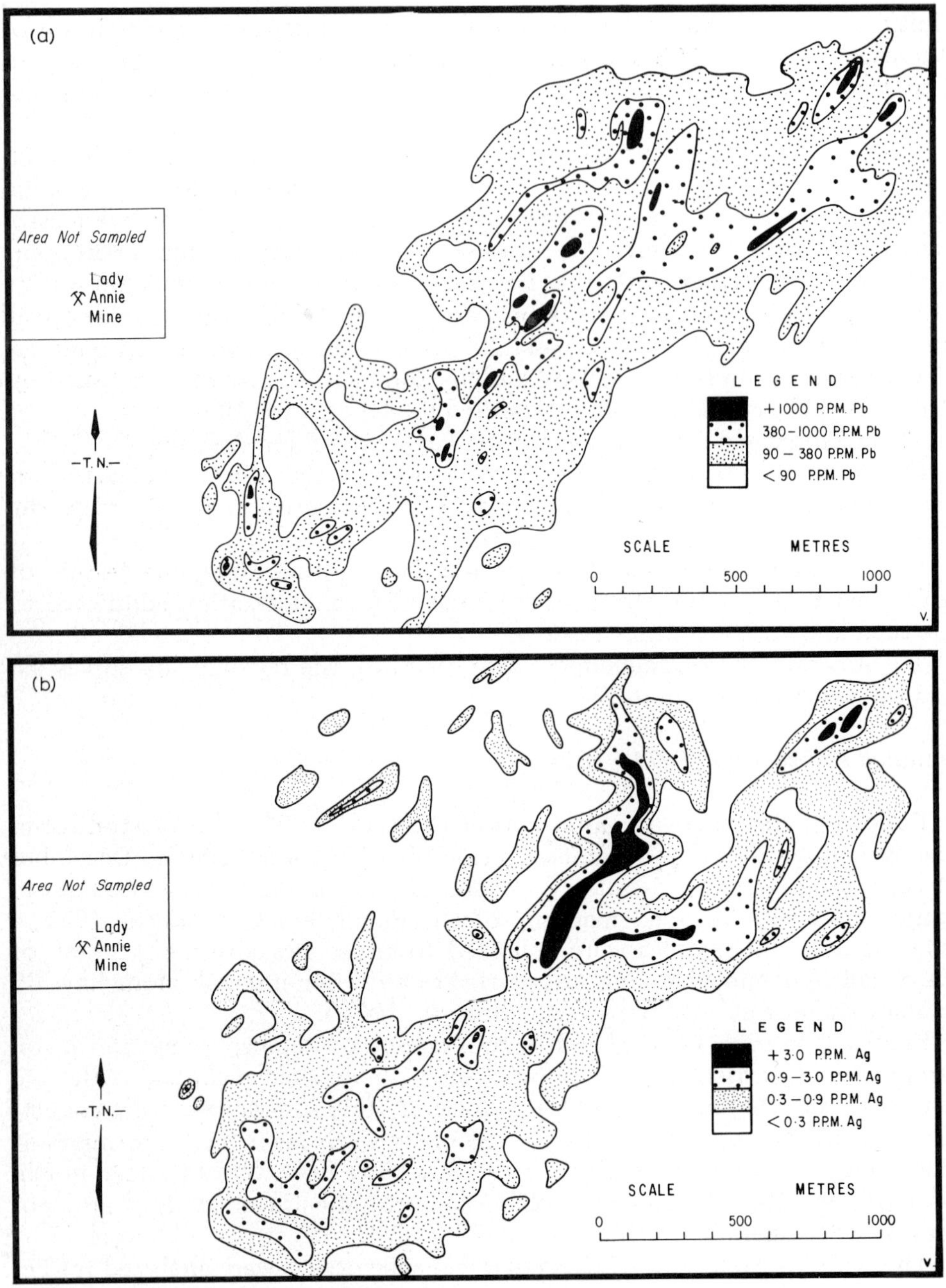

Fig. 3. Lady Loretta detailed soil survey plan. (a) Pb; (b) Ag.

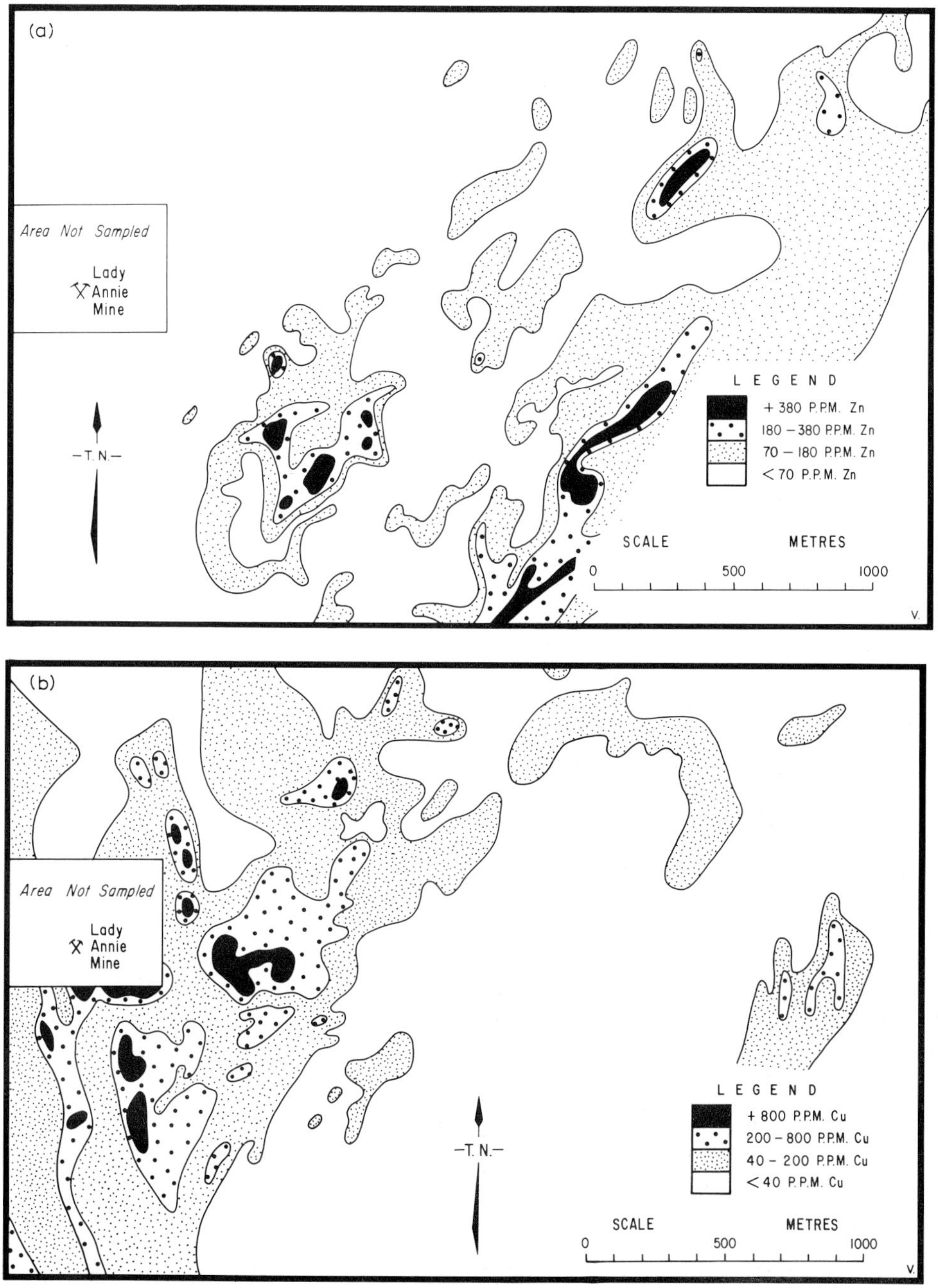

Fig. 4. Lady Loretta detailed soil survey plan. (a) Zn; (b) Cu.

scale cumulative frequency curves for the soil sample results from the Big Syncline (Pb-Ag-Zn domain) and Carlton Fault/Lady Annie (Cu domain) areas (Alcock and Lee, 1974a). Threshold and anomalous values for Pb, Ag and Zn within the Big Syncline are shown in Table II.

TABLE II

Threshold and anomalous values for Pb, Ag and Zn, in soils from Big Syncline, Lady Loretta area

Element	Threshold value (ppm)	Anomalous value (ppm)
Pb	60	+ 380
Ag	0.24	+ 0.9
Zn	35	+ 180

Cu is not anomalous in the Lady Loretta area.

The Lady Annie and Lady Loretta areas are located in two distinctly different geochemical domains. The distribution of Pb, Ag, Zn and Cu in the soils of the two domains, and their relationship to topography (Fig. 1a) and geology (Fig. 1b) may be summarized as follows:

Lead. The areal distribution of Pb (Fig. 3a) is related to the Ore Horizon and associated limonitic (after pyrite) sediments above and below this unit (Fig. 1b) and to the topographic highs associated with this stratigraphic unit (Fig. 1a). There has been little lateral dispersion of Pb in the soil although there is evidence from drilling of considerable movement vertically down the profile (i.e. in bedrock) within the Ore horizon.

Silver. The areal distribution of Ag (Fig. 3b) is closely related to that of Pb and to the Ore Horizon and associated pyritic beds (Fig. 1b). Ag shows slightly more lateral dispersion than Pb. This is best illustrated on the eastern limb of the Big Syncline where two Ag anomalies occur on the eastern and western flanks downslope from the Pb anomaly and Ore Horizon outcrop.

Zinc. The areal distribution of Zn (Fig. 4a) reaches maximum concentration downslope from the Pb-Ag anomalies and Ore Horizon outcrop. The distribution of Zn is erratic and it cannot be reliably used as a direct indicator of Zn-Pb-Ag mineralization in this environment because of its high mobility. Zn also shows considerable movement vertically down profile (i.e. in bedrock) within the Ore Horizon, being leached in excess of 100 m below surface.

Copper. The areal distribution of Cu within the Pb-Zn-Ag domain of the Lady Loretta area, southeast of the Carlton Fault zone (Fig. 1b and 4b), shows no anomalous values. There is a sharp change in values, however, to the northwest of the Carlton Fault zone. Cu values broadly indicate secondary Cu mineralization within stratigraphic units and/or fault structures in the Lady Annie region (Cu domain).

CONCLUSIONS

The stratiform Zn-Pb-Ag mineralization at Lady Loretta gives rise to well defined Pb and Ag anomalies in the skeletal soil overlying the Ore Horzion.

Geochemical soil sampling was responsible for delineating the sub-outcropping massive sulphide mineralization and led directly to the discovery of the deposit.

A new look at old data (in this case drill core) can often be worthwhile.

ACKNOWLEDGEMENTS

This paper briefly presents some of the geochemical results obtained at Lady Loretta during an extensive exploration and evaluation programme carried out by numerous Placer personnel.

The authors gratefully acknowledge the permission to publish given by Placer Exploration Limited, Triako Mines N.L. and Mount Isa Mines Limited. In addition they wish to thank Mr. R.J. Voss for drafting the illustrations, Miss E. Fry for typing the manuscript, and their colleagues for reviewing the original draft.

REFERENCES

Alcock, P.J. and Lee, M.F., 1974a. Progress report on the geology and geochemistry of the Big Syncline, Lady Loretta. Placer Prospecting (Australia) Pty. Ltd., Rep. No. 12/74 (unpublished).

Alcock, P.J. and Lee, M.F., 1974b. Aspects of the geology and exploration of the Lady Loretta lead-zinc-silver deposit, Northwest Queensland. Australas. Inst. Min. Metall., N.W. Qld. Branch, Regional Meet., August, 1974, pp. 207—215.

Alcock, P.J., Graylin, R.K., Dowling, J.F. and Cox, R., 1975. Progress report on exploration of Authority to Prospect No. 1171M (N.W. Qld.) as at 12th March, 1975. Placer Exploration Ltd., Rep. No. 11/75 (unpublished).

Arthur, E.W., 1972. Lady Annie soil sizing tests. Placer Prospecting (Australia) Pty. Ltd., Rep. dated 9/4/72 (unpublished).

Carr, M.J., 1971. Geological notes and recommendations on Authority to Prospect No. 963 M. Placer Prospecting (Australia) Pty. Ltd., Rep. No. 38/71 (unpublished).

Carter, E.K., Brooks, J.H. and Walker, K.R., 1961. The Precambrian mineral belt of northwestern Queensland. Bull., Bur. Miner. Resour., Geol. Geophys., Aust., 51.

Curtis, R., 1969. Report on Lady Annie Mine investigations for the period April 1968 to March 1969. Placer Prospecting (Australia) Pty. Ltd., Rep. No. 29A/69 (unpublished).

De Keyser, F., 1958. Geology and mineral deposits in the Paradise Creek area, northwest Queensland. Rec., Bur. Miner. Resour., Geol. Geophys., Aust., 100.

Graylin, R.K., Alcock, P.J., Lee, M.F., Lee, M.K., Storey, L.O., Challis, A. and Newton, A.W., 1972. Supplementary geological report and revised mineral inventory, Lady Annie, North Queensland for period May 1970—December 1971. Placer Prospecting (Australia) Pty. Ltd., Rep. No. 1/72 (unpublished).

Lee, M.F., 1972. A fold study of the Lady Loretta area, North West Queensland. B. Sc. (Hons) Thesis, University of Adelaide (unpublished).

Lewis, R.W., 1970. Geology report and revised geological mineral inventory as of May 1970, Lady Annie Mine, North Queensland. Placer Prospecting (Australia) Pty. Ltd., Rep. No. 33A/70 (unpublished).

Lewis, R.W., 1975. Lady Annie secondary copper deposit. In: C.L. Knight (Editor), Economic Geology of Australia and Papua New Guinea, 1. Metals. Australasian Institute of Mining and Metallurgy, Melbourne, Vic., pp. 1023—1033.

Loudon, A.G., Lee, M.K., Dowling, J.F. and Bourn, R., 1975. Lady Loretta silver-lead-zinc deposit. In: C.L. Knight (Editor), Economic Geology of Australia and Papua New Guinea, 1. Metals. Australasian Institute of Mining and Metallurgy, Melbourne, Vic., pp. 377—382.

Rennie, C.C., Loudon, A.G., Lee, M.K., Dowling, J.F., Lee, M.F. and Kowalczyk, P., 1973. Lady Loretta geological summary report and ore reserves. Placer Prospecting (Australia) Pty. Ltd., Rep. No. 10/73 (unpublished).

Stace, H.C.T., Stubble, G.D., Brewer, R., Northcote, K.H., Sleeman, J.R., Mulcahy, M.J. and Hallsworth, G.G., 1968. A Handbook of Australian Soils. Rellim Technical Publications, Glenside, S.A., 435 pp.

Journal of Geochemical Exploration, 8 (1977) 203—217

EXPLORATION GEOCHEMISTRY OF THE STUART SHELF, SOUTH AUSTRALIA

J.H. RATTIGAN, R.W. GERSTELING and D.G. TONKIN

Pacminex Pty. Limited, CSR Minerals and Chemicals Division, Sydney, N.S.W. (Australia)

(Revised version accepted February 23, 1977)

ABSTRACT

Rattigan, J.H., Gersteling, R.W. and Tonkin, D.G., 1977. Exploration geochemistry of the Stuart Shelf, South Australia. J. Geochem. Explor., 8: 203—217.

Exploration for base metals over 8000 km^2 of a stable tectonic block, with a sub-horizontal Precambrian sedimentary sequence, has involved sampling stream sediments, surface rocks, lake muds, drill cuttings, core, groundwaters and plants. The land surface is undulating with elevated country separated by soil plains and playas. Fixed dunes and other superficial strata mantle bedrock. Cu ore is mined from the region. Recent exploration led to the discovery of the hidden Cattle Grid sulphide orebody located at a quartzite-sandstone interface more than 25 m below the surface. Other mineralization is found in clastic and carbonate rocks in different stratigraphic situations.

Sampling of stream sediments and surface rocks was of limited use because of limited outcrop, particularly over stratigraphic zones of exploration interest. Sampling of lake muds revealed geochemical contrast, but anomalies drilled have only shown pyrite. Regular dispersion of anomalous Cu in groundwaters about the Cattle Grid body has not been firmly demonstrated. The most effective geochemical technique for detection of primary and secondary metal dispersions is profile analysis of drill samples. Shallow drill sampling and biogeochemistry have application in specific geological situations.

INTRODUCTION

Exploration geochemistry over the Stuart Shelf in South Australia has been carried out by several companies for more than a decade. The terrain has little outcrop, restricted drainage, leached sand cover and high salinity creating unusual exploration problems. Geochemical techniques adapted to the terrain, following orientation work over concealed mineralized bodies, have proved of value in the continuing search for orebodies.

This paper concerns the recent work by CSR Limited, tenement holders on the shelf, and operator through Mt. Gunson Mines Pty. Ltd., of a mining venture on the Cattle Grid orebody. The area tested involved 8000 km^2

located south and north of Mt. Gunson (Fig. 1). The purpose has been to detect dispersion haloes that might lead to concealed orebodies and this is the aspect discussed. Considerable data that are accumulating on rock and water geochemistry will assist in studies on ore genesis but this aspect will not be covered in this publication.

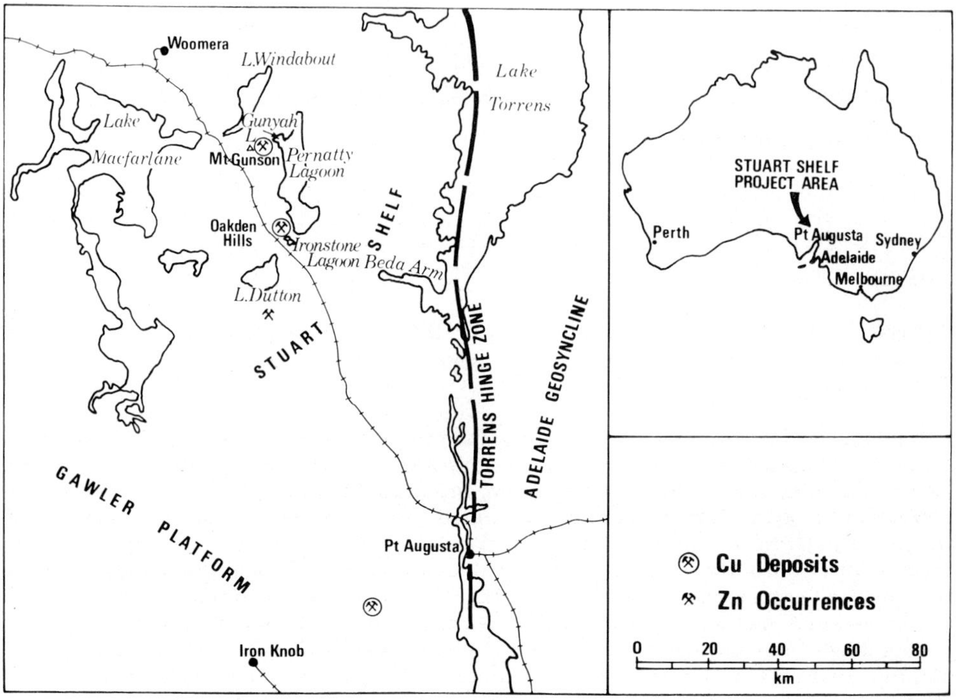

Fig. 1. Locality map, Stuart Shelf, South Australia.

DESCRIPTION OF THE AREA

The area, in central South Australia, is part of an extensive partly dissected plateau west of Lake Torrens. Rainfall is low with the annual range being 110—200 mm. Drainage is internal and streams ephemeral. The few drainage channels have short courses draining into saline lakes. Evaporation rate is high (to 2500 mm per annum), particularly in summer.

The landforms (Johns, 1968) derive from an extensive tableland, now dissecting, with prominent residuals preserved as mesas and buttes as high as 260 m at Mt. Gunson, and higher in areas to the south. Elevated country has some outcrop but is commonly strewn with lag gravels or covered with fixed, red dunes with relief to 10 m. Lower areas consist of sand and loam plains and large salt lakes including Pernatty Lagoon, Lake Windabout, Ironstone Lagoon, Lake MacFarlane and Lake Dutton. The lakes are mud-filled and encrusted with halite. Standing-water distribution varies with rainfall.

Tablelands have shrub steppe vegetation dominated by saltbush (*Atriplex vesicaria*). Lowland plains have savannah woodland of myall (*Acacia sowdenii*) with bluebush (*Kochia astrostricha*) or shrub steppe of bluebush and saltbush. Fixed dunes have wattle (*Acacia ligulata*), mulga (*A. aneura* and *A. Linophylla*) and native cypress pine (*Callitris collumelaris*).

GEOLOGY

The Stuart Shelf. This tectonic unit occupies the eastern margin of the Gawler Platform (Parkin, 1969). The sub-horizontal, relatively thin, Proterozoic (Adelaidean) sediments have remained relatively undisturbed since deposition.

Despite the stability of the shelf, the sequence is marked by lithofacies variation and disconformities created by transgression and regression of seas over surfaces which had erosional or minor tectonic irregularities. Several hiatuses are inferred (Thomson, 1976). One major structural uplift is the Pernatty Culmination of Johns (1974) near Mt. Gunson.

The Adelaidean sequence. The sequence (Table I) is believed to represent Adelaidean time from 1400 to 560 m.y. Mapping of lag gravel and limited outcrop areas was partly successful in identifying the bedrock geology but

TABLE I

Stratigraphic sequence, Stuart Shelf

Cainozoic-Tertiary	dune sands, redistributed sands, alluvial and colluvial sandy and loamy soils, calcrete, silcrete and ferricrete, gypseous and halitic muds	
Adelaidean	Tent Hill Formation	Simmens Quartzite Corraberra Sandstone Tregolana Shale Whyalla Sandstone
	Yudnapinna Beds	
	Tapley Hill Formation	(Woocalla Dolomite Member)
	McLeay Beds	
	Pandurra Formation	
	Roopena Volcanics	
Carpentarian	Gawler Volcanics	

knowledge of the sequence and the distribution of units has been greatly increased by exploration drilling. The Adelaidean sequence described by Parkin (1969) and Thomson (1976) rests on the acid Gawler Range Volcanics. In the area studied, the overlying basal Adelaidean units are mafic lavas of the Roopena Volcanics dated near 1340 m.y. The succeeding sediments include the Pandurra Formation, a widely developed sandstone, conglomeratic in part, with thin red shale members. The formation is silicified near the surface and its top is in places marked by red, haematitic breccia believed to relate to weathering effects on an old land surface. The Pandurra Formation is intruded by dolerites and is disconformably transgressed by younger Adelaidean units. The McLeay Beds are localized deposits of possibly glacigene shales and sandstones. The Tapley Hill Formation, including the Woocalla Dolomite Member, consists of massive dolomite, laminated dololutites and black shales. The Yudnapinna Beds are carbonate-bearing siltstones.

The Tent Hill Formation, the youngest Adelaidean of the region, is represented by the Whyalla Sandstone, Tregolana Shale (red lutites), Corraberra Sandstone and Simmens Quartzite.

Superficial sediments. The Adelaidean succession is mantled by fixed, red dunes and sands and loams with thicknesses as much as 19 m. Irregular silcrete and calcrete zones are developed as a capping. Lake sediments are gypseous muds encrusted with halite. Alluvial sands, gravels and silts occur along the short streams.

MINERALIZATION

Mineral occurrences in the region are discussed by Gersteling and Heape (1975) and Johns (1968). A feature of the region is the occurrence of minor mineralization in many units of the sequence including the Pandurra Formation, McLeay Beds, Tapley Hill Formation, Whyalla Sandstone and Tregolana Shale. Generally, the anomalous zones are small and not of economic interest. The only ore currently mined is from the Cattle Grid deposit located at the Pandurra Formation/Whyalla Sandstone interface west of Pernatty Lagoon. This is flat-lying, strata-controlled but epigenetic ore lying beneath the present water table and beneath 15—35 m of Whyalla Sandstone and 10 m of Quaternary sand (Van Herk et al., 1975). The dominantly chalcocite-bornite sulphide ore also carried significant Zn, Pb, Co and Bi. Other smaller, similar shallow deposits were formerly mined near and on Gunyah Lake. Above the water table, copper sulphides are oxidized and oxidized mineralization is not considered as ore. A second style of mineralization, never worked to any significant degree, occurs in shales of the Tapley Hill Formation. Cu (Gersteling and Heape, 1975), Pb and Zn (Johns, 1974), Mn and Ba are known from this unit.

METHODS AND MATERIALS STUDIED

Analytical methods

Inorganic solid materials were routinely analyzed by atomic absorption spectrometry (AAS) for Cu, Pb and Zn following hot concentrated perchloric acid leach for one hour on 0.25 g of sample. In orientation work, designed to determine if any pathfinder elements might usefully be used in exploration, AAS analyses for Fe, Co, Ni, Bi, Mn and V were also made. The prepared samples were analyzed in commercial laboratories to avoid contamination, possibly significant at the trace level, in a Mt. Gunson Mine laboratory used for assaying ore and mill products. In work prior to 1970, some semi-quantitative analyses by field colorimetric methods using biquinoline were used on stream sediment samples, but these methods were abandoned when a larger regional programme was initiated. Bore hole waters were tested for Cu, Pb and Zn by AAS following solvent extraction of buffered waters, whilst total dissolved solids (TDS) was determined by gravimetric methods.

Plant samples were washed thoroughly in demineralized water, ashed and the ash analyzed. The plants were tested as possible accumulators of Cu and for pathfinder elements. Eighteen samples were selected for emission spectrographic analysis of a wide range of elements. All subsequent samples were analyzed for Cu, Zn, Mn and Fe by AAS after hot, concentrated perchloric acid leach for one hour on 0.25-g samples. W and Mo were determined, following potassium pyrosulphate fusion of 0.2-g samples, by colorimetric methods using dithiol.

Stream sediment sampling

Few areas of the Stuart Shelf lend themselves to this method. Over the last decade, standard methods of sampling were used on short streams draining elevated areas underlain by the Pandurra Formation, Tapley Hill Formation (Woocalla Dolomite Member) and Whyalla Sandstone. The earliest known surveys were commenced in 1966 near Mt. Gunson and samples were analyzed by field biquinoline methods.

In 1971, orientation about shallow Cu mineralization established that the minus 60-mesh BSS fraction yielded the highest metal values and strongest contrast. Orientation showed also that the optimum time for sample collection was shortly after rainfall, when windborne material had been flushed along channels and higher absolute metal values and contrast were found in channel sediment compared with samples taken in dry periods. This study suggested that Cu may relate to the release in solution of metal from the normally alkaline source sediment by acid rainwaters and sorption of ions on fine sediment particles newly deposited from run-off. At periods between rain in this arid region, windborne material is not always clearly distinguishable from waterborne material and may have a diluent effect in samples of sediment of mixed provenance taken from stream channels.

This study suggested that Cu and the pathfinder Zn, were the most likely indicator elements for Cu bodies. Surveys carried out on the basis of this orientation were largely confined to a small area (200 km^2) over the Pernatty Culmination, where short streams are actively eroding silicified Pandurra Formation. One programme covered streams draining the Whyalla Sandstone, west of Beda Arm.

Rock chip sampling

Windblown sands and playas cover most of the stratigraphic zones of exploration interest. Outcrop sampling was consequently restricted. In 1967, 126 samples were collected by previous titleholders from two areas of outcropping Whyalla Sandstone. Each consisted of a composite of chips taken from areas up to 23 m^2 about stations.

Since 1971, rock sampling has been carried out in other areas. The Pandurra Formation was tested only in areas with induced polarization and electromagnetic geophysical anomalies. Elsewhere sampling was restricted by availability of outcrops which were mainly Whyalla Sandstone and silcrete cappings. The material was routinely sampled, crushed, split and analyzed by AAS methods for Cu, Pb and Zn.

Salt lake mud sampling

Extensive areas are covered by salt lakes. Because Cu deposits are located on the margins of Gunyah Lake and Ironstone Lagoon, exploration of such water-logged, mud-covered areas received priority. It was expected that sulphide deposits, if any, would be preserved unoxidized at shallow depths under lakes. Systematic prospecting by heavy drill rigs or geophysical methods was hampered by this terrain and the first indication of one shallow deposit, East Lagoon, was malachite-stained mud in a trench.

The most effective systematic method proved to be hand augering to mud base and analysis of samples from bottom mud and from weathered bedrock. In trial auger sampling, assays for Cu, Pb, Zn, Ni, Co, Mo and V were made. Later, in systematic sampling, analyses were made only for Cu, Pb and Zn.

Sampling was carried out over Gunyah Lake, Pernatty Lagoon, Ironstone Lagoon and other lakes. An orientation survey was made over mineralization covered by Tregolana Shale in Gunyah Lake. Another area of 3.5 km^2 on the north fringe of Gunyah Lake, where muds were undisturbed and not contaminated by mining, was selected for detailed sampling on a 100-m $\times$ 100-m grid. The area was underlain by Pandurra Formation, Whyalla Sandstone and Tregolana Shale. Sampling over Ironstone Lagoon was carried out on a 100-m × 200-m grid over an area of some 3 km^2. Bedrock was Pandurra Formation with some outliers of the Woocalla Dolomite Member with copper-bearing basal shales, perched along the northern shoreline. Pernatty Lagoon, 200 km^2 in area, was initially sampled at 200-m intervals along lines spaced at 2.5 km. This revealed only one area of interest underlain by dolomite which was subsequently sampled at 100-m × 100-m spacings.

Deep profile lithological sampling

Vertical drilling was required to investigate the stratigraphy and geochemistry of the flat-lying sedimentary strata. 8000 km^2 of the Stuart Shelf has been probed by deep stratigraphic drilling, at first on a broad irregular pattern determined by the distribution of target stratigraphic units inferred from geological mapping and airborne electromagnetic and magnetic geophysical surveys. Subsequently when zones of interest were confirmed a closer spacing was adopted commonly at 1-km centres. The rock profile penetrated in deep drilling was systematically sampled, from the drill collar downwards, once contamination from unconsolidated superficial strata had been minimized by casing seals. Sampling was carried out normally over 3-m intercepts but also over lesser intercepts dictated by geological boundaries or mineralization. In open holes, cuttings from the percussion holes were collected and split to a convenient size. In core drilling, the sample was split and one-half crushed and analyzed. Cu, Pb and Zn were the main elements determined, but Mn and V were also determined in some units.

Sampling of cuttings came from 450 scout drill holes ranging from 15 to 150 m depth. Another 500 holes have been drilled at closer spacing to delineate orebodies and areas of mineralization.

Penetration sampling of superficial cover

In selected areas, shallow drilling was used to sample both the basal layer of the superficial sand and top of bedrock. Samples from mechanical auger, rotary or percussion drilling were analyzed by AAS for Cu, Pb and Zn.

Ground water sampling

No systematic regional groundwater sampling has been carried out. In the vicinity of Mt. Gunson, a cylinder and ball valve sampling tool was used to collect a number of down hole waters (Creelman, 1974). In routine exploration, samples were taken only where strong water flows were encountered. The samples were analyzed for Cu, Zn and TDS. Some samples were also analyzed for Pb and U.

Biogeochemical testing

As early as 1966, exploration teams experimented with biogeochemistry in attempts to find a cheap method to explore the difficult terrain of the Stuart Shelf. This early work established that samples from *Acacia aneura* and *A. linophylla* might be indicators of near-surface mineralization above shallow deposits.

After discovery of the deeper Cattle Grid body, and development of concepts on circulating groundwaters as transporters of Cu, orientation tests were made to check biogeochemical response over deeply hidden mineralization. Twigs and leaves were collected from myall, mulga and cypress pine tress along traverses over mineralized and unmineralized country about the Cattle Grid orebody. Areas over mineralized and unmineralized shales were also tested.

Where the value in ashed plants over mineralization appeared to be significant for any element, statistical tests were applied to determine the level of significance. Sampling of *Acacia* species is being used as a regional exploration tool where drill hole geochemical control is lacking.

RESULTS

General

Mean metal values in unmineralized rock units from 450 holes drilled since 1971 are shown in Table II. These serve as reference background levels. Newly discovered deposits provided ideal orientation for the effectiveness of some of the geochemical techniques employed, using basic data available from some 500 ore delineation drill holes.

TABLE II

Arithmetic mean of metal values in unmineralized rock units

Unit	Number of samples	Cu (ppm)	Pb (ppm)	Zn (ppm)
Sand cover	108	11	28	33
Tregolana Shale	202	18	34	66
Whyalla Sandstone	584	12	22	33
Yudnapinna Beds	377	14	25	41
Tapley Hill Formation	264	102	170	259
McLeay Beds	46	17	43	73
Pandurra Formation	71	24	32	35

Stream sediment sampling

A 1966 survey near Mt. Gunson gave only low values for metals. However, confidence in the work was not strong as (1) most samples were collected from areas downstream from dolomite outcrop from which alkaline waters were derived, (2) field biquinoline analytical methods were unreliable, and (3) difficulty was experienced in distinguishing between windblown material and fluvial sediment.

In 1971, the application of this technique to the limited area where it could be sensibly applied, namely the 200-km^2 flanking the western side of Pernatty Lagoon, gave no anomalies of significance. Samples showed low values, ranging from 10 to 40 ppm for Cu, 20—30 ppm for Pb and 15—60 ppm for Zn. These were well below any threshold expected to relate to significant mineralization. The conclusion was reached that the Pandurra Formation, eroded of its uppermost members, was unprospective. Samples from streams heading in Whyalla Sandstone gave values of 10—30 ppm for Cu and no anomalies.

Rock chip sampling

In surveys commencing in 1967, two anomalous areas were detected, one a Cu and one a broad Pb-Zn anomaly. The latter, over shales, was regarded as the more promising. Drilling confirmed the presence of weak, but uneconomic , Pb-Zn subsurface mineralization.

In recent surveys, metal contents were generally in the ranges 10—30 ppm Cu, 20—30 ppm Pb and 5—35 ppm Zn. However, values as high as 500 ppm Cu were obtained in the periphery of the Cattle Grid body and may relate to groundwater migration. A maximum of 320 ppm Cu was encountered over the anomalous area previously located in 1967 but the significance has yet to be investigated by drilling.

Salt lake mud sampling

Surveys established contrasts in base metal concentrations (Fig. 2) from locality to locality, and generally uniform values through the mud profile except for a slightly lower value in the top 50 cm of mud. Basement lithologies and rock boundaries were also established for stratigraphic control of data. Summary data are in Table III.

The Gunyah Lake orientation suggested Tregolana Shale provided an impervious barrier to upward metal migration from mineralized bedrock. Higher mean values over the Tregolana Shale at the subcropping Tregolana Shale/ Whyalla Sandstone contact were attributed to channeling of migrating waters along the contact and "spilling over" with precipitation in mud near the contact.

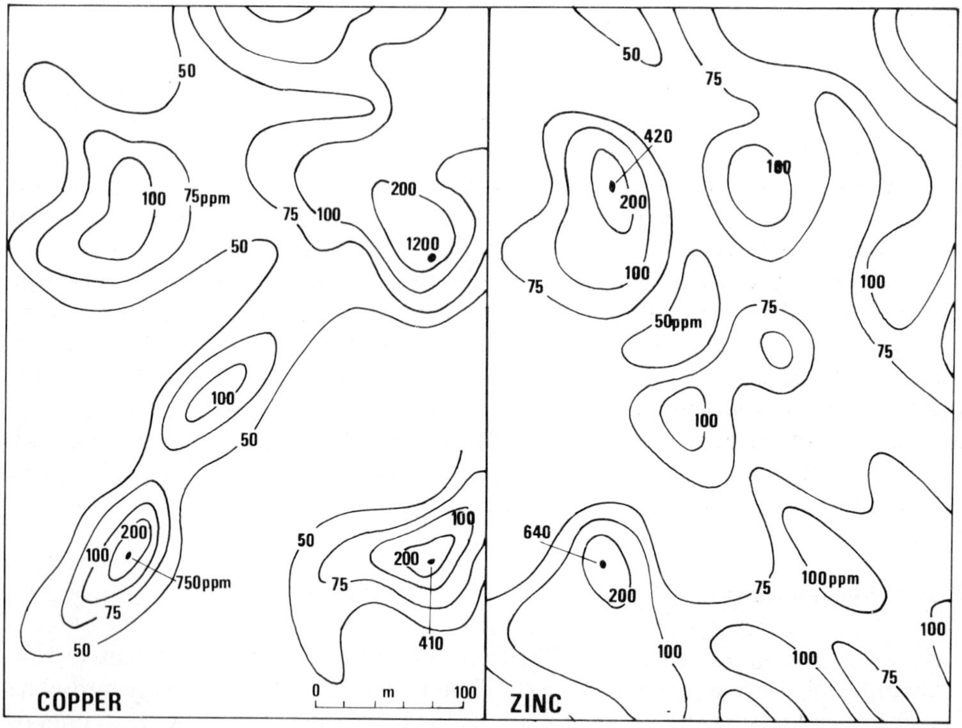

Fig. 2. Geochemical patterns for Zn and Cu in bottom lake muds in the same gridded area of Pernatty Lagoon.

TABLE III

Bottom mud geochemistry—arithmetic mean values

Bedrock type	Number of samples	Cu (ppm)	Pb (ppm)	Zn (ppm)
Gunyah Lake				
Tregolana Shale	93	92	16	90
Whyalla Sandstone	82	38	8	29
Pandurra Formation	31	37	13	32
Ironstone Lagoon				
Pandurra Formation	180	40	13	70

At Ironstone Lagoon, five spot Cu values ranging from 110 to 380 ppm Cu on the lake margin were attributed to outwash from known Cu occurrences. In Pernatty Lagoon, the most accessible of weak Cu anomalies were drilled but no Cu mineralization was intersected. Dolomite (Tapley Hill Formation, Table II) has high metal background which is believed to account for the false geochemical anomalies in overlying muds.

Deep profile penetration sampling

The initial geochemical indications to significant mineralized bodies, in units penetrated by bores to date, have been values that are strongly anomalous, with respect to the mean values in Table II. One feature which must be discounted in assessing sample values is the high background metal content in the Tapley Hill Formation compared with other units.

Secondary dispersion from the Cattle Grid body was studied and Figs. 3 and 4A show the response in overlying Whyalla Sandstone at or near the water table about 15 m below surface. This dispersion is believed to relate to cupriferous groundwater migration and precipitation of Cu near the water table in the permeable strata which characterizes the sequence over Cattle Grid.

Hidden mineralization in lutites (Tapley Hill Formation) or mineralization covered by Tregolana Shale shows no dispersion in overlying sediments (Fig.

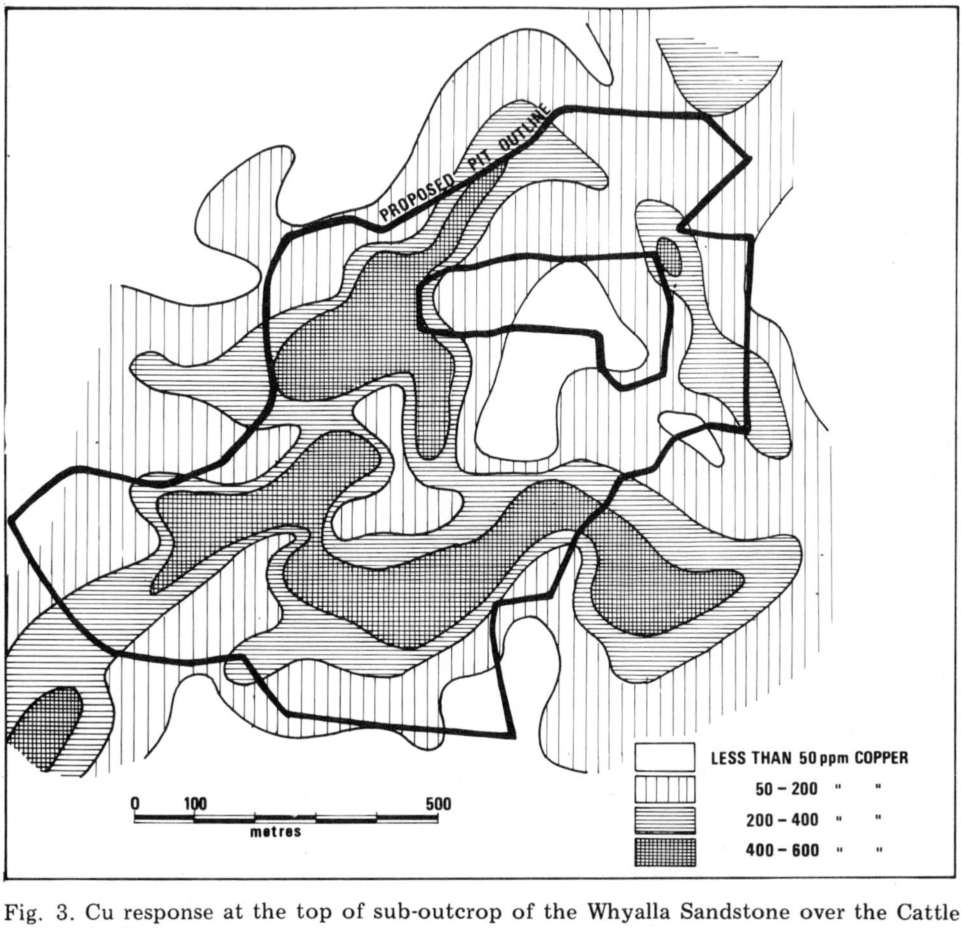

Fig. 3. Cu response at the top of sub-outcrop of the Whyalla Sandstone over the Cattle Grid orebody.

4B and C). This lack of dispersion is believed to be due to impervious seals to groundwater migration where lutites are met in the sequence, but in part may be due to the high pH associated with waters of carbonate environments which inhibit solution and dispersion of Cu.

A general conclusion is that except for direct sampling of primary mineralized haloes about ore, profile sampling for secondary dispersions is best applied to permeable sequences that allow upward migration of cupriferous waters from mineralized bodies.

Penetration sampling of superficial cover

Study of the shallow bedrock orientation data (Figs. 3 and 4) led to the drill testing of selected areas under deep sand cover. North of Cattle Grid, and at Oakden Hills, rock beneath sand cover showed values in Whyalla Sandstone in the ranges 5—25 ppm Cu, 20—50 ppm Pb and 20—50 ppm Zn. No anomalous areas were outlined. Spot samples in Tregolana Shale containing up to 75 ppm Cu, 90 ppm Pb and 420 ppm Zn were regarded as falsely anomalous because they were erratically distributed, had no vertical or lateral continuity and did not reflect mineralization in the underlying Whyalla Sandstone.

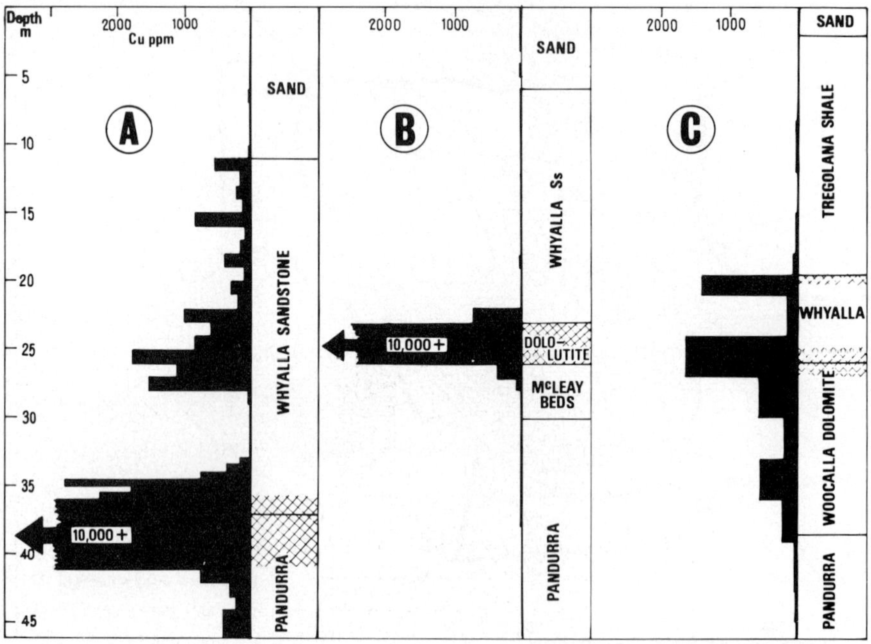

Fig. 4. Contrasts in geochemical profiles in different rock sequences penetrated in drill holes. A. Sandstone mineralization with permeable cover — Cattle Grid. B. Lutite mineralization with permeable cover — Mt. Gunson. C. Sandstone mineralization with impermeable cover — Oakden Hills.

Whereas this short penetration method is cost effective, it is regarded as useful only over areas where subcropping Whyalla Sandstone rests directly on the Pandurra palaeo-surface, the host to Cattle Grid ore.

Groundwater testing

Summary data are shown in Table IV. Salinity of groundwaters varied with the nature of aquifers. Waters from thick superficial clastic sequences are less saline than those in Adelaidean rocks. Whereas mine waters issuing from mineralized bodies and other waters (Creelman, 1974) in the Mt. Gunson area are highly anomalous in Cu, no regular pattern has been demonstrated for metals in bedrock waters in the periphery of Cattle Grid. Cu contents as high as 580 ppb have been found in formation waters in unmineralized terrain, whereas values of 98 ppb or less are recorded near Cattle Grid. More comprehensive work remains to be done on hydrogeochemistry.

TABLE IV

Summary of hydrogeochemical assays

Assays	TDS* (ppm)	Cu (ppb)	Zn (ppb)	Pb (ppb)	U_3O_8 (ppb)
Maximum value	109,500	580	1600	250	75
Minimum value	3730	<2	<2	<2	5
Mean**	31,472	92	222	26	26
Number of samples	86	92	86	23	6

* Total dissolved solids.
** Arithmetic.

Biogeochemical surveys

Whereas orientation in 1966 programmes gave anomalies for Cu over shallow mineralization, drilling of similar responses in an outlying area intersected no mineralization.

Statistical tests on 1975 data indicate that mulga trees accumulate Cu, Zn, Mn and Fe and that myall may be sensitive to Mo. Mulga leaves showed anomalous Fe and Mn over sandstone mineralization whereas twigs also showed anomalous response for Cu and Zn. Systematic surveys were therefore restricted to twig samples.

The recent orientation showed that mulga twigs, over sandstone and sand-covered Cu mineralization, respond to Cu, Zn and Fe (Fig. 5A). Over lutite mineralization, or mineralization overlain by shale (Fig. 5B), these metals do not respond significantly. Sampling is not yet of adequate intensity to

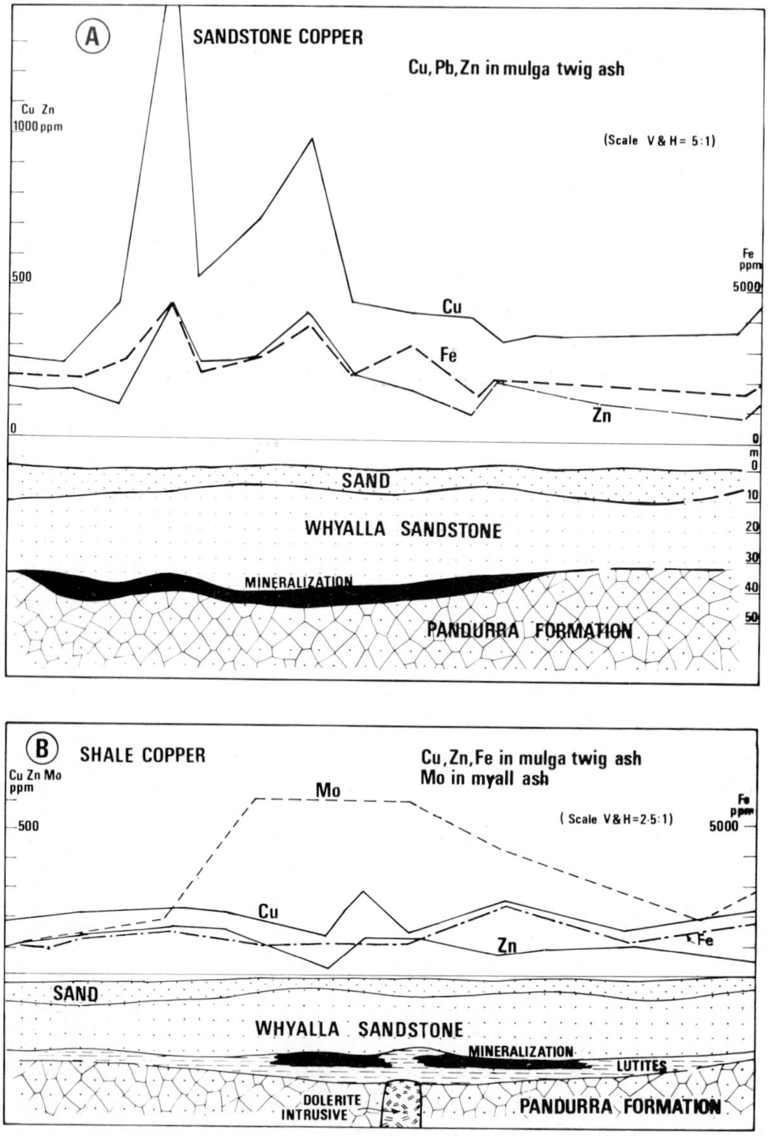

Fig. 5. Geochemical response in plants on section lines over (A) Cattle Grid orebody, and (B) lutite mineralization near Mt. Gunson.

attribute statistical significance to Mo pathfinder response. The reasons for lack of response over calcareous lutite mineralization is attributed to the impervious nature and alkaline character of shale matrix or cover, preventing dissolution and migration of metal ions in groundwater.

In view of the occurrence of highly anomalous metals in groundwater over non-mineralized terrain, caution should be used in interpreting biogeochemical data.

CONCLUSIONS

Geochemical techniques have proved an effective direct exploration method in exploring for hidden mineralization in arid, saline terrain where bedrock is masked by superficial strata. Deep drilling of the sequence and sampling the rock profile has been the most successful technique in locating primary and secondary metal dispersions. Shallow drilling to bedrock and biogeochemical surveys both show positive orientation responses to mineralization overlain by sandstone, but the application of these methods is restricted by the geology of the terrain. In areas where lutites are host to mineralization, or overlie it, there is no response. Lake mud sampling is believed to suffer from similar problems. Groundwater sampling, as presently known, gives false anomalies but needs more study.

ACKNOWLEDGEMENTS

The Management of CSR Minerals and Chemicals Division are thanked for permission to publish. Many geologists and geochemists of that Division, and of Mt. Gunson Mines Pty. Ltd., contributed to data collection and collation and their work is gratefully acknowledged.

REFERENCES

Creelman, R.A., 1974. The geology and geochemistry of the Pernatty Lagoon copper deposits, South Australia. M.Sc. (Honours.) Thesis, Macquarie University, Sydney, N. S.W. (unpublished).

Gersteling, R.W. and Heape, J.M., 1975. The Cattle Grid orebody, Mt. Gunson, South Australia. Proc., Australas. Inst. Min.. Metall. Conf., Adelaide, S.A., June, 1975, pp. 103—112.

Johns, R.K., 1968. Geology and mineral resources of the Andamooka-Torrens area. Geol. Surv. S. Aust., Bull., 41: 1—103.

Johns, R.K., 1974. Base metal mineralization in the Pernatty Lagoon region. Geol. Surv. S. Aust., Rep. Invest., 42: 1—28.

Parkin, L.W. (Editor) 1969. Handbook of South Australian Geology. Geological Survey of South Australia, Adelaide, S.A.

Thomson, B.P., 1976. Precambrian geology and tectonics of the Stuart Shelf and Torrens Hinge zone. In: B.P. Thomson, B. Daily, R.P. Coats and B.G. Forbes .Late Precambrian and Cambrian Geology of the Adelaide Geosyncline and Stuart Shelf, South Australia. 25th Int. Geol. Congr., Canberra, A.C.T., August, 1976, Excursion Guide No. 33A, pp. 1—11.

Van Herk, H., Houston, M.L. and Dudgeon, R.N., 1975. Planning, development and extraction of the Cattle Grid ore deposit, Mt. Gunson. Proc., Australas. Inst. Min. Metall. Conf., Adelaide, S.A., June, 1975, pp. 113—124.

Journal of Geochemical Exploration, 8 (1977) 219—234

GEOCHEMICAL DISPERSION PATTERNS ASSOCIATED WITH THE LAKE YINDARLGOODA SULPHIDE MINERALIZATION, WESTERN AUSTRALIA

G.H.W. FRIEDRICH and S.M. CHRISTENSEN

Institut für Mineralogie und Lagerstättenlehre, Technische Hochschule, Aachen (Federal Republic of Germany)
Geologisches Landesambt Schleswig-Holstein, Kiel (Federal Republic of Germany)

(Revised version accepted March 9, 1977)

ABSTRACT

Friedrich, G.H.W. and Christensen, S.M., 1977. Geochemical dispersion patterns associated with the Lake Yindarlgooda sulphide mineralization, Western Australia. J. Geochem. Explor., 8: 219—234.

Geochemical investigations have been carried out in the Lake Yindarlgooda area, 40 km east of Kalgoorlie, Western Australia, where sulphide deposits and black shales occur in a sequence of metavolcanics and metasediments of the Eastern Goldfields greenstone belt. The studies were made in order to understand the processes of element migration and the formation of dispersion patterns in semi-arid regions having partly lateritic weathering conditions. Outcrop exposure in the vicinity of the mineralized zones is limited to the siliceous caprock and gossans.

Mineralogical data, and geochemical data on the distribution of Cu, Ni, Co, Zn, Pb and other elements, have been obtained from unweathered and weathered rocks, and different horizons of the overlying soil in the area surrounding the mineralized zones of Lake Yindarlgooda, Scotia and Ringlock. The results indicate, in principle, two different kinds of weathering environment: (1) playa lakes, with the development of solonchaks; and (2) Tertiary peneplains outside the playa lakes, with lateritic soils, solonized brown soils or desert loams.

The trace element distribution patterns in the solonchaks are determined only by the initial element content of the alluvial sediments. Therefore, sampling of solonchaks generally cannot be recommended for geochemical exploration surveys, but sampling of weathered bedrock below the solonchaks is highly recommended in the playa lake environment. By sampling of the bedrock, the non-outcropping continuation of the Lake Yindarlgooda sulphide body has been precisely delineated.

In areas of the Tertiary peneplains outside the playa lakes, lateritic soils, solonized brown soils or desert loams cover the weathered bedrock. Comparative geochemical investigations in these areas in the vicinity of the nickel sulphide deposits at Scotia and Ringlock have shown that the migration of elements in solution is restricted, a result of the very low water table. In this area the main process of dispersion is mechanical. Therefore, the secondary dispersion haloes of Scotia and Ringlock are narrow, restricted to a few tens of metres, and sampling at closer intervals is needed to locate the orebodies.

INTRODUCTION

The type and extent of secondary dispersion patterns depend principally on climatic conditions during oxidation and migration of elements. Preliminary studies of weathering and soil formation, and especially of the interrelation between secondary dispersion, grain size distribution and mineralogical composition of soils, are the bases for the interpretation and evaluation of geochemical data obtained from soil samples.

In order to understand the processes of element migration and the formation of secondary dispersion patterns in semi-arid regions having partly lateritic weathering conditions, geochemical investigations were carried out in the Lake Yindarlgooda, Scotia and Ringlock areas, 40 km east and 70 km north of Kalgoorlie.

REGIONAL GEOLOGY

The area of investigation is situated between latitudes 30° and 31°S and longitudes 121°30′ and 122°30′E in Western Australia (Fig. 2). It forms part of the NNW-trending Eastern Goldfields greenstone belt of Archaean age, which is 300 km long and up to 150 km wide (Fig. 1).

The greenstone belt is composed of metavolcanics intercalated with geosynclinal metasediments, predominantly basic to intermediate intrusives, intermediate to acid extrusive rocks, tuffs and clastic sediments. The rock sequence comprises three volcanic cycles, each of which starts with basic volcanics and ends with acid volcanics. During periods of low volcanic activ-

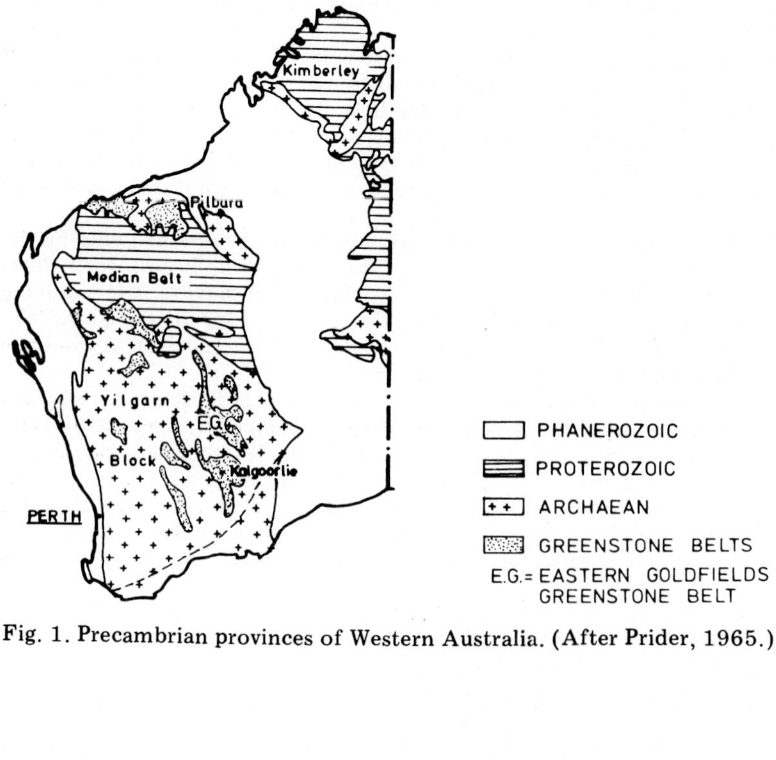

Fig. 1. Precambrian provinces of Western Australia. (After Prider, 1965.)

ity at the end of each volcanic cycle, graded sediments, arkosic sandstone, shale and tuffaceous sediments were deposited in local basins between centres of active volcanism. In one of these basins occurs the mineralization of Lake Yindarlgooda. Pyrite and small amounts of base metal sulphides were deposited together with chert bands and black shales. The pentlandite-pyrrhotite mineralization at Scotia, Ringlock and Carr Boyd Rocks are associated with ultramafics within the lowermost volcanic cycle, which comprises the Moreland and Gindalbie Formations.

The whole Archaean rock sequence was subjected to a low-grade regional metamorphism resulting in silicification and sericitization of acid volcanics and chloritization, with accompanying albitization, of basic volcanics. Pyroclastics have been altered to sericite-biotite-chlorite schists. Clastic sediments were silicified and carbonatized. The emplacement of the youngest igneous rocks, the post-granite Widgiemooltha dyke suite, is controlled by a 70°-trending fracture system. Overlying the Archaean rocks are remnants of flat-lying Cainozoic sediments. Classifications by volcanic cycles, stratigraphy and ages of the rock sequences in the study area are given by Turek (1966), Kriewaldt (1967), Sofoulis et al. (1969) and Williams (1970). The principal data of those studies are summarized in Table I and Figs. 2 and 3.

PHYSIOGRAPHY

The area of investigation lies within a vast inland plateau with a subdued topography. Relief differences between hills, which grade gently into broad alluviated valleys, seldom exceeds 100 m. The present climate is semi-arid with an aridity index of 8.23. The annual evaporation rate of 2580 mm is

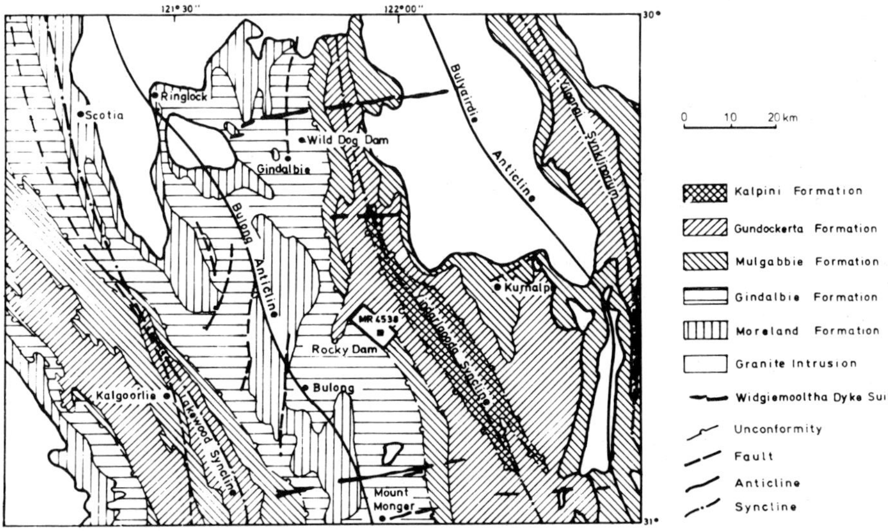

Fig. 2. Generalized geological map of the study area (after Williams, 1970), showing location of Ministerial Reserve Area MR 4538.

TABLE I

Stratigraphy, age, thickness and lithology of the Archaean in the area of investigation (after Turek, 1966; Kriewaldt, 1967; Sofoulis et al., 1969 and Williams, 1970)

		Formation	Age (m.y.)	Thickness (m)	Lithology
Proterozoic		Widgie-mooltha dyke suite	2420 ± 30		Quartz dolerite, gabbro, norite granophyre
Archaean	Volcanic cycle III	Granite Intrusion	2612 ± 13		Intrusion of granite, pegmatite; metamorphism
		Kalpini Formation		4500	Basic to intermediate extrusive and intrusive rock; ultramafic intrusive rocks, clastic sediments
	Volcanic cycle II	Gundockerta Formation	ca. 2615	6000	Fine- to medium-grained sediments, partly derived by erosion from:
				3000	acid to intermediate extrusive and intrusive rock, minor basic extrusives, pyroclastic rocks;
		Mulgabbie Formation	2660	4000	basic extrusive and intrusive rocks, ultramafic intrusives; minor intermediate to acid extrusive and intrusive rocks; *banded iron formation; pyrite mineralization Lake Yindarlgooda*
	Volcanic cycle I	Gindalbie Formation	2660	4500	Clastic sequence; fine- to medium-grained sediments mainly derived from:
				7500	intermediate to acid extrusives and intrusives, minor basic extrusive rocks, pyroclastics;
		Moreland Formation		5000	basic extrusive and intrusive rocks, ultramafic extrusives; *nickel sulphide mineralization Scotia*
		Gneiss	2800		Granitic gneisses

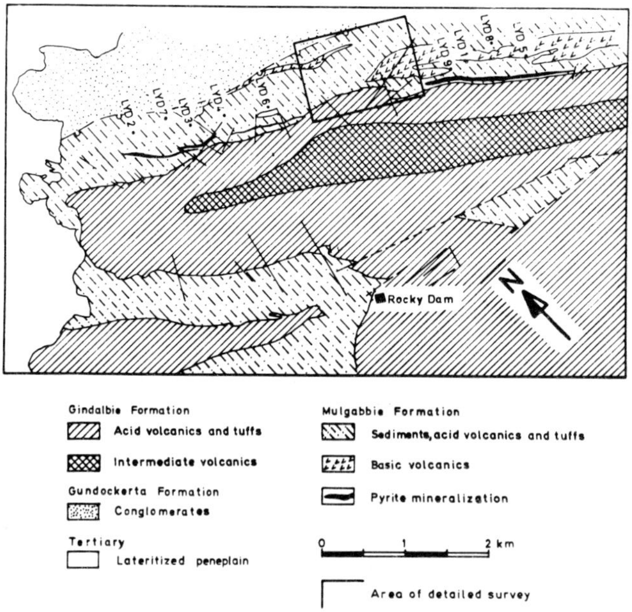

Fig. 3. Geological map of the Lake Yindarlgooda area showing locations of drill holes LYD 1—9 and area of detailed survey (see Fig. 7).

ten times higher than the average annual precipitation.

The physiography is characterized by two distinct land surfaces, known as the "old" and "new" plateaux (Jutson, 1934). The old plateau represents a peneplain formed in the Mesozoic and Tertiary and subjected to intense lateritization during the Pliocene. It is now represented by flat-topped hills capped with indurated laterite, sand and gravel plains forming isolated remnants restricted to the elevated headwater divide regions between the major drainage basins.

During the late Tertiary or early Quaternary, the lateritization slowed dramatically or ceased, due to the onset of more arid climatic conditions. Renewed erosion under arid climatic conditions formed the new plateau. It is represented by claypans and playa lakes, which evolved from trunk valleys of the early Tertiary river system. These valleys originally drained southeastwards towards the Nullarbor plain. The present drainage, however, is internal and there are no permanent rivers in the area. The drainage lines on the present plateau are poorly defined and sheet flooding into the internal drainage basins, which are occupied by playa lakes, is common after heavy rains.

Different soil types are related to the geomorphological units described above. Besides fossil laterites and gravel plains, polycyclic soils such as calcareous lateritic soils, solonized brown soils and desert loams can be found on the Tertiary peneplain and in the transition zone to the recent base of erosion. In the playa lakes, solonchaks are present. These developed from alluvial sediments.

In general, two different types of weathering environment can be distinguished:

(1) Predominantly chemical weathering in the playa lakes.

(2) Predominantly physical weathering on the Tertiary peneplains with lateritic soils, solonized brown soils and desert loams.

MINERALOGY AND GEOCHEMISTRY OF OXIDIZED BEDROCK AND SULPHIDES

In the Lake Yindarlgooda area, chemical weathering predominates, due to the high water table of the salt lake, about 1 m beneath the surface whereas outside the playa lakes the water table is deeper. Groundwater fluctuations during the pre-Holocene led to the formation of an oxidation zone of more than 100 m depth.

Processes of hydration, hydrolysis of silicates, solution and oxidation have resulted in an almost complete decomposition of quartz, silicates and sulphides, forming clay minerals and hydrous oxides of Fe, Al and Mn. Microscopic studies of diamond cores drilled down to the pyrite mineralization at Lake Yindarlgooda indicate a distinct secondary enrichment of clay minerals, chlorite and limonite in the oxidation zone (Table II).

The feldspar content of a tuff in Lake Yindarlgooda drill hole (LYD) 7 decreases from 60 to 70% in unweathered material to 20—25% in weathered material of the oxidation zone. The difference is equivalent to the increase of sericite and clay minerals. Carbonates and sulphides are completely decomposed. The quartz content is also lower than in the unweathered tuff. An almost complete decomposition of epidote, carbonate, amphibole and quartz can be observed in basic volcanics (basalts), drilled by LYD 6. The unweathered basalt contains 15—20% epidote, 5% carbonate, 50% amphibole and 10—15% quartz, whereas the weathered equivalent has 30—40% chlorite, 40—50% clay minerals and 10% limonite. The extremely high degree of decomposition is caused by sulphuric acid originating from weathering of pyrite in nearby black shales, and the high salt concentration dissolved in groundwater of the playa lake.

In spite of the high content of clay minerals and hydrous oxides of Fe and Mn in the weathered rocks, only a slight enrichment of trace elements in the oxidation zone in the vicinity of the pyrite mineralization, which is situated at the northwest margin of Lake Yindarlgooda, was noted. Table III presents the values of Co, Cu, Ni, Pb and Zn in weathered and unweathered basic volcanics and tuffs drilled by LYD 6 and LYD 7. Only Zn is enriched in the oxidation zone, due to the high mobility of Zn and the high Zn content of pelitic sediments in the hanging wall and footwall of the tuffs and basic volcanics.

The oxidation of the stratabound sulphides of Lake Yindarlgooda led to the formation of siliceous cap rocks and gossans. They originated by oxida-

TABLE II

Mineralogical composition (in wt.%) of weathered rocks (oxidation zone) and unweathered rocks, Lake Yindarlgooda area

Drill hole and depth (m)	Rock type	Feldspar	Sericite	Chlorite	Clay	Epidote	Carbonate	Amphibole	Limonite	Quartz
LYD 6	Basic volcanics:									
25—55	weathered	10	—	30—40	40—50	—	—	—	10	2— 5
60—90	unweathered	5—10	—	5— 8	—	15—20	5	50	—	10—15
LYD 7	Tuffs:									
3—45	weathered	20—25	30	2	30—40	3— 5	—	—	3	5
70—90	unweathered	60—70	20	—	—	3— 5	5	—	—	5—10

TABLE III

Co, Cu, Ni, Pb and Zn contents (in ppm) in weathered and unweathered basic volcanics (drill hole LYD 6) and tuffs (drill hole LYD 7), Lake Yindarlgooda area

Drill hole and depth (m)	Rock type	Co	Cu	Ni	Pb	Zn
LYD 6	Basic volcanics:					
20—55	weathered	20	17	35	15	279
60—90	unweathered	15	30	25	20	90
LYD 7	Tuffs:					
3—45	weathered	95	125	155	12	972
70—90	unweathered	50	120	100	10	174

tion of two sulphide bodies with different chemical composition, a massive sulphide zone in the north and a disseminated one in the south of the Lake Yindarlgooda area.

The oxidation led to the formation of Fe^{III}-sulphate, which hydrolized to FeOOH and H_2SO_4. Sulphuric acid increased the hydrolitic decomposition of silicates forming silicic acid and clay minerals. The gossans show, in part, a strong silicification, suggesting that silicic acid is preferentially adsorbed by iron oxides. Comparison of the Co, Cu, Ni, Pb and Zn content in the weathered mineralized zone and the unweathered mineralized zone shows that these elements are neither concentrated nor impoverished in the gossans, suggesting that the migration of ions was rather limited. The gossans which developed from two different types of sulphide bodies (massive pyrite and disseminated pyrite) are characterized by approximately the same element distribution as the corresponding unweathered sulphide bodies (Table IV).

TABLE IV

Co, Cu, Ni, Pb and Zn contents (in ppm) of gossans and primary sulphides of the two main mineralization types (disseminated pyrite and massive pyrite), Lake Yindarlgooda area

Type of mineralization		Co	Cu	Ni	Pb	Zn
Disseminated pyrite	Gossan	45—150	50— 850	25—330	20—200	255—3630
	Primary sulphides	103—230	435—1390	125—270	32—250	805—5400
Massive pyrite	Gossan	15— 95	10— 900	5—150	4—400	11— 650
	Primary sulphides	56—106	26— 116	38—110	12— 55	200— 660

MINERALOGY AND GEOCHEMISTRY OF SOILS

Outcrop exposure in the vicinity of the mineralized zones is limited to siliceous caprocks and gossans. Consequently, studies on the use of soil geochemistry for exploration of non-outcropping deposits beneath the playa lakes have been carried out.

Solonchaks and bedrocks were sampled at varying depth by use of a special soil drill. The trace element concentrations were determined by atomic absorption spectrophotometry, a 25% HNO_3 attack being employed. For the determination of the absorbed portion of the total element content 1 g of the minus 80-mesh sieved solonchak samples were leached with 10 ml of a 1% EDTA solution.

The solonchaks of the playa lakes developed from alluvial sediments show a zonation, which can be attributed to a sedimentary stratification during the deposition of transported weathering material from the drainage area of the internal basins of the playa lakes. Only in the uppermost horizon have soil formation processes led to an enrichment of water-soluble salts, a result of capillary rise and subsequent evaporation of saline groundwater. Based on grain size analysis, four different horizons can be distinguished (Fig. 4).

The profile consists of a lowermost gravel horizon, characterized by a high content of iron and manganese hydroxides, and three overlying horizons, with different grain size composition. The grain size fractions are given in Fig. 4. With increasing depth the silt and clay fractions decrease and the sand fraction increases. The relatively high content of the fine sand fraction in the uppermost horizon can be attributed to deposition by windblown sands.

The mineralogical composition of the solonchaks depends on the zonation within the profile (Table V). Besides halite, gypsum and bassanite are enriched in the uppermost horizon. The gravel horizon contains about 60% quartz whereas only 30—35% SiO_2 has been found in the silty clay horizon, which contains up to 40% kaolinite and 10% illite. No other clay mineral could be found.

Considering the total element content, the portion of adsorbed trace elements is nearly negligible, due to the limited sorption capacity of kaolinite, which forms 80% of the clay mineral fraction. The total element content is generally much higher than the portion bound by adsorption. Only for Cd, Mn and Co are considerable amounts bound by adsorption. The grain size composition does not affect the distribution of the adsorbed portion of Fe, Cu, Ni and Zn (Fig. 5). This leads to the conclusion that the trace element distribution of the solonchaks is not influenced by secondary dispersion processes from the underlying rocks, but only by the primary element content of the alluvial sediments. The Cu, Ni, Pb and Zn values in the solonchaks are, without exception, background values and do not depend on the primary trace element content of black shales, tuffs and volcanics beneath the playa lake sediments. The trace element content of different rock types

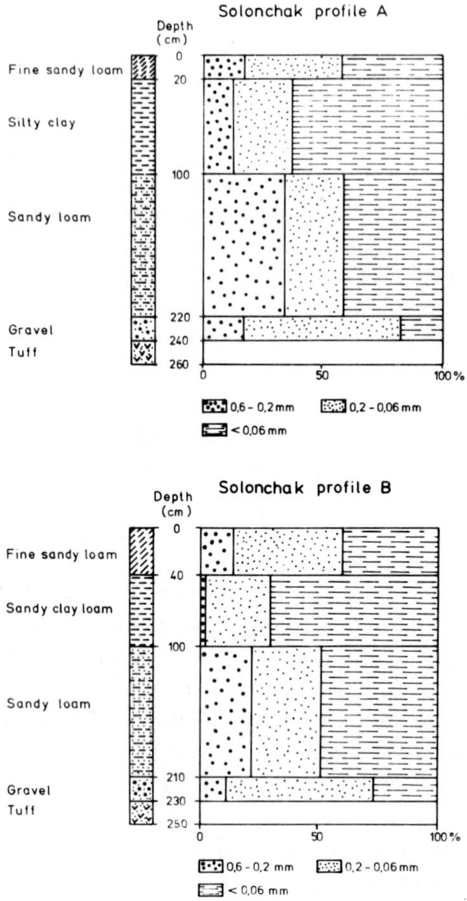

Fig. 4. Solonchak profiles *A* and *B* showing the distribution of three grain size fractions (in wt. %) in four horizons, Lake Yindarlgooda area.

TABLE V

Mineralogical composition (in wt.%) of solonchaks, Lake Yindarlgooda area

	Quartz	Halite	Feld-spar	Gyp-sum	Bas-sanite	Kaolin-ite	Illite	Haematite
Profile A								
0— 20 cm	50	4	3	2	2	20	5	2
20—100 cm	30	3	3	—	—	40	10	3
100—220 cm	45	3	3	—	—	35	10	2
220—240 cm	55	2	4	—	—	20	5	4
Profile B								
0— 40 cm	50	9	3	—	8	15	5	2
40—100 cm	35	3	5	—	—	40	5	3
100—210 cm	50	3	2—3	—	—	30	10	2
210—230 cm	60	3	3	—	—	10	5	5

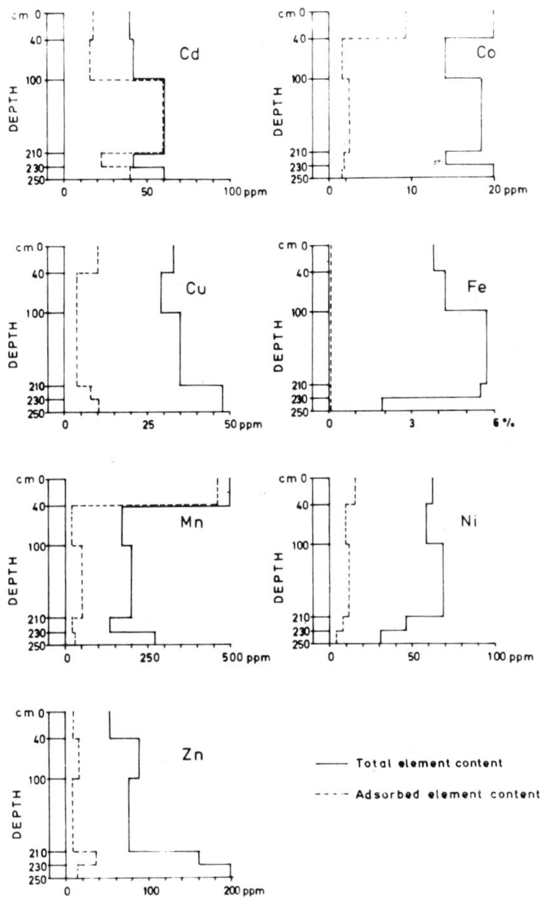

Fig. 5. Distribution of Cd, Co, Cu, Fe, Mn, Ni and Zn (total content and adsorbed element content) in five different horizons of solonchak profile A (minus 80-mesh fraction), Lake Yindarlgooda area.

is not reflected in the overburden. The solonchaks completely mask the bedrock (Fig. 6).

Therefore, sampling of solonchaks cannot be recommended for geochemical exploration surveys, but sampling of weathered bedrock below the solonchaks is highly recommended in a playa lake environment. By sampling of bedrock material with a special soil drill at varying depth (25—50 cm), the concealed continuation of the Lake Yindarlgooda mineralization has been precisely delineated (Fig. 7). The elongated Cu, Ni and Zn anomalies in the central part mark the northwest-trending mineralization. A second significant Ni anomaly occurs east of the mineralized zone, which probably can be attributed to an ultramafic rock. This anomaly was not drilled. The west-trending Pb anomaly in the north of the study area is associated with hydrothermal quartz veins.

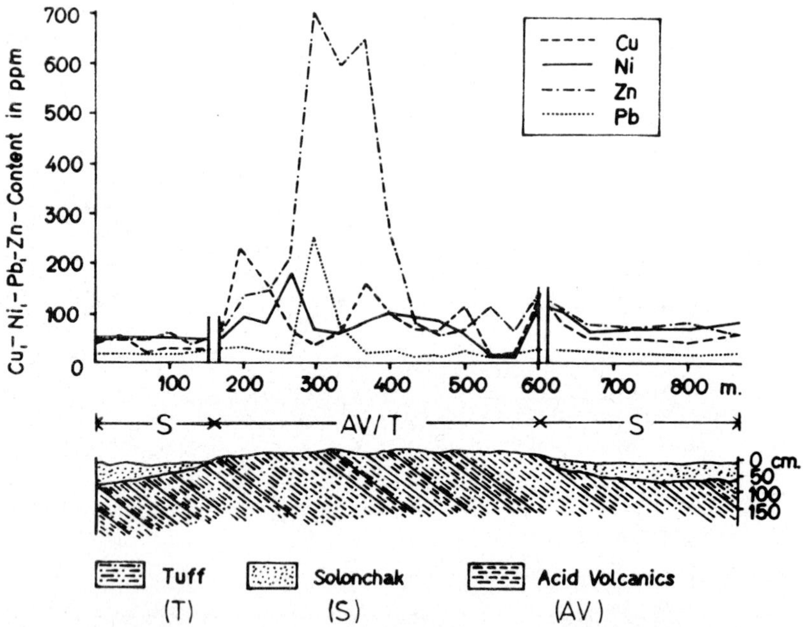

Fig. 6. Distribution of Cu, Ni, Zn and Pb in outcropping weathered bedrock compared with the corresponding element distribution in solonchaks above partly mineralized bedrock, SW-NE traverse line 286 X, Lake Yindarlgooda area.

In areas of Tertiary peneplains outside the playa lakes lateritic soils, solonized brown soils and desert loams cover the weathered bedrock. Comparative geochemical investigations in these areas, in the vicinity of the nickel sulphide deposits of Scotia and Ringlock, have shown that the migration of elements in solution is restricted as a result of the low groundwater table. Cu, Ni and Zn values obtained from B horizons of these soil types correspond to the element distribution in the underlying bedrock, whereas the trace element content of the A horizon varies considerably. Arid climatic conditions led to the enrichment of $CaCO_3$ and/or $CaSO_4$ in the B horizon. This "caliche" horizon (locally termed calcrete or kunkar) contains rock fragments of the weathered bedrock. The Ni and Cr distribution in the carbonate-rich B horizon and in the overlying A horizon above the nickel sulphide mineralization of Ringlock shows that the sampling of caliche is more suitable for geochemical exploration surveys than sampling of the uppermost horizon. The background-anomaly contrast is much more significant in caliche (Fig. 8). Anomalous Ni values are associated with the coarse grain fraction. This coincides with results obtained from geochemical investigations in the vicinity of the nickel sulphide deposits at Kambalda (Mazzucchelli, 1972) and Scotia. In these areas the main process of dispersion is mechanical. The secondary dispersion halo of Ni and Cu at Scotia is narrow, caused by limited mobility of the weathering products of the orebody cemented in the carbonate-rich B horizon.

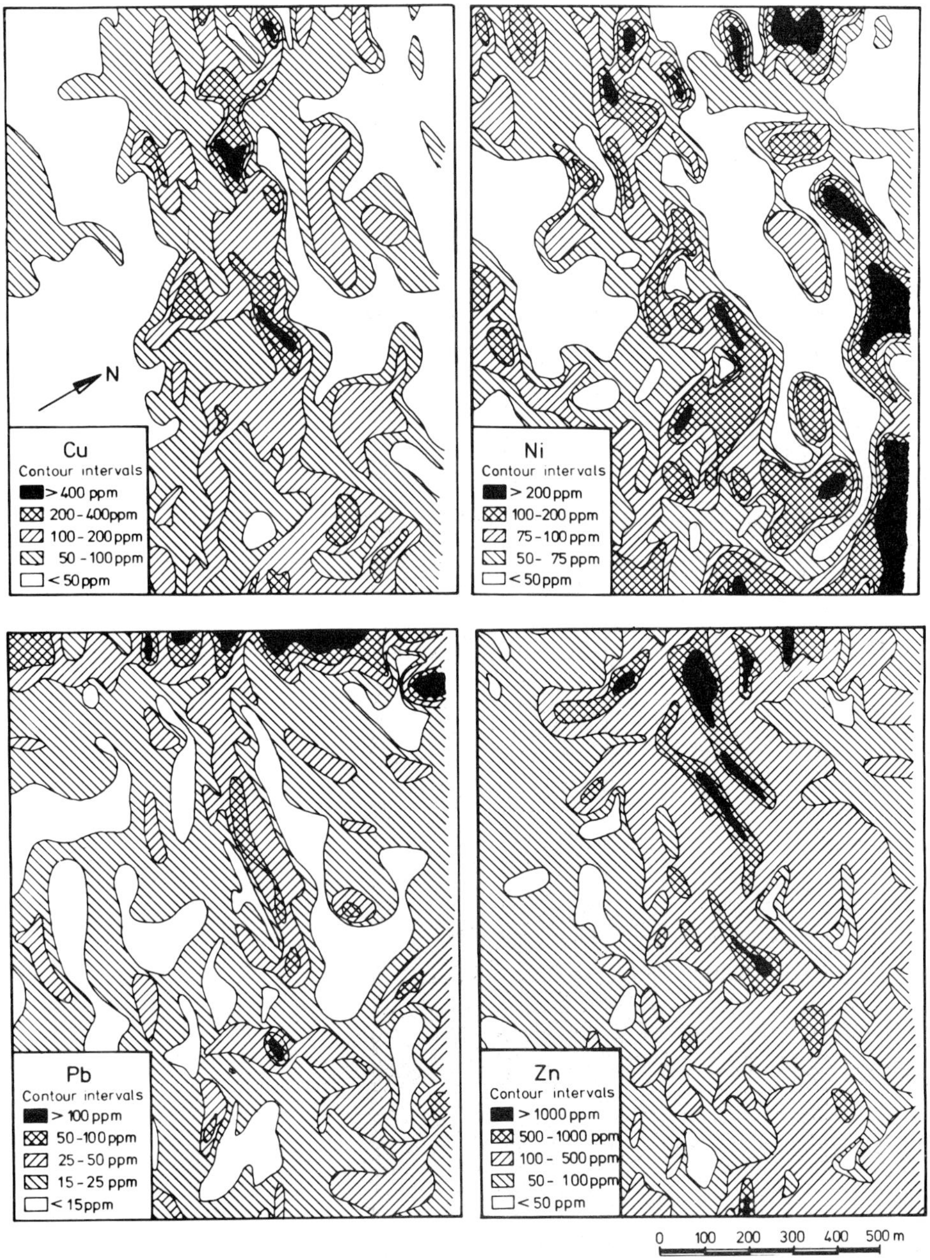

Fig. 7. Map showing distribution of Cu, Ni, Zn and Pb in weathered bedrock (beneath playa lake sediments), Lake Yindarlgooda area (see Fig. 3: area of detailed survey).

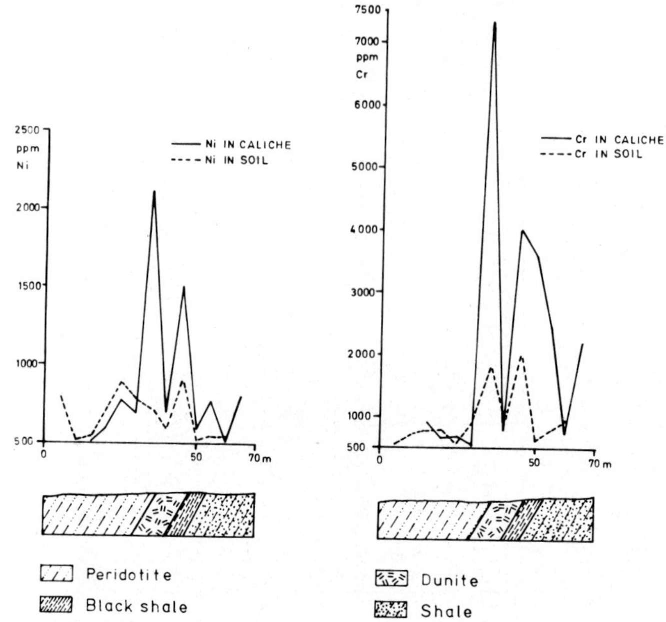

Fig. 8. Distribution of Ni and Cu in soil (A horizon) and caliche above the Ringlock sulphide mineralization (in dunite), Western Austrlia.

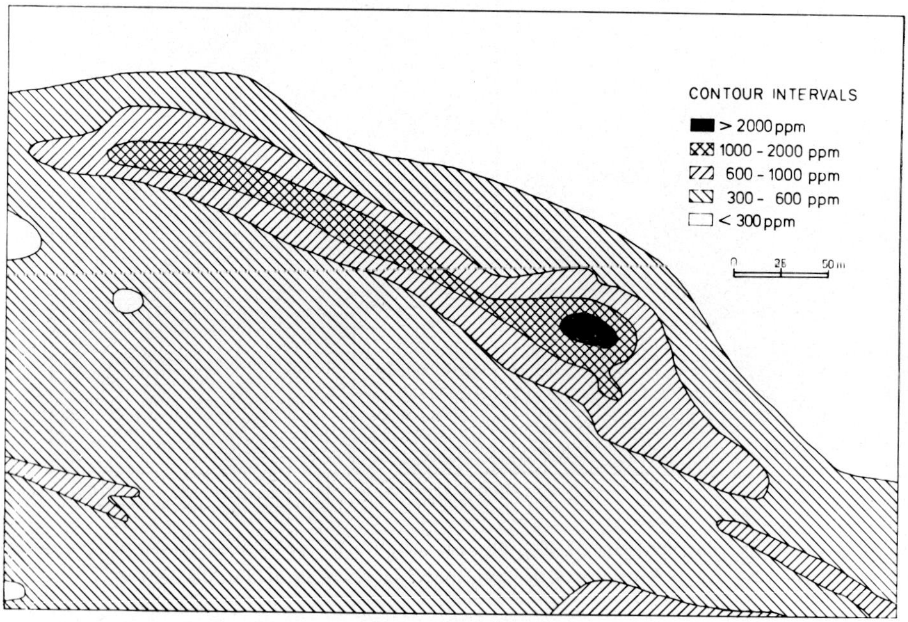

Fig. 9. Map showing the distribution of Ni in soils above the Scotia nickel sulphide deposit (sample depth: 15 cm, size fraction: minus 80-mesh). (After McKay, 1971).

Results of a systematic soil survey of the Scotia mine area by McKay (1971) show a symmetrical Ni anomaly above the mineralized ultramafic complex. The Ni values vary between 600 and 1200 ppm.

The ore zone at Scotia is marked by a Ni anomaly of 1200—3000 ppm in the overburden. The secondary dispersion is restricted to only 30—60 m (Fig. 9). Therefore, sampling at close intervals is needed to locate Scotia-type deposits in semi-arid regions.

CONCLUSIONS

Geochemical studies of secondary dispersion patterns in the Lake Yindarlgooda area show that the trace element distribution of the solonchaks of the playa lakes is determined only by the initial element content of the alluvial sediments. Anomalies in the underlying bedrock are not reflected in solonchaks. For geochemical exploration surveys, sampling of the bedrock beneath the playa lake sediments is necessary.

In areas outside the playa lakes, the main process of dispersion is mechanical. Secondary dispersion haloes are narrow and soil sampling at close intervals is needed to locate non-outcropping mineral deposits.

ACKNOWLEDGEMENTS

The authors would like to express their thanks to the Deutsche Forschungsgemeinschaft and Swiss Aluminium Mining Australia Ltd. for their support of this study. Thanks are also due to K.G. McKay of Great Boulder Gold Mines Ltd. for help and fruitful discussions. C.R.M. Butt kindly read the manuscript and made helpful comments.

REFERENCES

Christensen, S., 1974. Geochemisch-lagerstättenkundliche Untersuchungen im Bereich der Schwarzschiefer und stratiformen Sulfidervorkommen Lake Yindarlgooda, Scotia und Ringlock, West-Australien. Dissertation, RWTH Aachen, Aachen.

Friedrich, G., 1968. Anwendung und Ergebnisse moderner Methoden bei der geochemischen Exploration. In: Untersuchung und Bewertung von Erzlagerstatten, Heft 21 der GDMB, Clausthal-Zellerfield, pp. 58—91.

Friedrich, G. and Kulms, M., 1969. Eine Methode zur Bestimmung geringer Konzentrationen von Quecksilber in Gesteinen und Böden und ihre Anwendung bei der geochemischen Exploration von Erzlagerstätten. Erzmetall, 22: 74—84, 214—218.

Horwitz, R.C. and Sofoulis, J., 1965. Igneous activity and sedimentation in the Precambrian between Kalgoorlie and Norseman, Western Australia. Proc. Australas. Inst. Min. Metall., 214: 45—59.

Horwitz, R.C., Kriewaldt, M.J.B., Williams, I.R. and Doepel, J.J.G., 1967. A zone of Archean conglomerates in the Eastern Goldfields of Western Australia. W. Aust. Geol. Surv., Annu. Rep. 1966, pp. 53—56.

Jutson, J.T., 1934. The physiography (geomorphology) of Western Australia. W. Aust. Geol. Surv., Bull., 95, 2nd ed.

Kriewaldt, M.J.B., 1967. Explanatory notes on Kalgoorlie 1:250,000 Geological Sheet, Western Australia. W. Aust. Geol. Surv. Rec., No. 1967/10.

Mazzucchelli, R.H., 1972. Secondary geochemical dispersion patterns associated with the nickel sulphide deposits of Kambalda, W.A. J. Geochem. Explor., 1: 103—116.
McKay, K.G., 1971. Scotia geochemical orientation. Great Boulder Mines, Kalgoorlie, W.A. (unpublished report).
Prider, R.T., 1965. Geology and mineralisation of the Western Australia Shield. In: Geology of Australian Ore Deposits, 8th Commonw. Mining Metall. Congr. Australasian Institute of Mining and Metallurgy, Melbourne, Vic., pp. 56—65.
Sofoulis, J., Williams, X.K. and Rowston, D.L., 1969. Investigation of Ministerial Reserve 4538H, Lake Yindarlgooda, Bulong District, W.A. W. Aust. Geol. Surv. Rec., No. 1968/17.
Turek, A., 1966. Rubidium-strontium isotopic studies in the Kalgoorlie-Norseman area, Western Australia. Ph.D. Thesis, Australian National University, Canberra, A.C.T.
Williams, I., 1970. Explanatory notes on the Kurnalpi 1:250,000 Geological Sheet, Western Australia. W. Aust. Geol. Surv. Rec., No. 1970/1.

Journal of Geochemical Exploration, 8 (1977) 235—245

CHEMICAL CONSTRAINTS ON THE WEATHERING OF SERPENTINITES CONTAINING NICKEL-IRON SULPHIDES

E.H. NICKEL and M.R. THORNBER

Division of Mineralogy, Commonwealth Scientific and Industrial Research Organization, Wembley, W.A. (Australia)

(Revised version accepted December 30, 1976)

ABSTRACT

Nickel, E.H. and Thornber, M.R., 1977. Chemical constraints on the weathering of serpentinites containing nickel-iron sulphides. J. Geochem. Explor., 8: 235—245.

A review of mineralogical observations and results of Eh-pH measurements made on weathered profiles over serpentinites containing Ni-Fe sulphides in several areas of Western Australia permits some conclusions to be drawn relating to the chemical conditions prevailing during, and as the result of, weathering.

At depth, conditions of relatively low Eh and high pH prevail. Eh increases and pH decreases upward in response chiefly to a series of anodic oxidative changes in the sulphides. Sulphides decompose at the water table, resulting in a sharp drop in pH which, in turn, promotes the decomposition of serpentine and carbonates. Above the water table, where the mineralogy is dominated by goethite, the pH increases again, promoting the local deposition of carbonates. Near-surface silicification is attributed to deposition of silica from silicic acid generated by the decomposition of serpentine during an earlier phase of weathering.

INTRODUCTION

The weathering of nickel sulphide ores in the Western Australian environment has been investigated at several localities, including Kambalda (Keele and Nickel, 1974; Nickel et al., 1974; Thornber, 1975), Mt. Windarra (Watmuff, 1974), Spargoville (Wilmshurst, 1975) and Perseverance (Nickel et al., 1977a). These published studies, as well as additional information derived from unpublished observations made on samples from the Mt. Keith deposit and from laboratory experiments, have led to a general understanding of the mineralogical and chemical changes that take place in the weathering profile. This paper attempts to collate and analyze these data so as to provide a general picture of the chemical conditions that prevail throughout the weathering profile of serpentinites containing Ni-Fe sulphides.

The major nickel sulphide deposits of Western Australia occur in ultramafic belts within the Yilgarn block — an Archaean craton occupying the southwestern part of Western Australia. Geological descriptions of the deposits can be obtained from a number of papers in the recent monograph edited by Knight (1975).

Most of Western Australia, including the area providing material for this study, has been subjected to deep lateritic weathering. Although the region is now semi-arid, the bulk of the lateritic weathering is thought to have taken place under more humid comditions between the Eocene and Middle Miocene epochs (Johnstone et al., 1973), and possibly as late as the Pliocene (Stephens, 1971). Subsequent uplift, accompanied by erosion, has caused the upper parts of the laterite profile to be stripped away in places; in others, appreciable thicknesses of lateritic cover, sometimes exceeding 100 m (Loftus-Hills, 1975), are retained. The relatively wet climate responsible for the development of the laterites was followed by more arid conditions, resulting in a lowering of the water table and the development of geomorphological features sush as sand plains, endorheic drainage, and salt playas (Mulcahy, 1967). In recent times, the surface has been further modified by the development of siliceous hardpan over a large area (Bettenay and Churchward, 1974).

This complex weathering and erosional history has left a complicated imprint on the rocks, making it difficult to interpret the chemistry involved in the weathering of such material. However, the authors feel that the available data do permit some useful conclusions to be drawn.

MINERALOGICAL AND CHEMICAL VARIATIONS IN THE WEATHERED PROFILE

The unweathered host rocks are serpentinized ultramafics now consisting largely of serpentine with varying amounts of talc-carbonate alteration (Burt and Sheppy, 1975; Martin and Allchurch, 1975; Roberts, 1975; Ross and Hopkins, 1975). Locally, as at the Perseverance deposit, some olivine has survived serpentinization. The primary sulphides consist largely of pyrrhotite and pentlandite, with relatively minor pyrite; in one of the Kambalda deposits (Otter shoot), millerite is one of the principal sulphides (Keele and Nickel, 1974). In some of the deposits, notably at Kambalda, the sulphides occur chiefly as massive sheets and lenses; in others, such as Mt. Keith, the sulphides are sparsely disseminated; in still others, sulphides occur in both massive and disseminated forms, e.g. Perseverance and Mt. Windarra.

In all deposits there is a sharp mineralogical and chemical discontinuity between the oxide (or gossan) zone above the water table, and the sulphide-serpentine assemblage below. However, the effects of supergene alteration have modified the sulphide assemblage far below the water table; at the Perseverance deposit, for example, supergene alteration extends down to about 400 m (Nickel et al., 1977a), whereas the water table is at about 35 m.

Changes in the sulphide mineralogy (below the water table) have been well documented (Nickel et al., 1974, 1977a; Watmuff, 1974; Thornber, 1975).

Pentlandite is converted to violarite, and pyrrhotite is replaced largely by pyrite and/or marcasite and, to a lesser extent, by violarite. These replacements have been interpreted as anodic reactions in an electrochemical model in which the principal cathodic reaction is the reduction of oxygen at the water table. Toward the upper part of the sulphide zone, as the water table is approached, Ni enrichment occurs by the partial exchange of Ni for Fe in violarite, of Ni for Mg in serpentine, and the occasional deposition of supergene millerite. At the water table, sulphides and serpentine decompose.

The chemical expression of the weathering profile over a disseminated portion of the Perseverance deposit is shown in Fig. 1, compiled from data recently published by Nickel et al. (1977a). The decomposition of the sulphides is marked by the disappearance of S above about 35 m. The behaviour of the other elements will be dealt with in the following discussion. Chemical profiles over other nickel sulphide bodies have not been equally well documented, but preliminary data from the Mt. Keith deposit (Butt and Sheppy, 1975; C.R.M. Butt, personal communication, 1976) indicate a general similarity to the Perseverance data if it is assumed that the top 40 m of the Mt. Keith profile represent transported material.

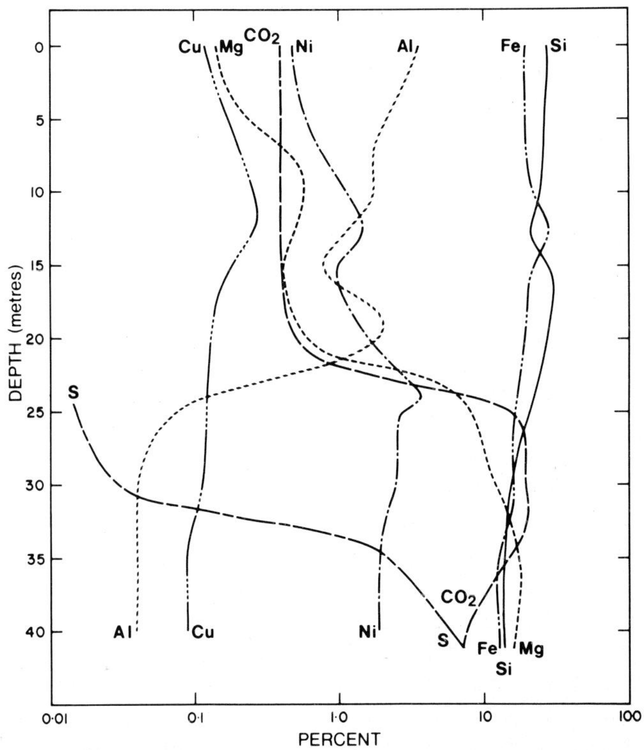

Fig. 1. The distribution of some of the chemical components in the weathered profile over the Perseverance deposit. (Adapted from data in Nickel et al., 1977a.)

THE CHEMICAL ENVIRONMENT

Below the water table

Below the water table conditions generally become increasingly reducing and alkaline with depth. Measurements made on material from Kambalda indicate pH values between 8 and 9, and Eh values as low as − 0.3 volt, at depths of 100 m or more below the water table. Thornber (1972) first suggested that the oxidation of Ni-Fe sulphide orebodies can be interpreted on an electrochemical model, and this suggestion has been borne out by subsequent mineralogical studies (Nickel et al., 1974; Watmuff, 1974; Thornber, 1975). For the purpose of this paper, the significance of this electrochemical concept is that the cathodic reduction of oxygen at the water table induces a series of anodic reactions in the sulphides far below the water table. The electrons generated in the anodic reactions are conducted toward the surface by the sulphides, while charge balance is maintained by the net upward movement of cations in solution. Electronic conductivity is prevented in the case of sparsely disseminated sulphides, with the result that disseminated sulphides are generally not weathered as deeply as massive ones. Where there is deep weathering of disseminated sulphides below the present-day water table, previous oxygen access at depth, i.e. by lowering of the water table, is indicated.

Eh measurements have shown that the oxidation potential in a nickel sulphide deposit undergoing weathering increases upward (Thornber, 1975), and mineralogical observations have confirmed that the original sulphide assemblage changes to a more highly oxidized one toward the water table (Nickel et al., 1974, 1977a; Watmuff, 1974; Thornber, 1975).

Quantitative interpretation of oxidative changes in the sulphide minerals on the basis of thermodynamic considerations is difficult for two reasons. Firstly, room-temperature thermodynamic data are, in some cases, non-existent, or if they do exist, they are not sufficiently precise to establish the stability conditions of a particular mineral phase. Secondly, phase stabilities in relation to the variables of Eh and pH are usually based on assumed metal activities of 10^{-6} *M*, whereas electrochemical experiments show that metals in solution at this activity play no detectable part in many sulphide reactions. In view of these difficulties, mineralogical observations combined with experimental data can be used to augment thermodynamic data in the construction of empirical Eh-pH diagrams that help to elucidate the chemical constraints imposed by the weathering environment.

If it is assumed that the oxidative changes in the iron sulphides are consistent with the mineral associations that can be calculated from thermodynamic data in the aqueous Fe-S-O system as shown in Fig. 2a (Nickel et al., 1977a) then a combination of mineralogical observations and experimental measurements allows this diagram to be extended and adjusted to fit mineralogical observations and experimental data for the Ni-Fe sulphide assemblages, as shown in Fig. 2b. In Fig. 2b the pentlandite-violarite reaction is shown as coinciding with the pyrrhotite-pyrite reaction. In fact, mineralogical observations have shown that the pentlandite-violarite reaction proceeds while pyrrhotite is still stable with respect to pyrite, so that the lines representing these reactions should actually be separated, with the pentlandite-violarite line at a lower Eh. However, because of the uncertainties noted above, these lines have not been shown separately. Also, the conversion of pentlandite to violarite, and that of pyrrhotite to pyrite, clearly takes place in the presence of water, so that the boundary between Fe_7S_8 and FeS_2 in Fig. 2a, shown to be

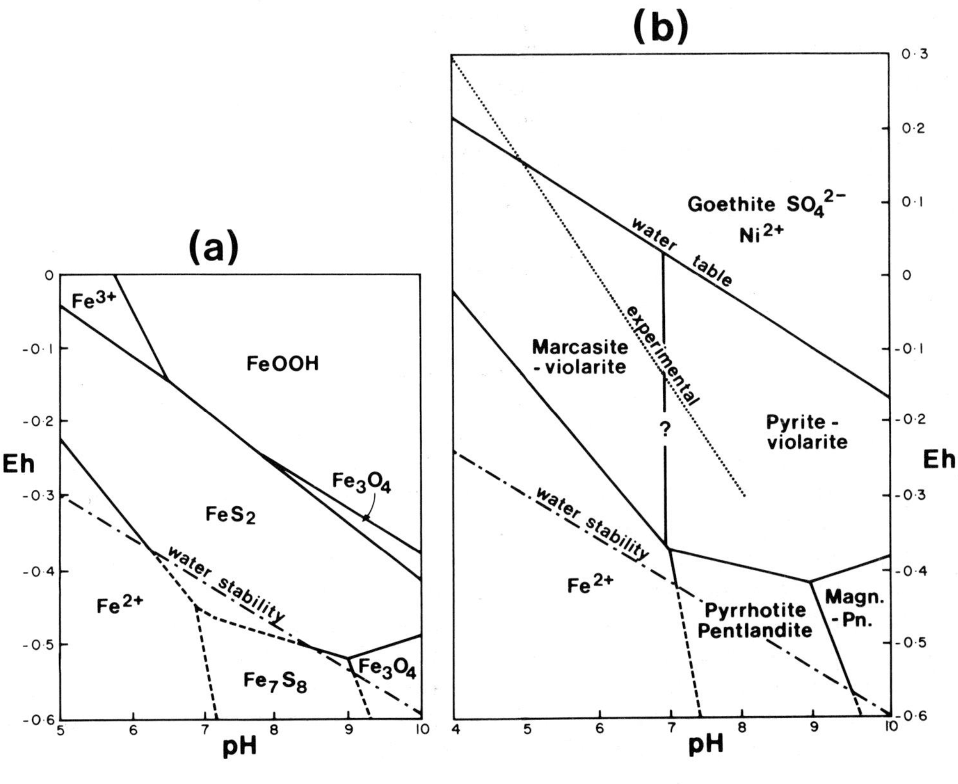

Fig. 2. (a) Eh-pH diagram for Fe phases in part of the aqueous Fe-S-O system at 25° C and atmospheric pressure. Boundaries between solid and liquid phases are drawn at $[Fe^{2+}] = 10^{-6}\,M$. Activities of the combined sulphur species are also calculated at $10^{-6}\,M$. (From Nickel et al., 1977a.) (b) An Eh-pH diagram based on mineralogical observations and laboratory measurements on natural mineral assemblages. The slopes of the lines used in Fig. 2a have been retained, but the positions have been altered to conform with the empirical data. The dotted line denoted by "experimental" represents the results of measurements made on aqueous electrolyte in contact with pentlandite-pyrrhotite and violarite-pyrite assemblages. Abbreviations used are: Pn = pentlandite; Magn = magnetite; and Vl = violarite.

largely within the field of water instability, is obviously too low. In Fig. 2b, therefore, the boundary has been raised so as to fall entirely within the field of water stability.

With regard to the upper part of Fig. 2b, we were able to show electrochemically that violarite is unstable with respect to sulphate at about 0.2 volt and pH = 4. This point has been used to position the potential of the water table shown in Fig. 2b. The dotted line, which represents a combination of the measurements made of the mixed potential and pH of the aqueous electrolyte in contact with pentlandite-pyrrhotite, violarite-pyrite assemblages at various stages of weathering, comes to within 0.1 volt of that point, which confirms the approximate positioning of the water table.

The marcasite/pyrite ratio may well be an indicator of pH conditions during the replacement of pyrrhotite by disulphides, since the results of Rickard's (1969) experimental work have shown that the formation of marcasite is favoured by low pH. This conclusion is supported by the supergene sulphide assemblages at Kambalda, where the more acid Lunnon shoot contains an appreciably higher marcasite/pyrite ratio than the more alkaline Durkin and Otter shoots. The Mt. Keith deposit also contains a high proportion of supergene marcasite, although groundwater pH measurements have not been made at the appropriate depth there to confirm this correlation.

Since marcasite formation is favoured by low pH, and pyrite by high, it should be possible to separate the dominance fields of these minerals by a vertical line, as has been done in Fig. 2b. The nature of Rickard's data does not permit this line to be positioned precisely, so the line was drawn arbitrarily at a pH of 7, which falls between the measurements made in the Lunnon and Durkin shoots (Nickel et al., 1974).

It should be stressed that Fig. 2b is not intended to represent equilibrium states for the mineral phases designated, but is rather a representation of the chemistry of a dynamic system as observed in the field and in the laboratory. It can, perhaps, be construed as a cautionary exercise in the too-rigid application of thermodynamic data to weathering processes.

The serpentine below the water table remains relatively stable, although it becomes enriched in Ni toward the top due to downward-percolating nickel-bearing solutions. Carbonates, predominantly magnesite, commonly occurring with the serpentine, also undergo chemical changes, showing enrichment in both Fe and Ni (Thornber and Nickel, 1976).

At the water table

Sulphides, serpentine, and carbonates decompose in the vicinity of the water table. Although oxidative potential is the principal factor in sulphide decomposition (Fig. 2a and b), it should not be a significant factor for serpentine since it does not contain appreciable amounts of oxidizable components. The controlling factor in the decomposition of serpentine appears to be pH, since thermodynamic calculations have shown that it becomes unstable relative to H_4SiO_4 and Mg^{2+} in solution at a pH of about 8 (Nickel et al., 1977a). Although it may be argued that ionized silicate species, such as $H_7SiO_6^-$, may also affect the stability of serpentine, such species are not likely to be of significance at moderate values of pH; they can be expected to assume increasing importance at pH values in excess of 9, but such high values do not appear to occur in weathering profiles examined to date.

At high carbonate ion activities, serpentine can be expected to be replaced by magnesite, which can persist to somewhat lower pH values. The layer of magnesitic carbonate that sometimes overlies the decomposed serpentine, as at the Perseverance deposit, giving rise to a carbonate peak in the profile just above the water table (Fig. 1) implies that the carbonate ion activity must have been particularly high in view of the low pH that would be expected at this position (see below).

The serpentine develops a characteristic bleached appearance toward the water table. Although the colour change has not yet been linked unequivocally to a well-defined change in chemical composition, and no change in X-ray diffraction pattern can be discerned, electron microprobe analyses indicate that the pale colour is associated, in some samples at least, with a slightly lower Mg/Si ratio. Leaching experiments on serpentine and olivine have the same effect when Mg^{2+} is dissolved from the surface. Since Mg^{2+} is not a chromatophoric ion, the colour change is likely to be an optical effect due to light scattering from serpentine in the early stages of decomposition. Since the first stage in the dissolution of a silicate is replacement of the bases (in this case Mg^{2+} by H^+; Luce et al., 1972), loss of some of the Mg^{2+} ions might produce sufficient disruption of the structure to cause the inferred light scattering.

The decrease in pH at the water table can be attributed chiefly to the decomposition of the sulphides and the hydrolysis of Fe^{3+}. The reactions involving the oxidation of pyrite and the formation of goethite can be written as:

$$2FeS_2 + 2H_2O + 7O_2 \rightarrow 2Fe^{2+} + 4SO_4{}^{2-} + 4H^+$$

and $4Fe^{2+} + O_2 + 6H_2O \rightarrow 4FeOOH + 8H^+$

and these combine to give:

$$4FeS_2 + 10H_2O + 15O_2 \rightarrow 4FeOOH + 8SO_4{}^{2-} + 16H^+$$

These reactions are mainly responsible for the depression of the pH where the sulphides have ready access to air and water.

This being the case, one would expect that sulphide-bearing serpentine should decompose more rapidly than serpentine essentially devoid of sulphides. Paradoxically, however, evidence obtained from the study of the Perseverance deposit (Nickel et al., 1977a) indicates that barren serpentinite adjoining sulphide-bearing serpentinite is more deeply weathered than the latter. A possible explanation for this anomalous behaviour might be greater permeability in the barren serpentinite, but this possibility has not yet been evaluated. However, in any case the deeper weathering of the serpentinite is due to unexpected behaviour of one or more of the mineral phases, laboratory experiments on the artificial weathering of sulphide-bearing and barren serpentinites have been initiated, and may ultimately shed some light on this problem.

Above the water table

Chemical conditions above the water table are much more difficult to define because of the absence of a permanent aqueous phase. The oxidation potential is undoubtedly high because of easy access of atmospheric oxygen to the decomposed and frequently porous mineral assemblage. The pH over weathered ultramafic rock is generally in the range between 7 and 8.5 (C.R. M. Butt, personal communication); although lower values are certainly not uncommon, they are usually due to remnant sulphides or decaying organic matter.

Trace element data through complete weathered profiles in Western Australian serpentinites are still rather sparse, and there is as yet insufficient information to interpret with confidence trace element distribution in terms of mineralogy and chemistry. Nevertheless, a few of the more obvious correlations are coming to light. For example, at the Perseverance deposit, the sharp Ni peak at a depth of 24 m (Fig. 1) is due primarily to eardleyite, a hydrous carbonate of Ni and Al, belonging to the hydrotalcite group (Nickel et al., 1977a, 1977b). It occurs at the interface between an underlying carbonate layer and an overlying zone of Al enrichment. Obviously, the existence of eardleyite is predicated on the availability of both Al and CO_2, and the pH would presumably have to be reasonably high. The mineral was probably precipitated when downward-moving nickel-rich solutions encountered this favourable environment.

It is reasonable to expect that it will eventually be possible to correlate most trace element concentrations with particular chemical factors. Laboratory experiments have shown that co-precipitation with siderite of metals follows the decreasing sequence Cu^{2+}, Mn^{2+}, Co^{2+}, Ni^{2+}, Ca^{2+}, and Mg^{2+} (Thornber and Nickel, 1976). Further experiments on the precipitation of goethite in the presence of trace amounts of Cu, Co and Ni have shown that Cu is the most strongly co-precipitated, followed by Co, and then Ni. The pH at which the goethite forms appears to affect the amount of trace element that is combined.

Unlike most laterite profiles where there is depletion of silica toward the surface, the Perseverance and Mt. Keith deposits exhibit a substantial degree of near-surface silicification. These deposits lie within the area affected by surface silicification, termed the "Wiluna hardpan", evidently a late, post-lateric development (Stephens, 1971; Bettenay and Churchward, 1974), it might be thought that the silicification of the gossans at Perseverance and Mt. Keith derive from the same origin. However, the silicification at Perseverance, Mt. Keith and over dunitic ultramafic bodies generally, extends to greater depths than the more superficial Wiluna hardpan, suggesting that the silica was derived from weathering of the ultramafic bodies and deposited at an earlier, higher water table. Subsequent erosional stripping has exposed the silicified layer at the present-day land surface. Retention of intricate sulphide textures in the Perseverance gossan (Nickel et al., 1977a) tends to confirm this explanation for the silicification.

Since the silicification is developed preferentially over serpentinite, it can be assumed that the silica was derived from the silicic acid liberated by the decomposition of the serpentine. The low Al content of ultramafic bodies precludes the formation of appreciable amounts of alumino-silicates such as clay minerals, and the silicic acid content of the water therefore builds up until the saturation limit (120—140 ppm) is reached. Eh and pH, at least at the levels found in normal soil profiles, appear to have little effect on silica precipitation.

CONCLUSIONS

Mineral stability data and direct Eh-pH measurements enable some aspects of the chemical environment in sulphide-bearing serpentinite to be inferred. An attempt to summarize these data quantitatively, and to hazard a guess at the original residual profile before erosional stripping, appears in Fig. 3. Although the numerical values given in this illustration can only be approximations, derived as they are from measurements made under existing conditions and with interpolation of inferences drawn from known and inferred

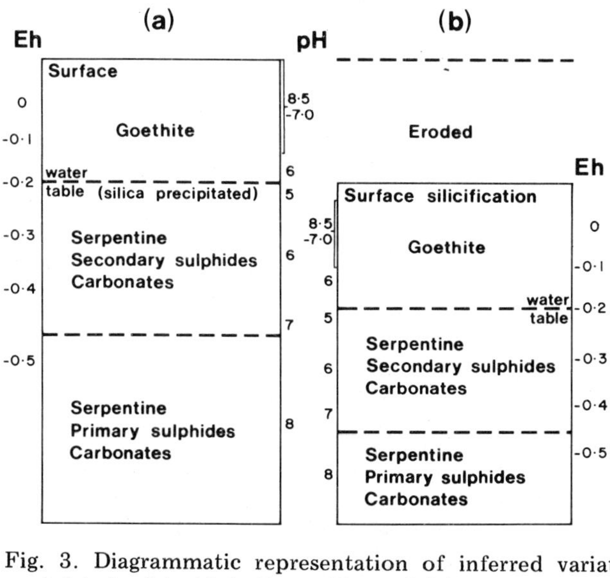

Fig. 3. Diagrammatic representation of inferred variations in Eh and pH through (a) postulated original laterite profile; and (b) present profile.

mineral stabilities, all indications are that the general trend is for Eh to increase and pH to decrease toward the surface. The pH is anomalously low at the water table, where the decomposition of the sulphides and the hydrolysis of Fe must reduce the pH to values below that in underlying and overlying parts of the profile. The extent to which the pH is depressed at the water table will, of course, be affected by a variety of factors, including the rate at which the sulphides are oxidizing and the presence or absence of mineral buffers, such as carbonates.

Individual weathering profiles will, of course, vary depending on particular circumstances. Thus, for example, the magnesite layer overlying the decomposed serpentine-sulphide assemblage at Perseverance has not been recognized elsewhere, and may be the result of a set of circumstances peculiar to that area. However, much more information is required before this and other problems in interpreting the mineralogy and chemistry of various weathering profiles can be solved.

ACKNOWLEDGEMENTS

The authors are grateful to W.E. Ewers and C.R.M. Butt, both of the Division of Mineralogy, C.S.I.R.O., who read the manuscript and made helpful suggestions. The illustrations were prepared by C.R. Steel.

REFERENCES

Bettenay, E. and Churchward, H.M., 1974. Morphology and stratigraphic relationships of the Wiluna hardpan in arid Western Australia. *J. Geol. Soc. Aust.*, 21: 73—80.

Burt, D.R.L. and Sheppy, N.R., 1975. Mount Keith nickel sulphide deposit. In: C.L. Knight (Editor), *Economic Geology of Australia and Papua New Guinea,* 1. Metals. Australasian Institute of Mining and Metallurgy, Parkville, Vic., pp. 159—168.

Butt, C.R.M. and Sheppy, N.R., 1975. Geochemical exploration problems in Western Australia exemplified by the Mt. Keith area. In: I.L. Elliott and W.K. Fletcher (Editors), *Geochemical Exploration 1974.* Elsevier, Amsterdam, pp. 391—415.

Johnstone, M.H., Lowry, D.C. and Quilty, P.G., 1973. The geology of southwestern Australia — a review. *J. R. Soc. W. Aust.*, 56: 5—15.

Keele, R. and Nickel, E.H., 1974. The geology of a primary millerite-bearing assemblage and supergene alteration at the Otter Shoot, Kambalda, Western Australia. *Econ. Geol.*, 69: 1102—1117.

Knight, C.L. (Editor), 1975. *Economic Geology of Australia and Papua New Guinea,* 1. Metals. Australasian Institute of Mining and Metallurgy, Parkville, Vic., 1126 pp.

Loftus-Hills, G.D., 1975. Ora Banda nickel deposits, Western Australia. In: C.L. Knight (Editor), *Economic Geology of Australia and Papua New Guinea,* 1. Metals. Australasian Institute of Mining and Metallurgy, Parkville, Vic., pp. 1010—1011.

Luce, R.W., Bartlett, R.W. and Parks, G.A., 1972. Dissolution kinetics of magnesium silicates. *Geochim. Cosmochim. Acta,* 36: 35—50.

Martin, J.E. and Allchurch, P.D., 1975. Geology of the Perseverance nickel deposit. In: C.L. Knight (Editor), *Economic Geology of Australia and Papua New Guinea,* 1. Metals. Australasian Institute of Mining and Metallurgy, Parkville, Vic., pp. 149—155.

Mulcahy, M.J., 1967. Landscapes, laterites and soils in southwestern Australia. In: J.N. Jennings and J.A. Mabbutt (Editors), *Landform Studies from Australia and New Guinea*: Australian National University Press, Canberra, A.C.T., pp. 211—230.

Nickel, E.H., Ross, J.R. and Thornber, M.R., 1974. The supergene alteration of pyrrhotite-pentlandite ore at Kambalda, Western Australia. *Econ. Geol.*, 69: 93—107.

Nickel, E.H., Allchurch, P.D., Mason, M.G. and Wilmshurst, J.R., 1977a. Supergene alteration at the Perseverance nickel deposit, Agnew, Western Australia. *Econ. Geol.*, 72: 184—203.

Nickel, E.H., Davis, C.E.S., Bussell, M., Bridge, P.J., Dunn, J.G. and MacDonald, R.D., 1977b. Eardleyite as a product of the supergene alteration of nickel sulfides in Western Australia. *Am. Mineral.* (in press).

Rickard, D.T., 1969. The chemistry of iron sulphide formation at low temperatures. *Stockholm Contrib. Geol.*, 20: 67—95.

Roberts, J.B., 1975. Windarra nickel deposits. In: C.L. Knight (Editor), *Economic Geology of Australia and Papua New Guinea*, 1. Metals. Australasian Institute of Mining and Metallurgy, Parkville, Vic., pp. 129—143.

Ross, J.R. and Hopkins, G.M.F., 1975. Kambalda nickel sulphide deposits. In: C.L.Knight (Editor), *Economic Geology of Australia and Papua New Guinea*, 1. Metals. Australasian Institute of Mining and Metallurgy, Parkville, Vic., pp. 100—121.

Stephens, C.G., 1971. Laterite and silcrete in Australia. A study of the genetic relationships of laterite and silcrete and their companion minerals and their collective significance in the formation of the weathered mantle, soils, relief and drainage of the Australian continent. *Geoderma*, 5: 5—52.

Thornber, M. R., 1972. Pyrrhotite — the matrix of nickel sulphide mineralization. Conf., Australas. Inst. Min. Metall., Newcastle, N.S.W., May-June, 1972, pp. 51—58.

Thornber, M.R., 1975. Supergene alteration of sulphides, II. A chemical study of the Kambalda nickel deposits. *Chem. Geol.*, 15: 117—144.

Thornber, M.R., and Nickel, E.H., 1976. Supergene alteration of sulphides, III.The composition of associated carbonates. *Chem. Geol.*, 17:45—72.

Watmuff, I.G., 1974. Supergene alteration of the Mt. Windarra nickel sulphide ore deposit, Western Australia. *Miner. Deposita*, 9: 199—221.

Wilmshurst, J.R., 1975. The weathering products of nickeliferous sulphides and their associated rocks in Western Australia. In: I.L. Elliott and W.K. Fletcher (Editors), *Geochemical Exploration 1974*. Elsevier, Amsterdam, pp. 417—436.

Journal of Geochemical Exploration, 8 (1977) 247—258

YILGARN NICKEL GOSSAN GEOCHEMISTRY — A REVIEW, WITH NEW DATA

P.G. MOESKOPS

Australian Mineral Development Laboratories, Frewville, S.A. (Australia)

(Revised version accepted February 10, 1977)

ABSTRACT

Moeskops, P.G., 1977. Yilgarn nickel gossan geochemistry — a review, with new data. J. Geochem. Explor., 8: 247—258.

Recent chemical investigations by the author on 40 Ni ores, 103 true gossans and 69 false gossans (collected from 32 localities, including most producing nickel mines) have largely supported the findings of previous authors who have published on nickel gossans from the Western Australian Yilgarn Shield and have indicated that, in the early stages of gossan recognition, particular attention should be paid to "signatures" based on Cu, Ni, Zn, Mn and acid-soluble Cr ($Cr^{(s)}$).

Compared with ores at depth, gossans are typically depleted in platinoids (mainly Pd and to a lesser degree Pt). This may be explainable in terms of low-temperature Eh-pH relations for the system Pd-Pt-H_2O-Cl. Furthermore, true gossans often contain a lower content of $Cr^{(s)}$ than false gossans. This probably occurs because (1) nickel sulphide ores are commonly depleted in total Cr, and hence $Cr^{(s)}$, with respect to their host ultramafic rocks and (2) under the highly acid conditions which prevail during the early stages of gossan formation, $Cr^{(s)}$ (mainly present in secondary ferrochromite rims on refractory chromite grains and in associated secondary silicates) is removed by leaching.

INTRODUCTION

The Western Australian Yilgarn Shield (Block), of Archaean age, covers some 800,000 km^2 and lies roughly between latitudes 25° and 35° S and longitudes 116° and 124° E (Fig. 1). The shield is composed largely of granitoid rocks which form a "sea" containing numerous discontinuous, semi-connected, N-NNW elongated greenstone belts. In detail, these belts consist largely of folded and regionally metamorphosed cyclic volcanic successions. Larger greenstone belts, such as that which encompasses the Kalgoorlie region, contain evidence of up to three volcanic cycles, with each cycle representing an upward change from quiescent sub-aerial basic volcanism, with associated chemical sedimentation, to later, more explosive, acid volcanism,

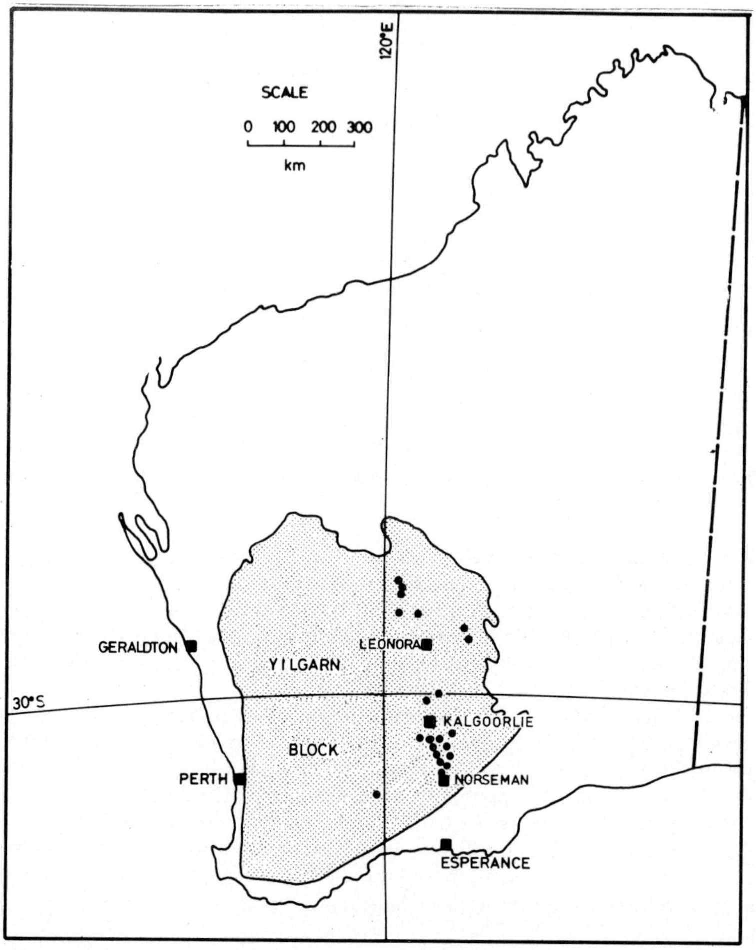

Fig. 1. Yilgarn Block of Western Australia, showing major nickel sulphide occurrences (small black circles).

with associated clastic sedimentation. The southern part of the Yilgarn Block is cut by a suite of ENE-trending Proterozoic dykes.

A high proportion of the Yilgarn is covered by extensive soil and scree plains and by Tertiary weathering crusts. The weathering crusts account for much of the relief seen in the region. Climatically, the Yilgarn is predominantly arid with the annual rainfall ranging from about 540 mm in the extreme southwest to about 160 mm, or less, in the extreme northeast. Temperatures range form 0° to 45° C and vary markedly both diurnally and seasonally. The drainage is internal and extensive playa systems abound throughout the area.

Known nickel sulphide deposits (Fig. 1) are restricted almost entirely to ultramafic rocks associated with stratigraphically lower basic volcanic zones

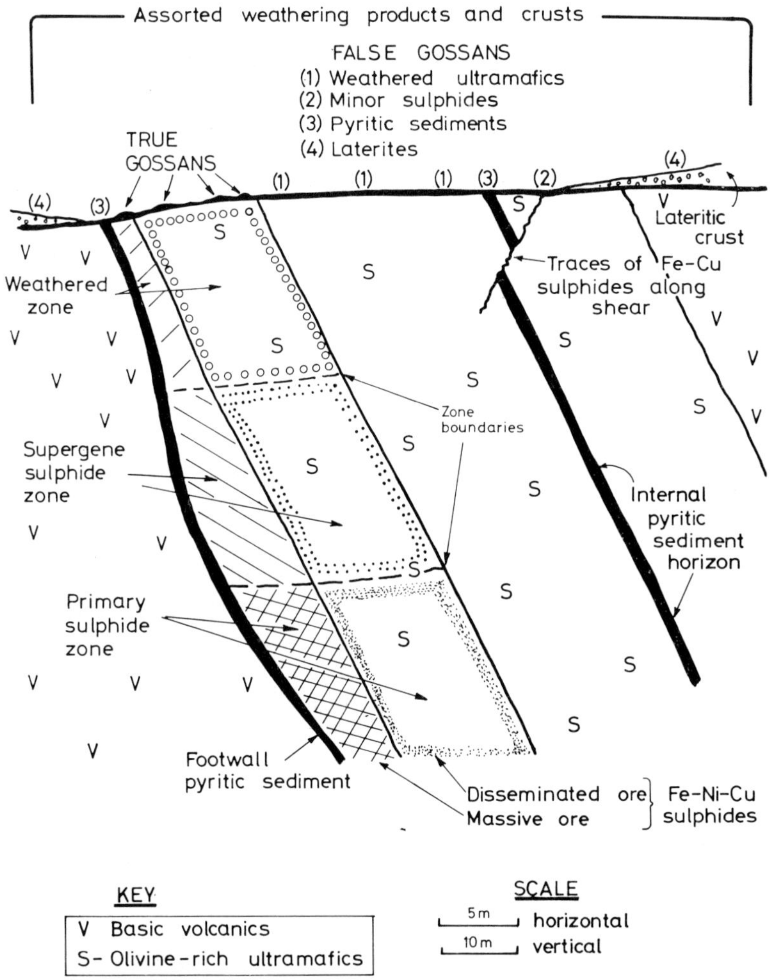

Fig. 2. Schematic, vertical section through typical Yilgarn nickel sulphide deposit, showing mineralized zones and relationships between true and false gossans.

of greenstone belts. These ultramafics occur largely as thick, highly continuous sheets; nickel sulphides are concentrated locally in shoots which typically occur at, and directly above, the stratigraphic bases (i.e. footwalls) of sheets. The mineralization typically consists of the simple assemblage pyrrhotite-pentlandite-pyrite-chalcopyrite. In most deposits (Fig. 2), two mineralized zones are evident; a thin (1—3 m) massive sulphide basal zone overlain by a thick (up to 30 m) disseminated sulphide zone. Massive sulphide zones commonly rest on thin "footwall" metasedimentary horizons (usually pyritic cherts and shales containing minor chalcopyrite and/or sphalerite) and the lowermost few centimetres are commonly enriched in chalcopyrite. Boundaries between massive and disseminated zones, and between disseminated zones and overlying barren ultramafics, are usually sharp. All shoots are generally oxidized to a vertical depth of about 40 m, with a zone of supergene sulphides developed between the oxidized and primary zones (Fig. 2).

To date over twenty major deposits of the type described above have been discovered with reserves estimated to be in excess of 10^8 tonnes at an average grade slightly above 2.4% Ni. Other types of nickel sulphide deposits such as the large, low-grade, Mt. Keith-type deposits and some of the more unusual deposits, such as those located at Mt. Martin and Carr Boyd Rocks, are not included in this reserve estimation.

NICKEL GOSSANS AND OTHER WEATHERING CRUSTS

Virtually all Yilgarn nickel sulphide deposits were discovered as a direct consequence of gossan searching, hence the importance of the "gossan search" approach can hardly be over-emphasized. Unfortunately, as noted by Clema and Stevens-Hoare (1973), discovery and recognition of true nickel gossans is hampered not only by the lack of outcrop but also by the wide variety of other types of ironstones (false gossans) in the region. False gossans include normal weathering crusts (e.g. laterites, ferruginous silcretes) as well as those produced by weathering of non-economic sulphides (e.g. pyritic shales, pyrite-chalcopyrite-filled shears in ultramafics). "Nickel gossans", "false gossans" and "laterites" are defined as follows:

Nickel gossans are accumulations of limonitic and/or jasperoidal material derived largely, or entirely, by leaching of economic, or sub-economic, nickel sulphide mineralization and associated minerals.

False gossans are accumulations of limonitic and jasperoidal material derived from any source other than economic or sub-economic nickel sulphide mineralization. False gossans superficially resemble true gossans and include gossanous material derived from non-economic sulphides, typically pyrite, and *laterites* which are accumulations of secondary oxides and hydrated oxides of Fe, Al, Cr etc. Laterites develop in arid environments as resistate mineral assemblages as a result of weathering. Yilgarn laterites tend to be more ferruginous (i.e. less siliceous) than nickel gossans and are commonly quite magnetic.

Relations between true and false gossans are illustrated schematically in Fig. 2.

REVIEW OF LITERATURE

Cochrane (1973) determined the Cu, Pb, Zn, Ni, Co, $Cr^{(s)}$ (i.e. perchloric-nitric acid-soluble chromium), Mn, Ag, Mo and Fe contents of 73 true and 10 false gossans from 23 localities in the Yilgarn, including all mines operating at the time. The data suggest that:

(1) Unlateritized true gossans generally yield high Cu (500—5000+ ppm) and Ni (1000—5000+ ppm) associated with low $Cr^{(s)}$ (500 ppm), Mn (500 ppm) and Zn (100 ppm).

(2) False gossans, derived by non-lateritic weathering of unmineralized ultramafic rocks, generally yield similar, or lower, Ni values to those observed in true gossans but are strongly depleted in Cu (100 ppm). Lateritic variants are commonly enriched in $Cr^{(s)}$ (2000—10,000+ ppm) and Zn (250+ ppm).

(3) Lateritic false gossans derived from various rock types are characterized by variable Mn (20—1200 ppm), Ni (15—3500 ppm), $Cr^{(s)}$ (30—10,000 ppm) and Zn (20—500± ppm) and are generally more ferruginous (i.e. less siliceous) than true gossans.

Clema and Stevens-Hoare (1973) devised successful empirical scattergrams which make use of various combinations and ratios of Cu, Ni, Zn, Pb, Mn and $Cr^{(s)}$(see Fig. 3), to distinguish true from false gossans. Joyce and Clema (1974) applied the statistical technique of principal component analysis to derive a simpler scattergram (Fig. 4) based only on Cu, Ni and $Cr^{(s)}$ values. Bull and Mazzucchelli (1975) applied the multivariant statistical technique of discriminant analysis (details of discriminant equation not given), using the five elements Ni, Cu, Co, Zn and $Cr^{(s)}$, to distinguish largely between true and false gossans, and also noted Te to be a useful confirmatory element.

Results of orientation studies over particular deposits have been reported by Butt and Sheppy (1975; Mt. Keith), Cox (1975; Pioneer) and Wilmshurst (1975; Perseverance and Spargoville). Butt and Sheppy (1975) noted that values of 4000 ppm Ni and 300 ppm Cu in weathering crusts (mainly lateritic) at Mt. Keith are usually, but not always, indicative of disseminated Fe-Ni-Cu sulphides at depth. They concluded that parameters other than simple Cu and Ni values and ratios are necessary even when searching for large, disseminated, Mt. Keith-type Fe-Ni-Cu sulphide deposits. (N.B. Gossans on disseminated sulphides were excluded from previous gossan studies.) Cox (1975) re-emphasized the usefulness of soil sampling in areas where soils are shallow (see Mazzucchelli, 1972) as a method of locating gossan float and sub-outcrop. Wilmshurst (1975) reported on extensive geochemical investigations carried out on the Perseverance and Spargoville Ni deposits and noted that:

(1) Fresh Fe-Ni-Cu sulphides generally possess Ni/Cu ratios of about 15; in the case of true gossans the ratio is commonly lower and in the case of false gossans it is commonly higher.

(2) High Zn values, even where associated with moderately high Cu values, usually characterize false gossans (typically laterites).

(3) True gossans are most clearly characterized by the presence of traces of Pd and Pt (originally noted by Keays, 1972).

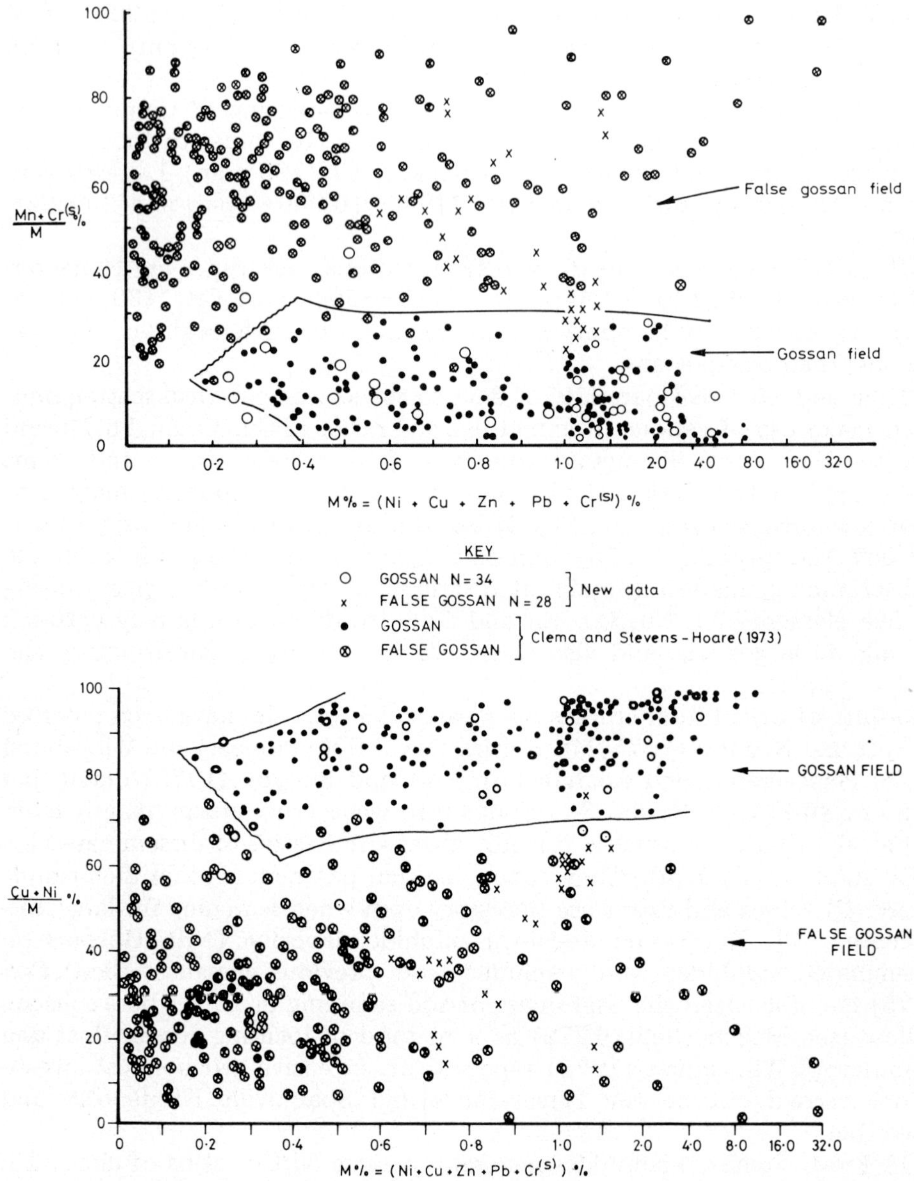

Fig. 3. Signature diagrams of Clema and Stevens-Hoare (1973) designed to distinguish true from false massive nickel sulphide gossans. New data are also plotted onto the diagrams.

(4) High As values in true gossans (commonly associated with high Au and Zn values) are related either to the presence of nickel arsenides, associated with the nickel sulphides at depth, or to the presence of arsenopyrite in adjacent metasediments.

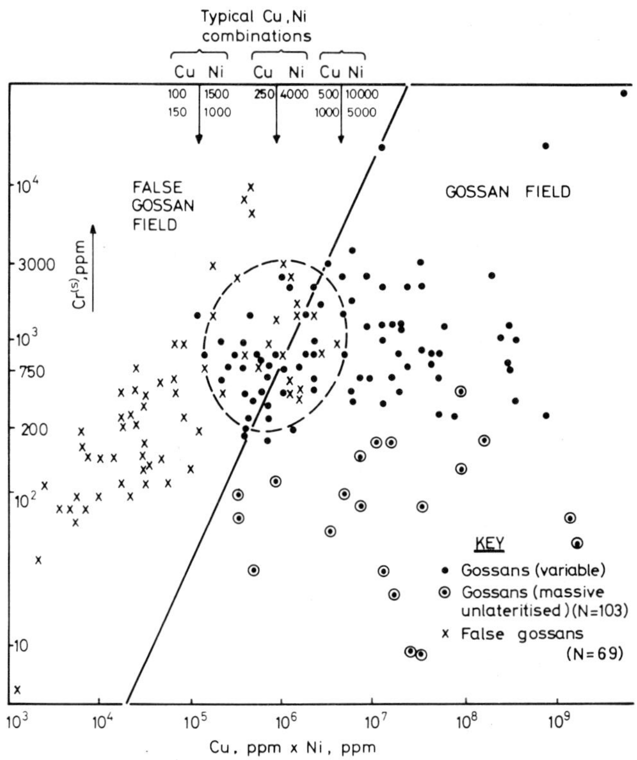

Fig. 4. New data plotted onto the simplified gossan scattergram of Joyce and Clema (1974). The main area of overlap (i.e. uncertainty) is indicated. (Note: four coincident gossan points in gossan field; two coincident false gossan points in false gossan field).

NEW DATA

Sample collection and analysis

Samples were collected by the author and his associates during the period 1966–1974 from most, but not all, known nickel sulphide gossans and environs. At each true gossan outcrop (confirmed as such by drilling or mining), two or three samples were collected over a narrow stratigraphic interval (about 2–10 m). False gossan samples were collected in a similar manner from unmineralized basal contact areas (confirmed as such by drilling) of mineralized ultramafic tracts. Most false gossans consisted of ferruginized (mainly lateritized) and silicified ultramafic rocks; however, about 10 samples were collected from weathered pyritiferous "footwall" sediments and from weakly mineralized (pyrite-chalcopyrite) shears in ultramafics. Wads were not sampled. Very few "grab" samples (sensu stricto) were collected; most samples consisted of about a dozen "chips" collected from areas of 1–10 m^2. In all, 103 representative true gossans and 69 representative false gossans were collected from 32 localities. In addition, 40 representative primary sulphide ore (i.e. unweathered and non-supergene modified) samples were also collected. The samples were analyzed for a wide range of elements (Table I).

Results and discussion

The data in Table I indicate that true gossans, compared with false gossans, (1) are generally, but not always, less ferruginous (i.e. more jasperoidal) (2) are generally, but not always, enriched in Ni and Cu; (3) generally possess lower Ni/Cu ratios (mostly 2—6; 5—20 in false gossans; (4) generally yield low Zn, Mo, Mn values; (5) are generally somewhat enriched in Bi; and (6) are *invariably* enriched in Pt, Pd and Se.

The true and false gossans also can be distinguished fairly well using the three earlier-mentioned scattergrams (Figs. 3, 4) and other diagrams such as the triangular plot Cu-Ni-Zn (Fig. 5). The author found these diagrams to be generally about 80—90% successful in distinguishing true and false gossans. The data presented in Table I indicate that Pt, Pd and Se values must be determined, however, to distinguish true gossans positively.

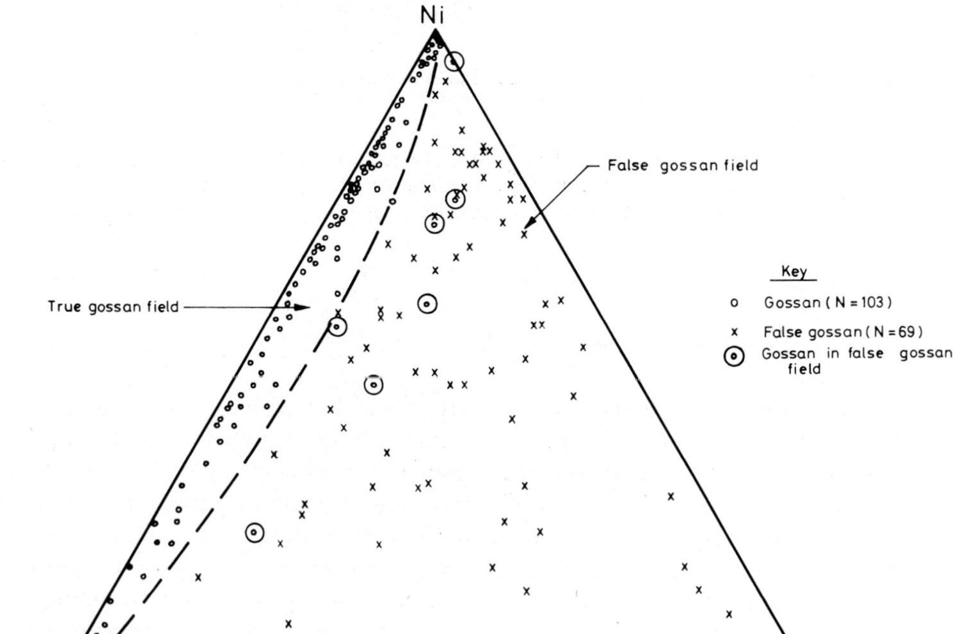

Fig. 5. Triangular diagram based on weight percent Ni, Zn and Cu, showing subjectively assigned true and false gossan fields. The indicated field boundary successfully classifies 93% of true gossans and all false gossans studied by the author.

TABLE I

Summary of new geochemical data on nickel sulphide ores, true gossans and false gossans

Element (ppm)		Fresh nickel sulphide ores				True gossans				False gossans			
or oxide (wt.%)		*N*	*V*	*NR*	*M*	*N*	*V*	*NR*	*M*	*N*	*V*	*NR*	*M*
SiO_2			n.d.	—	—	26	5—70	50—60	52	26	4—70	40—50	42
ΣFe_2O_3			n.d.	—	—	26	9—60	40—50	44	26	12—78	50—60	56
MgO			n.d.	—	—	26	0—12	1—3	1.5	26	0—15	1—4	2
Al_2O_3			n.d.	—	—	26	0—2	0.5—1	0.8	26	0—5	0.5—2	0.8
Cu	(5)	40	200—9500	2000—5000	2100	103	110—13,000	800—3000	1000	69	30—1200	80—150	90
Ni	(5)	40	12,000—85,000	25,000—40,000	31,000	103	950—25,000	1500—5000	3000	69	50—10,000	800—2500	1500
Zn	(5)	40	20—300	60—120	80	103	20—1100	50—200	140	69	50—2200	200—1500	800
Pb	(5)	40	20—70	30—60	40	34	20—80	30—40	35	28	5—40	10—20	30
Co	(5)	40	150—3500	400—1800	1200	103	40—1200	200—400	280	28	10—800	200—800	480
$Cr^{(s)}$	(5)	40	80—3000	300—700	600	103	10—11,000	200—2000	750	69	40—10,000	100—1800	600
Mn	(5)	40	50—1400	250—600	450	103	100—400	200—400	600	28	200—5000	800—2000	1400
Mo	(5)	40	<5	<5	<5	103	<5—120	<5	—	20	<5—20	∿10	10
Pt	(0.02)	20	0.07—2.2	0.08—0.80	0.45	20	0.05—0.28	0.1—0.2	0.15	20	⩽0.02	—	—
Pd	(0.02)	20	0.15—1.7	0.5—1.2	0.70	20	0.08—0.48	0.1—0.3	0.15	20	⩽0.02	—	—
Au	(0.02)	20	<0.02—0.52	∿0.02	0.02	20	0.0—1.7	⩽0.15	—	20	⩽0.02	—	—
Ag	(2)	20	<2—15	∿3.0	3.0	50	<3—11	⩽3	2	20	⩽0.02	—	—
As	(5)	20	<5—1500	<5—10	8	20	5—60	20—30	25	20	10—15	20—30	25
Bi	(5)	20	25—65	30—50	40	20	15—180	20—60	45	20	0—30	∿15	15
Se	(2)	20	5—100	10—40	25	20	10—200	20—40	35	20	0—2	⩽2	—
Sb	(4)	20	<4—25	<4	<4	20	<4—410	20—40	30	—	n.d.	—	—
Ni/Cu		40	1.6—57.4	8—20	14.8	103	0.4—11.9	2—6	3.7	69	1.5—31.7	5—20	14.9

$Cr^{(s)}$ refers to acid-soluble ($HClO_4$-HNO_3) chromium; figure in parentheses following elements represents detection limit.

Analytical techniques (initial sample weights 500—1000 g): Si, Fe, Mg, Al, As, Bi, Se, Sb — X-ray fluorescence (±2%); Cu, Ni, Zn, Pb, Co, $Cr^{(s)}$, Mn, Mo, Ag — atomic absorption spectrometry (±5%); Pt, Pd, Au — fire assay/atomic absorption spectrometry (±5%).

Symbols: *N* = number of samples analyzed, *V* = variation, *NR* = normal range (i.e. 15—85 percentiles) values rounded, *M* = median value.

n.d. = not determined.

Two further points are worthy of mention, namely the behaviour of Pd relative to Pt in gossans, and the significance of $Cr^{(s)}$ in gossan interpretation.

Examination of median Pd and Pt values in true gossans and fresh parent ores (Table I) suggests that relative to ores, Pd in gossans is more strongly depleted than Pt; this is possibly explainable in terms of Eh-pH relations for the system Pd-Pt-H_2O-Cl at 25°C (Fuchs and Rose, 1974; see Fig. 6). Under the highly acidic and oxidizing conditions which prevail within the zone of oxidation (e.g. Nickel et al., 1974; Blain and Brotherton, 1975;) Pd is more leachable than Pt. Nevertheless, it must be remembered that Pt is commonly less abundant and more erratically distributed than Pd in nickel sulphide ores and hence, by inference, in nickel gossans. Wilmshurst (1975) noted Pd/Pt ratio changes of between 1 and 8 over relatively short stratigraphic intervals (10—15 m) in sulphides at Spargoville, south of Kalgoorlie. In view of these findings, the significance of Pd and Pt values and ratios in nickel gossans should be regarded with considerable suspicion.

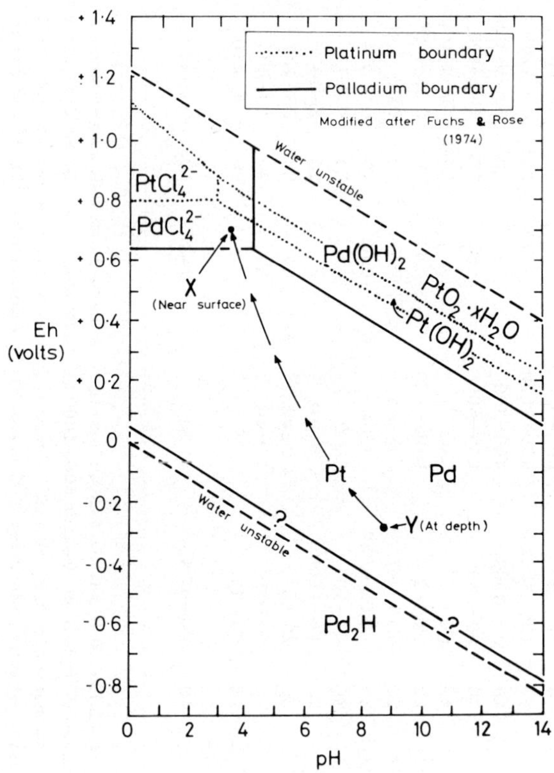

Fig. 6. Eh-pH diagram for the system Pd-Pt-H_2O-Cl at 25°C. The boundary conditions are: Pt = 10^{-12} *M* (1×10^{-4} ppb), Pd = 10^{-12} *M*, and Cl = 3×10^{-4} *M* (10 ppm). (After Fuchs and Rose, 1974.)

It seems that *unlateritized* sulphide gossans derived from extensive massive sulphide ores are generally strongly depleted in $Cr^{(s)}$ (mainly $\leqslant$200 ppm; see Fig. 4). This is possibly not fortuitous; during the early stages of gossan formation, oxidizing ores are extensively leached by large volumes of percolating dilute H_2SO_4 (e.g. Nickel et al., 1974). Mineragraphic investigations by the author and by Ewers et al. (1976) have indicated that most $Cr^{(s)}$ is tied up in secondary ferrochromite which occurs as rims about refractory chromite grains. These rims are more highly reflecting under the microscope, and may be readily dissolved in the laboratory using dilute $HClO_4$ or H_2SO_4. For example, in one test run, a relatively pure separate composed of ferrochromite-rimmed chromite grains was heated (80°C) and stirred continually in excess 1*M* H_2SO_4 solution for about eight hours; the original separate assayed 14.6% Cr, whereas the final residue assayed 24.8% Cr and weighed 50% less. Mineragraphic examination of the original separate and the final residue indicated that all ferrochromite (rims) had been removed (i.e. about 50% of the original weight, containing a calculated 4.5% Cr). It therefore seems possible that in certain field situations, $Cr^{(s)}$ may be dissolved out of true gossans by percolating acidic groundwaters and may possibly be precipitated in nearby false gossans. It must be remembered, however, that massive sulphide ores, compared with associated disseminated sulphide ores and unmineralized ultramafic rocks, are generally somewhat depleted in total $Cr^{(s)}$, hence low $Cr^{(s)}$ values in gossans may be largely due to this inherent feature. The nature and origin of the $Cr^{(s)}$ in true and false gossans needs to be researched in more detail.

CONCLUSIONS

In conclusion, although preliminary discrimination of true gossans can be made largely as suggested in previously published literature, using Cu, Ni, $Cr^{(s)}$, Mn, Zn and Pb values, data presented here indicate that Pt, Pd and Se values are very useful at the confirmatory stage. Te may also be determined at the confirmatory stage, although this element was not determined by the writer. As mentioned earlier, Bull and Mazzucchelli (1975) found it to be particularly useful. Needless to say, textural (i.e. mineragraphic) studies constitute a valuable adjunct to geochemical studies in positively identifying true gossans.

REFERENCES

Blain, C.F. and Brotherton, R.L., 1975. Self potentials in relation to oxidation of nickel sulphide bodies within semi-arid climatic terrains. Trans. Inst. Min. Metall., 84-B: 123—127.

Bull, A.W. and Mazzucchelli, R.H., 1975. Application of discriminant analysis to the geochemical evaluation of gossans. In: I.L. Elliott and W.K. Fletcher (Editors), Geochemical Exploration 1974. Elsevier, Amsterdam, pp. 221—226.

Butt, C.R.M. and Sheppy, N.R., 1975. Geochemical exploration problems in Western Australia exemplified by the Mount Keith area. In: I.L. Elliott and W.K. Fletcher (Editors), Geochemical Exploration 1974. Elsevier, Amsterdam, pp. 391—416.

Clema, J.M. and Stevens-Hoare, N.P., 1973. A method of distinguishing nickel gossans from other ironstones of the Yilgarn Shield, Western Australia. J. Geochem. Explor., 2: 393—402.

Cochrane, R.H.A., 1973. A guide to the geochemistry of nickeliferous gossans and related rocks from the Eastern Goldfields. In: Report of the Department of Mines, Western Australia, for the year ending December, 1972. Rep. Department of Mines, Perth, W.A. Div. IV, pp. 115—121.

Cox, R., 1975. Geochemical soil surveys in exploration for nickel-copper sulphides at Pioneer, near Norseman, Western Australia. In: I.L. Elliott and W.K. Fletcher (Editors), Geochemical Exploration 1974. Elsevier, Amsterdam, pp. 437—460.

Ewers, W.E., Graham, J., Hudson, D.R. and Rolls, J.M., 1976. Crystallization of chromite from nickel-iron-sulphide melts. Contrib. Mineral. Petrol., 54: 61—64.

Fuchs, W.A. and Rose, A.W., 1974. The geochemical behaviour of platinum and palladium in the weathering cycle in the Stillwater Complex, Montana. Econ. Geol., 69: 332—346.

Joyce, A.S. and Clema, J.M., 1974. An application of statistics to the chemical recognittion of nickel gossans in the Yilgarn Block, Western Australia. Proc. Australas. Inst. Min. Metall., 252: 21—24.

Keays, R.R., 1972. Platinoids as discriminators of gossans derived from nickeliferous sulphides. Geol. Soc. Aust. Joint Spec. Groups Meet., Canberra, A.C.T., 1972 (unpublished).

Mazzucchelli, R.H., 1972. Secondary geochemical dispersion patterns associated with the nickel sulphide deposits at Kambalda, Western Australia. J. Geochem. Explor., 1: 103—116.

Nickel, E.H., Ross, J.R. and Thornber, M.R., 1974. The supergene alteration of pyrrhotite-pentlandite ore at Kambalda, Western Australia. Econ. Geol., 69: 93—107.

Wilmshurst, J.R., 1975. The weathering products of nickeliferous sulphides and their associated rocks in Western Australia. In: I.L. Elliott and W.K. Fletcher (Editors), Geochemical Exploration 1974. Elsevier, Amsterdam, pp. 417—436.

Journal of Geochemical Exploration, 8(1977)259—281 259

SOME ASPECTS OF THE USE OF GEOCHEMISTRY IN THE SEARCH FOR NICKEL SULPHIDES IN LATERITIC TERRAIN IN WESTERN AUSTRALIA

B.H. SMITH

Western Mining Corporation, Kalgoorlie, W.A. (Australia)

(Revised version accepted March 1, 1977)

ABSTRACT

Smith, B.H., 1977. Some aspects of the use of geochemistry in the search for nickel sulphides in lateritic terrain in Western Australia. J. Geochem. Explor., 8: 259—281.

Extensive lateritization of the rocks of the Archaean Shield of Western Australia, often to depths in excess of 100 m, constitutes a considerable barrier to geochemical exploration for nickel sulphides. Some of the chemical and mineralogical factors that affect the interpretation of geochemical data for laterites are the subjects of studies reported here.

In the mottled clay zone of laterite profiles, Cu is mainly associated with goethite, while Ni is associated with smectite. Ni and Cu change from an association with magnesium silicates to one with iron oxides as weathering progresses, with Cu moving more readily to the Fe association than Ni.

In a mineralized environment, where nickel sulphides are present with iron sulphides and minor amounts of copper sulphides, the association of Ni and Cu with iron oxide, both adjacent to, as well as at the sites of mineralization, could therefore be expected to occur lower in the weathering profile than in a barren situation. Electron microprobe studies confirmed this, showing that higher concentrations and proportions of Ni are present in iron-rich phases in the saprolite zone of a mineralized environment than are present in similar phases in a barren environment. However, the Ni contents of iron-rich phases in the higher ferruginous zone of the laterite profiles of the two environments were similar.

The dissolution of "active Fe" from samples from both mineralized and barren lateritized environments using sodium dithionite also released Ni and Cu and this supported the mechanism of an association with iron oxide in mineralized localities. The Ni extracted also gave good discrimination between a sulphide origin and a silicate origin. Discrimination was less definite when total Ni values were used. Cu is strongly held by iron oxide, so total values of Cu as well as that extracted by dithionite gave good discrimination.

The significance of the distribution of Cr, Co, Mn, Mg and Ca in the interpretation of data is also discussed using an example of a laterite profile developed over low-grade nickel sulphide mineralization near Kalgoorlie.

INTRODUCTION

The Archaean Shield of Western Australia covers an area of approximately 800,000 km^2 and has from time to time been the scene of intense mineral exploration activity for Au and base metals. The shield has been levelled to a peneplain, termed the Great Plateau of Western Australia (Jutson, 1914), which has a highly weathered mantle, often in excess of 100 m deep, and a diffuse drainage system.

Lateritization has been extensive and there has been much debate as to how this is related to epeirogenic uplift of the shield. Woolnough (1918) and Jutson (1914) proposed at least two periods of lateritization, one evidenced by a higher "Old Plateau" and the other by a lower "New Plateau", generally at a height differential of about 20 m. Thus, the New Plateau results from rejuvenation of the drainage upon uplift and erosion by streams. As an alternative, Prider (1966) and Playford (1954) suggested a single lateritic event following uplift, with the different levels of the laterite surface due simply to the topography developed by the drainage system.

There has also been debate as to whether laterites are mainly formed on in-situ material (Prider, 1966) or include detrital products of earlier laterites (Woolnough, 1918; Mulcahy, 1960, 1967; Mulcahy and Bettenay, 1972; Mulcahy et al., 1972). It suffices to say that the weathering, erosional and depositional history of the area is complex and so far no precise sequence of events has emerged.

Furthermore, there is debate about the absolute age of lateritization. It is generally thought to be of late Miocene or Pliocene age (Mulcahy 1959; Prider, 1966) although a Permian to Jurassic age (Daily et al., 1974) must be considered if the laterites on the shield are coeval with laterites in some other parts of Australia (David, 1950).

The shield laterites are products of considerable chemical weathering and, being quite different from the parent rocks, they therefore constitute a formidable barrier to geochemical exploration for base metals. The geology of the Archaean Shield, also known as the Yilgarn Block, is summarized in Fig. 1, while the areal extent of lateritization is shown in Fig. 2. The area of laterite soils shown also includes eroded laterite profiles where some of the original horizons have been removed. It has been estimated (C.S.I.R.O., 1975) that only 20% of the total area in the Kalgoorlie region considered to have mineral potential can be effectively covered by surface geochemical techniques. This is even after special geochemical techniques, such as the grouping of data according to weathering history prior to statistical analysis, have been brought to bear on the problem.

The object of the study was to gain a better understanding of the chemical and mineralogical changes that occur with lateritization of ultrabasic rocks, both from situations barren of nickel sulphides and those that are mineralized. It was hoped that as a result, some reduction might be made in the present requirement of having to drill to fresh rock on a detailed grid pattern

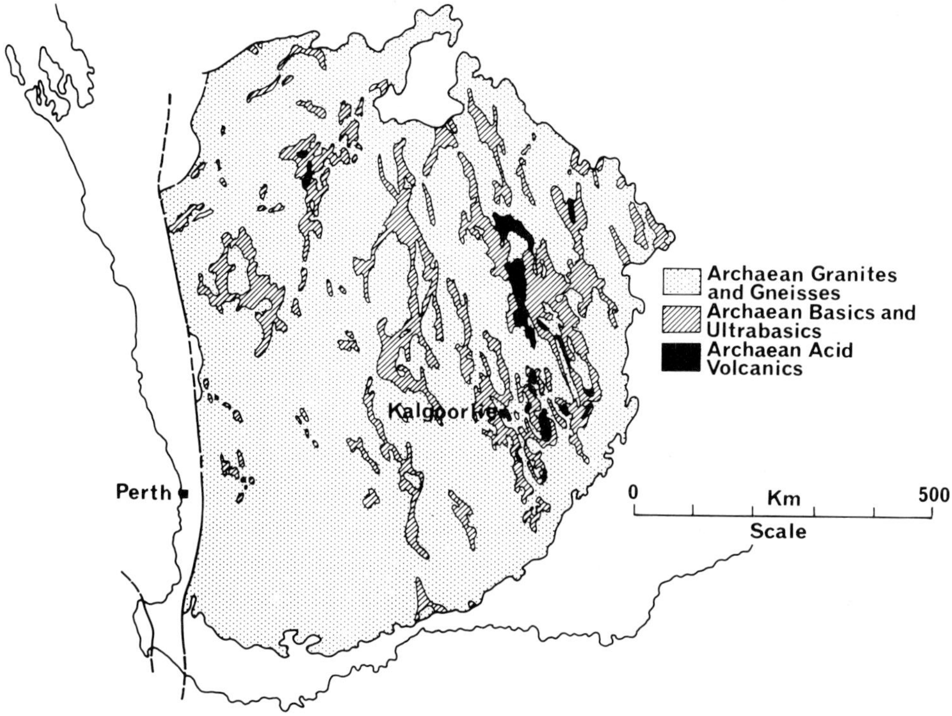

Fig. 1. The geology of the Yilgarn Block, Western Australia.

to locate mineralization. The results discussed here have relevance to laterites in the southeast sector of the shield in an area within a 200-km radius of Kalgoorlie. However, some sections of the work may have a wider application.

METHODS AND MATERIALS STUDIED

The laterite profiles discussed are considered to be representative of those commonly encountered during exploration for nickel sulphides in laterite-covered terrain in the area described above.

The changes in the morphology, chemistry and mineralogy through a laterite profile formed on ultrabasic rocks barren of nickel sulphides are outlined by reference to an example from near Ora Banda, approximately 60 km northwest of Kalgoorlie. Mineralogical determinations were made by X-ray diffraction techniques using Co radiation. Chemical analysis was carried out by atomic absorption spectrophotometry (AAS) after digestion of the finely ground samples in a refluxing mixture of HNO_3, $HClO_4$ and HF and boiling to dryness prior to dissolution in HCl.

The relationship of elements to mineralogy within particular horizons in other laterite profiles also formed on ultrabasic rocks barren of nickel sulphides was studied by carrying out partial mineralogical separations by me-

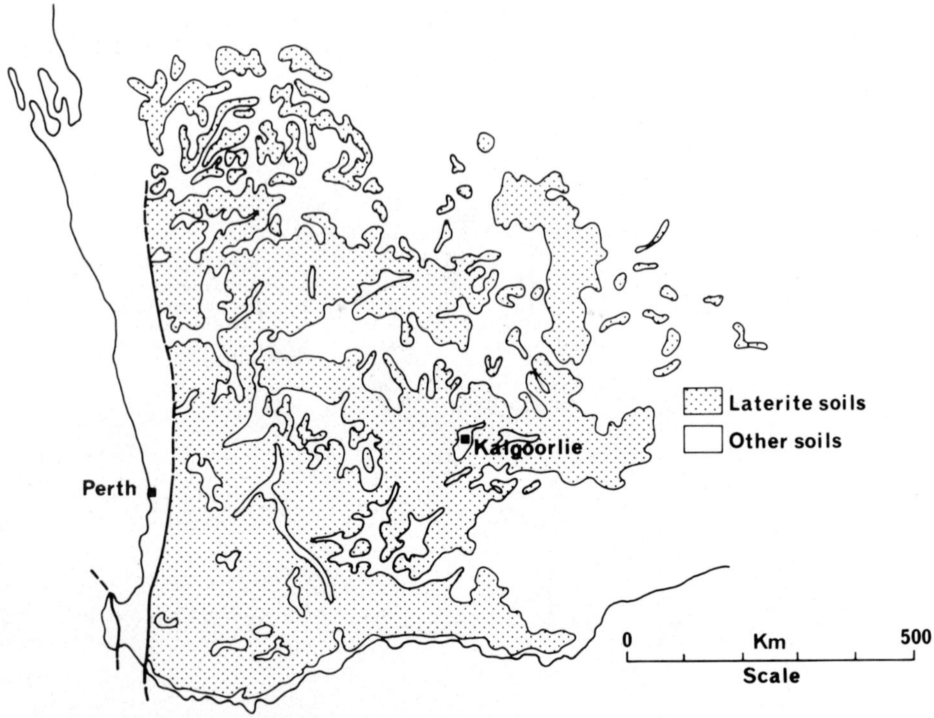

Fig. 2. The areal extent of lateritization of the surface of the Yilgarn Block, W.A.

chanical means. It was found that the minerals present are preferentially distributed in various particle size ranges. Three 50-g samples from four different horizons in each profile were vigorously agitated with a propeller-type stirrer in 500 ml of 4×10^{-3} M NaOH and 0.005% (w/v) sodium hexametaphosphate. Four fractions, > 200 μm, 200—20 μm, 20—2 μm and < 2 μm were separated from each sample by sedimentation and sieving. The fractions were chemically analyzed using HNO_3-$HClO_4$-HF digestion followed by AAS and the minerals present were identified and estimated by X-ray diffraction (XRD).

The association of elements and minerals was also studied at two different horizons in laterite profiles developed on mineralized and unmineralized ultrabasic rocks using an electron microprobe analyzer (K. Norrish, personal communication, 1976). Approximately 50 random points on each sample were scanned for Ni, Co, Fe, Si, Mn, Al, Cr and Mg. In addition, some discrete mineral phases were analyzed.

The relation of Fe to both Cu and Ni through a laterite profile developed on mineralized ultrabasic rock from near Widgiemooltha 100 km south of Kalgoorlie and a profile developed on unmineralized ultrabasic rock from near Ora Banda was studied by extraction of the more chemically active fraction of Fe with sodium dithionite solution using the method of Coffin

(1963). The Ni, Cu and Fe extracted were analyzed by AAS after oxidation of any citrate and partially oxidized dithionite products left in solution by using HNO_3-$HClO_4$ mixture.

RESULTS AND DISCUSSION

The discussion of the results of these studies on laterite profiles emphasises factors considered to have relevance to the interpretation of geochemical data generated in the search for nickel sulphide mineralization in ultrabasic rocks.

Morphology and mineralogy

General characteristics of a laterite profile

Figs. 3 and 4 show the main characteristics of a laterite profile from Ora Banda formed on ultrabasic rocks. These are discussed briefly so that later observations can be seen in the context of a complete laterite profile. The most commonly observed features of laterites developed on ultrabasic rocks are the remobilization and enrichment of Fe in a dominant ferruginous zone (or plinthite zone when massive), so that none of the original rock tex-

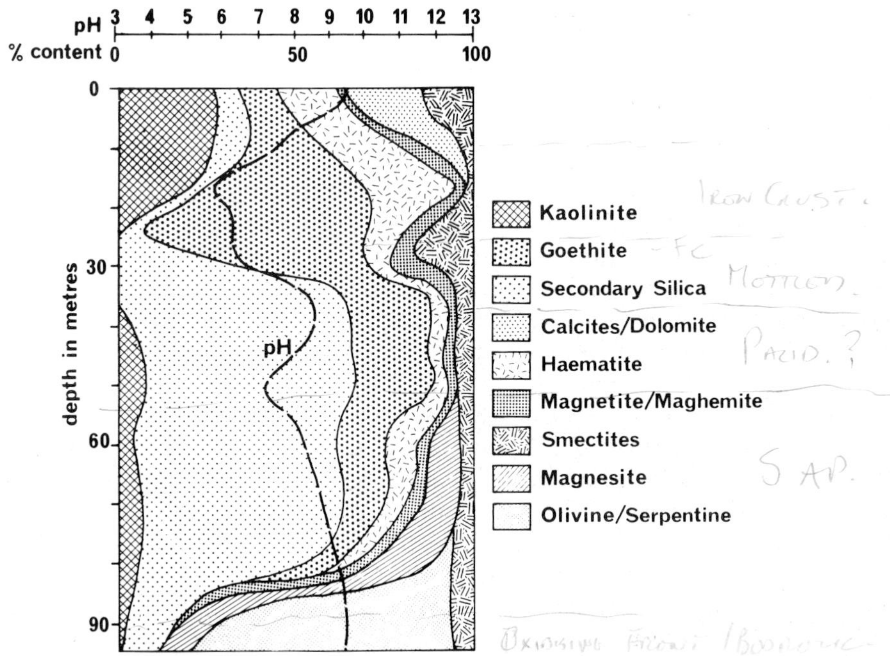

Fig. 3. The mineralogy and pH of an unmineralized laterite profile (SG41) at Ora Banda, W.A.

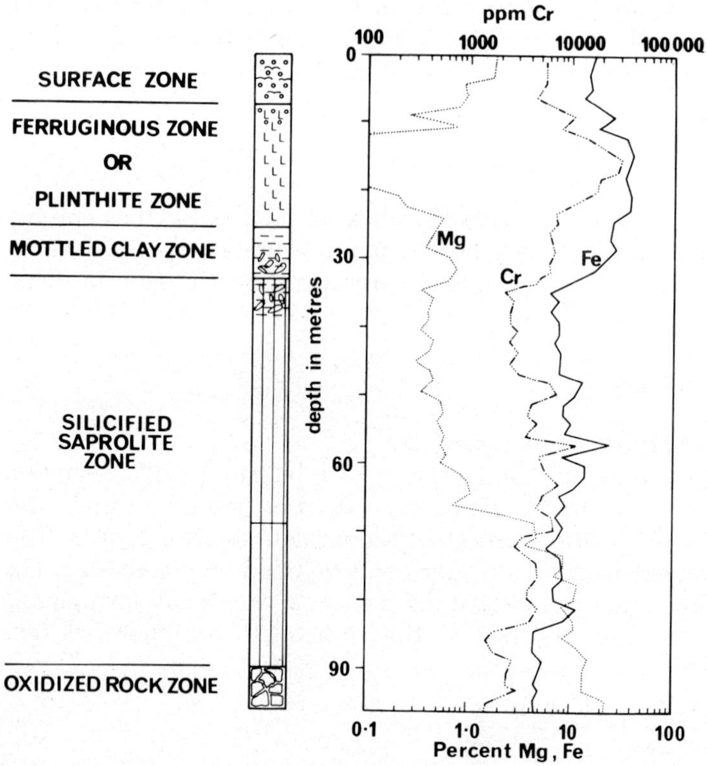

Fig. 4. Mg, Cr and Fe concentrations through an unmineralized laterite profile (SG41) at Ora Banda, W.A.

tures remain, the presence of mottled clays immediately below this and, lower in the profile, deposition of secondary silica and formation of clays with variable morphological features. This is followed, at depth, by saprolitic clays showing the morphology of the original rock-forming minerals, with a slow gradation to oxidized rock. The water table at the time of formation of the laterites is assumed to have fluctuated about the boundary of the ferruginous zone and mottled clay zone on the basis of observations on laterites presently forming in the tropics (Van Schuylenborgh, 1969).

The major characteristics of the mineralogy shown in Fig. 3 are the presence of large amounts of goethite in the upper part of the profile, with haematite increasing in proportion towards the surface. Secondary silica in the lower part of the profile occurs as opaline silica, chalcedony and cristobalite. Magnetite, with the possibility of maghemite as it oxidized, remains in constant proportion through to the surface of the profile. Smectite increases in proportion near the inferred ancient water table, while smectite, calcite and dolomite at the surface have probably been recently deposited by wind and water. Calcite and dolomite as well as salt are commonly deflated from bare playa areas under the present arid conditions, while flash storms in summer transport clay material.

The pH (Fig. 3) is poised between 8 and 9 in the surface horizons by the recent additions of carbonates, while values between 5 and 6 in the ferruginous zone reflect the considerable leaching that has occurred in the past. Lower in the profile, the inferred ancient water table is indicated by a sharp rise in pH to 8, with a further gradual increase to 9 down the profile to the oxidizing rock zone.

Common variants of laterite profiles

Multiple ferruginous zones. Enrichment of Fe at separate levels through a laterite profile is frequently observed and is thought to be caused by changes in the average water table depth for long periods of time. These variations in the water table could result from climatic changes or from alterations in the local base levels of erosion. At present, water tables are generally much lower than during the major laterite-forming period. Minor enrichment of Ni can also occur at these subsidiary ferruginous zones.

Lateritization of a previously existing saprolite zone that has been silicified. In contrast to normal laterite profiles, textures of the original rock are sometimes preserved in the surface horizons in material known locally as "cap-rock". This material is massive and has a secondary silica content that may be in excess of 50% SiO_2, compared with values of less than 20% SiO_2 normal for the upper horizons of laterites. The simplest explanation of the formation of these "cap-rocks" is that they are the remnant silicified saprolite horizons of earlier laterite profiles that have been exposed by erosion. Commonly, lateritization has again progressed to great depths below this impervious "cap-rock", with the formation of large solution cavities, recementing of angular material that has fallen into these cavities, and re-establishment of the normal sequence of lateritic horizons.

Chemical profiles

Laterite development makes exploration geochemistry difficult because of the very large losses of some rock-forming components. In the profile from Ora Banda (SG 41), weathering has extended to a depth of more than 90 m. In Fig. 5 an attempt has been made, using Cr as a base, to give an indication of the relative losses and gains of some elements that are of importance in exploration geochemistry. A similar calculation was made for a laterite profile at Greenvale, Queensland, by Zeissink (1969). Such a compilation suffers from two major errors; the erratic distribution of Cr in the original rock both in silicates and spinels, and any loss of Cr caused by the alteration of these minerals, as well as the elimination of Cr from iron oxides in the upper part of the ferruginous zone. Measurements of the weight of element per unit

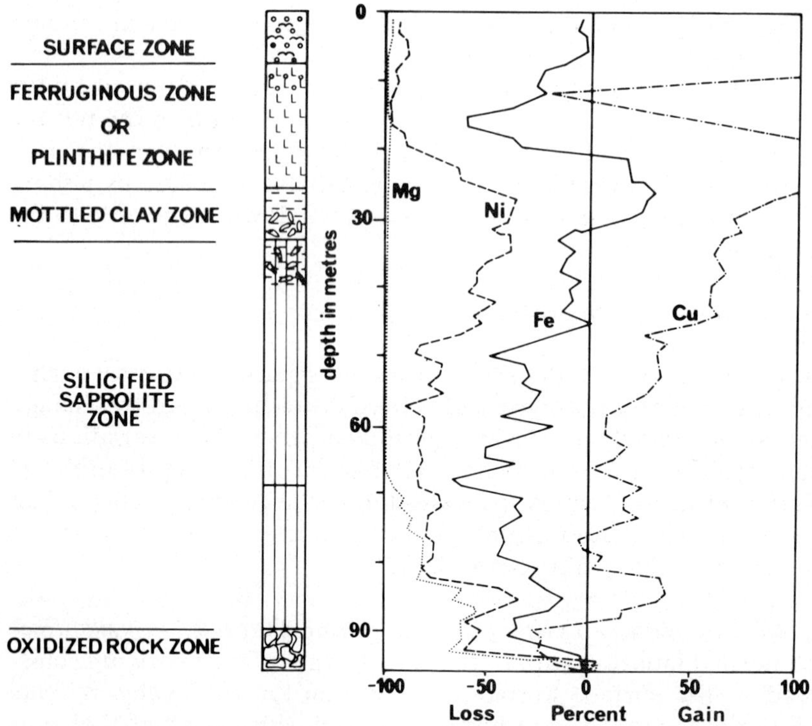

Fig. 5. Percent change in initial content of Mg, Fe, Ni and Cu relative to Cr of a laterite profile (SG41) at Ora Banda, W.A.

volume have also been used to calculate losses. However, in most ferruginous zones there is little or no preservation of original rock textures, so such measurements can involve even larger errors. The important features to recognize in Fig. 5 are the large relative losses of Mg and Ni low in the profile, with a smaller loss of Ni relative to Cr at the mottled clay zone. Cu shows continuous enhancement, through most of the profile relative to Cr, with a maximum in the lower ferruginous zone.

Values of up to 4% Ni in mottled clay zones have, in the past, provided the impetus for much effort in follow-up drilling for nickel sulphides thought to be responsible for these anomalies. However, such Ni values are simply the product of weathering processes and rarely bear any relation to Ni mineralization. Other base metals may also be concentrated at this mottled clay horizon. Iron oxide is implicated here because of its obvious presence in this horizon and the cyclical dissolution and reprecipitation of Fe that occurred during formation of the profiles on reduction and oxidation. Goethite has been shown to be an efficient adsorber of base metals (Grimme, 1968; Forbes et al., 1976) and freshly precipitated material would have a high surface activity. However, clay minerals such as smectite are also present at this horizon and the relative importance of these and iron oxides in retaining Ni and Cu is discussed later.

Relationship of nickel and copper distribution in a laterite profile to mineralogy

The broad relation between Ni and smectite in the upper horizons of laterite profiles is evident from the profiles of the mineralogy and chemistry of the Ora Banda profile that are shown in Figs. 3 and 6. Smectite, together with calcite, dolomite and some of the kaolinite above 20 m, are considered to be exotic, being additions from recent aeolian and fluvial processes. There is also an association of Ni and Cu with both goethite and haematite evident from the same figures. A similar association of Ni and Cu with iron oxides has also been found by Zeissink (1969) for laterite profiles at Greenvale, Queensland, by Wilmshurst (1975) for oxidized mineralized ultrabasic rocks at Spargoville, Western Australia, by Butt and Sheppy (1975) for lateritized ultrabasic rocks from Mt. Keith, Western Australia, and by McKenzie (1975) for some soils in South Australia.

The relationships of Ni and Cu to mineralogy were initially studied in the mottled clay zone of laterite profiles. A sample at this zone was taken from eight separate laterite profiles located from 80 km south to 250 km north of

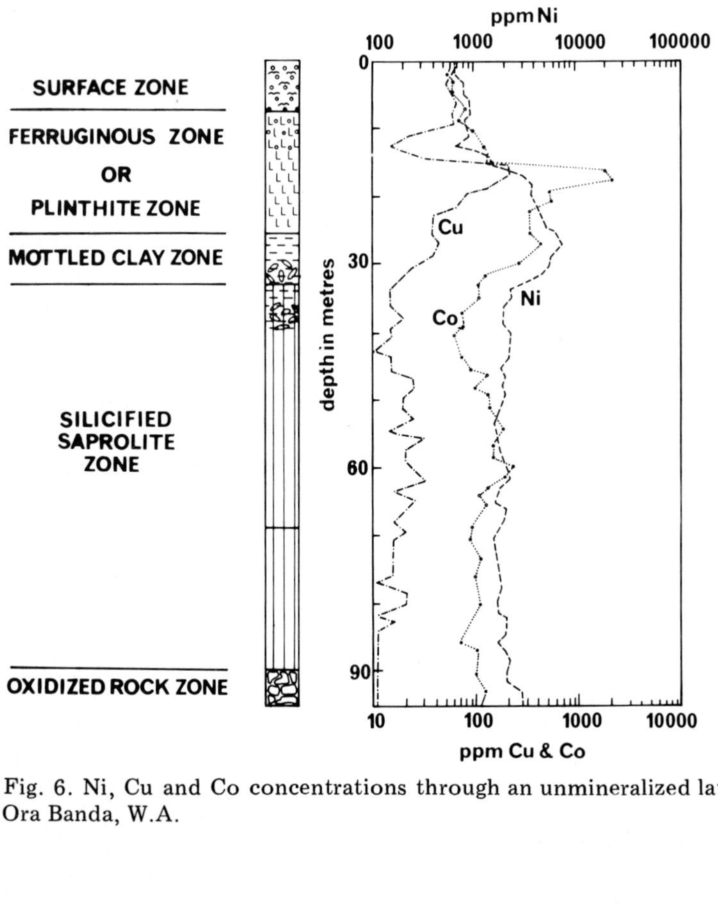

Fig. 6. Ni, Cu and Co concentrations through an unmineralized laterite profile (SG41) at Ora Banda, W.A.

Kalgoorlie, each profile being developed from olivine-rich rock barren of Ni mineralization having approximately the same silicate Ni content of about 2500 ppm in fresh rock. Each sample was separated into four size ranges, > 200 μm, 200—20 μm, 20—2 μm and < 2 μm. This resulted in a partial mineralogical separation, as the component minerals were preferentially distributed in the various size ranges. Smectite is mainly in the < 2-μm fraction, goethite and haematite had a more random distribution with a tendency to concentrate in the 2—20-μm and 20—200-μm size ranges, while secondary silica was highest in the largest size range. The correlations of Ni and Cu with the mineralogy as estimated by XRD were calculated on the log-normalized data. The results are shown in Table I. As there are sampling problems, due to the high variability of the profiles and errors due to the difficulties of XRD estimation of clay material, the results can only indicate trends.

A relationship of Ni with smectite and of Cu with goethite in the mottled clay zone of barren profiles is shown by the values in Table I and this relationship was further studied by analysis of a wider range of material also from barren profiles. However, rather than use costly XRD analysis for each sample to estimate the proportions of smectite and goethite, only total Mg and Fe analyses were carried out. It was found that these two elements were dominantly contained in the two minerals of interest and the minerals were assumed to have a constant content of the elements. In this study, any change in the relationship of Ni and Cu with Fe and Mg was examined by sampling zones above and below the mottled clay zone.

Initially, three samples were taken from each of four adjacent zones in seven different profiles. Each sample was then fractionated into the four different size fractions used previously, so there were finally twelve separate samples from each of four weathering zones in seven profiles. The composition of laterites is variable on a detailed scale, e.g., when comparing adjacent 1-cm cubes. Thus, variability of the initial three samples in each zone of each profile, combined with preferential distribution of minerals in the four size fractions, gave a range of compositions for the final twelve samples that could reveal the correlations of Ni and Cu with Fe and Mg in the normalized data.

The results from one profile are shown in Fig. 7 and similar changes in correlation were observed in the other six profiles. Over the section of the

TABLE I

Correlation coefficients for Ni and Cu with minerals in the mottled clay zone

	Smectite	Goethite
Ni	0.62***	0.23
Cu	− 0.21	0.65***

* $P < 0.05$, ** $P < 0.01$, *** $P < 0.001$.

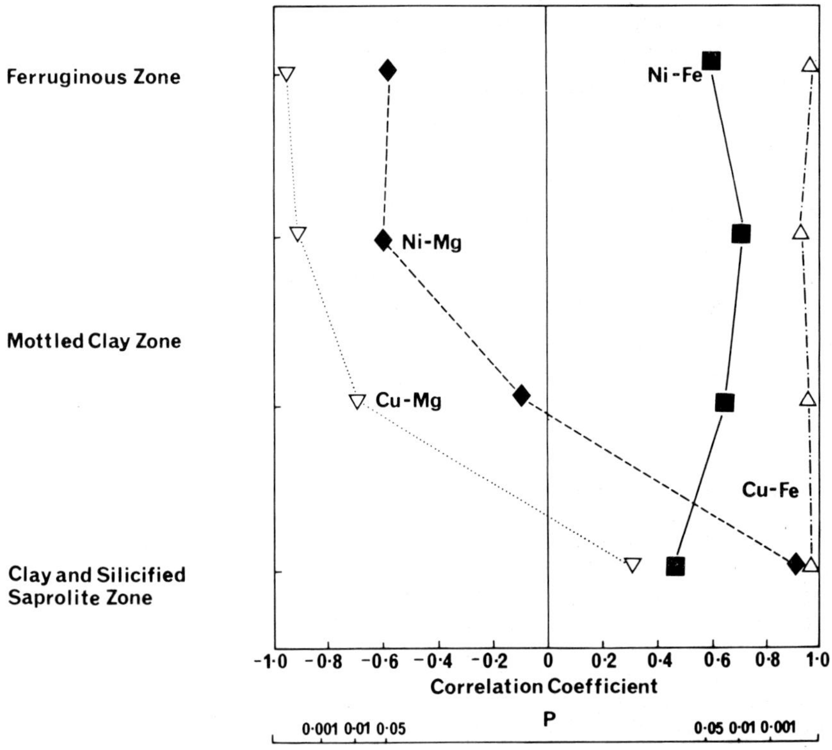

Fig. 7. The changes in the correlations of Ni and Cu with Fe and Mg in the transition from the silicified saprolite zone to the ferruginous zone of a laterite profile.

laterite profiles studied, from the silicified saprolite zone to the ferruginous zone, Ni changes its association with magnesium silicates to one with iron oxides, while Cu apparently remains associated with iron-oxides.

Electron microprobe studies

The association of Ni with Fe was further studied using an electron microprobe (K. Norrish, personal communication, 1976). Analysis of the Cu-Fe relationship was not attempted as the detection limit for Cu is only about 300 ppm and it was assumed that Cu was evenly distributed throughout the samples. The Cu content of the samples ranges from 40 to 320 ppm.

It was argued that, where Ni mineralization (usually of a pentlandite, pyrrhotite, pyrite and chalcopyrite association) has been lateritized, Ni would be dominantly associated with iron oxides in a lower weathering zone in the profile. This is in contrast to the situation where Ni is initially only present in olivine (Rajamani et al., 1975). This early Ni-Fe association would occur in haloes around mineralization as well as at the site of mineralization due to the outward movement of Fe from the reducing and acidic environment of the weathering sulphides. This hypothesis was tested by scanning samples from the ferruginous zone or saprolite zone of three profiles, one known to be related to nickel sulphide mineralization, and two others from areas barren of mineralization. The samples from the two barren associations were of interest because of their high Ni contents of about 1%. The concentrations of Ni in various mineral phases of the samples that were identified from their major element compositions are given in Table II. These, combined with the results from semi-quantitative estimation of the mineralogy by XRD, were used to apportion total Ni to the various minerals.

Comparison of the two samples from saprolite zones, sample B from a barren environment and sample D from a mineralized environment, supports the above hypothesis. The concentrations of Ni in iron-rich phases, which are

TABLE II

Ni content of various minerals in the ferruginous zone and saprolite zone of laterite profiles from barren and nickel sulphide associations

Sample No.	Profile zone, association, hole No. and depth (m)	Component minerals	Mineral content of sample (%)	Ni content of mineral (%)	Ni contained in mineral in 100 g of sample (g)	Ni distribution (%)
A	ferruginous	smectite	30	1.1	0.33	30±7
	barren	chlorite	30	0.6	0.18	20±5
	SED 1	secondary silica	25	0	0	0
	12.5—13	manganese oxide	trace	10	0.05	5±3
		goethite	15	3.0	0.45	45±10
		total			1.0	
B	saprolite	smectite	35	0.2	0.07	5±3
	barren	chlorite	30	2.9	0.87	80±15
	XV 33	chrysotile	15	0.2	0.03	1±1
	6—8	goethite	20	0.8	0.16	15±5
		total			1.1	
C	ferruginous	talc	60	0.2	0.12	50±12
	nickel sulphide	chrysotile	10	0.2	0.02	10±5
	35600S	secondary silica	10	0	0	0
	12—14	goethite	20	0.5	0.10	40±10
		total			0.24	
D	saprolite	talc	70	0.2	0.14	30±7
	nickel sulphide	smectite	10	0.2	0.02	5±3
	35600S	chrysotile	10	0.2	0.02	5±3
	22—24	goethite	10	2.4	0.24	60±12
		total			0.42	

dominantly goethite, are 0.8 ± 0.5% Ni and 2.4 ± 0.5% Ni for the barren and nickel sulphide associations respectively. Despite the higher proportion of goethite in the barren associate B, the goethite in this sample accounts for only 15% of the total Ni, compared with 60% for sample D. However, most of the Ni in sample B in silicate form is contained in chlorite, while the other silicate minerals in samples B and D have consistently low concentrations. The reason for the high concentration of Ni in chlorite is not known. The dominant association of Ni with goethite in D, the sample related to nickel sulphides, over that with magnesium silicate minerals is also reflected in the correlation coefficients of semi-quantitative element analyses carried out at 50 or more random points on the samples with the electron microprobe. For the nickel sulphide association, Ni-Fe r=0.95***, and for the barren association, Ni-Fe r=0.52*** and Ni-Mg r=0.67*** (*** $P < 0.001$).

In samples A and C from the ferruginous zones, the proportions of the total Ni accounted for by goethite and magnesium silicate phases are similar for both the nickel sulphide and barren associations. Also, goethite in both samples has relatively higher concentrations of Ni than the accompanying magnesium silicates. In sample A, manganese oxide contains 10% Ni; however, it only accounts for a small proportion of the total Ni present. High concentrations of Ni in manganese oxides of laterites have also been found by Turner (1968) and Hudson (1970).

In these two samples from the ferruginous zone, the Fe points that are greater than 30% Fe_2O_3 show a decrease in Ni as Fe increases, the correlation coefficient being $r = -0.93$***. High Fe values are found in the larger crystals, some microns in diameter, and it is probable that during crystal growth, impurities, such as Ni, are expelled.

Extraction of iron with sodium dithionite

The association of Ni with Fe in the saprolite zone in areas related to mineralization as shown by the electron microprobe results, suggested that selective dissolution of Fe might discriminate samples of a nickel sulphide association from those of barren situations. Total Ni values do not necessarily give such a discrimination, due to large absolute losses of Ni from the profile on lateritization. Accordingly, samples from a profile known to be associated with mineralization (BD204), and samples from a barren profile (SER-E), were extracted with sodium dithionite in citrate buffer (Coffin, 1963). The results are shown in Figs. 8 and 9.

The barren profile is complicated by an acid intrusive in the original lithology and also three separate zones of Fe enrichment. This illustrates the variability of individual profiles and is a major problem when working with lateritized materials. Only the Ni and Cu extracted are shown in the figures for clarity. However, the Fe results show that dithionite extracted the highest amount, and the highest proportion of Fe from the clay zones and the base of the ferruginous zones in each profile. The total Fe values shown in

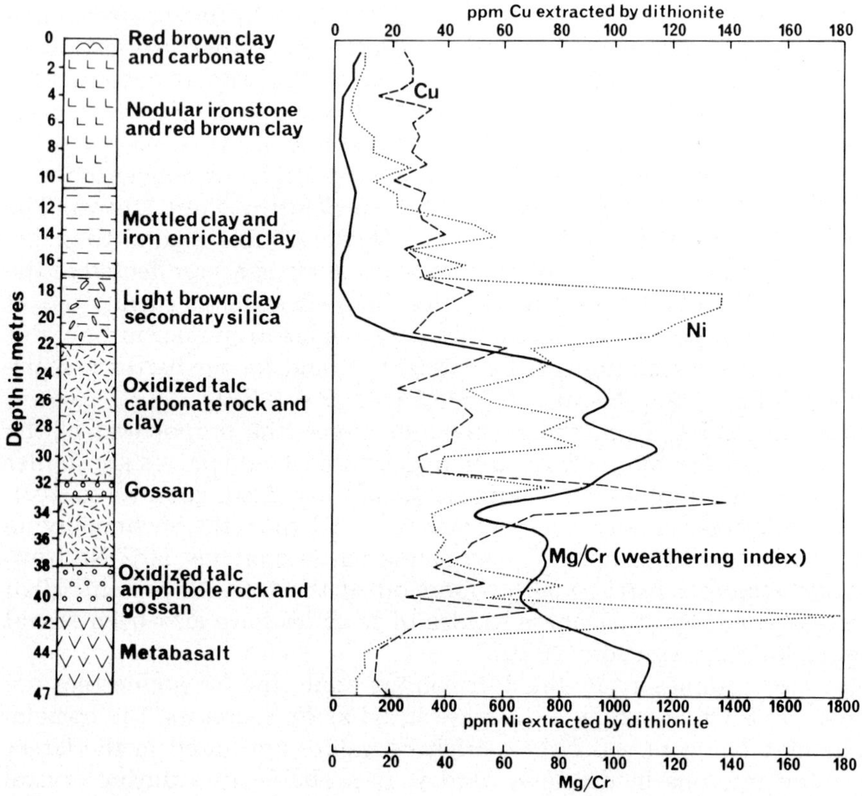

Fig. 8. Ni and Cu extracted by sodium dithionite from a laterite profile (BD204) developed on mineralized ultrabasic rocks.

Figs. 10 and 11 have the same trends and range of values in both profiles with additional peaks in the lower section of the mineralized profile where gossanous material occurs. The amount and proportion of Fe extracted decreased in the top of the ferruginous zones, probably due to the greater crystallinity of the iron oxides.

As shown in Figs. 8 and 9, the amounts of Ni extracted were greatest from the iron-rich clay zones in both profiles as well as from the gossan zones of the mineralized profile. No distinction can be made between a mineralized and unmineralized association from the extracted Ni in the upper parts of the profiles, particularly as the barren profile is complicated by the three zones of Fe enrichment. However, greater concentrations of Ni were extracted from the lower half of the mineralized profile than the same portion of the barren one. The presence of a Ni halo in weathered ultrabasic rocks that hosted mineralization was of interest, so attention was focussed on horizons adjacent to the gossan zones of the mineralized profile and away from the zone of acid intrusive rocks in the barren profile. In the barren profile, the Ni extracted ranged between 50 and 350 ppm from total contents of 1000—

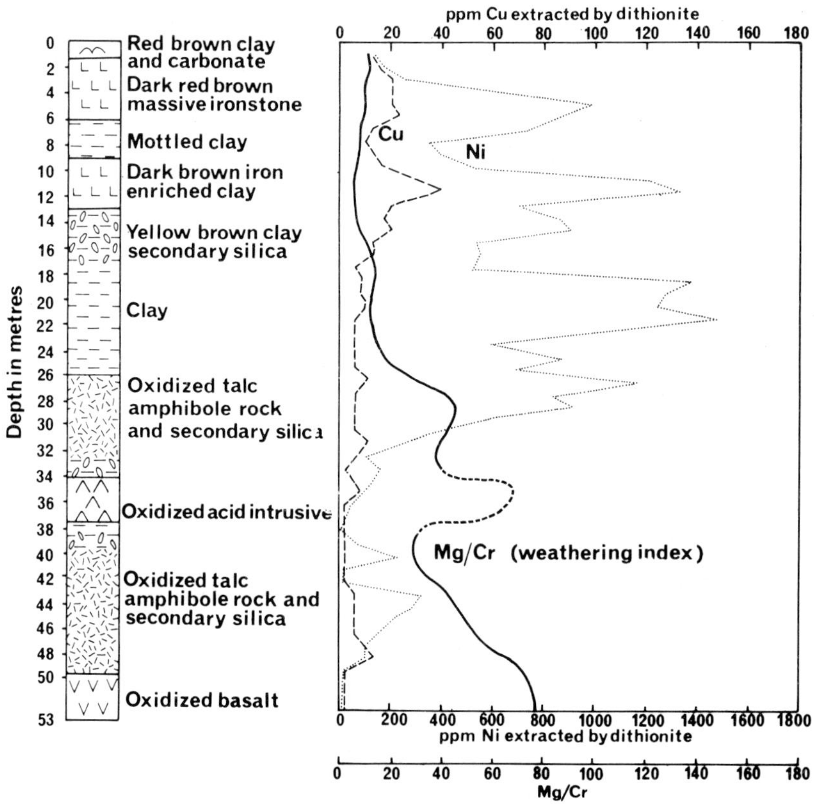

Fig. 9. Ni and Cu extracted by sodium dithionite from a laterite profile (SER-E) developed on barren ultrabasic rocks.

2500 ppm Ni, while in the mineralized profile, the Ni extracted ranged between 400 and 900 ppm from total contents of 2500—4000 ppm Ni. Therefore, Ni released by extracting "active" Fe with dithionite gives a better discrimination of mineralized environments from barren environments than when using total Ni.

A weathering index was used to indicate the point in a laterite profile below which dithionite extraction values may give discrimination. There are a number of possibilities for such an index, e.g. the concentration of Mg or Fe. It was found that the ratio Mg/Cr was the best index, since a marked increase in the ratio occurs at the point in the profile below which extraction values are useful. The ratio is plotted in Figs. 8 and 9.

In contrast to Ni, the total Cu contents of the oxidized rock and clay are good discriminators between the mineralized and unmineralized associations. However, the distinction is not so clear cut in the upper sections of the profiles due to a relative concentration of Cu in the ferruginous zones of the barren profile. Below the critical weathering horizon given by the Mg/Cr

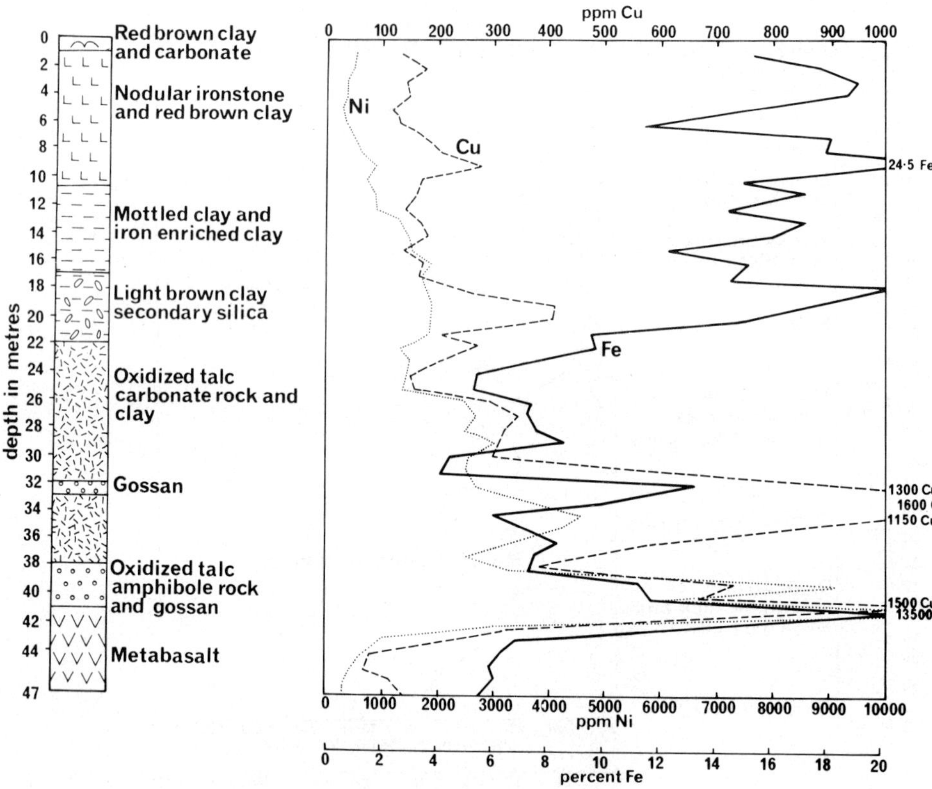

Fig. 10. Ni, Cu and Fe concentrations through a laterite profile (BD204) developed on mineralized ultrabasic rocks.

ratio, the total values for Cu range between 150 and 350 ppm in samples adjacent to the gossan zones in the mineralized profile, and from <10 to 50 ppm in the barren profile. The dispersion of Cu is limited by the strong association of Cu and iron oxides and a similar observation was made on laterite profiles at Mt. Keith, Western Australia, by Butt and Sheppy (1975) and at Greenvale, Queensland, by Zeissink (1970). The relative strength of this association is shown by a lower proportion being extracted by dithionite from total Cu than total Ni. Thus, the extraction values for Cu do not give any better discrimination than that possible from total values.

Implications for base metal exploration of the observations on laterites

In the light of these results, it is worthwhile now discussing the main geochemical features that are used to distinguish a laterite profile of mineralized association from that of unmineralized association. Some of these features have also been discussed by Butt and Sheppy (1975) and Wilmshurst (1975)

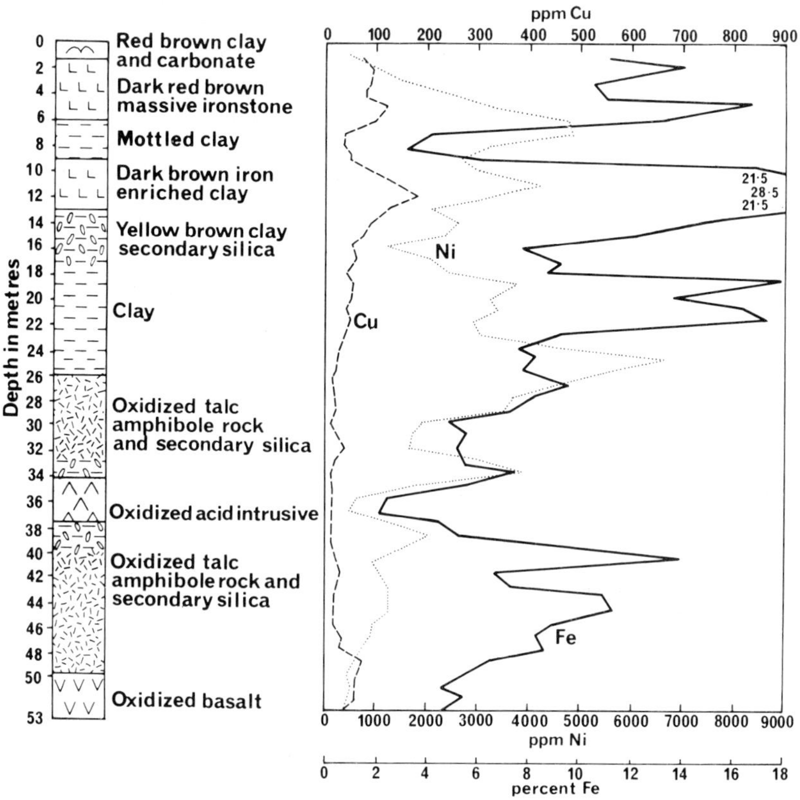

Fig. 11. Ni, Cu and Fe concentrations through a laterite profile (SER-E) developed on barren ultrabasic rocks.

and the use of Pt, Pd and Ir as pathfinders in lateritized material has been particularly successful (Keays and Crockett, 1970; Wilmshurst, 1975; Travis et al., 1976). The comments made here do not include any discussion of these pathfinder metals. Reference is made to a laterite profile developed on significant nickel sulphide mineralization approximately 80 km northeast of Kalgoorlie in support of some of these features. The geology and geochemical dispersion patterns of Ni and Cu are shown in Figs. 12, 13 and 14.

In the example, a sequence of ultramafic, basaltic and sedimentary rocks have been overturned and they now dip steeply to the west. The rocks are strongly foliated and have been partially invaded by the enclosing granites, while metamorphism has been pervasive. Lateritic weathering extends to depths of 10—30 m and the greatest depth of weathering is at the contact of the ultramafic and basaltic rocks. Colluvium covers the area to a depth of about 3 m and a weakly cemented ferruginous zone has developed. There is no preservation of the textures of the original lithologies above about 10 m.

The main geochemical features that are of use in the search for nickel sulphides in lateritic terrain are as follows.

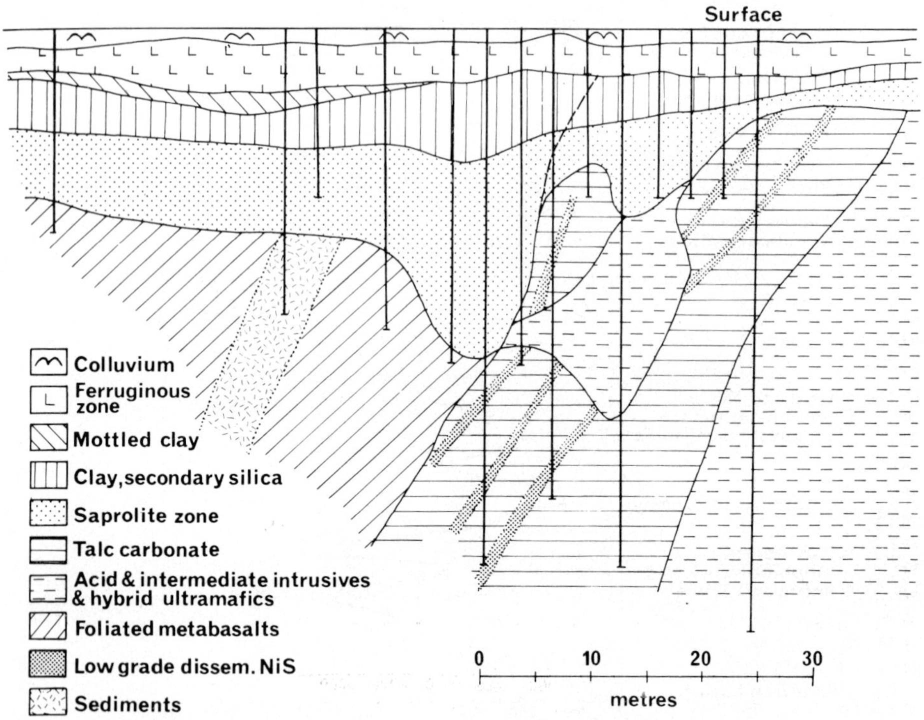

Fig. 12. Geology of the weathered zone over nickel sulphide mineralization 80 km northeast of Kalgoorlie.

Nickel. High Ni concentration is not a reliable indicator of the presence of mineralization. Ni is rapidly depleted with the onset of lateritic weathering and it can also concentrate to high values in clays derived from rocks that are barren of mineralization. The cross-section in Fig. 13 shows a rapid depletion of Ni values in the saprolite zone from original values of > 5000 ppm to values < 500 ppm.

Copper. As Cu has a low mobility in the weathered zone due to an association with iron oxides, the dimensions of any secondary halo around mineralization are usually limited, and in the example the halo of the mineralization near the contact has a width of only 10 m (Fig. 14). Other Cu haloes have been generated by a copper-rich sediment in the footwall and additional zones of disseminated nickel sulphide mineralization well away from the contact. High Cu values of sedimentary origin may cause confusion when coupled with the high Ni concentrations that can accumulate in lateritic profiles from purely silicate-Ni sources. High Zn values may assist in determining a sedimentary derivation for this Cu. However, Zn too can accumulate in the same profile zone as Cu and Ni.

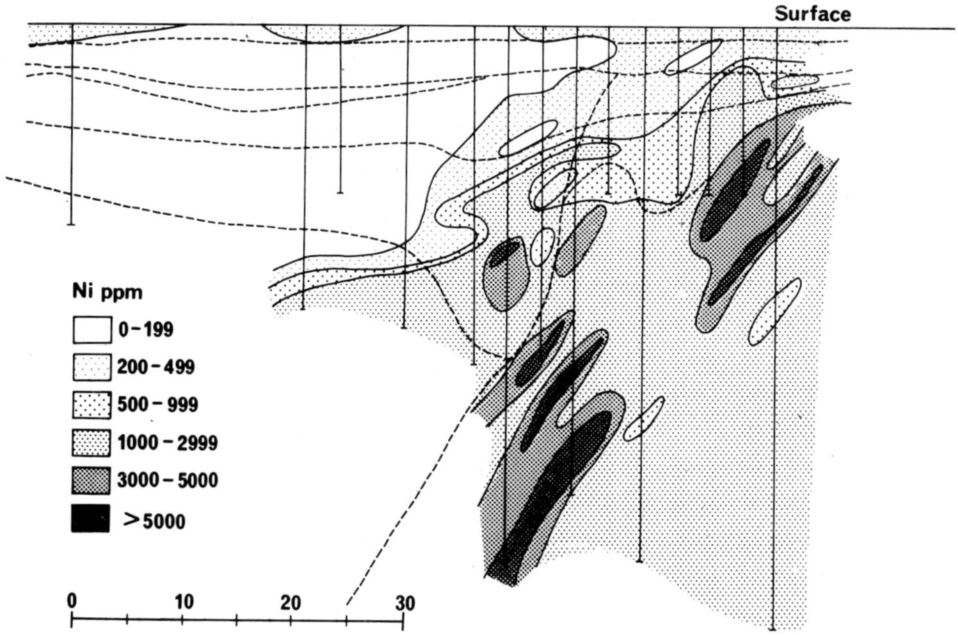

Fig. 13. The distribution of Ni in the weathered zone over nickel sulphide mineralization 80 km northeast of Kalgoorlie.

Furthermore, the relative enrichment of Cu in the upper horizons of the laterite profile from rocks that are barren of mineralization can also cause confusion. However, the values of Cu from such concentration rarely exceed 500 ppm and high values usually only have a sporadic distribution. Values of $>$ 100 ppm Cu are considered as anomalous but only assume significance where they occur below the mottled clay zone, and particularly where they are fairly constant with depth.

Chromium. This element has a limited mobility in the weathering zone due to the relative stability of the chromium-containing spinels as well as to the combination of Cr with iron oxides (K. Norrish, personal communication, 1976). Cr is often used as a base to calculate the relative enrichment (or depletion) of other elements and is also used to determine an ultrabasic origin for highly weathered material.

Magnesium. The concentration of Mg can be a useful indicator of the rock origin of weathered material as well as of the degree of weathering. However, the ratio of Mg/Ca is a more accurate indicator of origin, since account is then taken of any loss of Mg by weathering. Ca has a similar mobility to Mg in the weathering zone.

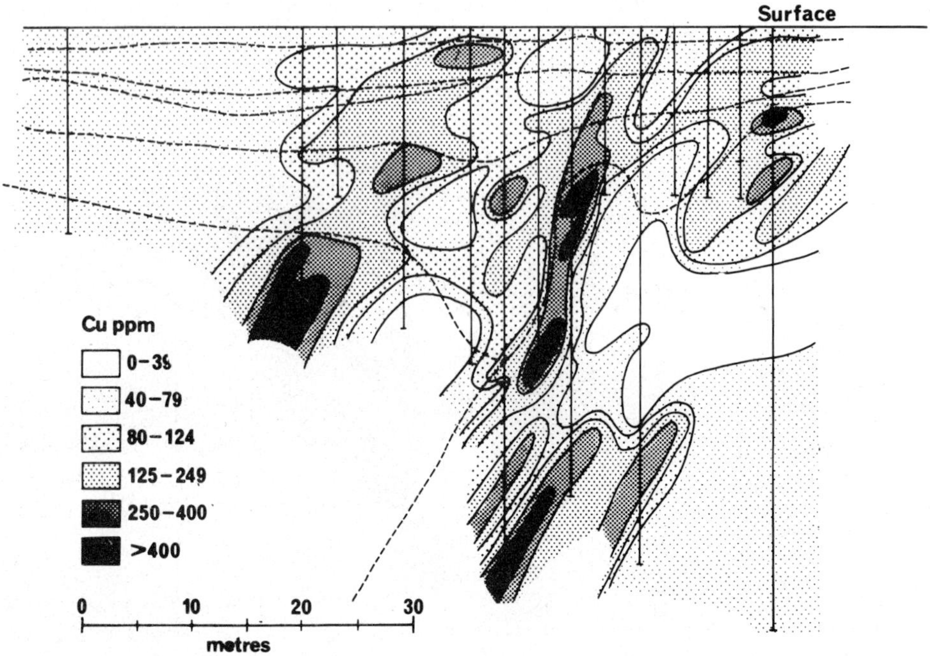

Fig. 14. The distribution of Cu in the weathered zone over nickel sulphide mineralization 80 km northeast of Kalgoorlie.

Iron. The Fe concentration is useful in distinguishing the various weathering zones of a laterite profile. Many laterite profiles are extremely complex due to erosion of some horizons and additions of detrital material followed by further lateritic weathering. The Fe concentration can indicate the nature of the chemical weathering that has taken place in each horizon.

Manganese. This element concentrates in the ferruginous zone and mottled clay zone and usually has a much more erratic distribution than Fe. Manganese oxides often have Ni concentrations of the order of 5—10% (K. Norrish, personal communication, 1976; Turner, 1968; Hudson, 1970). However, manganese oxides are usually only present in small proportions, the ratio Fe/Mn being $\gg$ 100:1, so this association of Ni is not important.

Cobalt. This element has a very similar chemistry to Ni. Thus the ratio Ni/Co is fairly constant, both in fresh ultrabasic rocks and in highly weathered material. This ratio is usually between 10:1 and 30:1, while the ratio in mineralized associations is often higher, with values of 20:1 to 50:1. However, the usefulness of this ratio can be negated due to the specific association of Co with manganese oxides (McKenzie, 1967, 1972). Where manganese oxides are present, the ratio may have very low values even in mineralized associations.

CONCLUSIONS

The work showed that during lateritic weathering, Ni, from barren ultrabasic rocks, is firstly associated with smectite, while Cu is associated with goethite. With further weathering, an association of Ni with goethite occurs, while that of Cu and goethite continues. At a later stage, as the iron oxide crystals grow in size, the concentrations of some elements that were previously associated with iron oxides, such as Ni, Cu and Cr, are reduced to low levels.

The work also showed that associations of Ni and Cu with iron oxides dominate where Ni mineralization occurs, both in gossan zones and in haloes surrounding mineralization. The use of a moderate extractor for Fe such as sodium dithionite in citrate buffer also released some Ni and Cu, and was successful in improving the discrimination between mineralized and barren associations over that of total Ni values. Extracted Cu values gave no improvement over total Cu values, because Cu is strongly associated with iron oxides and is not depleted during weathering to the same extent as Ni. Thus, the dimensions of a Cu halo around Ni mineralization are usually limited and the halo had a 10-m diameter in an example of lateritized nickel sulphides from northeast of Kalgoorlie.

The studies demonstrate the need for careful observations on samples from laterite profiles. Such observations can reveal the nature of the lateritic weathering that has affected the sampled horizon and so the main chemical controls on the component elements and minerals of the sample can be determined.

Clearly, there are great challenges for the exploration geochemist working with the lateritized rocks of the Archaean Shield of Western Australia. Some advance has yet to be made over the present costly requirement to drill as deep as the saprolite zone in order to be confident about an interpretation. The shield has been proven to be highly mineralized and therefore exploration in the area has a long-term future. Further advances in geochemistry could be made by using semi-quantitative estimations in addition to the present qualitative estimations of geochemical factors shown to be important in the interpretation of element data, e.g., how much goethite?, how much smectite?, etc. Then, multivariate analysis could play a more useful role in data interpretation and some additional successes in the delineation of mineralization in this difficult lateritic environment might be expected.

ACKNOWLEDGEMENTS

The permission of Western Mining Corporation to publish this paper is gratefully acknowledged. Dr. K. Norrish of C.S.I.R.O. Division of Soils, Adelaide, carried out the electron microprobe analyses and contributed some valuable discussion. Dr. R. Gilkes of the University of Western Australia also contributed some discussion. The chemical and X-ray diffraction analyses were carried out at the WMC Laboratories, Kalgoorlie. Mrs. M. Liddicoat is thanked for technical assistance, Miss A. Otter for preparation of the diagrams and Mrs. B. Chisholm for typing.

REFERENCES

Butt, C.R.M. and Sheppy, N.R., 1975. Geochemical exploration problems in Western Australia exemplified by the Mt Keith area. In: I.L. Elliott and W.K. Fletcher (Editors), Geochemical Exploration 1974. Elsevier, Amsterdam, pp. 391—415.

Coffin, D.E., 1963. A method for the determination of free iron in soils and clays. Can. J. Soil Sci., 43: 7—17.

C.S.I.R.O., 1975. Understanding the near-surface environment of the Kalgoorlie region to aid mineral exploration and water management. Report of W.A. State Committee of C.S.I.R.O., Perth, W.A., 24 pp. (unpublished).

Daily, B., Twidale, C.R. and Milnes, A.R., 1974. The age of the lateritized summit surface on Kangaroo Island and adjacent areas of South Australia. J. Geol. Soc. Aust., 21: 387—392.

David, T.W.E., 1950. In: The Geology of the Commonwealth of Australia, 2. Arnold, London, 618 pp.

Forbes, E.A., Posner, A.M. and Quirk, J.P., 1976. The specific adsorption of divalent Cd, Co, Cu, Pb and Zn on goethite. J. Soil Sci., 27: 154—166.

Grimme, H., 1968. Die Adsorption von Mn, Co, Cu and Zn durch Goethit aus verdünnten Lösungen. Pflanzenernähr. Bodenkd., 121: 58—65.

Hudson, R., 1970. A preliminary examination of Ora Banda laterite. Report to Western Mining Corporation, Kalgoorlie, W.A., 6 pp. (unpublished).

Jutson, J.T., 1914. An outline of the physiographical geology (physiography) of Western Australia. Bull. Geol. Surv. W. Aust., 61: 366 pp.

Keays, R.R. and Crocket, J.H., 1970. A study of precious metals in the Sudbury nickel irruptive ores. Econ. Geol., 65: 438—450.

McKenzie, R.M., 1967. The sorption of cobalt by manganese minerals in soils. Aust. J. Soil Res., 5: 235—246.

McKenzie, R.M., 1972. The sorption of some heavy metals by the lower oxides of manganese. Geoderma, 8: 29—35.

McKenzie, R.M., 1975. An electron microprobe study of the relationships between heavy metals and manganese and iron in soils and ocean floor nodules. Aust. J. Soil Res., 13: 177—188.

Mulcahy, M.J., 1959. Topographic relationships of laterite near York, Western Australia. J. R. Soc. W. Aust., 42: 44—48.

Mulcahy, M.J., 1960. Laterites and lateritic soils in southwestern Australia. J. Soil Sci., 11: 206—225.

Mulcahy, M.J., 1967. Landscapes, laterites and soils in southwestern Australia. In: J.N. Jennings and J.A. Mabbutt (Editors), Landform Studies from Australia and New Guinea. Australian National University Press, Canberra, A.C.T., pp. 211—230.

Mulcahy, M.J. and Bettenay, E., 1972. Soil and landscape studies in Western Australia, 1. The major drainage divisions. J. Geol. Soc. Aust., 18: 349—357.

Mulcahy, M.J., Churchward, H.M. and Dimmock, G.M., 1972. Landforms and soils on an uplifted peneplain in the Darling Range, Western Australia. Aust. J. Soil. Res., 10: 1—14.

Playford, P.E., 1954. Observations on laterite in Western Australia. Aust. J. Sci., 17: 11—14.

Prider, R.T., 1966. The lateritized surface of Western Australia. Aust. J. Sci., 28: 443—451.

Rajamani, V., Brown, G.E. and Prewitt, C.T., 1975. Cation ordering in Ni-Mg olivine. Am. Mineral., 60: 292—299.
Travis, G.A., Keays, R.R. and Davison, R.M., 1976. Palladium and iridium in the evaluation of nickel gossans in Western Australia. Econ. Geol., 71: 1229—1243.
Turner, A.R., 1968. The distribution and association of nickel in the ferruginous zones of the laterites of the Giles Complex. Amdel Bull., 5: 76—93.
Van Schuylenborgh, J., 1969. Weathering and soil-forming processes in the tropics. In: Natural Resources Research, XI. Soils and Tropical Weathering. UNESCO, Paris, 57 pp.
Wilmshurst, J.R., 1975. The weathering products of nickeliferous sulphides and their associated rocks in Western Australia. In: I.L. Elliott and W.K. Fletcher (Editors), Geochemical Exploration 1974. Elsevier, Amsterdam, pp. 417—436.
Woolnough, W.G., 1918. The Darling Peneplain of Western Australia. J. Proc. R. Soc. N.S. W., 52: 385—396.
Zeissink, H.E., 1969. The mineralogy and geochemistry of a nickeliferous laterite profile (Greenvale, Queensland, Australia). Miner. Deposita, 4: 132—152.
Zeissink, H.E., 1971. Trace element behavior in two nickeliferous laterite profiles. Chem. Geol., 7: 25—36.

Journal of Geochemical Exploration, 8(1977)283—303

THE APPLICABILITY OF SOME GEOCHEMICAL ANALYTICAL TECHNIQUES IN DETERMINING "TOTAL" COMPOSITIONS OF SOME LATERITIZED ROCKS

A.Z. GEDEON, C.R.M. BUTT, K.A. GARDNER and M.K. HART

Division of Mineralogy, Commonwealth Scientific and Industrial Research Organization, Wembley, W.A. (Australia)

(Revised version accepted March 9, 1977)

ABSTRACT

Gedeon, A.Z., Butt, C.R.M., Gardner, K.A. and Hart, M.K., 1977. The applicability of some geochemical analytical techniques in determining "total" compositions of some lateritized rocks. J. Geochem. Explor., 8: 283—303.

Analytical techniques used in studies of deep weathered and lateritized profiles have been investigated. Hydrofluoric-perchloric (HF), nitric-perchloric (NP) and perchloric (PE) acid digestions were done on batches of selected samples. Atomic absorption spectrometry on a single solution was used to obtain multi-element analysis. The elements studied were: Cr, Co, Cu, Fe, Mn, Ni, V, Zn and Pb. In addition, the applicability of background correction for non-atomic absorption was evaluated.

The mineralogy of the starting materials and residues after acid digestions was studied by X-ray diffraction. HF digestion left little or no residue. NP and PE digestion often left residues, which could contain goethite, haematite, serpentine, talc, muscovite, chromite and other minerals, depending upon the starting material.

The precision of the acid digestions was best for HF and least satisfactory, in general, for the PE digestion. For all elements, precision was closely related to concentration, with poorer results at the extremes of the concentration range. Cr, V and Pb caused the most concern.

Accuracy of the acid digestion was estimated by using XRF results as a base for comparison. The PE digestion tended to give very low extractions, often with marked differences between sample types. The NP digestion gave better extractions, but many minerals were still only partly dissolved. HF digestion was the best of the three methods, with extractions close to 100%, except for Cr and V.

Background correction, preferably simultaneous, was found to be beneficial, particularly at low concentrations.

INTRODUCTION

Over much of the Archaean Shield of Western Australia, rocks are commonly weathered to depths of 80 m or more. The deep weathering is the product of several weathering episodes, possibly extending back to the Mesozoic, the most important of which have been lateritization under warm humid conditions during the Tertiary and further alteration during subsequent arid and semi-arid periods. The weathering has resulted in very considerable compositional changes, even in single profiles. Thus, as a serpentinized dunite weathers, the principal mineral assemblage changes from serpentine-magnesite to silica-goethite-kaolinite and the chemical composition changes from 40% MgO, 35% SiO_2, 8% FeO + Fe_2O_3 and 0.1% Al_2O_3, to a variety of materials which, in extreme samples, may contain 90% SiO_2 or 75% Fe_2O_3 or 30% Al_2O_3, all with less than 0.01% MgO.

The deep weathering profiles can be subdivided into broad zones, but these have no practical significance in sampling. Thus, during exploration, it is rarely possible to collect directly comparable samples. This introduces many interpretational problems (Butt and Sheppy, 1975) but also stresses the importance of the analytical methods employed being able to produce reliable and consistent data throughout the matrix and concentration range.

Analysis should be total rather than partial, because for the most part active chemical dispersion processes have ceased; the dispersed metals tend to be held within the lattices of fairly resistant minerals, such as haematite and goethite, rather than only being adsorbed on clays and amorphous hydrous iron oxides.

The analytical methods commonly employed are the acid digestion procedures based on perchloric acid, namely hydrofluoric-perchloric acids (HF), nitric-perchloric acids (NP) and perchloric acid alone (PE), followed by atomic absorption spectrometry (AAS). They allow multi-element analysis over a wide concentration range from a single digestion and are also suitable for batch operation. Clearly the weaker digestions are not claimed to be truly "total" attacks, yet their precisions and accuracies are often assumed to be constant. Although they are least effective for Cr analysis, NP-soluble Cr is used as one criterion for distinguishing gossans (Joyce and Clema, 1974; Moeskops, 1977).

Previous studies have examined the effectiveness of these procedures in analyzing fresh and weathered rocks from temperate regions (for example, Foster, 1973). As our own experience led us to doubt their reliability in analyzing lateritized rocks of highly variable composition, we have tested them further, using a wide range of sample types and compositions. The elements considered were Cr, Co, Cu, Fe, Mn, Ni, V, Zn and Pb. In addition, the applicability of simultaneous and non-simultaneous background correction for non-atomic absorption has been evaluated. The digestion results were compared with X-ray fluorescence (XRF) analysis which provided an independent comparison base not so readily given by a further wet chemical technique. The XRF itself was checked against international standard samples. The use of XRF does not imply that it is necessarily better than the acid digestion-AAS procedures, although it may be assumed to give an approximation to a total analysis.

MATERIALS AND METHODS

Samples

Samples were selected for the study on the basis of their varying chemical and mineralogical compositions. They represent several zones in profiles through: (1) fresh and weathered ultramafic rocks (groups A—J); and (2) fresh and weathered basic and acidic volcanic rocks and alluvium (groups K—Q).

The chemical compositions, summarized in Table I, show a very wide variability for each sample type. In part, the wide ranges were chosen deliberately, to provide samples of varying matrix and metal contents to test the analytical methods. Nevertheless, they are typical of those encountered in individual profiles. Samples (200—500 g) were crushed, homogenized and a 10-g split ground to minus 200-mesh BSS (75 μm) using a Retsch agate micro-rapid mill.

Acid digestion and analysis

The analyses were conducted over a total period of seven months. Each sample was analyzed twice using each digestion, the time between duplicate analyses being at least four weeks. Digestions were carried out in Teflon crucibles, placed in wells in a pre-heated aluminium block, using the methods summarized in Table II. The final solutions were made up in glass test tubes and elemental determinations were made using a Varian-Techtron AA5 atomic absorption spectrometer. Instrument settings were standardized, although slight adjustments were made to obtain maximum absorbance. The unknowns were compared with a nine-step multi-element standard, each step spiked to the equivalent of 15% Fe_2O_3, 10% Al_2O_3 and 8% MgO in a hypothetical solid.

In each case, four readings, each with integration for one second, were recorded on standards, samples and deionized water. Using the same instrument setting, each batch was run through three times, taking readings in (1) uncorrected mode (UC) with the single-element hollow-cathode lamp; (2) simultaneously corrected mode (SC) with the single-element hollow-cathode and the hydrogen lamps intermittently pulsed; and (3) non-simultaneously corrected mode (NS) with the hydrogen lamp; readings were subtracted from those of (1) above to obtain corrected values.

Background corrections for Fe and non-simultaneous correction for V were not possible with our instrument. Each solution was read for 9 elements so that, including the correction modes, this involved each solution being read at least 24 times. The procedure took up to five working days per batch of 50—60 samples.

The recorded readings were averaged by an off-line data handling computer program; drift correction was made by subtracting water readings. For

TABLE I

Compositions of samples used in analytical tests (ranges and arithmetic means, determined by XRF analysis)

Group*	Sample type	Number of samples	Fe_2O_3 (%)	SiO_2 (%)	Al_2O_3 (%)	MgO(%)	Cr(ppm)
A	Pisolites with clay	4	37.8—63.3 46.5	17.5—23.9 19.9	4.5—24.0 17.2	0.4— 0.7 0.5	5850—127 8930
B	Pisolites	3	11.0—59.9 37.2	13.1—45.0 25.3	12.5—27.7 20.3	0.5— 0.7 0.6	1930—124 7080
C	Clays — oxide-rich	5	19.2—63.7 40.8	15.9—48.1 30.8	1.7—22.4 9.1	0.8— 5.8 2.4	8830—351 17400
D	Clays — oxide-poor	2	9.3, 13.3	38.8, 44.1	25.1, 25.8	1.8, 2.4	14400, 163
E	Clays — with silica	5	13.0—37.6 23.7	31.6—46.4 39.3	0.2— 1.3 0.5	1.6—22.9 15.9	1450—161 5200
F	Silica pans	5	5.3—14.5 9.6	63.6—86.8 76.6	0.1— 0.4 0.2	0.7—11.3 4.9	901—970 3010
G	Saprolitic serpentinite	10	6.0—19.7 10.7	21.5—49.8 37.1	0.1— 6.7 1.9	12.4—36.8 26.6	783—504 1850
H	Fresh serpentinite	8	6.3— 8.5 7.4	22.8—39.0 31.4	0.1— 2.4 0.60	31.6—41.2 36.4	855—155 1230
I	Weathered talc-carbonate	6	6.5—32.8 18.8	34.9—53.9 45.2	2.1— 8.0 5.0	17.0—24.1 20.0	1580—659 4170
J	Saprolite talc-carbonate	4	8.2—10.7 9.6	39.1—51.9 47.1	2.8— 5.8 4.5	27.0—29.7 28.3	1140—181 1510
K	Kaolinitic rhyodacite and alluvium	12	2.5—22.1 7.9	44.3—67.5 55.5	12.6—28.4 21.2	0.2— 2.6 1.0	14—653 188
L	Ironstones	3	17.3—42.9 31.1	20.9—39.7 29.4	18.3—26.8 22.1	0.3— 0.7 0.5	296—113 703
M	Saprolitic mica schist	7	2.0—21.9 11.5	45.8—66.7 56.5	12.5—26.0 18.3	0.4— 0.9 0.6	186—659 423
N	Saprolitic amphibolite	5	6.5—17.7 11.8	48.4—59.1 53.6	13.2—24.2 18.2	0.6— 4.2 2.0	18—359 232
O	Fresh amphibolite	2	9.9, 9.9	49.6, 50.7	14.7, 15.6	8.2, 8.5	194, 21
P	Rhyodacite saprolitic — fresh	2	7.1, 7.2	58.2, 60.1	14.2, 14.5	3.0, 3.6	101, 10
Q	Saprolites tuffs	5	9.6—13.1 11.8	51.4—65.7 59.2	10.5—20.3 13.2	1.3— 5.2 4.00	201—25 1310

*Groups A—J: derived from ultramafic rocks; groups K—Q: derived from other volcanic rocks.

each of the elements and reading modes, daily calibrations were obtained by a least-square fitting of a 3rd-order polynomial equation to the standards' concentration and absorbance values. The concentration values of the unknowns were calculated from the polynomial equation. No extrapolation was used. Samples that exceeded the readings for the highest standard were diluted 10- to 100-fold, to bring absorbance within the calibration range.

X-ray fluorescence analysis

X-ray fluorescence analysis was done using a Philips PW 1220C spectrometer. Au radiation was used for the determination of all trace elements but

Co(ppm)	Cu(ppm)	Mn(ppm)	Ni(ppm)	V(ppm)	Zn(ppm)	Pb(ppm)
70–500	1–1345	130–2355	250–5460	85–2600	25–200	
205	565	745	2200	1375	95	
25–130	1–85	65–295	245–380	320–805	9–13	
85	45	155	305	600	12	
100–2375	290–1840	165–12300	3885–14600	60–1200	165–380	
740	1145	3165	11345	305	285	
54, 61	59, 82	450, 550	875, 2030	695, 830	13, 60	
275–7690	750–1880	880–43890	15150–28475	16–40	105–535	
2950	1100	13062	19350	25	170	
65–315	165–555	110–1215	3195–8725	6–17	60–95	
195	395	610	5885	11	70	
145–880	190–740	570–2120	5200–20645	5–105	20–120	
324	425	1340	8730	35	65	
100–210	245–365	640–1380	3380–6655	5–45	15–45	
150	265	910	5490	15	30	
45–150	190–740	40–160	1290–4470	75–130	40–80	
85	440	75	2440	105	65	
120–485	35–285	1070–7585	3315–7690	85–140	40–75	
270	185	3155	5395	120	60	
5–70	1–1515	25–690	7–930	70–345	10–1020	3–70
30	305	140	200	195	210	25
35–85	30–755	75–300	25–150	390–1345	18–480	20–2585
60	330	155	80	915	185	875
10–50	50–1850	30–245	20–60	140–400	50–385	11–3190
25	650	120	35	215	160	950
20–50	30–235	100–1240	105–210	145–485	40–215	9–18
35	110	605	155	395	110	12
50, 55	5, 60	1200, 1325	95, 105	210, 240	75, 80	6, 9
31, 36	< 2	705, 725	55, 70	100, 110	21, 23	4, 6
30–140	45–730	410–1390	180–1375	150–250	40–260	5–20
100	265	730	990	215	210	10

Pb, for which Mo radiation was used. Cr radiation was used for major element determinations. Trace elements were determined from pressed powder samples backed with boric acid. Synthetic standards were made by grinding Spec-pure (Johnson-Matthey) oxides with an arbitrary matrix of 85% Si_2O and 15% Fe_2O_3. For Co, Cu, Pb, Ni and Zn, the Compton scattered line from the X-ray tube was used to correct for matrix effects; for Cr, Mn and V, calibration was based on the arbitrary matrix. V was corrected for interference by Ti. Fe and other major elements were determined using fused glass discs and results calculated by the methods of Norrish and Hutton (1969). Duplicate analyses were performed using newly prepared pressed powders and fused glass discs.

TABLE II

Acid digestion methods

	Hydrofluoric-perchloric (HF)	Nitric-perchloric (NP)	Perchloric (PE)
Sample weight	0.3 g	0.3 g	0.3 g
Reagents	5 ml 50% HF, 1 ml 72% $HClO_4$	3 ml acid mixture: 3 vols. 68% HNO_3 1 vol. 72% $HClO_4$	4 ml 72% $HClO_4$
Digestion	fume dry at 220° C; add 2 ml conc. HCl, fume dry	fume dry at 220° C	1 hour at 180° C
Leach	6 ml 5*M* HCl at 90° C for 30 minutes; dilute to 30 ml with water	6 ml 5*M* HCl at 90°C for 30 minutes; dilute to 30 ml with water	dilute to 30 ml with water
AAS standards	multi-element, 1*M* HCl	multi-element, 1*M* HCl	multi-element 1*M* $HClO_4$

TABLE III

Predominant minerals in starting materials and in residues following nitric-perchloric acid digestions, determined by X-ray diffraction. No data available for groups O and Q

Group	Sample type	Initial mineralogy	Residue after NP digestion
A	Pisolites with clay	He, Qz, Sp (Go) (Tc) (Ka)	He, Qz, (Go) (Sp)
B	Pisolites	He, Qz, Sp (Go)	Qz, Sp, Go
C	Oxide-rich clays	Go, Ka, Sp, Sa, Tc, He, (Do) (Qz)	He, Qz
D	Oxide-poor clays	Sa, Ka, Qz, Go	Qz
E	Clays with silica	Sp, Qz, (He) (Tc)	He, Qz
F	Silica pans	Qz (Sp) (He)	Qz, (He)
G	Saprolitic serpentinite	Sp, Tc, Sa, Cl	Tc, Sp
H	Serpentinite	Sp, Mg, Dol, Qz	Go, (Qz)
I	Weathered talc-carbonate	Go, Tc, Ka, Qz, (Sp)	Tc, Qz, (Go) (Sp)
J	Saprolitic talc-carbonate	Tc, Sp, Mg, Cl, (Go) (Do)	Tc, (Go)
K	Kaolinitic rhyodacite and alluvium	Qz, Ka, (Go) (Mu)	Qz, Ka (Mu)
L	Gossanous ironstones	Mu, Go, Ka, (Mo)	Mu, Ka, Go
M	Saprolitic mica schist	Qz, Mu, Ka, Fd, (Go) (Mo) (An)	Mu, Ka, Fd, Qz, (An)
N	Saprolitic amphibolites	Mo, Fd, Qz, Cl, Am, Go	Qz, Fd, (Am).
P	Rhyodacites	Qz, Fd, (Am)	Qz, Fd, (Am).

Key: Am = amphibole; An = andalusite; Cl = chlorite; Do = dolomite; Fd = feldspars; Go = goethite; He = haematite; Ka = kaolinite; Mg = magnesite; Mo = montmorillonite; Mu = muscovite (and/or fuchsite); Pn = pentlandite; Qz = quartz; Sa = saponite; Sp = serpentine; Tc = talc.

X-ray diffraction analysis

X-ray diffraction analyses of pressed powders of the starting materials and the leach residues were made with a Siemens diffractometer using Ni-filtered Cu radiation at an excitation voltage of 40 kV and a current of 20 mA.

MINERALOGY OF SAMPLES AND LEACH RESIDUES

The mineralogy of the starting materials and the residues left after NP digestion are compared in Table III. Residues were often strongly coloured by iron oxides. The results indicate that only montmorillonite, saponite and carbonates are fully dissolved, whereas quartz, talc, muscovite, feldspars, amphiboles and kaolinite are often little affected. Serpentine, goethite and haematite seem to be partially dissolved, the amounts varying from sample to sample.

Some important minor minerals were not detected by diffractometry. Chromite (and/or stichtite) and magnetite are present in all samples derived from ultramafics and are retained, in part, in the residues. "Fresh" serpentinites contained pentlandite or violarite, although the latter would certainly be altered to goethite prior to analysis. Ilmenite is common in the basic volcanic materials. Some samples derived from the acid volcanics contain the chromium-bearing mica fuchsite, which is only partly dissolved; Pb in these samples can be associated with partially soluble heavy minerals.

The residues left after PE digestions were more voluminous, and the changes in mineralogy caused by the extraction were less marked.

Little or no residue was left after HF digestions; where present it consisted of quartz and, occasionally, chromite and graphite.

ASSESSMENT OF ANALYTICAL RESULTS

Data treatment

The data have been assessed by comparing analytical pairs (Thompson and Howarth, 1973). All data were used — no outliers were omitted. *Precision* was estimated by comparing duplicate analyses of each sample by each analytical method. *Accuracy* was estimated by comparing means of duplicate pairs for each digestion and instrument reading mode with XRF analyses.

For this assessment, the following parameters were calculated:

(1) for the whole data set, for each method:

correlation coefficient; slope (s) of regression line ($y = sx$)

(2) for comparing analytical pairs by each method and means of pairs by different methods:

relative difference (RD) = $|x_1 - x_2| / [(x_1 + x_2)/2]$ %

where x_1 = analysis 1, x_2 = analysis 2;

mean RD = $(RD_1 + ... + RD_n)/n$

probable RD = RD has a 90% probability of being below a stated value

(3) for determining accuracy:

extraction = AAS value/XRF value %

Precision of the analytical procedures

Reproducibility or precision is indicated by RD. In addition the *correlation coefficient* indicates the ability of the method to duplicate the variance of the data set and the *slope* indicates the overall effect of variations in digestion and instrument conditions, etc., between determinations. If identical values were obtained the slope would be one.

The slopes and correlation coefficients (Table IV) suggest that the most consistent results are given by HF digestion for elements other than V and perhaps Pb and Zn. Overall, the PE digestion is the least satisfactory. The poor slopes given by Cr, especially by NP, and the erratic slopes for Zn and V, probably reflect inconsistent digestion of chromite and magnetite respectively. Furthermore, for NP digestion after fuming to dryness, difficulties can be experienced in redissolving iron and manganese oxides and associated elements.

Nevertheless, although poor slopes may indicate inconsistent conditions between batches, correlation coefficients may remain satisfactory. Thus, although the absolute values between the two runs may be significantly different, relative values may be very close, e.g. whilst PE digestion gives a slope $\leqslant$ 0.92 for Cr, Co and Mn, correlation coefficients are close to unity, indicating duplication of the overall variance of the data.

The ranges in element concentration for which probable RD are <20% and 10% are shown in Table V. In addition, the range for which mean RD $\leqslant$10% is also shown. An important feature shown by the ranges, which is common to all elements and procedures, is that precision is closely related to concentration, with poorer results obtained at the extremes.

Chromium. No procedure, other than XRF, has probable RD <10% over a significant range. Several procedures give probable RD <20% but for NP and, especially PE, the total range of the data set, as determined by the method, is much reduced. The digestion giving most precise results over the widest range is HF-UC.

TABLE IV

Slopes and correlation coefficients as a measure of the precision of duplicate analyses

	Cr		Co		Cu		Fe		Mn		Ni		V		Zn		Pb	
	*s**	*r**	*s*	*r*	*s*	*r*	*s*	*r*	*s*	*r*	*s*	*r*	*s*	*r*	*s*	*r*	*s*	*r*
HF-perchloric																		
UC	*0.92*	*0.9760*	0.98	0.9990	*0.97*	*0.9951*	1.01	0.9972	*0.99*	*0.9978*	1.03	0.9964	*0.88*	*0.9312*	0.94	0.9914	*0.91*	*0.9980*
SC	*0.89*	*0.9854*	1.14	1.0	*0.98*	*0.9950*	—	—	*0.99*	*0.9976*	0.95	0.9948	*0.54*	*0.8835*	1.01	0.9953	*0.89*	*0.9985*
NS	*0.90*	*0.9735*	1.00	0.9988	*0.95*	*0.9934*	—	—	*0.96*	*0.9981*	1.07	0.9973	—	—	0.95	0.9892	*0.94*	*0.9983*
Nitric-perchloric																		
UC	*0.70*	*0.9111*	1.01	0.9994	*0.95*	*0.9863*	0.96	0.9065	*1.12*	*0.9983*	0.88	0.9934	*0.93*	*0.9557*	0.99	0.9837	*1.05*	*0.9975*
SC	*0.74*	*0.9419*	0.92	0.9994	*0.95*	*0.9860*	—	—	*1.03*	*0.9928*	0.96	0.9911	*1.11*	*0.5984*	1.01	0.9900	*1.02*	*0.9986*
NS	*0.70*	*0.9092*	0.98	0.9993	*0.97*	*0.9823*	—	—	*1.13*	*0.9983*	0.89	0.9910	—	—	1.00	0.9861	*1.03*	*0.9973*
Perchloric																		
UC	*0.92*	*0.9870*	0.91	0.9997	*0.97*	*0.9939*	1.04	0.9911	*0.87*	*0.9974*	1.06	0.9664	*1.39*	*0.9084*	1.49	0.9632	*1.08*	*0.9946*
SC	*0.74*	*0.9816*	0.90	0.9995	*0.95*	*0.9927*	—	—	*0.87*	*0.9979*	1.03	0.9649	*0.46*	*0.3911*	1.14	0.9566	*1.08*	*0.9976*
NS	*0.87*	*0.9813*	0.84	0.9997	*0.98*	*0.9938*	—	—	*0.86*	*0.9975*	1.05	0.9676	—	—	1.45	0.9666	*1.05*	*0.9935*

* s = slope of linear regression line ($y = sx$); r = correlation coefficient.

AAS reading modes for non-atomic absorption; UC, uncorrected; SC, simultaneously corrected; NS, non-simultaneously corrected.

TABLE V

Precision of analytical procedures, based on a comparison of duplicate analyses. Results indicate total ranges determined by each procedure, ranges within which relative differences between duplicates have 90% probabilities of being below 20% and 10% (pRD), and ranges for which mean relative difference is 10% or less (mRD)

	Range		Cr(ppm)	Co(ppm)	Cu(ppm)	Fe(%)	Mn(ppm)	Ni(ppm)	V(ppm)	Zn(ppm)	Pb(ppm)
HF-perchloric											
UC	Total		15–26400*	5–9260	4–1680**	0.08–47.65	14–2580*,**	5–15200**	0–1875**	2–890	2–3180
	pRD	<20%	140– 1150	38–9260	21–1680	1.20–47.65	14–2580	302–15200	— —	17–890	194–3180
	pRD	<10%	— —	— —	153– 763	1.20–47.65	303–2580	3050– 8680	— —	42– 93	— —
	mRD	⩽10%	— —	38–9260	54–1680	1.20–47.65	14–2580	210–15220	— —	2–890	— —
SC	Total		6–22800*	1–9750	2–1550**		9–2550*,**	0–15800**	0– 892*	3–860	0–3070
	pRD	<20%	78– 1050	25–9750	67–1550		9–2550	67–15800	— —	17–860	158–3070
	pRD	<10%	— —	— —	111–1550		275–2550	1490– 7630	— —	124–860	— —
	mRD	⩽10%	77– 1050	25–9750	64–1550		10–2550	69–15800	— —	24–860	— —
NS	Total		7–26400*	1–9050	4–1660**		11–2510*,**	3–15000**		1–865	1–3150
	pRD	<20%	19– 1070	57–9050	148–1660		11–2510	251–15000		17–865	— —
	pRD	<10%	— —	— —	456– 753		296– 930	720– 8240		42– 90	— —
	mRD	⩽10%	85– 1070	44–9050	70–1660		11–2510	71–15000		17–865	— —
Nitric-perchloric											
UC	Total		10–12800	1–9570	6–1650**	0.05–41.62	6–6070**	12–21900	0–1340**	4–580	0–2830
	pRD	<20%	871– 1460	68–9570	141– 389	— —	228– 692	5230–14000	— —	— —	— —
	pRD	<10%	— —	— —	229– 389	— —	— —	— —	— —	— —	— —
	mRD	⩽10%	— —	79–9570	200–1650	— —	— —	— —	— —	— —	— —
SC	Total		6–14800	0–8860	2–1490**		2–5940**	6–24500	14– 577*	3–535	0–2800
	pRD	<20%	— —	73–8860	192–1490		253–5940	2630–24500	— —	68–535	— —
	pRD	<10%	— —	— —	— —		— —	— —	— —	— —	— —
	mRD	⩽10%	— —	107–8860	192–1490		253–5940	4650–24500	— —	68–535	— —
NS	Total		1–12500	0–9330	1–1620**		4–6030**	4–19400		3–555	0–2830
	pRD	<20%	853– 1320	58–9330	196– 690		240– 678	4430–19400		240–555	— —
	pRD	<10%	— —	58– 171	— —		— —	— —		— —	— —
	mRD	⩽10%	— —	58–9330	185–1620		— —	4580–13500		— —	— —

Perchloric											
UC	Total		*3–4340***	3–8820	*0–1190***	0.01–32.05	*0–45800*	5–21400	*1–313***	2–705*	*9–2100***
	pRD	<20%	*327–1010*	105–8820	*69–1190*	3.11–32.05	*18–45800*	79–21400	– –	– –	– –
	pRD	<10%	– –	144– 437	*282–1190*	– –	– –	3160–21400	– –	– –	– –
	mRD	⩽10%	*100– 660*	45–8820	*40–1190*	2.76–32.05	*22– 380*	54–21400	– –	– –	– –
SC	Total		*1–4000***	0–8710	*0–1170*		*1–46100*	1–20900	*9–391***	1–650	*0–2040***
	pRD	<20%	– –	56–8710	*65–1170*		*40–46100*	23–20900	– –	29–87	– –
	pRD	<10%	– –	– –	– –		– –	2910–20900	– –	– –	– –
	mRD	⩽10%	– –	56–8710	*65–1170*		*45– 850*	23–20900	– –	26–130	– –
NS	Total		*1– 4250***	0–8620	*0–1190***		*0–45900*	1–21500		0–675*	*3–2080***
	pRD	<20%	– –	60– 793	*65–1190*		*25–45900*	72–21500		– –	– –
	pRD	<10%	– –	– –	*279–1190*		– –	3170–21500		– –	– –
	mRD	⩽10%	– –	60– 427	*65–1190*		*41– 250*	72–21500		– –	– –
XRF	Total		*0–35100*	4–7800	*0–1770*	1.45–23.17*	*27–44500*	5–27700	*3–2590*	8–1010	*3–3210*
	pRD	<20%	*0–35100*	109– 274	*0–1770*	1.45–23.17	*27–44500*	5–27700	*7–2590*	11–1010	*9–3210*
	pRd	<10%	*0–35100*	– –	– –	1.45–23.17	*85–44500*	102–27700	*40–2590*	63–1010	*31–3210*
	mRD	⩽10%	*0–35100*	101–7800	*56–1770*	1.45–23.17	*27–44500*	5–27700	*19–2590*	8–1010	*9–3210*

*Incomplete sample set. For Cr: groups K–Q missing; Fe: only 15% repeated; Mn: groups A–J missing; V: groups A–J missing; Zn: groups A–J missing; all Pb: groups A–J missing.

**Top concentration in range given by different samples.

AAS reading modes for non-atomic absorption: UC, uncorrected; SC, simultaneously corrected: NS, non-simultaneously corrected.

Cobalt. Only two procedures have probable RD <10% over a significant range, but all HF and some NP and PE procedures have mean RD ⩽10% over wide ranges. For the weaker attacks, however, the precision is poor for samples with low Co contents. Thus NP-NS has probable RD <10% for samples containing over 50 ppm and comprising only 30% of the data set, whereas for HF-UC, probable RD is <10% for samples with over 40 ppm Co — comprising 80% of the data set.

Copper. Many procedures give quite satisfactory results, with wide ranges with probable and mean RD <10%. The best results are given by HF-SC, with mean RD ⩽5 for over 60% of the data set. The weaker acids give poor results for samples with low Cu contents and the ranges with mean RD ⩽10% comprise less than 40% of the data set. XRF results were affected by a change of X-ray tube-target between runs.

Iron. Results for NP and PE digestions are often poor, the result of inconsistent solution of crystalline iron oxides. Residues are frequently coloured and contain goethite and magnetite.

Manganese. The results are incomplete, with duplicate results by HF of the ultramafic-derived samples being unavailable. Nevertheless, the HF procedures are the most satisfactory, particularly in those samples containing more than 220 ppm Mn.

Nickel. Most HF and PE procedures have mean RD ⩽6% over wide ranges; however, sporadic poor results in samples containing 500—1500 ppm tend to limit to high concentrations the ranges with probable RD ⩽10%.

Vanadium. Poor results are obtained from all digestion procedures, due to a combination of difficult instrument conditions and inconsistent solution of vanadium-bearing minerals.

Zinc. The most useful results are given by HF digestions, with probable RD <8% for samples containing ⩾20 ppm Zn.

Lead. Only the volcanic-derived samples were analyzed for Pb, with the best results given by HF-SC. Most samples contain less than 100 ppm Pb, held in heavy minerals largely insoluble to acid digestion.

In general, the most precise results are obtained by XRF for all elements tested other than Co and perhaps Cu. Of the digestions, HF procedures give the most precise results and PE procedures the least. The latter also have the disadvantage that they severely limit the ranges, as a result of incomplete solution of several minerals. Results of NP procedures for Co, Cu, Fe and Mn were poorer than expected and this appears to be due in part to varying solubility of the residues after the initial digestion.

Accuracy of the analytical procedures

X-ray fluorescence. Since XRF analysis was being used as a measure of the accuracy of the digestion procedures, the accuracy of XRF itself was judged against published results for various international standards. Comparative results are given in Table VI. In most cases, the results obtained in this study were within 10% of the published mean values and always within the published ranges. Since the ranges upon which such means were calculated are often wide and the number of analyses few, evaluation of results is subjective. Nevertheless, the performance of this instrument is mostly satisfactory, although some Zn results tend to be a little low.

Forty of the test samples were also analyzed for Cr by a commercial laboratory, which gave results approximately 5% lower including a result of 8250 ppm for standard UM-2. Judgment as to which results are better is again subjective. However, the general agreement with published standard data, and the mostly good precisions, indicate the XRF results are a suitable basis for evaluation of the acid digestion procedures.

Acid digestions. Estimates of accuracy of the digestion procedures were obtained by comparing means of duplicate results with the XRF data (Tables VII and VIII). Here the *correlation coefficient* indicates the ability of the two methods to reproduce the variance of the data and the *slope* signifies the fraction extracted, i.e., accuracy. Not surprisingly, these results show that extractions decline HF > NP > PE. The decline in correlation coefficient implies that not only does the overall efficiency of extraction decline, but that a greater fluctuation is superimposed on the data, probably reflecting the varying susceptibility of the constituent minerals to acid attack. The least accurate results are given by Cr, due to the poor solubility of chromite under these conditions, and, particularly for PE procedures, by Fe and V, both of which are retained in undissolved iron oxides. With few exceptions, only the HF procedures have significant ranges for which probable RD between acid digestion and XRF is below 30%. These ranges illustrate also that, as with precision, accuracy is in part a function of concentration, with poorer results being obtained at the extremes of the range.

The effect of sample type on accuracy

Variations in the quality of results due to differences in sample type are difficult to assess. Within-group variations in extraction can be as pronounced as those between groups. This is often a response to the variations in concentration of each metal in each group for, as discussed previously, accuracy and precision tend to be a function of concentration. Thus, if the trace metal contents of samples from a given geological type vary over a wide range, poorer results will be given by the samples having the extreme concentrations. However, some trends are apparent. Mean extractions of each metal

TABLE VI

Comparison of means for International Standards

Sample	Source	Method	Cr (ppm)	Co (ppm)	Cu (ppm)	Fe (%)	Mn (ppm)	Ni (ppm)	V (ppm)	Zn (ppm)	Pb (ppm)	Number of determinations
UM 2	1	XRF	8505	158	1110	10.06	620	3100	161	45	46	2
	2	ES	10300	181	959	10.07	620	3850	—	32	—	—
SY-3	1	XRF	0	42	0	—	2280	5	38	245	128	1
	2	ES	n.d.	n.d.	18	4.9	2320	n.d.	<30	180	120	—
AGV-1	1	XRF	4	22	64	—	784	13	106	86	—	1
	3	—	12	14	60	6.76	763	19	125	84	35	—
GXR 2	1	XRF	31	8	98	1.86	1350	21	58	622	638	2
	4	XRF	—	10	116*	—	—	25	—	650*	708	2–8
GXR 5	1	XRF	106	36	463	3.62	402	77	63	58	12	2
	4	XRF	—	28	505	—	—	88	—	68	41	2–8
GXR 6	1	XRF	93	31	82	5.78	1340	27	220	138	88	2
	4	XRF	—	32	90	—	—	32	—	147	92	2–8

Sources: 1 = this study; 2 = Faye, 1973; 3 = Flanagan, 1973; 4 = Allcott and Lakin, 1974.
XRF = X-ray fluorescence; ES = emission spectrometry.
*Median value. — = not available; n.d. = not detected.

TABLE VII

Slopes and correlation coefficients, comparing means of acid digestion and XRF results, as a measure of accuracy

	Cr		Co		Cu		Fe		Mn		Ni		V		Zn		Pb	
	*s**	*r**	*s*	*r*	*s*	*r*	*s*	*r*	*s*	*r*	*s*	*r*	*s*	*r*	*s*	*r*	*s*	*r*
HF-perchloric																		
UC	*0.70*	*0.9731***	1.15	0.9954	*0.90*	*0.9809*	0.91	0.9876	*0.92*	*0.9971***	1.00	0.9807	*0.77*	*0.9879*	1.12	0.9728	*1.32*	*0.8909*
SC	*0.59*	*0.9759***	1.03	0.9934	*0.84*	*0.9836*	—	—	*0.92*	*0.9971***	1.02	0.9866	*0.63*	*0.9611*	1.01	0.9834	*1.20*	*0.8922*
NS	*0.67*	*0.9704***	1.00	0.9942	*0.87*	*0.9822*	—	—	*0.91*	*0.9970***	0.97	0.9784	—	—	1.05	0.9765	*1.20*	*0.8926*
Nitric-perchloric																		
UC	*0.39*	*0.9352*	1.06	0.9947	*0.83*	*0.9574*	0.77	0.9706	*0.80*	*0.9927*	0.87	0.9775	*0.50*	*0.9810*	0.83	0.9027	*1.14*	*0.8842*
SC	*0.42*	*0.9325*	0.90	0.9934	*0.76*	*0.9614*	—	—	*0.70*	*0.9666*	0.91	0.9782	*0.37*	*0.8742*	0.76	0.8967	*1.09*	*0.8851*
NS	*0.38*	*0.9361*	0.96	0.9928	*0.79*	*0.9615*	—	—	*0.77*	*0.9927*	0.81	0.9729	—	—	0.79	0.9047	*1.12*	*0.8851*
Perchloric																		
UC	*0.18*	*0.6379*	0.90	0.9927	*0.51*	*0.7903*	0.28	0.5723	*0.78*	*0.9941*	0.76	0.9242	*0.18*	*0.5524*	0.60	0.6612**	*0.87*	*0.8684*
SC	*0.16*	*0.6274*	0.88	0.9941	*0.51*	*0.7943*	—	—	*0.76*	*0.9924*	0.74	0.9187	*0.39*	*0.4037*	0.52	0.6836**	*0.82*	*0.8610*
NS	*0.18*	*0.6346*	0.84	0.9910	*0.51*	*0.7898*	—	—	*0.78*	*0.9940*	0.77	0.9218	—	—	0.56	0.6690**	*0.84*	*0.8700*

* *s* = slope of linear regression line ($y=sx$); *r* = correlation coefficient.

** All data, but single AAS determination.

AAS reading modes for non-atomic absorption: UC, uncorrected; SC, simultaneously corrected; NS, non-simultaneously corrected.

TABLE VIII

Accuracy of HF digestion, with XRF as comparison base. Results indicate ranges within which relative differences between AAS and XRF procedures have 90% probabilities of being below 30%, 20% and 10% (pRD). For NP and PE, pRD is always >20%; ranges with pRD <30% were recorded for uncorrected Cu, Fe, Mn and Zn data with NP digestion, and for Mn with PE digestion. Corrected data from NP digestion for Co, Cu and Zn also had ranges with pRD <30%

	Range	Cr(ppm)	Co(ppm)	Cu(ppm)	Fe(%)	Mn(ppm)	Ni(ppm)	V(ppm)	Zn(ppm)	Pb(ppm)
XRF	Total	0—35100	4—7800	0—1770	1.45—23.17*	27—44500	5—27700	3—2590	8—1010	3—3210
HF-perchloric										
UC	pRD <30%	215— 783**	172—7800	44— 884	1.43—44.58	59— 1468**	265—18900	— —	31— 264	— —
	pRD <20%	252— 653	194—7800	59— 267	4.64— 5.78	111— 397	265— 5340	— —	51— 148	— —
	pRD <10%	— —	— —	78— 249	— —	— —	— —	— —	— —	— —
SC	pRD <30%	58— 783**	68—7800	37— 267		25— 614**	159—27000	— —**	16—1010	— —
	pRD <20%	— —	— —	31— 205		59— 397	159— 4720	— —	49— 148	— —
	pRD <10%	— —	— —	— —		— —	173— 3180	— —	— —	— —
NS	pRD <30%	95— 783**	45—7800	31— 636		59— 614**	102—18900		16—1010	— —
	pRD <20%	95— 194	141—7800	31— 222		59— 108	265— 5390		49— 148	— —
	pRD <10%	— —	— —	— —		— —	— —		— —	— —

*Incomplete sample set.

**Single AAS determination.

AAS reading modes for non-atomic absorption: UC, uncorrected; SC, simultaneously corrected; NS, non-simultaneously corrected.

TABLE IX

Mean extraction (percent) obtained by acid digestion procedures, sorted according to sample type. Results determined by comparing simultaneously corrected atomic absorption results with XRF results. For sample description and composition, refer to Table I

Sample type	Cr			Co			Cu			Fe			Mn			Ni			V			Zn			PB		
	HF	NP	PE	HF	NP	PE	HF	NP	PE	HF	NP	PE	HF	NP	PE	HF	NP	PE	HF	NP	PE	HF	NP	PE	HF	NP	PE
A	55	55	<10	70	50	25	90	65	<10	95	75	12	100	70	30	110	75	30	*	40	25	115	50	15			
B	50	45	<10	55	45	<10	120	80	20	95	75	<10	95	85	50	115	45	15	65	30	20	145	80	25			
C	55	40	25	100	80	65	95	80	70	100	85	*	100	80	65	110	95	55	85	70	90	120	80	45			
D	65	30	<10	90	55	40	105	50	<10	85	75	20	80	55	55	95	40	10	80	*	30	125	60	15			
E	65	60	30	95	85	80	80	75	65	95	90	45	90	95	85	100	90	80	*	*	>150	105	95	75			
F	70	65	30	95	90	85	80	80	65	95	95	40	85	90	80	95	95	75	*	*	>150	100	100	70			
G	55	55	35	100	90	80	80	75	75	95	90	50	85	80	20	100	100	95	*	50	>150	100	90	65			
H	60	55	45	115 •	110	115	75	80	80	90	90	45	65	75	65	125	125	125	*	>150	>150	110	120	90			
I	55	30	<10	80	80	<10	85	65	10	90	65	<10	95	45	<10	100	55	10	75	*	*	85	40	<10			
J	60	60	40	100	80	80	90	85	75	90	70	50	85	75	75	105	80	70	20	*	110	90	60	35			
K	105	35	20	75	55	45	90	60	35	80	65	15	110	40	35	105	35	15	60	20	25	110	50	25	95	35	30
L	75	40	10	65	25	45	105	75	40	90	80	10	95	55	35	>150	50	35	55	35	< 10	90	85	30	55	45	40
M	90	25	<10	70	65	95	95	80	35	85	90	25	110	70	65	135	85	45	50	20	35	110	90	35	85	65	40
N	90	35	15	80	60	75	90	70	45	80	70	*	100	50	40	135	60	45	75	45	20	110	50	35	25	30	30
O	110	25	20	90	30	55	100	80	80	85	40	25	100	30	25	105	35	45	95	30	30	100	45	30	—	—	—
P	120	70	40	145	75	130	—	—	—	85	60	40	105	45	45	70	45	55	60	50	35	125	80	75	—	—	—
Q	85	75	45	85	80	70	80	75	80	80	85	35	85	70	65	95	85	70	50	55	20	105	90	85	—	—	—

*Highly variable data; — values close to detection limit.
HF = hydrofluoric-perchloric acid digestion; NP = nitric-perchloric acid digestion; PE = perchloric acid digestion.

have been calculated by comparing simultaneously corrected acid digestion data with XRF data (Table IX). Such a comparison, of course, assumes that XRF is not susceptible to such variations itself — which, in one case (see below) is not certain. The results again demonstrate how extraction varies according to acid strength and with the element analyzed, but also suggest that it varies with sample type. The effect of sample type is most pronounced with PE digestion, with extractions varying from <10% to >50% for each element. The lowest extractions are mostly from highly ferruginous samples — not unexpectedly, since goethite and haematite are found in residues — and in strongly weathered talc-carbonate rocks, residues of which contain goethite, haematite and talc. Higher extractions are obtained from samples in which metals are held in more soluble minerals such as poorly crystalline iron and manganese oxides, saponite, pentlandite and serpentine. Much of the variation tends to be eliminated with the stronger acids, particularly HF, which dissolves most minerals although for Cr, even HF results are unsatisfactory. The better results seem to be associated with those samples in which Cr is not associated with chromite, such as in fuchsite-bearing mica schist derivatives and in talc-carbonate rocks and silicified serpentinites in which Cr is associated with iron oxides.

Very high extractions (⩾110%) of Co and Ni are recorded for each digestion procedure from fresh serpentinites (group H). Most of the Ni and Co is associated with pentlandite and a closer inspection of the individual results shows that higher extractions are obtained from the sulphide-rich samples. A similar disparity was noted between our XRF results and the published values for the nickel-sulphide-bearing ultramafic standard UM-2 (Table VI). The XRF may be at fault, but the reason is unclear.

The effect of background correction on precision and accuracy

Reference to Tables IV and V indicates that precision, as shown by slopes, correlation coefficients and the ranges for which relative differences are low, is usually only marginally improved by background correction, if at all. Marked improvements would not necessarily be expected, in any case, since in obtaining the additional information, greater instrument noise, which contributes to imprecision, would be expected.

Correction does, however, improve results, especially at lower concentrations. When our HF-UC and HF-SC values for standard AGV-1 are compared with the "1972 values" given by Flanagan (1973) it can be seen that the HF-SC are in all cases in closer agreement (Table X). In general, extraction as estimated by uncorrected data tends to be higher at lower concentrations. If the apparent extraction is plotted against concentration, there is usually an inflection point above which extraction is more or less constant but below which it increases rapidly with decreasing concentration. The effect of background correction is to lower the apparent extraction. Below the inflection point, this effect is usually quite significant but, above it, only marginal.

TABLE X

Comparison of HF-digestion values with "1972 values" (Flanagan, 1973) for standard AGV-1 (all data in ppm)

	Cr	Co	Cu	Mn	Ni	Zn
HF-UC	59	52	67	855	85	94
HF-SC	23	13	57	835	25	90
"1972 value"	12	14	60	763	19	84

AAS reading modes for non-atomic absorption: UC, uncorrected; SC, simultaneously corrected.

Govett and Whitehead (1973) noted that background correction did give a significant reduction at high concentrations but, in their study, they did not use spiked standards. Inflection-point concentration expressed as parts per million in solution and mean extractions above and below that point, have been calculated for the uncorrected data for the HF digestion (Table XI). The mean extraction below the inflection point is markedly reduced by correction and the ratio of extractions above and below the point is closer to unity. The net effect, therefore, is to standardize the extraction throughout the range; this makes the data more self consistent and maintains the true contrast which exists in the data set. Essentially similar effects are expected for the weaker acids but, since extractions are generally lower for these, a greater proportion of the data fall below the inflection point concentration.

TABLE XI

Mean extraction (percent) for uncorrected and corrected data above (a) and below (b) inflection point concentration, HF digestion. Inflection point concentration expressed as ppm in solution

	Cr inflection point 2.9 ppm		Co 1.0 ppm		Cu 0.77 ppm		Mn 1.60 ppm		Ni 9.2 ppm		Zn 0.45 ppm		Pb 0.34 ppm	
	b	a	b	a	b	a	b	a	b	a	b	a	b	a
UC	189	65	330	113	165	88	116	88	268	102	145	109	427	116
SC	110	58	86	97	130	83	105	87	123	103	116	100	149	85
NS	138	62	101	98	136	86	107	86	182	98	121	103	64	82

AAS reading modes for non-atomic absorption: UC, uncorrected; SC, simultaneously corrected; NS, non-simultaneously corrected.

CONCLUSIONS

The applicability of a single digestion, followed by AAS, for multi-element analysis of lateritized rocks has been examined for perchloric (PE), nitric-perchloric (NP) and hydrofluoric-perchloric (HF) acids, in terms of precision, accuracy and the effect of sample type.

The analytical scheme adopted has tended to maximize certain errors — for example, those due to evaporative loss during the period required for multi-element analysis and variations in conditions between batches — but, in so doing, has given an appropriate assessment of overall performance.

The samples chosen had very wide ranges of mineralogical and chemical compositions and these ranges appear to have exceeded the optimum operating regions of the standardized procedures followed. Precision and, to some extent, accuracy tend to be poor at the extremes of concentration range. The lower limit for which acceptable precisions can be obtained depends upon the solubility of the host mineral but nevertheless, at low concentrations in solution, precision suffers as the signal/noise ratio decreases.

The HF digestion gives the most precise results over the widest ranges. Extractions, except for Cr and V, tend to be close to 100%. The NP digestion gives better extractions than the PE digestion but both leave many minerals only partly dissolved. In general, the weaker digestions give moderately precise, if inaccurate, results for Co, Cu, Fe, Mn, Ni and Zn, especially from the weathered ultramafic rocks in which the concentrations are within the optimum range. However, since the efficiency of digestion can vary between batches, the precision can be poor.

All three digestions are clearly unsatisfactory for Cr and V. For both metals, the low solubility of the main host minerals (chromite, fuchsite, haematite and magnetite) is a problem and for V, instrument conditions are difficult. More accurate results may be expected from a fusion, followed by either AAS or colorimetry, but this is less suited to batch analysis.

Even when acceptable precisions can be obtained from the weaker acids for certain elements, the results may nevertheless be unsatisfactory, since extractions are not only often very poor but tend to vary with sample type. Thus, if varying sample types are compared, which is usual during exploration in a lateritic environment and inevitable in studies of whole profiles, then the validity of conclusions based on the weaker attacks is open to question. Since secondary minerals such as haematite and goethite are not being fully dissolved, then secondarily dispersed metals such as Cu, Ni and Zn, held within these minerals, are not being released and the data obtained are incomplete and perhaps distorted. Where the amounts of metals released by the weaker acids are used as criteria for interpretation, for example in distinguishing between true and false gossans, then the conditions under which the analysis is performed must be controlled as rigorously as possible, so as to obtain truly comparable data.

ACKNOWLEDGEMENTS

We are grateful to Mr. P. Wilson and Dr. E.H. Nickel for obtaining and interpreting the X-ray diffraction results, Mr. E.P.A. Osborn for assistance with computer programming, Mr. C.G. Pope for the sample preparation and other members of the CSIRO Division of Mineralogy for assistance and helpful discussion.

REFERENCES

Allcott, G.H. and Lakin, H.W., 1974. Statistical summary of geochemical data furnished by 85 laboratories for six geochemical exploration reference samples. U.S. Geol. Surv., Open-File Rep.

Butt, C.R.M. and Sheppy, N.R., 1975. Geochemical exploration problems in Western Australia, exemplified by the Mt. Keith area. In: I.L. Elliott and W.K. Fletcher (Editors), Geochemical Exploration 1974. Elsevier, Amsterdam, pp. 391—415.

Faye, G.H., 1973. Standard reference ores and rocks available from the Mines Branch as of October 1973. Dep. of Energy, Mines and Resources, Ottawa, Ont., Inf. Circ., IC 309.

Flanagan, F.J., 1973. 1972 values for international geochemical reference samples. Geochim. Cosmochim. Acta, 37: 1189—1200.

Foster, J.R., 1973. The efficiency of various digestion procedures on the extraction of metals from rocks and rock-forming minerals. Bull. Can. Inst. Min. Metall., 66(736): 85—92.

Govett, G.J.S. and Whitehead, R.E., 1973. Errors in atomic absorption spectrophotometric determination of Pb, Zn, Ni and Co in geologic materials. J. Geochem. Explor., 2: 121—131.

Joyce, A.S. and Clema, J.M., 1974. An application of statistics to the chemical recognition of nickel gossans in the Yilgarn Block, Western Australia. Proc. Australas. Inst. Min. Metall., 252: 21—24.

Moeskops, P.G., 1977. Yilgarn nickel gossan geochemistry — a review, with new data. J. Geochem. Explor., 8: 247—258.

Norrish, K. and Hutton, J.T., 1969. An accurate X-ray spectrographic method for the analysis of a wide range of geological samples. Geochim. Cosmochim. Acta, 33: 431—453.

Thompson, M. and Howarth, R.J., 1973. The rapid estimation and control of precision by duplicate determinations. The Analyst, 98: 153—160.

Journal of Geochemical Exploration, 8(1977)305—323

GEOCHEMICAL EXPLORATION UNDER EXTREMELY ARID CONDITIONS

E.M. EL SHAZLY[1], N. BARAKAT[2], E.A. EISSA[2], H.H. EMARA[3], I.S. ALI[4] S. SHALTOUT[4] and F.S. SHARAF[4]

[1] *Atomic Energy Establishment, Academy of Scientific Research and Technology, Cairo (Egypt)*
[2] *Faculty of Science, Ain Shams University, Cairo (Egypt)*
[3] *Mineral Research Laboratories, Cairo (Egypt)*
[4] *Military Research Laboratories, Cairo (Egypt)*

(Revised version accepted March 29, 1977)

ABSTRACT

El Shazly, E.M., Barakat, N., Eissa, E.A., Emara, H.H., Ali, I.S., Shaltout, S. and Sharaf, F.S., 1977. Geochemical exploration under extremely arid conditions. J. Geochem. Explor., 8: 305—323.

Geochemical exploration has been carried out in several areas in the southern Eastern Desert of Egypt between latitudes 22° N and 25° N. The region where these areas are located is characterized by an extremely arid environment in which the scarce rainfall usually leads to the formation of rare torrents pouring in the valleys (wadis). The latter are filled with Quaternary alluvium and they are incised in mountainous country comprising igneous-metamorphic basement rocks of essentially Precambrian age.

Bedrock and alluvium sampling have been used during exploration for Cu, Ni, Pb, Zn, Co, Be, Li and B. Bedrock sampling has been found applicable for all these elements, irrespective of their oxidation stage or their partial leaching and mobilization in the oxidation zone. Geochemical anomalies for all the investigated elements are well expressed, depending on the closeness of spacing of the sampling pattern. Chemical characteristics of the arid environment do, however, influence the mobilization or fixation of certain elements in the oxidation zone.

On the other hand, the process of rock weathering in this environment, the chemical and physical nature of the alluvium accumulating in the valleys and the leaching action of the torrents in these valleys, lead to the removal of the chalcophile elements Cu, Ni, Pb, Zn and Co, and eventually to the ineffectiveness of alluvial geochemical exploration. These elements can usually only be detected a short distance from their source, normally within about 100 m. Under the same conditions, however, the oxyphile elements Be, Li and B are preserved and could be traced by alluvial geochemical prospecting for long distances from their source, although they are also partly mobilized in the weathering and transportation processes.

INTRODUCTION

Geochemical exploration, utilizing alluvium and rocks, has been conducted in Abu Swayel and Haimur in the south Eastern Desert of Egypt (Fig. 1) for Cu, Ni, Pb, Zn and Co. The results from the two areas have been found to be comparable and, accordingly, emphasis here will be given to the first locality. The rock exposures in these two areas are of igneous-metamorphic basement rocks which are incised by valleys (wadis) filled with Quaternary alluvium of generally coarse nature comprising gravel and sand but with some silt (El Shazly et al., 1965).

Geochemical exploration, using alluvium, for Be, Li and B has been carried out in Wadi Sikait area in the south Eastern Desert (Fig. 1). The alluvium-filling Wadi Sikait and its tributaries is of Quaternary age; it is distinguished into a widely spread younger alluvium at the present wadi level and an older alluvium at a higher level. The alluvium is coarse, ranging from gravel to sand to minor silt, and is derived from basement igneous-metamorphic rocks essentially of Precambrian age (El Shazly and Hassan, 1970).

The prospected areas are all located in the southern part of the Eastern

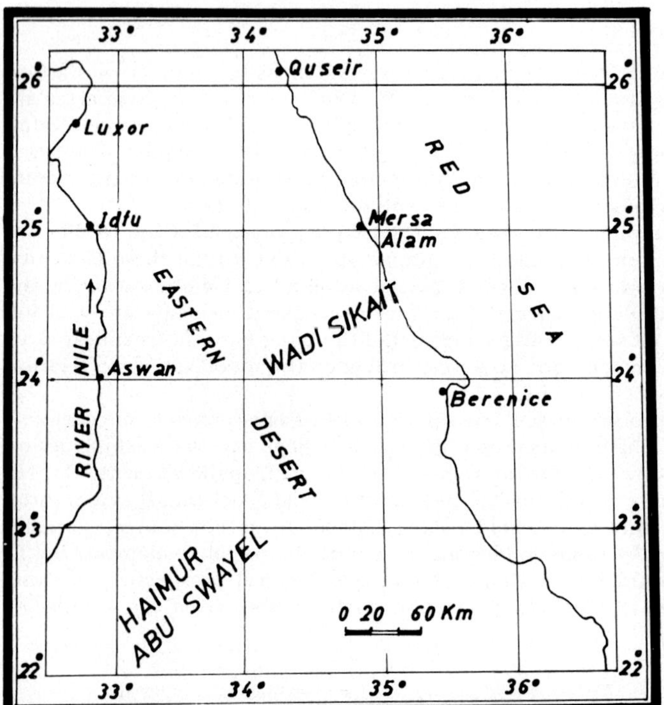

Fig. 1. Location of prospected areas in the south Eastern Desert of Egypt.

Desert of Egypt, where the dominating climatic conditions are of extreme aridity, with scarce rainfall and high evaporation; it is cool in winter but hot to very hot in other seasons.

SAMPLING AND ANALYSIS

Samples of alluvium and rocks were collected from the Abu Swayel and Haimur areas in a square grid pattern of 50 m (Fig. 2) and to a depth of 30 cm. At Wadi Sikait the predominating younger alluvium has been sampled

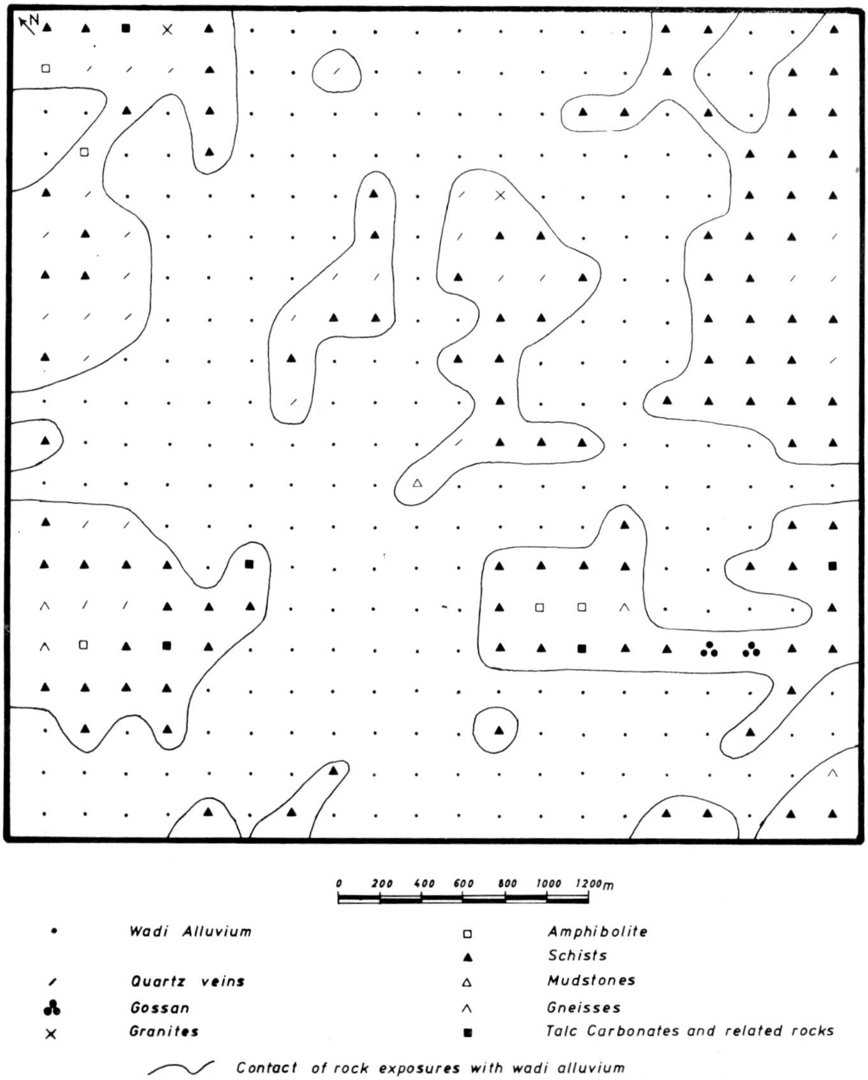

Fig. 2. Map of distribution of alluvium and rocks, Abu Swayel area.

on a 50-m square grid pattern (Fig. 11) and to a depth of 30 cm. Each alluvium sample was mixed, quartered and passed through a 2-mm sieve in preparation for quantitative analysis. The samples of alluvium and rocks from Abu Swayel and Haimur areas were analyzed for Cu, Ni, Zn, Pb and Co, while the samples of alluvium from Wadi Sikait area were analyzed for Be, Li and B. The analytical techniques and the limits of detection of Cu, Ni, Zn, Pb, Co, Be, Li and B are given in Table I.

TABLE I

Techniques of analysis and detection limits for Cu, Ni, Pb, Zn and Co

Element	Analytical technique	Detection limit (ppm)
Cu (rocks and alluvium)	polarographic	5
Cu (water)	colorimetric	0.01
Ni (rocks and alluvium)	colorimetric	2
Ni (water)	colorimetric	0.02
Zn (rocks and alluvium)	polarographic	5
Zn (water)	colorimetric	0.002
Pb (rocks and alluvium)	polarographic	5
Pb (water)	colorimetric	0.001
Co (rocks and alluvium)	colorimetric	1
Co (water)	colorimetric	0.005
Be (alluvium)	spectrographic	0.1
Li (alluvium)	spectrographic	0.1
B (alluvium)	spectrographic	0.05

ABU SWAYEL AND HAIMUR AREAS

Regional setting

The Abu Swayel and Haimur areas are mainly underlain by Precambrian rocks of the igneous-metamorphic basement, which is the northern extension of the African Shield. The oldest rocks are some 1200—1300 m.y. old and are regionally metamorphosed geosynclinal sediments and volcanics. They are mainly represented by schists, gneisses, meta-mudstones, amphibolites, talc carbonates and related rocks (El Shazly et al., 1965) and are intruded by granite bodies and quartz veins.

Old mines exploiting Cu from Cu-Ni mineralization are known in both the Abu Swayel and Haimur areas. The locality of the gossan of the Cu-Ni mineralization at Abu Swayel is shown in Fig. 2. In the underground mine workings, chalcopyrite, pentlandite and pyrrhotite are the main sulphide minerals replacing the amphibolite country rocks, which are hydrothermally altered and sheared in a NW-SE direction. The foliation of the metamorphic rocks also trends approximately NW-SE.

Wadis, filled with alluvium of Quaternary age, are found incised in the basement rocks. The material constituting the alluvium is derived from the basement rocks, the erosion occurring especially during the pluvial periods. During the present dry period, rare torrents, resulting from the scarce rainfall, are still active. Acacia trees are the most noticeable shrubs growing in the wadis (El Shazly et al., 1970).

Geochemical results

The ranges and means of Cu, Ni, Zn, Pb and Co in the rocks and alluvium are given in Tables II and III for the Abu Swayel and Haimur areas respectively. The geochemical anomalies are shown on the maps in Figs. 3, 4, 5 and 7 for Cu, Ni, Zn and Pb in the rocks of Abu Swayel area, and in Figs. 6 and 8 for Zn and Pb in the alluvium of the same area. Histograms showing the frequency distributions of Cu, Ni, Zn, Pb and Co in the alluvium of the Abu Swayel area are in given in Fig. 9.

TABLE II

Mean Cu, Ni, Pb, Zn and Co contents in rocks and alluvium, Abu Swayel (includes only those samples in which the elements were detected)

Sample type (number of samples)		Cu (ppm)	Ni (ppm)	Zn (ppm)	Pb (ppm)	Co (ppm)
Schists	M*	228	94	158	117	10
(109)	R	30— 4000	12— 750	20—1430	15— 480	1—20
Gneisses	M	195	51	183	110	13
(4)	R	70— 380	25— 80	40— 440	10— 170	12—14
Amphibolites	M	1294	385	274	190	13
(5)	R	40— 5000	50— 1500	130— 520	16— 250	12—14
Gossans	M	119000	7550	6100	2100	20
(2)	R	113000—125000	4800—10300	4000—8200	2000—2200	20
Talc carbonates	M	695	126	290	117	12
(5)	R	150— 2600	32— 500	90— 552	80— 150	7—18
Granites	M		60		40	4
(2)	R		50— 70		40	4
Quartz	M	76	50	48	51	5
(29)	R	15— 200	15— 150	40— 150	10— 120	1—12
Alluvium	M	109	49	104	119	11
(240)	R	30— 290	18— 190	40— 220	20— 320	2—32

*M = mean; R = range.

TABLE III

Mean Cu, Ni, Pb, Zn and Co contents in rocks and alluvium, Haimur (includes only those samples in which the elements were detected)

Sample type (number of samples)	Cu	Cu (ppm)	Ni (ppm)	Zn (ppm)	Pb (ppm)	Co (ppm)
Schists	M*	137	63	73	56	3
(73)	R	20— 3660	2— 250	10— 260	20— 220	2— 7
Mudstones	M	96	20	102	60	4
(22)	R	30— 190	8— 60	30— 400	20— 100	2—40
Gneisses	M	30	14			2
(3)	R	30	8— 25			2
Amphibolites	M	113	22	94	49	4
(51)	R	30— 500	5— 40	20— 500	20— 120	2—10
Gossans	M	36980	4945	2467	640	11
(4)	R	140—139000	35—15000	600—6000	100—1300	7—15
Talc carbonates	M	138	31	413	85	3
(8)	R	100— 180	25— 40	150— 720	80— 100	2— 5
Alluvium	M	38	28	26	58	2
(45)	R	30— 120	10— 40	10— 50	20— 250	2— 3

*M = mean; R = range.

The rocks of the Abu Swayel area exhibit anomalies of Cu, Ni, Zn and Pb, expressed as contour lines of concentrations of 200, 1000, 2000 and 10000 ppm of each element (Figs. 3, 4, 5 and 7). These include 13 Cu anomalies, 7 Ni anomalies, 12 Zn anomalies and 13 Pb anomalies. Some of these anomalies are superimposed on the known Cu-Ni mineralization, but the majority are newly recorded. The most important of the latter actually prove to be complex anomalies of more than one element and in some cases incorporate the four elements Cu, Ni, Zn and Pb. Although the prospected areas are poor in Co (Tables II and III), relatively high Co values accompany the Cu-Ni anomalies.

In comparison, the alluvium of the Abu Swayel area does not exhibit separate anomalies of Cu and Ni and only 3 Zn anomalies and 8 Pb anomalies are recorded (Figs. 6 and 8). It may be noted that the anomalies of Zn and Pb in the alluvium are of small extent and possess only low values of 200—300 ppm. The greater number of Pb anomalies recorded in the alluvium, in comparison to other elements, may be explained by the relative difficulty of leaching Pb in the oxidation zone due to the formation of lead sulphate and carbonate; nevertheless, in the arid environment, where the chloride ion is abundant, the element is partly mobilized as lead chloride.

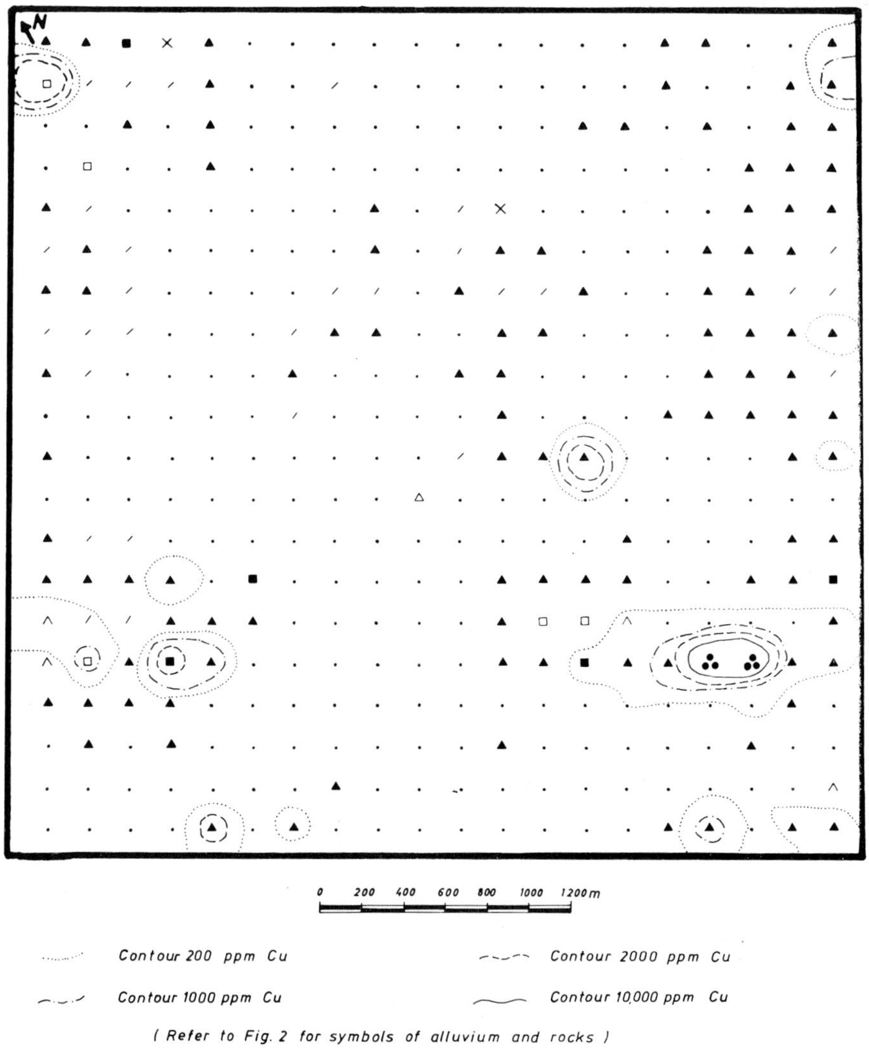

Fig. 3. Map showing contours of Cu anomalies in rocks, Abu Swayel area.

The histograms of Cu, Ni, Zn, Pb and Co (Fig. 9A to E) show the distribution of these elements in the different concentration ranges in the alluvium of Abu Swayel area. The distributions are usually approximately normal though with variable degrees of irregularity. The presence of a considerable number of low Ni values may be due to the presence of this element in silicates in the ultrabasic and basic rocks of Abu Swayel area, as well as its occurrence in the form of pentlandite in the sulphide mineralization.

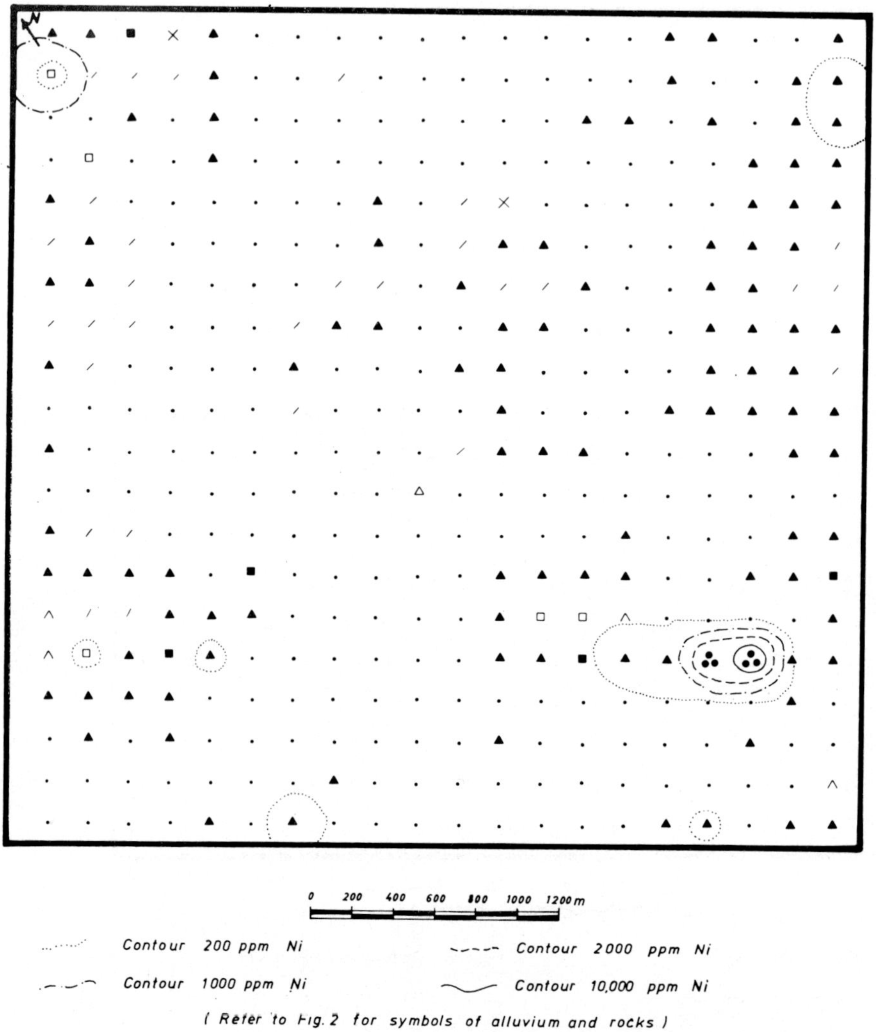

Fig. 4. Map showing contours of Ni anomalies in rocks, Abu Swayel area.

WADI SIKAIT AREA

Regional setting

Wadi Sikait area has basement rocks of Precambrian age, represented in the prospected area by psammitic gneiss, schists (especially biotite-rich ones), serpentinized dunites and pegmatites (Fig. 10). These rocks represent

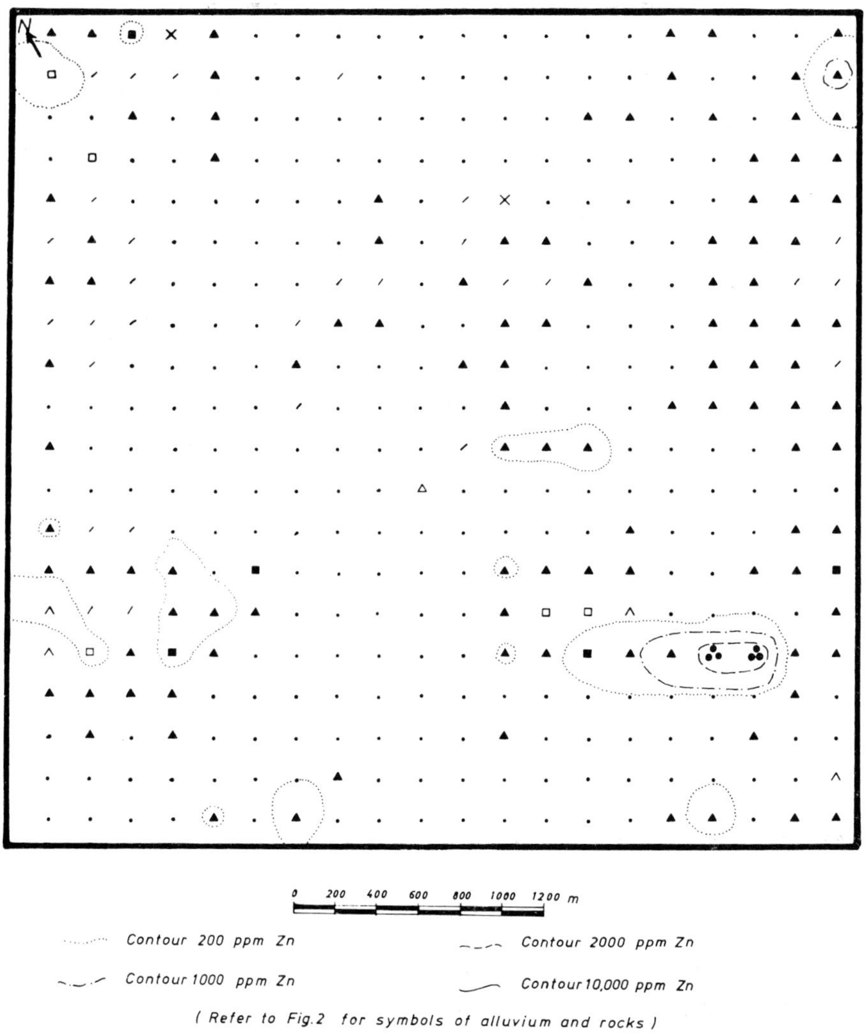

Fig. 5. Map showing contours of Zn anomalies in rocks, Abu Swayel area.

sediments and volcanics which have been regionally metamorphosed and folded. An anticlinal structure has been noted in the Wadi Sikait area (El Shazly and Hassan, 1970). Ancient emerald mines are associated with the schists in the prospected area.

The alluvium which covers Wadi Sikait and its tributaries is distinguished into younger alluvium, which is dominant and occurs at the present wadi level, and older alluvium which is found as relics at a higher level. Both

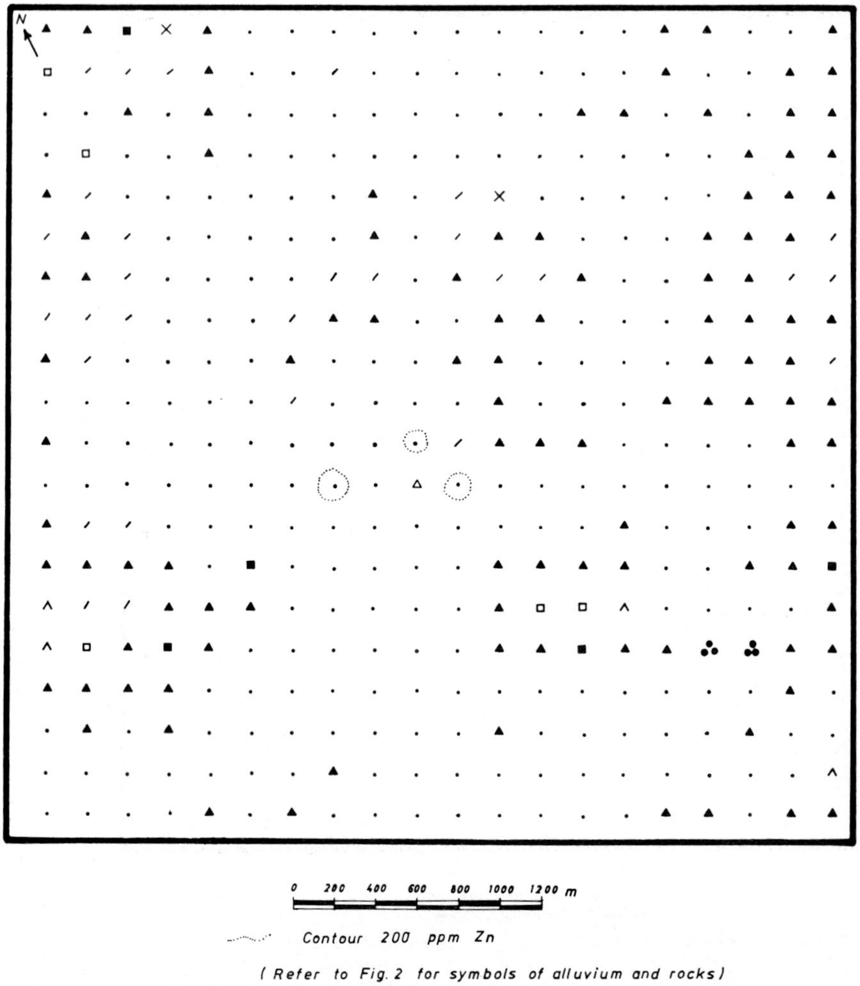

Fig. 6. Map showing contours of Zn anomalies in alluvium, Abu Swayel area.

types of alluvium are of Quaternary age and have been developed by the weathering of the basement rocks, especially during the previous pluvial periods. During the present dry period, rare torrents resulting from scarce though sudden rainfall are still passing through Wadi Sikait and its tributaries. The younger alluvium has been used in the prospecting for Be, Li and B.

Geochemical results

In spatial distribution, the highest Be concentrations, ranging from 20 to 40 ppm, are only encountered in the tributaries. In the main Wadi Sikait,

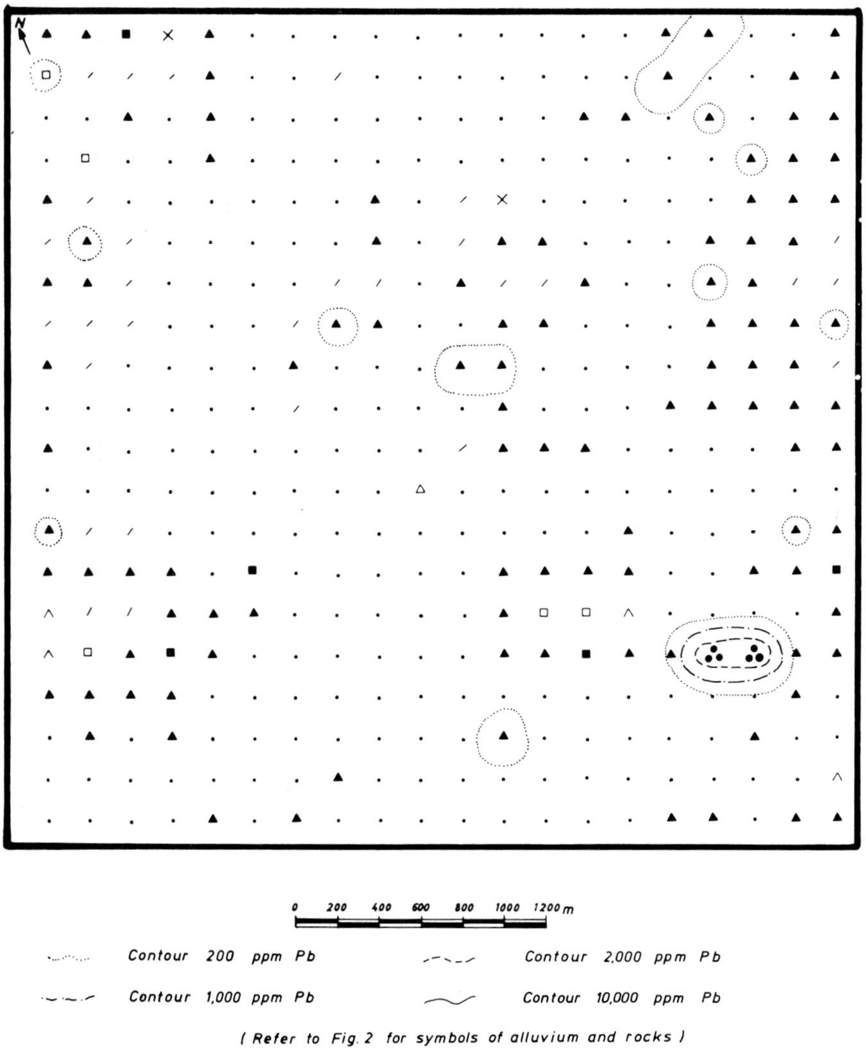

Fig. 7. Map showing contours of Pb anomalies in rocks, Abu Swayel area.

however, the concentration of Be in the alluvium is in the range 1.1—20 ppm (Fig. 11). The dilution of Be in the alluvium in this case is explained by the abundance of foreign materials in the long course of the main wadi.

The Be content of the alluvium has an overall range of 1.1—37.6 ppm with a mean of 9.1 ppm (detection limit 0.1 ppm). The histogram (Fig. 14A) shows a positively skewed frequency distribution, with the greatest number of samples (45.7%) in the first interval (1.1—5 ppm Be) and with progressively fewer samples in the higher concentration intervals.

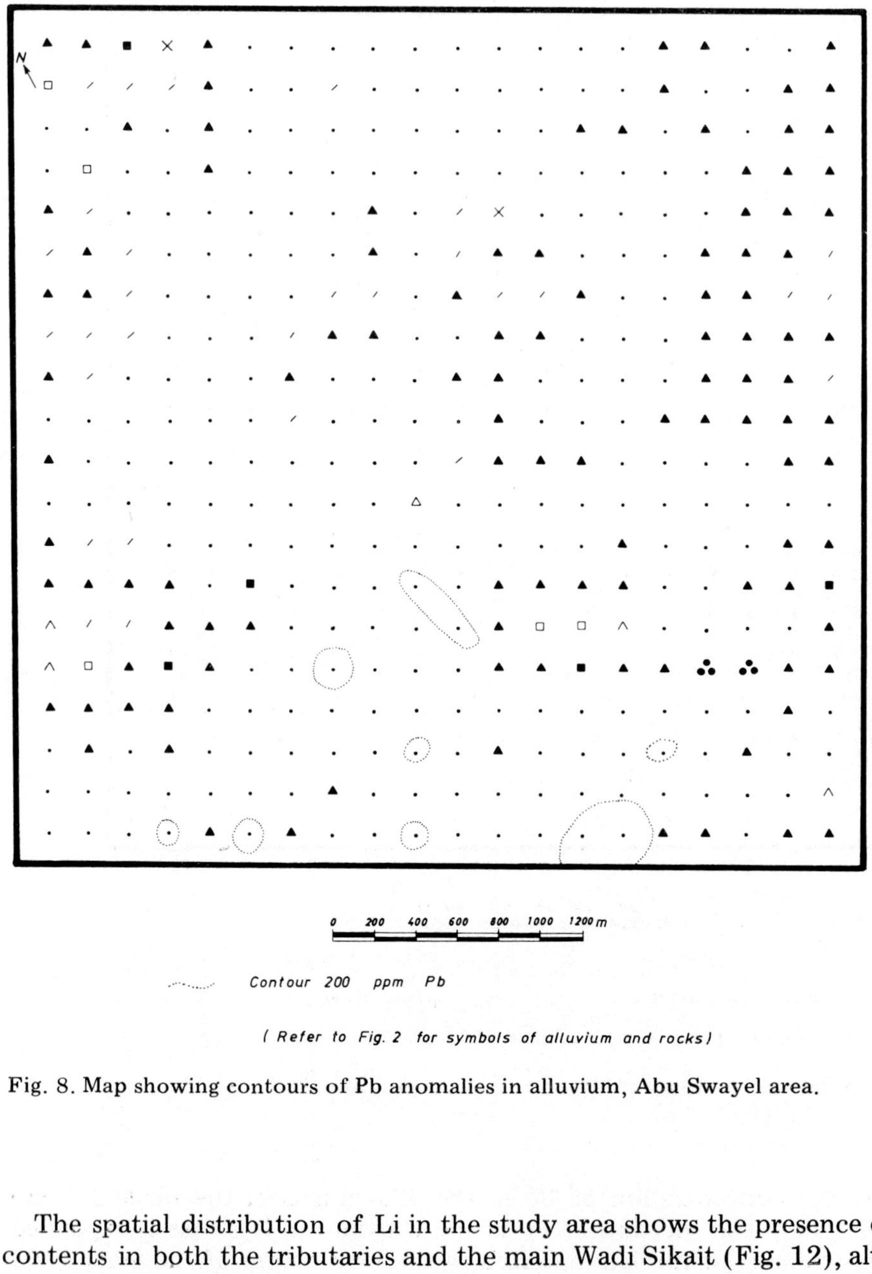

Fig. 8. Map showing contours of Pb anomalies in alluvium, Abu Swayel area.

The spatial distribution of Li in the study area shows the presence of high contents in both the tributaries and the main Wadi Sikait (Fig. 12), although the high values are more common in the tributaries. Li has been detected in the alluvium in values ranging from 14.7 to 107.6 ppm, with a mean of 50.8 ppm (detection limit 0.1 ppm). The mean Li content in the alluvium of Wadi Sikait is considerably higher than the average Li in soils of 30 ppm given by Vinogradov (Vinogradov, 1959; Hawkes and Webb, 1962). The

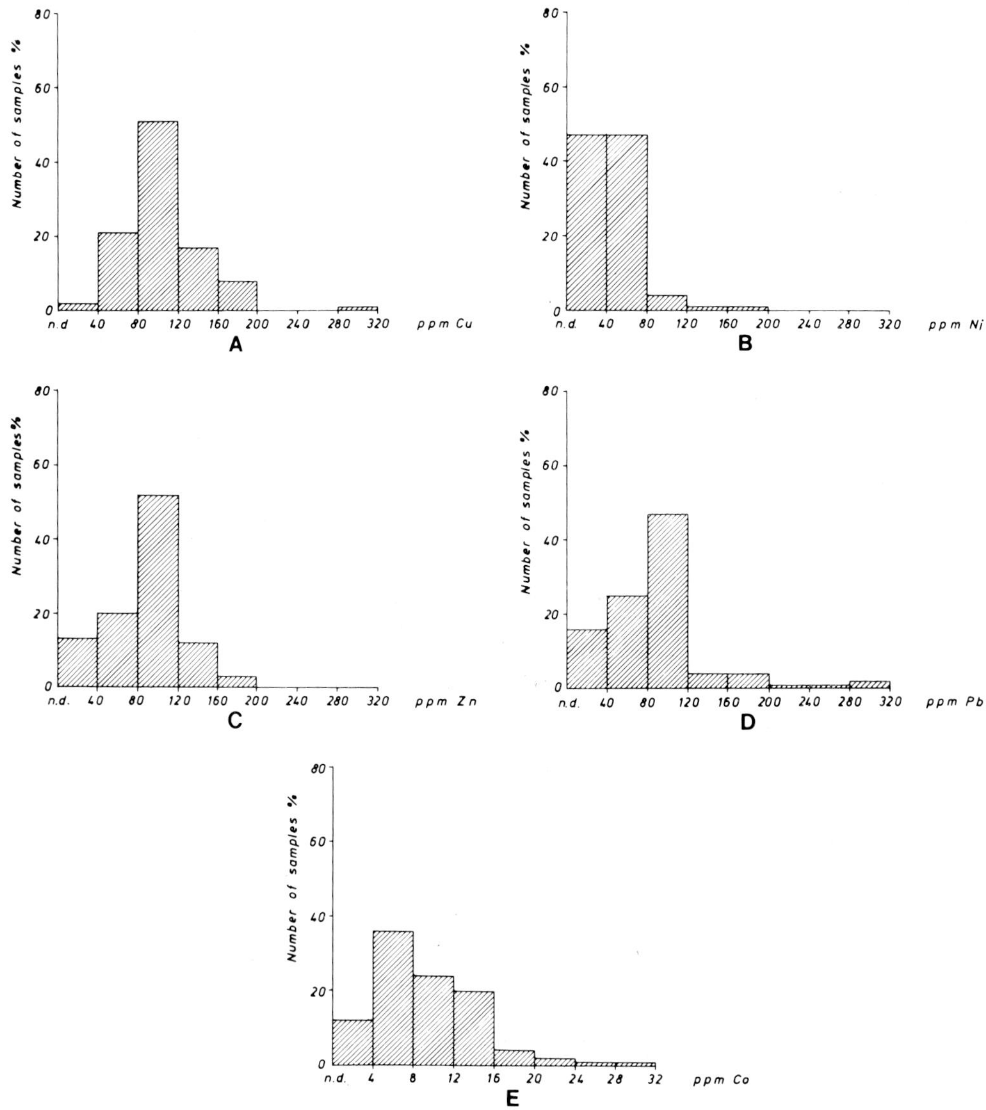

Fig. 9. Histograms showing frequency distributions of chalcophile elements in alluvium, Abu Swayel area. Number of samples: 240.

frequency distribution of Li in the alluvium of Wadi Sikait is approximately normal with 32.44% of the samples occurring about the mean in the 40—60 ppm range (Fig. 14B).

The spatial distribution of B (Fig. 13) illustrates that the values above the mean are mainly encountered in the tributaries and at their intersections with the main Wadi Sikait. B has been found in the alluvium samples in

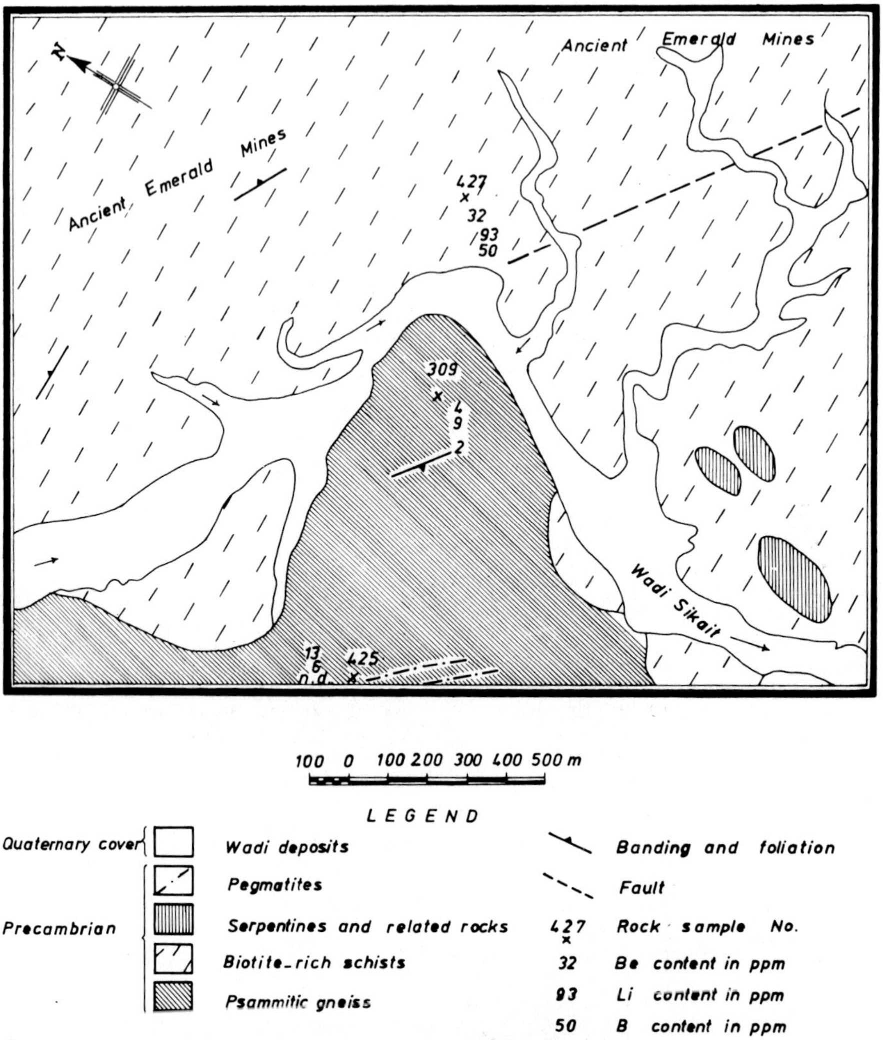

Fig. 10. Geological map of the prospected Wadi Sikait area.

concentrations ranging from 0.8 to 158.5 ppm with a mean of 16.6 ppm (detection limit 0.05 ppm). The mean B content is higher in the alluvium of Wadi Sikait area as compared to its average content in soils of 10 ppm reported by Vinogradov (Vinogradov, 1959; Hawkes and Webb, 1962). The histogram (Fig. 14C) shows a positively skewed frequency distribution with the greatest number of samples (77.05%) in the lowest concentration interval (0.79—20 ppm), with progressively smaller numbers of samples in higher concentration intervals.

In the alluvium, the correlation coefficient between B and Be (+0.31) is

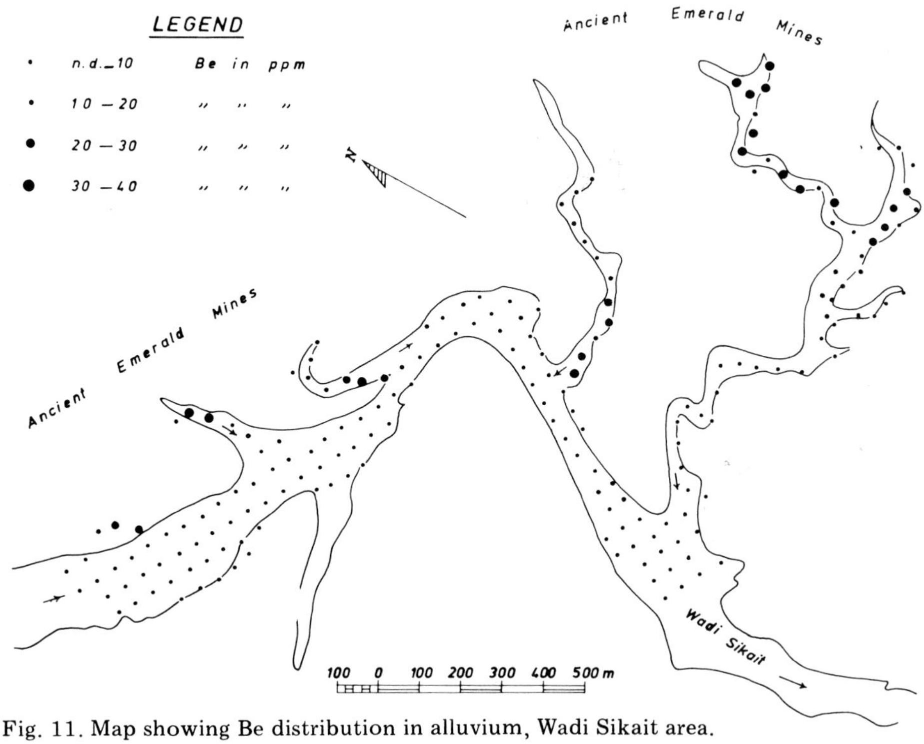

Fig. 11. Map showing Be distribution in alluvium, Wadi Sikait area.

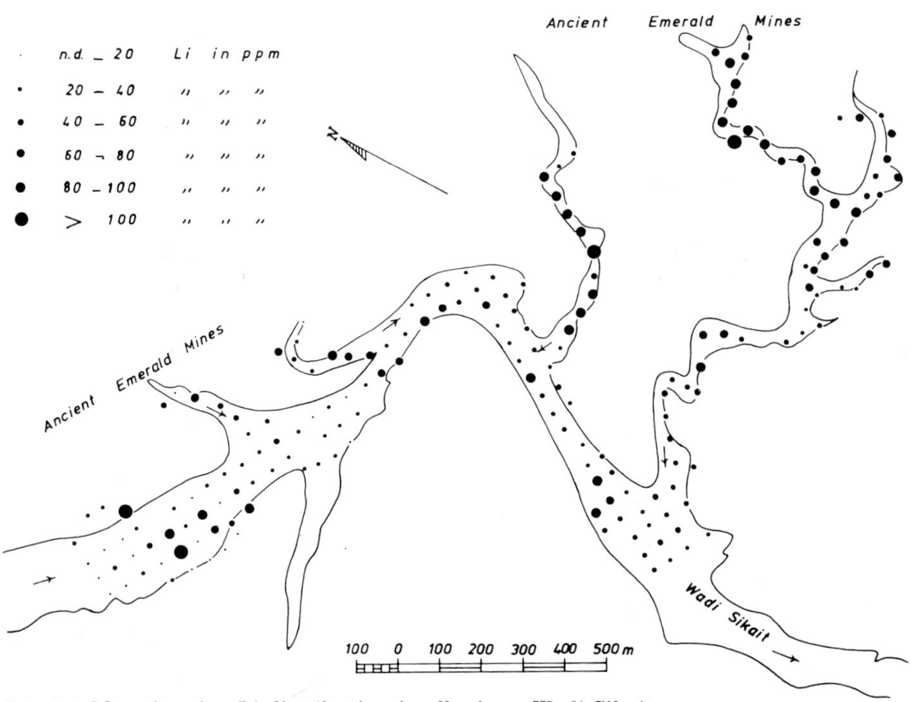

Fig. 12. Map showing Li distribution in alluvium, Wadi Sikait area.

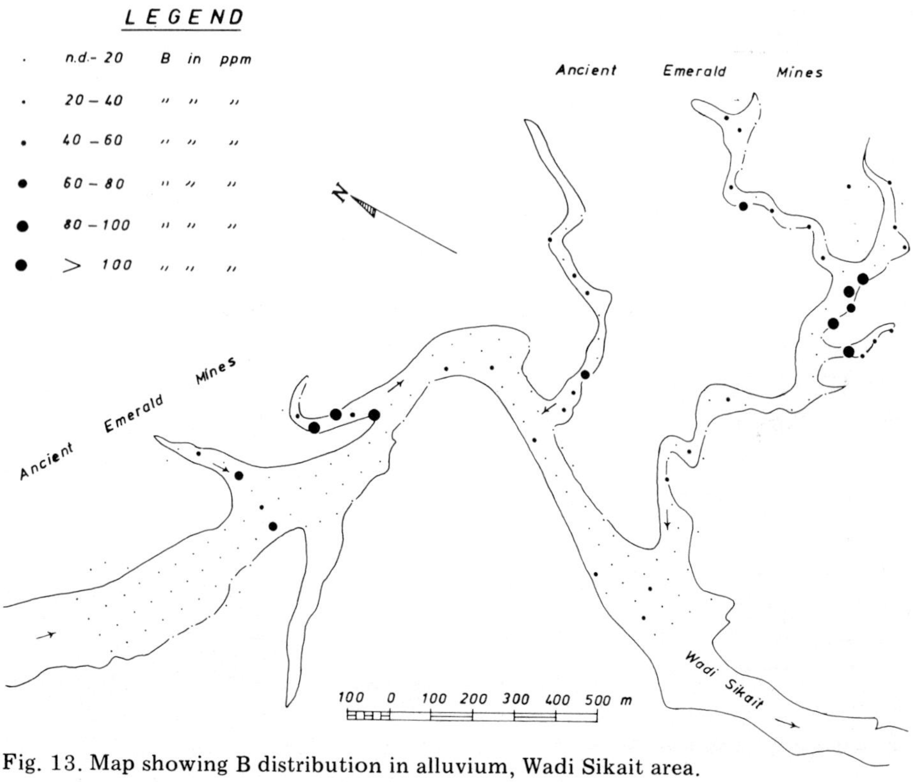

Fig. 13. Map showing B distribution in alluvium, Wadi Sikait area.

lower than that between Li and Be (+0.47). This difference in correlation coefficients is, however, not particularly significant.

SUMMARY AND CONCLUSIONS

The overall picture clearly indicates that alluvial geochemical prospecting is not applicable to the chalcophile elements (Cu, Ni, Zn, Pb and Co) in the prevailing extremely arid environment. There have been no anomalies recorded in the alluvium for Cu and Ni, and only small low-level anomalies registered for Pb and Zn which can be missed in prospecting. This is explained by the rapid leaching of the elements studied, especially in the presence of Cl^- and SO_4^{2-} in this very arid medium. The alluvium itself normally has a particle size coarser than the clay fraction and its capacity to adsorb the mobilized ions of the chalcophile elements is not sufficient to retain high concentrations of these elements at appreciable distances from the source.

The effect of weathering of the rocks is illustrated by the chemical analyses carried out on two samples of alluvium (Table IV) and one sample of

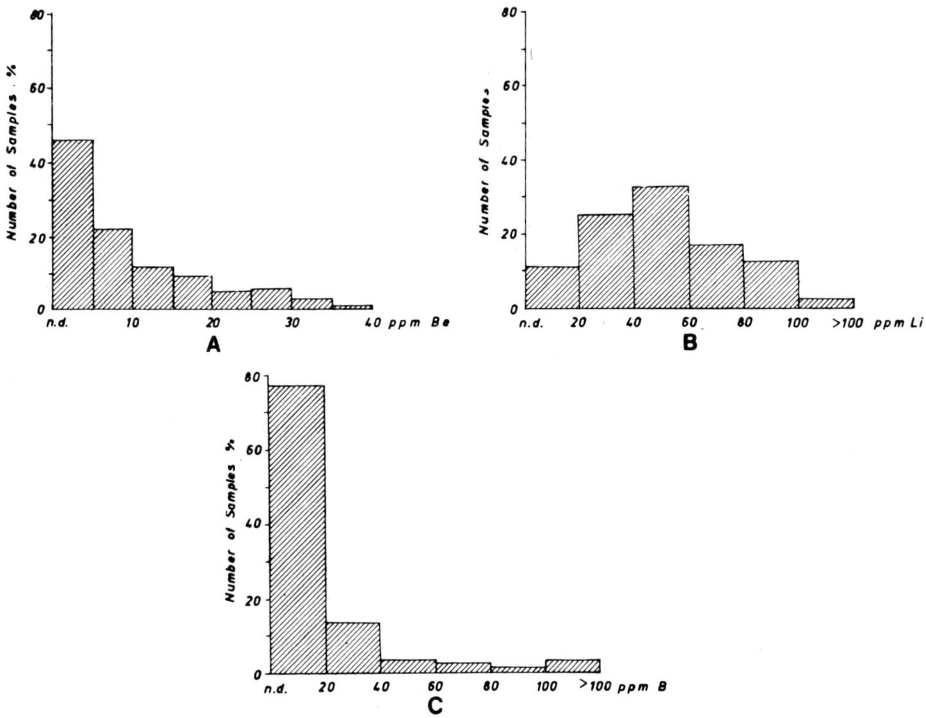

Fi[illegible] Histograms showing frequency distributions of oxyphile elements in alluvium, Wadi Sikait area.

groundwater from the fractured rocks of Abu Swayel shaft (Table V). Compared to the average chemical composition of crystal rocks, the alluvium is impoverished in alkalis Na and K, while the ferrous iron is largely oxidized to ferric iron. In the water sample, evidence for leaching of Ni, and to a lesser extent, of Cu and Co, is illustrated.

Bedrock prospecting for the investigated chalcophile elements proved very effective and it shows a considerable number of new anomalies — sometimes of high concentrations — for Cu, Ni, Pb and Zn. The most suitable rocks for prospecting are the amphibolites, talc carbonates and schists. However, this bedrock geochemical technique is suitable only for detailed prospecting and localization of the mineralization, not for regional and semi-detailed exploration, as closely spaced grid patterns have to be used. In spite of their mobility, these elements are preserved in quite high concentrations.

El Shazly et al. (1970) have demonstrated that Acacia trees are highly applicable to reconnaissance and semi-detailed exploration for chalcophile elements, as these elements are fixed by the trees at considerable distances from their source. In comparison, alluvial geochemical prospecting could

TABLE IV

Chemical analysis of alluvium, Abu Swayel and Haimur

Chemical component	Alluvium 75 A, Abu Swayel (wt.%)	Alluvium 131 H, Haimur (wt.%)
SiO_2	76.62	74.64
Al_2O_3	6.05	6.77
FeO	1.19	1.35
Fe_2O_3	4.10	3.76
CaO	2.24	3.43
MgO	1.34	1.81
Na_2O	1.01	1.28
K_2O	1.45	1.21
MnO	0.11	0.11
TiO_2	1.10	1.30
H_2O^-	0.38	0.44
Loss on ignition	3.67	2.70
SO_4^{2-}	0.98	1.14
Cl^-	0.06	0.03
Cu	0.012	0.004
Ni	0.009	0.0025
Zn	0.014	0.004
Pb	0.009	0.002
Co	0.0012	n.d.

TABLE V

Abridged chemical analysis of groundwater, Abu Swayel shaft

Anions		Major components		Metals	
Cl^-	886 ppm	Ca hardness	1000 ppm	Cu	0.3 ppm
SO_4^{2-}	887 ppm	Mg hardness	600 ppm	Ni	0.68 ppm
S^{2-}	2.02 ppm	SiO_2	27.5 ppm	Zn	n.d.
F	6.8 ppm	Na^+	940 ppm	Pb	n.d.
B	2.0 ppm	K^+	20 ppm	Co	0.02 ppm
		organic matter	673 ppm	Fe	0.1 ppm
Total dissolved solids			6764 ppm	pH 7.7	

not be applied on any scale, while bedrock geochemical techniques may only be used in detailed prospecting and final localization of the mineral concentration.

The results of this study indicate the suitability of applying alluvial prospecting techniques to Be, Li and B under extremely arid conditions.

This is due to their relative stability in weathering and transportation; however, B values become lower at considerable distances from the source area. Alluvial geochemical prospecting can be applied in the cases of these three oxyphile elements on regional and semi-detailed scales. For detailed exploration, analyses of the favourable bedrocks identified should be carried out in great numbers in the appropriate grid pattern. The present work has led to better delineation of beryl and tourmaline occurrences, and to the discovery of occurrences of lithium-bearing mica in the psammitic gneiss and schists in the prospected Wadi Sikait area.

It is of interest to compare the present findings with those reached by biogeochemical prospecting using Acacia trees in the same area. It has been concluded by El Shazly et al. (1970) that biogeochemical prospecting is of limited value in reconnaissance and semi-detailed applications for the three oxyphile elements in question, although in the case of Li the efficiency of biogeochemical techniques is comparable to alluvial geochemical methods. The balance between the success of the trends is based on the mobilization of the Be, Li and B in weathering and transportation, and the capacity of Acacia trees to absorb their released ions. The lowest mobility is apparently attained by Be and the highest by B, while the plants have very special aptitude to absorb B. Accordingly, Li holds an intermediate position which makes it feasible to follow it in the field by alluvial geochemical and biogeochemical techniques. Be is best suited for alluvial prospecting, while B, although relatively the most mobile, is indiscriminately absorbed by plants so that they do not reflect its content in the alluvium.

REFERENCES

El Shazly, E.M. and Hassan, M.A., 1970. Geology and mineralization of Wadi Sikait—Wadi Nagrus area, Eastern Desert. J. Geol. U.A.R., 14.

El Shazly, E.M., Farag, I.A.M. and Bassyuni, F.A., 1965. Contributions to the geology and mineralization of Abu Swayel area, Eastern Desert. J. Geol. U.A.R., 9: 45—67.

El Shazly, E.M., Barakat, N., Eissa, E.A., Emara, H.H., Ali, I.S., Shaltout, S. and Sharaf, F.S., 1970. The use of Acacia trees in biogeochemical prospecting. Can. Inst. Min. Metall., Spec. Vol., 11: 426—434.

Hawkes, H.E. and Webb, J.S., 1962. Geochemistry in Mineral Exploration. Harper and Row, New York, N.Y., 415 pp.

Vinogradov, A.P., 1959. The Geochemistry of Rare and Dispersed Chemical Elements in Soils (translated from Russian). Consultants Bureau, New York, N.Y., 2nd ed., 209 pp.

Journal of Geochemical Exploration, 8 (1977)325—335 325

SOME ASPECTS OF NATURAL GAMMA RADIATION IN ORE SEARCH

GEOFFREY H. SHERRINGTON

Geopeko Limited, Chatswood, N.S.W. 2067 (Australia)

(Revised version accepted March 9, 1977)

ABSTRACT

Sherrington, G.H., 1977. Some aspects of natural gamma radiation in ore search. J. Geochem. Explor., 8: 325—335.

Natural gamma radiation from the decay chains headed by ^{238}U, ^{235}U, ^{232}Th and ^{40}K can be detected and measured by a variety of instruments.

Hand-held total count scintillometers indicate a portion of the gross gamma radioactivity. Portable four-channel differential spectrometers allow some discrimination between contributions from the U, Th and K series. Similar discrimination is possible with downhole and airborne gamma-ray spectrometers. Laboratory spectrometers using high-resolution germanium detectors show the contribution of many individual isotopes.

The most obvious and widespread use for all of these instruments is in exploration for U deposits. Deposits which are radiometrically simple, such as Ranger 1, Northern Territory, need less complex instrumentation than deposits such as Mary Kathleen, Queensland, which has significant ^{232}Th series activity. Radiometric complexity is also caused by disequilibrium in the U decay series.

Exploration for base metal deposits can also be assisted by gamma spectrometry. The weak radiometric expression can show character attributable to recognizable events (such as K metasomatism) or can give an empirical pattern correlatable with lithology.

There is increasing attention being given to airborne gamma surveys as a guide to geological mapping.

A current resurgence of interest in natural gamma applications is being assisted by the development of better, more compact instrumentation.

INTRODUCTION

Four natural decay series are dealt with in this paper. They are headed by ^{238}U, ^{235}U, ^{232}Th and ^{40}K. These four series contain 47 radioisotopes, of which 36 are gamma emitters (Lederer et al., 1967). The number of individual gamma emissions from these 36 isotopes, so far detected, approaches 1000 in the energy range 0.025 — 3.183 MeV (Smith and Wollenberg, 1972).

The practical exploration methods described herein use only a few of these thousand possible emissions. This is because a given emission, to be useful, should give a high count rate and be well-resolved from other intense emissions. These requirements dictate the choice of measurement instrument. A compromise is involved since, although count rate increases with detector size, resolution decreases.

Detector volumes vary roughly from 1 cm^3, for laboratory semi-conductors, to 50,000 cm^3 for airborne scintillometers. In current terminology, resolutions vary from 0.5% to 10% for these examples, with counting times ranging from 1 day to 1 second respectively.

Within this range of options, other factors such as cost and portability assist in defining the optimum instrument for an application. Examples are given of several instruments and a variety of applications. Some of the applications are not novel and can be referenced in standard texts. The sole reference known by the author to cover the application of high-resolution spectrometry is that of Smith and Wollenberg (1972).

METHODS AND APPARATUS

*Total count ground surveys.*The type of instrument used was a hand-held scintillometer similar to the Scintrex BGS-1S. This has a sodium iodide (NaI) detector of about 15 cm^3 volume, a meter readout in counts per second (cps) and an energy range of about 0.08 MeV upwards.

This device is most commonly used on the ground for locating areas of anomalous radioactivity and often for determining gross radioactive distribution on a surveyed grid. In areas not noted for radioactive expression the background count rate is of the order of 20 cps.

Four-channel spectrometer ground surveys. A typical instrument is the Geometrics DISA-400. This has a NaI detector of about 350 cm^3 and sequential digital display of counts per selected sample period. The display shows uncorrected counts for total count, K, U and Th, represented by ^{40}K, ^{214}Bi and ^{208}Tl centred at 1.46, 1.76 and 2.62 MeV, with total count from 0.2 MeV upwards.

Counting times of one minute typically give background readings of 5000 for total count and 50—100 for the other channels.

Airborne survey. One instrument discussed herein comprises 50,000 cm^3 of NaI detectors in the main pack, a 4000-cm^3 crystal to monitor atmospheric Rn, a spectrometer of 256 channels for each crystal pack and a computer-based data acquisition system. The entire spectrum from 0.32 to 2.90 MeV is taken each second and put onto magnetic tape. It is one of the most advanced airborne spectrometers assembled to date and at the time of writing had not been flown.

Mention is made of a past airborne survey using 14,700 cm^3 of crystal, flying speed 170 km/hr, terrain clearance 100 m and line spacing 200 m.

High-resolution spectrometry. This spectrometer comprises a hyperpure planar germanium detector, cooled by liquid nitrogen. Active area is 80 mm^2

resolution at 122 keV is 600 eV full width at half maximum (FWHM).The detector is coupled to a 1024-channel multichannel analyzer (ORTEC 6200) with punched paper tape output.

Samples of 50 ±1 g are counted for periods between 100 seconds and 1 day, usually in the energy range 0—204.8 keV. The count rate for the most intense emission in the U spectrum (75 keV) is about 0.2 cps per percent uranium.

In this energy range, which is well below that commonly used with NaI detectors, there are some tens of emissions from isotopes in each of the ^{238}U and ^{232}Th series, which can be used to study disequilibrium.

APPLICATIONS AND RESULTS

Total count ground surveys

No actual examples of this common technique are given here. The instruments have widespread use in locating anomalous areas (particularly after an airborne survey). Of course, they are most useful in those areas where the radiation derives mostly from a single decay series. Work undertaken on several prospects in the Ranger areas, Northern Territory, is in this category. Several examples from this region have shown the close correspondence between airborne U channel response, portable U channel spectrometer response and total count.

Some differences arise between the total count instrument and the four-channel spectrometer, largely because of the setting of the lower-energy discriminator. The total count instrument, with its lower threshhold, displays more counts per unit crystal volume. The additional counts are at low energy and are susceptible to more selective attenuation by non-radioactive cover and the air.

Widespread use of multiple instruments has often demonstrated the need for frequent instrument calibration.

Four-channel spectrometer ground surveys

The two examples given here relate to base metal exploration. It is understood, however, that the most widespread use for the four-channel spectrometer is, and is likely to remain, in U exploration.

The four-channel spectrometer has been applied to base metal exploration for two purposes — first, as an aid to geological mapping and second, for seeking specific targets such as a zone of hydrothermal alteration characterized by an anomalous K distribution. A major limitation to its use in both cases is that less than a metre of colluvium or other transported cover will give a false impression of the bedrock distribution of radioactivity. In many exploration areas in Australia, outcrop is often sparse and weathering can extend to 150 m depth. Preliminary trials in the Northern Territory of Australia, over lateritized terrain, have shown rather featureless four-channel spectrometer response. The two examples below are drawn from temperate climates, the first at Parkes, N.S.W. (latitude 33°S) and the second at Mount Morgan (latitude 23°S).

In the Parkes example (Fig. 1), observation of gossanous material prompted a geochemical and geological mapping survey on an area designated here as Prospect 1. Minor mineralization was found in a sandstone outlier unconformably overlying a dipping sequence of trachyte-sediment-trachyte. The outlier was characterized radiometrically by low-U, high-K and moderately elevated Th, compared to the underlying sequences. The radiometric contrast, expressed as the ratio of U to K (uncorrected counts) is shown in Fig. 1, superimposed on the simplified geology, which was obtained by logging auger cuttings from 2 to 5 m depth. Discriminate analysis, incorporating radiometrics and geochemistry for Cu, Pb, Zn was moderately successful in distinguishing the four major rock units shown. This is an example of the use of four-channel radiometrics to assist in geological mapping.

In the Mount Morgan example (Fig. 2), Prospect 2 was around a small mine where minor tonnages of high-grade secondary Cu and Zn had been mined about 1920. Diamond drilling showed that a Devonian acid volcanic sequence was the host so that the target was a volcanogenic massive sulphide deposit similar to those described by Sangster (1972) from the Archaean in Canada. Zones of hydrothermal alteration have often been described

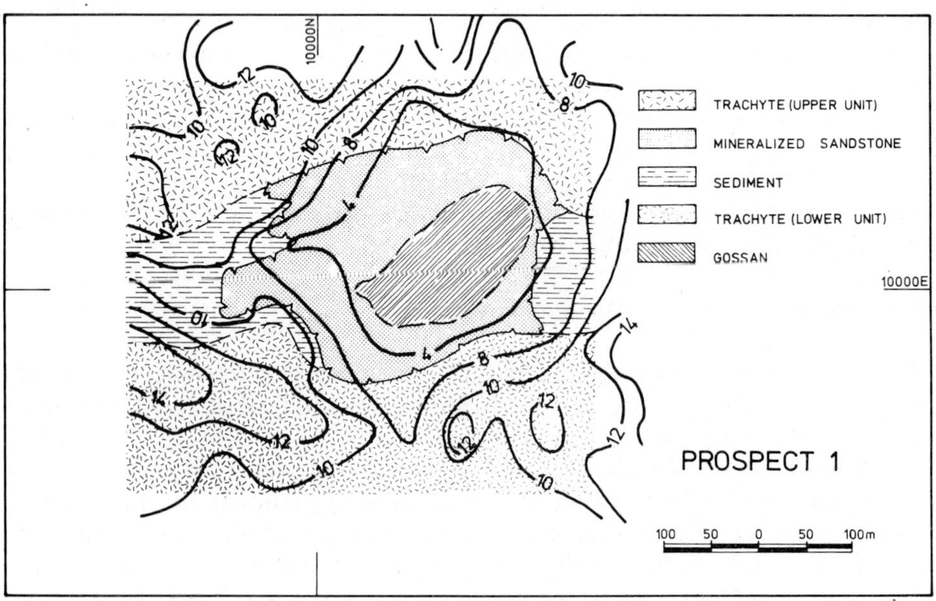

Fig. 1. Prospect 1 near Parkes, N.S.W. Radiometric contours are superimposed on simplified geology. The contours are the ratio of uncorrected U to K counts, one-minute count. Geometrics DISA-400 spectrometer. The sandstone/gossan unit carries uneconomic base metal mineralization.

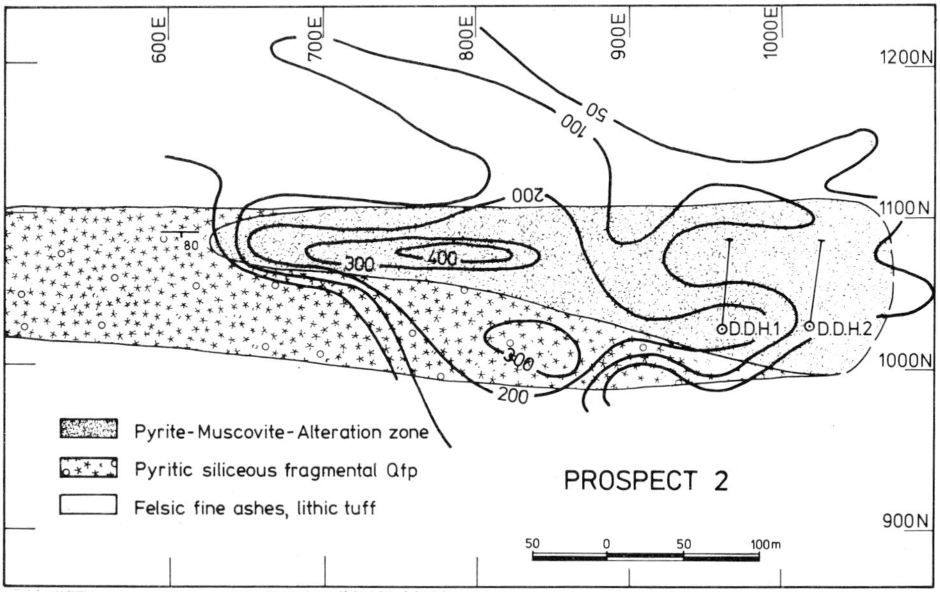

Fig.2. Prospect 2 near Mt. Morgan, Queensland. Radiometric contours for K are superimposed on simplified geology. The drill holes intersected uneconomic Zn-Cu-Ag-Au mineralization. Contour intervals are in uncorrected K counts per minute, DISA-400 instrument.

in association with this type of occurrence. Fig. 2 shows the simplified geology, with raw K radiometric counts superimposed upon it. It is evident that K is anomalous in the vicinity of mineralization found in the old mine and diamond drill holes. Several further K radiometric anomalies have been found in the general strike direction of this sequence and uneconomic base metal mineralization has been shown associated with some of them. The four-channel spectrometer is a very rapid and reliable device for exploration at this level.

Airborne survey

In the approximate time span of 1970—1976, two developments in airborne survey instruments for gamma radiation became available and advanced the method. One was the "square" NaI detector crystal (commonly described as 16 × 4 × 4 inches), with a single photomultiplier tube at one end. The other was the continued development of microprocessors and small electronic computing devices amenable to operation in a low-flying aircraft.

The square detectors, compared to the former cylindrical crystals, are more compact, require fewer photomultiplier tubes and seem more easily held to constant temperature. Crystal volumes of 3000 cubic inches (50,000 cm^3) are now being flown in the square configuration, in the U.S.A., Canada, Australia and (shortly) Iran.

The use of miniature electronic data processing equipment has allowed multichannel spectra to be recorded in flight. Given the approximate 10% resolution (at 0.662 MeV FWHM) of the crystal packs, a popular choice of channel numbers has been 256, with a one-second reading time. The abundance of data generated has required parallel buffer systems to allow information to be acquired at the same time as it is read onto magnetic tape.

Some current airborne systems use an "upward-looking" crystal assemblage to attempt to monitor atmospheric Rn and daughters. The theory behind this approach is riddled with missing parameters and an alternative approach is needed.

As more airborne surveys are completed, the need becomes more pressing to standardize the units of measurement. Units being adopted are parts per million thorium (ppm Th), equivalent uranium (eU) and percent potassium (%K); the conversion of count rates to these units depends upon the availability of good-quality, well-characterized test strips. Those anticipating substantial airborne gamma survey work should strive to assist in the preparation of test strips and test pads as a matter of some urgency.

The other area of airborne survey requiring attention is precise navigation and position recovery.

Downhole logging for ore reserve calculation

Downhole gamma logging for U has not been used quantitatively in the development of the Ranger 1 orebodies. A qualitative total-count log is taken to select the intervals required for X-ray fluorescence (XRF) analysis on half-core from diamond drilling.

Studies have been undertaken on the recovery of material from diamond and percussion drill holes placed a metre or so from each other. In one case, a 30-m intersection in a diamond hole returned an average of 0.314% U_3O_8 and the adjacent section in the percussion hole gave 0.434% U_3O_8, both by XRF analysis. In all cases studied, even with the inclusion of drill sludge weights and analysis and material balance work, results from percussion drilling are higher than those from diamond drilling. It is inferred that one or both of the drilling methods suffers U gains to, or losses from, the recovered material. It is likely that the same inference applies to the walls of the holes, upon which downhole gamma readings are taken. We do not currently know if downhole logging would show bias between diamond and percussion holes, nor do we know if this potential problem is unique to the Ranger orebodies.

We do consider it valuable to attempt quantitative logger calibrations by reference to synthetic boreholes rather than by comparison to material recovered from exploration drill holes.

High-resolution gamma spectrometry

Using the apparatus described above, it is possible to detect some twenty useful gamma emissions in each of the ^{232}Th and U series, when the latter two occur in their natural abundances of ^{238}U to ^{235}U.

By using standard gamma spectrometry methods, it is possible to:

(1) Measure the relative abundances of ^{238}U and ^{235}U, from the emissions at 92.8 keV (^{234}Th) and 143.8 keV(^{235}U).

(2) Perform gamma assays for ^{238}U which are essentially free from disequilibrium effects, again using ^{234}Th at 92.8 keV. The half-life of ^{234}Th is only 24 days so it accurately reflects the abundance of its parent, ^{238}U.

(3) Correct the gamma assays of (2) for sample density effects by using a second ^{234}Th emission at the lower energy of 63 keV. Changes in sample density affect the relative counts from each of these emissions.

(4) Estimate the relative abundances of ^{238}U and ^{232}Th in samples containing a mixture of both series.

(5) Study disequilibrium in the ^{238}U series.

At equilibrium in the ^{238}U series, all of the isotopes in the series have a fixed abundance relative to each other. Disequilibrium occurs when one or more isotopes are physically transported away from this environment. Those isotopes most likely to move furthest have the longest half-lives and include ^{238}U, ^{234}U, ^{230}Th and ^{226}Ra. The inert gas ^{222}Rn is also mobile because of its physical form.

Gamma spectrometry in the energy range 40—200 keV allows estimation of ^{238}U (through ^{234}Th), ^{230}Th (50 keV), ^{226}Ra (186 keV, not resolved from ^{235}U but capable of mathematical resolution) and ^{222}Rn (from essentially all isotopes below it in the series). A reported emission (Smith and Wollenberg, 1972) at 121 keV from ^{234}U lacks the intensity to be useful. It is thus mostly possible to determine at which isotope in the series geochemical differentiation occurs.

The isotopes ^{234}U, ^{230}Th, ^{226}Ra and ^{222}Rn, in the limited experience of the author, are able to migrate from the U source by distances which can assist in locating U mineralization. It is probable that a radioactive anomaly headed by ^{234}U or ^{230}Th will be close to the ^{238}U source, generally overlapping in part. ^{226}Ra can migrate and be immobilized at distances so far from the source that no useful directional or positional inferences are possible. Anomalies headed by ^{222}Rn are generally within a few tens of metres from the source.

A radioactive anomaly can have an excess of ^{238}U over subsequent isotopes. This can arise from recent or current addition of U to the anomaly or from depletion of daughter isotopes. When such an anomaly is found in the weathering profile, it is of interest to distinguish between the two possibilities by disequilibrium studies to attempt predictions about associated fresh rock mineralization. ^{230}Th is a key isotope in such an interpretation.

A more detailed discussion of disequilibrium is beyond the scope of this paper. The use of high-resolution spectrometry in such studies is illustrated in part in Fig. 3, which has five spectra labelled *A* to *E*. Each sample in these examples was counted overnight and each had about 1% eU except *E*, which was monazite with 6.8% ThO_2.

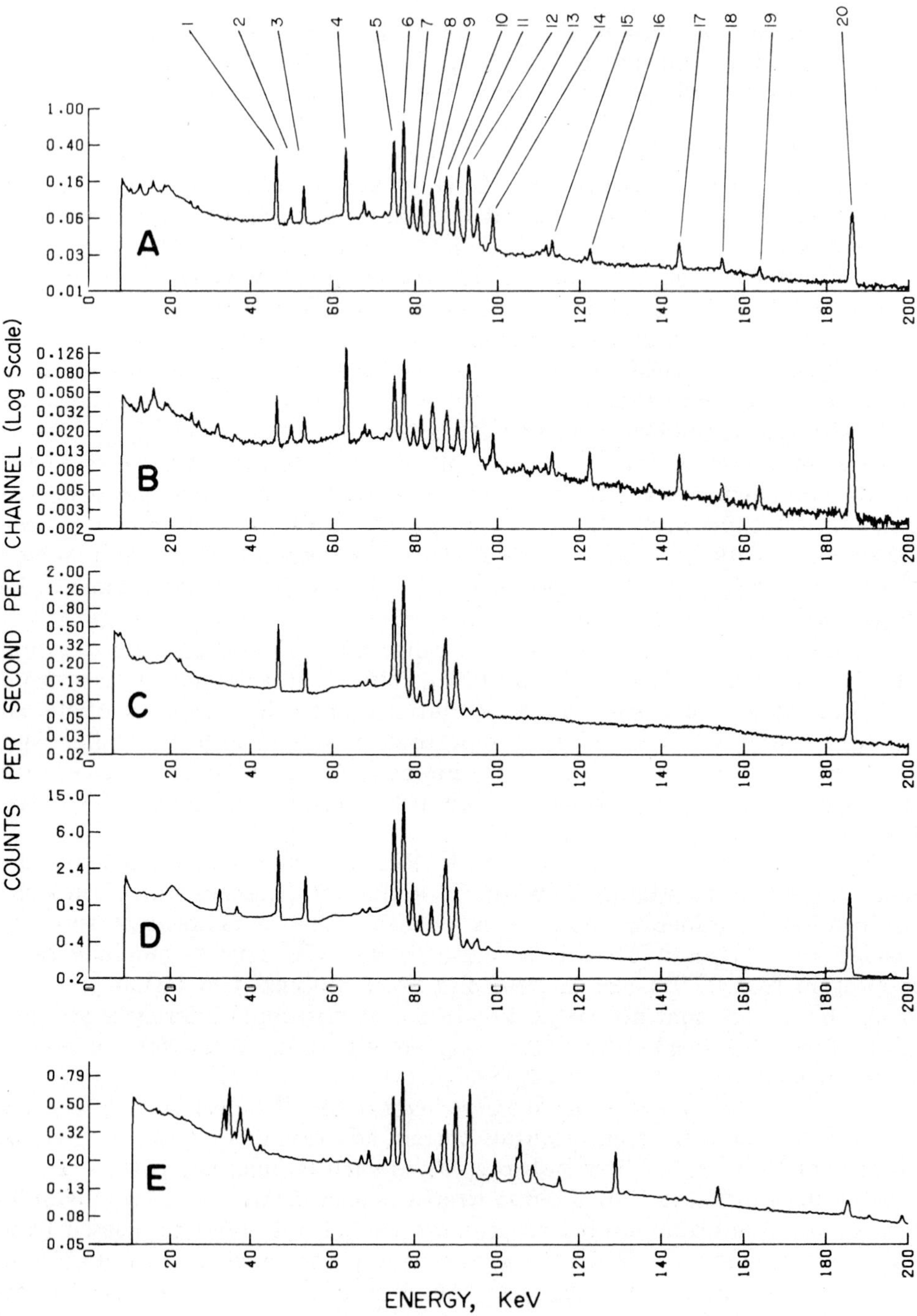

Fig. 3. Five spectra from high-resolution gamma spectrometer. From *A* to *E* they are: equilibrium U ore; ore with enrichment of U; rock with depletion of U; ^{226}Ra and daughters; monazite sand.

Spectrum *A* is that of a U ore in equilibrium. The 20 peaks numbered are given the assignments of Table I. The 50-keV peak assigned to ^{230}Th is in doubt to the extent that it may arise in the ^{235}U series. Several of the other peaks are unresolved multiple peaks, particularly in the abundant X-ray region of 75—110 keV. Some of the low-energy un-numbered peaks are fluorescent X-rays from Ba and rare earths.

In spectrum *A*, peaks are present for the major isotopes likely to migrate and cause disequilibrium, with the exception of ^{234}U. Spectrum *B* is an example of disequilibrium where the actual U content is about three times the radiometric grade. The interpretation is that some Ra and daughters are now being, or have recently been, removed from the sample location.

Spectrum *C* is from a "Ra hot spot" with below 5 ppm U but with 1% radiometric U. Again, Ra is the migrating isotope. In this case, it was fixed by crystalline sulphates in a surface evaporative environment.

Spectrum *D* was taken from some Ra point about 15 years old. The similarity with spectrum *C* is noted. Minor differences occur in the intensity of the Rn X-rays produced by decay of Ra, which can be ^{226}Ra (half-life 1620 years) or ^{223}Ra (half-life 11.7 days), as seen in spectra *A* and *D*, peaks 8 and 9.

Spectrum *E* is of monazite beach sand with 6.8% ThO_2 and an estimated

TABLE I

Assignment of 20 gamma emissions numbered in Fig.3

Peak Number	Energy (keV)	Assignment
1	46.52	^{210}Pb
2	50.0	^{230}Th
3	53.23	^{214}Pb
4	63.30	^{234}Th
5	74.8	Bi $K_{\alpha 2}$ X-ray
6	77.1	Bi $K_{\alpha 1}$ X-ray
7	79.3	Po $K_{\alpha 1}$ X-ray
8	81.3	Rn $K_{\alpha 2}$ X-ray
9	84.1	Rn $K_{\alpha 1}$ X-ray
10	87.3	Bi $K_{\beta 1}$ X-ray
11	89.8	Bi $K_{\beta 2}$ X-ray
12	92.3/92.8	^{234}Th
13	94.6	U $K_{\alpha 2}$ X-ray
14	98.4	U $K_{\alpha 1}$ X-ray
15	110	U K_{β} X-rays
16	122	^{57}Co reference
17	143.8	^{235}U
18	154.2	^{228}Ac
19	163.4	^{235}U
20	185.7, 186.0	^{235}U + ^{226}Ra

0.05% U. The emission at 129.1 keV, due to ^{228}Ac, is suitable for gamma assay of ^{232}Th, even in the presence of abundant U series isotopes.

A brief study of disequilibrium has been made in the No. 1 Orebody at Ranger 1. Short counting times led to imprecise results which confused the interpretation. However, it is clear that some disequilibrium exists in both primary and secondary ore and that it can be either uranium-enriched or uranium-depleted in nature. It appears that the overall radiometric grade is similar to the overall chemical grade, although differences of $\pm$10% occur in individual samples.

CONCLUSIONS

Base metal exploration

The weak radiometric character found with a portable four-channel spectrometer can be a useful aid in the geological mapping of a prospect. The instrument is able to detect some types of hydrothermal alteration. The best application is on outcrop or residual soil.

Some occurrences give gamma character which can be detected by airborne survey. Some porphyry Cu occurrences give a positive K anomaly, others do not.

We are currently unable to predict if a gamma survey will be useful on a prospect. However, past results have encouraged use of the portable instrument on a routine basis.

Uranium exploration

The trend in development of gamma sensors is towards greater portability and more advanced electronic circuitry. This is assisting the replacement, where desired, of total count instruments by spectrometers.

Gamma measurements are an important aspect of U exploration. Choice of the optimum instrument for a purpose involves resolution, count rate, portability, detector type and cost.

High-resolution gamma spectrometry has promise for gamma assays of U and Th, determination of isotopic abundances and disequilibrium studies. The work reported in this paper does not represent the limit of potential information — in fact, not all of the emissions detected have been unequivocally assigned. There is a requirement for more work into the geochemical mobility of selected natural isotopes and this can be assisted by such high-resolution spectrometry.

In all aspects of gamma measurement, it is desirable to reduce the data to the correct units of expression. This requires construction of test blocks, boreholes, test strips for airborne surveys, etc. A desirable by-product of such activity will be more frequent instrument calibration.

ACKNOWLEDGEMENTS

Geopeko Limited geologists G.J. Jones and R.R. Large collected the information used in Figs. 1 and 2, and T. Quinlan undertook computing.

The author thanks Peko-Wallsend Limited for permission to use the data reported.

Mention of a proprietary name does not imply endorsement of that name. Such names are used to describe a class of instrument whose specifications are available from the manufacturer.

REFERENCES

Lederer, C.M., Hollander, J.M. and Perlman, I., 1967. Table of Isotopes. John Wiley and Sons, New York, N.Y., 6th ed.

Smith, A.R. and Wollenberg, H.A., 1972. High-resolution gamma-ray spectrometry for laboratory analysis of the uranium and thorium decay series. In: J.A.S. Adams, W.M. Lowder and T.F. Gesell (Editors), The Natural Radiation Environment II. Proceedings of the Second International Symposium on the Natural Radiation Environment, August 7—11, Houston, Texas, Vol. I.

Sangster, D.F., 1972. Precambrian volcanogenic massive sulphide deposits in Canada: a review. Geol. Surv. Can., Paper, 72-22: 44 pp.

Journal of Geochemical Exploration, 8(1977)337—355

URANIUM AND COEXISTING ELEMENT BEHAVIOUR IN SURFACE WATERS AND ASSOCIATED SEDIMENTS WITH VARIED SAMPLING TECHNIQUES USED FOR URANIUM EXPLORATION

KAREN J. WENRICH-VERBEEK

U.S. Geological Survey, Denver Federal Center, Denver, Colo. 80225 (U.S.A.)

(Revised version accepted March 1, 1977)

ABSTRACT

Wenrich-Verbeek, K.J., 1977. Uranium and coexisting element behaviour in surface waters and associated sediments with varied sampling techniques used for uranium exploration. J. Geochem. Explor., 8: 337—355.

Optimum sampling methods in surface water and associated sediments for use in uranium exploration are being studied at thirty sites in Colorado, New Mexico, Arizona and Utah. For water samples, filtering is recommended to increase sample homogeneity and reproducibility because for most elements studied water samples which were allowed to remain unfiltered until time of analysis contained higher concentrations than field-filtered samples of the same waters. Acidification of unfiltered samples resulted in still higher concentrations. This is predominantly because of leaching of the elements from the suspended fraction. U in water correlates directly with Ca, Mg, Na, K, Ba, B, Li and As.

In stream sediments, U and other trace elements are concentrated in the finer size fractions. Accordingly, in prospecting, grain size fractions less than 90 μm (170 mesh) should be analyzed for U. A greater number of elements (21) show a significant positive correlation with U in stream sediments than in water. Results have revealed that anomalous concentrations of U found in water may not be detected in associated sediments and vice versa. Hence, sampling of both surface water and coexisting sediment is strongly recommended.

INTRODUCTION

Because of the urgent need to locate additional U deposits, the distribution of U in surface waters and associated sediments is being studied for use in exploration. Surface waters contain dissolved elements contributed by both surface and subsurface rocks; these elements occasionally concentrate in the bottom sediments and organic material. Although surface waters are not as good a geochemical evaluator of subsurface terrane as groundwater, sampling of surface waters is a cheaper, faster and more far-reaching method of exploration.

This research is part of a five-year study by the U.S. Geological Survey on the use of analyses of U carried in surface water (streams, lakes, swamps) as a tool for locating uranium-bearing terrane. Included in the study are the effects of climate, seasonal streamflow and variations in stream environment and associated sediments on the U concentrations in surface water. A parallel study is in progress to determine the extent to which the absorptive properties of modern decaying plant material and wood, soils and surface-water sediments can be used as a geochemical exploration tool in the search for U deposits.

Before the use of U in surface waters and stream sediments can be adequately tested as a tool for U exploration, methods of sampling need to be established. A study is in progress of 30 sites along streams that drain a broad variety of geologic terranes and represent portions of major drainage basins within the states of Colorado, New Mexico, Arizona and Utah. Some of the sites are near known U deposits whereas others represent areas not known to contain U deposits. The data and statistical analyses presented here represent 16 of these sampling sites. A detailed discussion on the behaviour of U at these sites in both water and stream sediments has been presented by Wenrich-Verbeek (1976).

SAMPLING PROCEDURES

Both the water and stream sediments were studied at each site. Multiple water samples were taken at each location and treated differently, utilizing all combinations of filtered and unfiltered, with acidified (by nitric acid) and unacidified samples. Filtration was done using both 0.45-μm nuclepore and millipore filters. This is the membrane size standardly used by Water Resources Division of the U.S. Geological Survey. Stream discharge, conductivity, pH, Eh and temperature were measured in situ. Details of the sampling equipment and field procedures are discussed by Wenrich-Verbeek (1976).

Multiple stream sediment samples were also taken at most sites; each sample represents composites of fifteen or more grab samples from the stream bed, both across the channel and parallel to the channel length. Occasionally samples were taken from specific locations within a stream meander. The samples were placed in polyethylene bags to avoid loss of the very fine clay fraction and dried at less than 43°C within two weeks of collection. The samples were sieved into the following seven size groups:

Group	Sieve mesh size (U.S. sieve series)	Grain size (μm)
1	>60	>250
2	60—100	149—250
3	100—140	105—149
4	140—200	74—105
5	200—230	63— 74
6	230—325	44— 63
7	<325	<44

CHEMICAL ANALYSES

Samples were analyzed by both the U.S. Geological Survey (USGS), Denver, Colorado and the U.S. Energy Research and Development Administration (ERDA), Oak Ridge Gaseous Diffusion Plant (ORL), Oak Ridge, Tennessee. Both laboratories analyzed the 95 water samples for U by use of direct and extraction fluorimetry (Barker et al., 1965) with limits of detection placed at 0.01 μg/l (USGS) and 0.03 μg/l (ORL). The agreement in the analytical results between laboratories for U in water is excellent (Wenrich-Verbeek, 1976). In addition, 48 trace and major elements were determined by emission spectroscopy or atomic absorption on each of the water samples. Similar analyses were made on the suspended fraction (residue upon filtration) of those samples unfiltered in the field but later filtered in the laboratory. The USGS laboratory determined U in the 213 stream sediment samples by delayed neutron activation (DNA), while ORL used fluorimetry on a sample leachate. The fluorimetry results were lower in 90% of the samples than the values obtained by the delayed neutron technique.

An attempt was made to determine parameters contributing to the lower values of the fluorimetry analysis through correlation coefficients between the various elements in the sediments and the ratio of fluorimetry to DNA. The fluorimetry determinations are expected to be somewhat lower, in that the sediment leach only incorporates the soluble U. U in the structure of such resistate minerals as zircon would not contribute to the analysis. Consequently, a significant correlation at the 99% confidence limit with such elements as Zr would be expected. Indeed, there is a significant negative correlation between the ratio of fluorimetry to DNA and Zr, which suggests some U is incorporated in zircon and does not contribute to the fluorimetry analysis. Cr, Ni, Cu, Sc, Rb and K also show correlations with the ratio. Sc, Rb and K have significant positive correlations with the ratio, whereas Cr, Ni and Cu, which are known quenchers — Cr and Ni are strong quenchers — of fluorescence (Barker et al., 1965, p. C5), show negative correlations. The negative correlations emphasize some of the problems involved with fluorimetry determinations for U in sediments. A regression analysis of U-fluorimetry/U-DNA shows Cr and Rb account for 35% of the variation. The other correlating elements explain an additional 11%. The remaining discrepancy between fluorimetry and DNA is probably a result of inconsistencies in the sediment leach associated with the fluorimetric determination. Although proponents of fluorimetry on sediments insist that total U is not desired since U tied up in such minerals as zircon has no exploration value, DNA may be preferable due to the analytical problems and inconsistancies in the leach associated with fluorimetry.

Thirty-eight other major and trace elements were determined for the stream sediments by atomic absorption, emission spectroscopy, and X-ray fluorescence.

DATA TREATMENT

Correlation and multivariate R-mode factor analyses were made on both the water and sediment data to detect correlations between U and associated elements and to determine factor groupings among the various elements. Scatter diagrams were made for each correlation as a check on the significance of the statistical results; that is, for assurance that one or two data points are not biasing the results.

Twenty-six geochemical variables in the water samples were subjected to four multivariate factor analyses, one for each of the different sample treatments. This not only facilitated grouping of the various geochemical parameters, but also underscored the effects that filtration, acidification or the lack thereof can have on water samples.

Identical statistical analyses were performed on the stream sediments. The multivariate analyses were run first on all 213 samples for geochemical variables and then separately by sample site to observe the influence of grain size on the correlations and factor groupings.

RESULTS AND INTERPRETATIONS

Water

Filtration and acidification

A minimum of four water samples were collected at each site and represent combinations of field-filtered, unfiltered, acidified, and unacidified samples. The unfiltered samples were later filtered in the laboratory at the time of analysis in order to detect changes in the dissolved fraction resulting from extended contact of the water with the suspended fraction. Most of the elements show behaviour similar to U for each of the four sample treatments. Table I portrays the variations in element concentrations with acidification and filtration using millipore filters.

The variation between sample treatments, shown in Figs. 1 through 4 and Table I, is a result of predominantly one geochemical feature, the leaching of the elements from the suspended fraction. A possible minor effect, but not shown to be conclusive, is the loss of elements by adsorption onto the sample container.

When unacidified- and acidified-unfiltered samples are compared (item 1, Table I), the elements Li, As, B, Ra, Fe, Mg, Al, Mn, Pb and U show a greater concentration in the acidified samples. Elements present in large concentrations such as Ca, K, Na and Ba, show no highly significant concentration differences between these two sample treatments. In the unacidified sample, however, most of the elements remained in the suspended material but were removed by filtration just prior to analysis. The major elements, in contrast, are readily leached merely on standing so that the addition of acid removes little more of their concentration from the suspended material.

TABLE I

Behaviour of the various elements with acidification and filtration

1. *Unfiltered:*		
Increased when unacidified	No significant difference in element concentration between unacidified-unfiltered and acidified-unfiltered samples	Increased when acidified
organic carbon	Ca, K, Na, Ba**	Li, As, B, Ra*, Mg, Al*, Mn*, Pb*, Fe*, U
2. *Filtered:*		
Increased when unacidified	No significant difference in element concentration between unacidified-filtered and acidified-filtered samples	Increased when acidified
none	organic carbon**, Li, Ba**, As, Mg, Ra, Ca, K, Na, Al**, U	B, Hg, Mn, Pb, Fe
3. *Unacidified:*		
Increased when unfiltered	No significant difference in element concentration between unfiltered-unacidified and filtered-unacidified samples	Increased when filtered
Ba, Mg, Mn, Pb, Ca, Fe	Li, As, Hg**, Ra**, K, Na, Al**, U, B**	organic carbon
4. *Acidified:*		
Increased when unfiltered	No significant difference in element concentration between unfiltered-acidified and filtered-acidified samples	Increased when filtered
organic carbon*, Li, Ba*, As, B, Mg, Ra*, Mn* Pb, Ca*, Fe*, K, Al, U	Hg*, Na	none

*Indicates a very large increase in concentration.

**Values for samples do not follow a straight line trend but show a scatter on both sides of the line.

This is supported by their relatively low concentrations in the suspended fraction of those samples that were later filtered in the laboratory. Also, most of these major elements are in sufficiently large concentrations in the water, that the adsorption onto the container walls has a negligible effect. This becomes even more apparent when these unfiltered-acidified and unfiltered-unacidified samples are compared to filtered samples (items 3 and 4, Table I). In the unfiltered samples Ca, Ba and K are more concentrated, which illustrates that these alkalies and alkaline earths are indeed being leached from the suspended material.

Comparing filtered-acidified with filtered-unacidified samples shows little or no difference for most of the elements, with the exception of B, Hg, Mn, Pb and Fe, which have an increased concentration in the filtered-acidified samples. Some elements listed as "no significant difference" (Table I), including U, show small differences between the two sample types; these differences are small increases in concentration in the acidified samples. These increased element concentrations in the acidified samples are attributed to the acidification minimizing loss of solutes by oxidation and/or precipitation and by adsorption to the container surface (Brown et al., 1970, p. 16).

Although U shows only weak increases in concentration in the unfiltered samples over the filtered samples (detailed diagrams of U variation with sample treatment are shown in Wenrich-Verbeek, 1976), many other elements are strongly affected by the long-term presence of the suspended material as shown in Table I (items 3 and 4) and Figs. 1–4.

One of the most consistent trends reflected by these various sample treatments is the large deviation of values for turbid samples from the line of

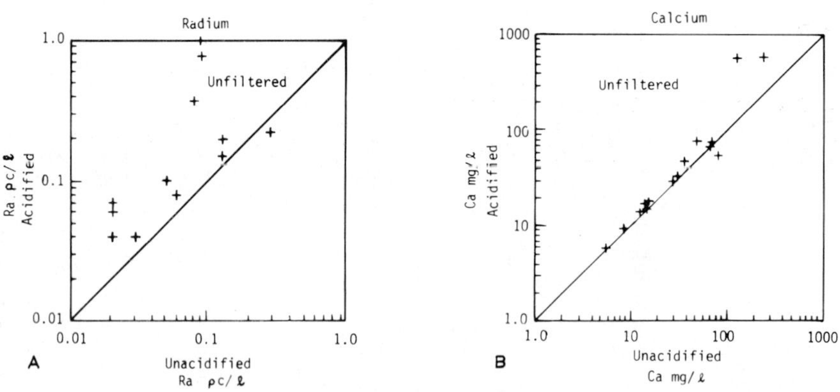

Fig. 1. A. Effect of acidification on Ra content of unfiltered samples. Similar results were obtained for Li, As, B, Al, Mn, Pb, Fe and U. The line represents points of equal concentration for the two sample types. B. Effect of acidification on Ca content of filtered samples. Similar trends showing no significant difference between treatments are exhibited by K, Na and Ba. The two highest samples with an increased concentration for the acidified samples are also the two most turbid samples. The line represents points of equal concentration.

equal concentration. Turbid samples consistently show increased concentrations of elements in the unfiltered samples. This is true even for those elements such as U, which are not intensely affected by sample treatment. Chi-square tests of U values for unfiltered-acidified, unfiltered-unacidified, and filtered-unacidified sample types show 60%, 70%, and 85% probabilities, respectively, of representing the true population of U values. The true population is considered to be that of the filtered-acidified type in which the U variations subsequent to sampling are at a minimum. Other elements have probabilities considerably lower; the element concentration from unfiltered (but later filtered in the laboratory) turbid samples frequently show proba-

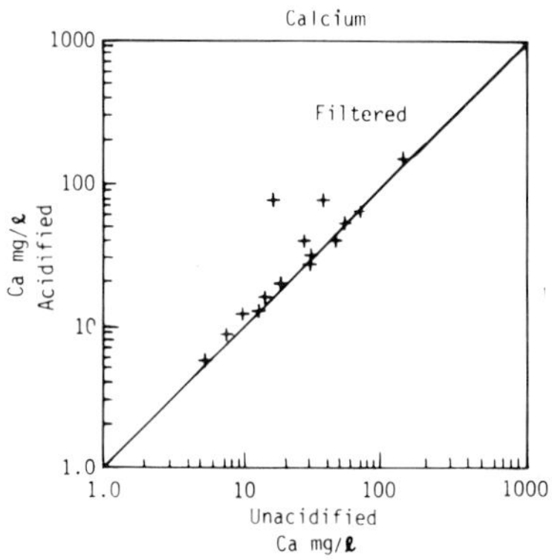

Fig. 2. Effect of acidification on filtered samples for Ca. Ca along with Li, Ba, As, U, Mg, Ra, K, Na and Al show essentially no or little variation with acidification, although there is a slight increase in concentration even by these elements in the acidified samples, as shown by this figure. The line represents points of equal concentration.

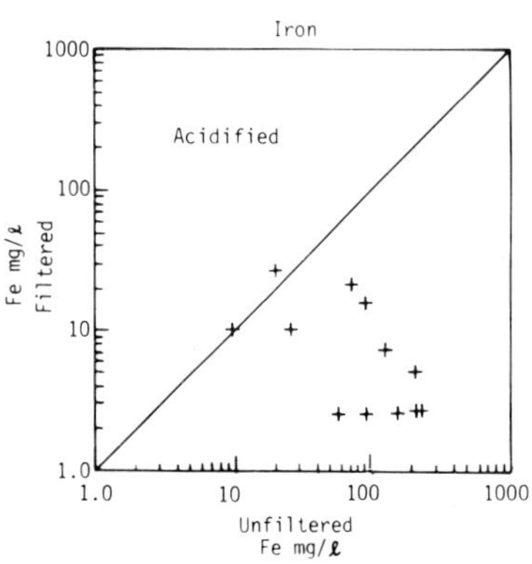

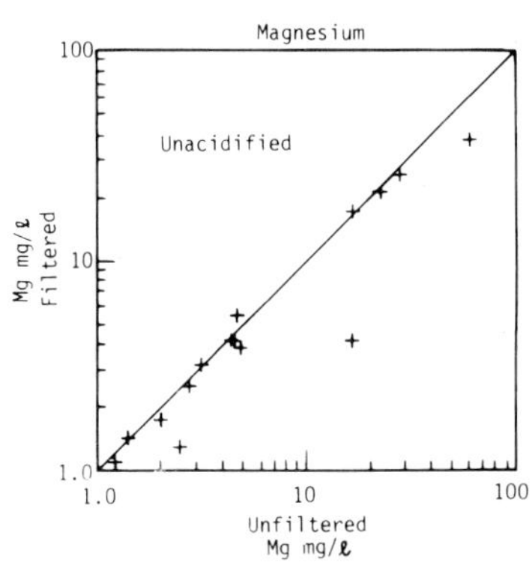

Fig. 3. Acidification results in more Fe concentrated in unfiltered than in filtered samples. Organic carbon, Li, Ba, As, B, Mg, Ra, Mn, Pb, Ca, Fe, K, Al and U behave similarly.

Fig. 4. When unacidified, samples show more Mg concentrated in unfiltered than in filtered samples. Similar trends are shown by Ba, Mn, Pb, Ca and Fe.

bilities less than 20% of representing the true population of element concentrations.

These probabilities underscore the problem of using unfiltered samples. If the sample contains a large suspended load, there is not only the resultant problem of element leaching from this material, if it is permitted to exist in contact with the water for a period of time, but also a sampling problem. It is much more difficult to collect a homogeneous sample of the suspended material than the already moderately homogeneous dissolved fraction of water. Suspended sediment content is positively related to the amount of surface runoff. A heavy rain upstream could thus result in inconsistent changes in the concentrations of U and other elements for unfiltered samples. Thus, if unfiltered samples are used, one of the advantages of water sampling over stream sediment sampling is lost, namely, that of a minimized problem in collecting a representative sample.

If the sample is field-filtered, the variation between acidified and unacidified samples is minimal. Unfiltered samples should never be acidified as it increases the leaching from the suspended material. Acidification of field-filtered samples is recommended to minimize ion loss by precipitation or onto the container, although the necessity of acidifying is minor compared to that of filtration.

Element behaviour with different types of filters

In addition to 0.45-μm millipore filters, nuclepore filters both with (0.45 μm) and without (0.40 μm) surfactant were used on the samples. Each nuclepore-filtered sample was done in replicate; one was acidified and the other remained unacidified.

Scatter diagrams show there to be no significant difference in element concentrations between nuclepore and millipore filters for Ra, Ca, Na, Li, Mg, K, and U. Thus, U (Fig. 5A) as with the previous filtration and acidification trends, behaves similarly to the alkalies and alkaline earths, especially Ca (Fig. 5B). However, B, Mn, Ba (Fig. 5C), and Al show higher concentrations in nuclepore-filtered samples, which suggests that their ions may interact more strongly with millipore filters, possibly due to adsorption by the millipore filters. In contrast Hg (Fig. 5D) and As show a lower concentration in nuclepore-filtered samples. Hg, in particular, shows a large decrease in concentration with passage of the water sample through nuclepore filters. Although more data are needed, there is no evidence that a surfactant coating has any effect on element concentration.

Correlations between elements and other geochemical parameters

Fewer elements correlate with U in water than in stream sediments; this is due partly to the lower concentration, frequently below the detection limit, of most elements in water and partly to the dependence of their solubility on other conditions, such as the presence of anions needed to form soluble complexes.

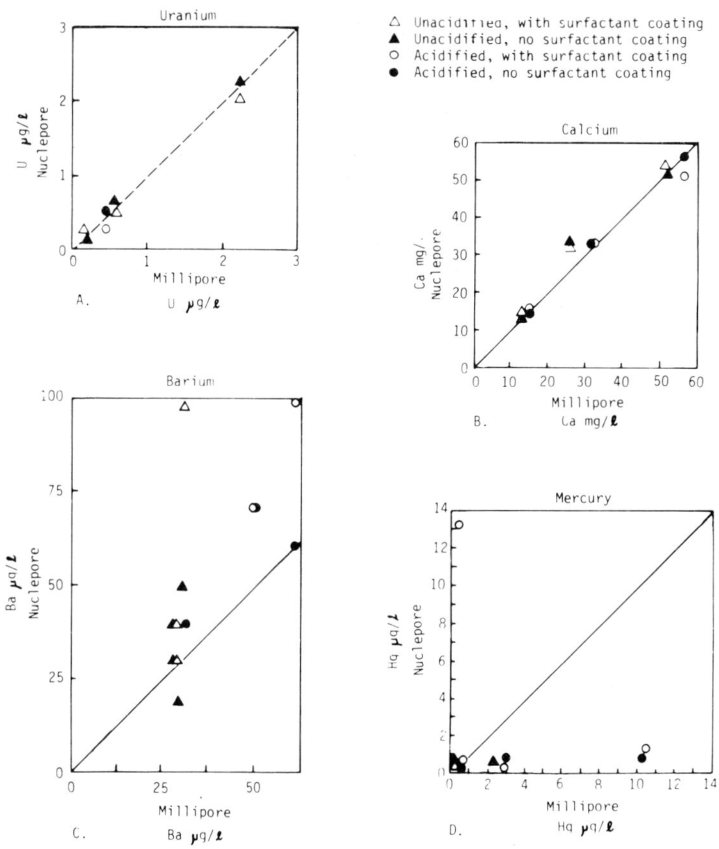

Fig. 5. Variation between nuclepore, with surfactant coating and without, and millipore filters. The line represents points of equal concentration.

The correlation coefficients between U and other elements vary somewhat with conditions of filtration and acidification (Table II). Unfiltered-unacidified samples have the largest number of elements correlating with U. These samples are influenced considerably by the suspended material as can be seen in Table I; some of the elements that correlate with U in these unfiltered-unacidified samples, but that do not correlate in the filtered-unacidified samples, show correlations with U in the bottom sediments — hence the correlation is one with the suspended sediment, not the water. The problems of acidification of unfiltered samples, as discussed earlier, are again reflected in the data. Although sixteen elements are significantly correlated in unacidified samples only five still retain these correlations in unfiltered-acidified samples. This is due to the inconsistent leaching of the elements from the suspended material by the acid, which introduces additional geochemical variables into the correlations; this is shown very nicely by the factor analysis grouping in Table III, where six factors were required for unfiltered-acidified samples instead of four or three to explain a similar percentage of the element variation. Correlations in Table II for filtered-unacidified samples are similar but slightly better than the filtered-acidified ones, this is probably a function of fewer data available for these filtered-acidified samples. Elements which show a strong positive correlation with U in water, for both the unfiltered- and filtered-unacidified samples are Ca, Mg, Na, K, Ba, B and Li. In addition, P, organic carbon, Cr, Cu, Ni, V, Th and Co correlate with the unfiltered-unacidified samples, but not with the filtered-unacidified samples (Ni, V, Th and Co) or data was insufficient (P, organic carbon, Cr, Cu). Arsenic correlates only in the filtered-unacidified samples. Additional data are needed to determine the strength of the correlations for these elements. In

TABLE II

Elements showing significant correlations with U in water

Element	Unfiltered-unacidified	Unfiltered-acidified	Filtered-unacidified	Filtered-acidified
P	×*	0	—	—
Organic carbon	×*	0	—	—
Ca	×*	×	×*	×
Mg	×*	×	×*	×*
Na	×*	×*	×*	×*
K	×*	×	×	×
As	0	0	×	0
Ba	×	0	×	0
Be	×*	0	—	—
B	×	0	×*	×
Cr	×*	0	—	—
Cu	×	0	—	0
Fe	0	—	0	0
Pb	0	0	0	—
Mn	0	0	0	0
Mo	0	0	0	—
Ni	×*	0	0	0
V	×	0	0	—
Zn	0	0	0	0
Al	0	0	0	0
Li	×	×*	×*	×
Ti	0	0	0	0
Zr	0	0	—	—
Hg	0	0	0	0
Th	×*	—	—	—
Co	×*	0	0	0

All correlations that were significant at the 95% confidence level are marked with an ×. Those marked by ×* are significant at the 99% confidence limit. 0 = no significant correlation. — = not determined, insufficient data.

addition, conductivity was measured directly in the water (unfiltered-unacidified) and shows a very strong positive correlation with U. This is not unexpected as conductivity is controlled by the concentration of the major ions, Ca, Mg, Na and K, which correlate with U in water.

Selected factor groups extracted from *R*-mode factor analysis of the data matrix are shown in Table III. The least number of factors (factor groups) are required and the best factor loadings are shown for the filtered-unacidified samples and are the worst for the unfiltered-acidified samples. The large number of factor groupings, 6, with poor factor loadings, and the large quantity of secondary relationships are a reflection of the acidification and consequent leaching of the suspended material increasing the number of geochemical processes influencing the element behaviour.

It might be noted that there are no secondary relationships in the factor analysis for filtered-acidified samples (Table IIIc); all elements fall into three discrete groups. All of the elements in each group are positively related with the exception of Zn and Mn. These two negatively related elements vary inversely with the other group 1 elements, U, Mg, Na, K, Li, Ca, Ba, As, B and Co.

Factor 1. It is apparent that U in water is most closely related to the alkalies and alkaline earths. This is expected, particularly in the case of Ca, which is related to the CO_3^{2-} anion that has a large influence on the solubility of U. Because of the close relation between the concentrations of Ca, Na, K, and Mg and U in water, these associated elements can be used to normalize the U concentration within an area. Evaporation of surface water may greatly increase the U concentration, or a sudden rain may dilute it (Wenrich-Verbeek et al., 1976). This becomes a significant problem when a large geographic area is studied requiring more than a few days of sampling. In this situation evaporation and increased discharge effects may enter into the resultant U concentrations. The major ions are also directly affected by these two factors, and hence their concentrations can be used to minimize evaporation and dilution effects. A cheaper and faster method of minimizing those effects is with the use of conductivity (measured in the field), which is controlled by the concentration of the major ions. Hence, conductivity, used to normalize the U concentration, is the most important additional parameter needed and can greatly increase the effectiveness of the use of surface waters as an exploration tool by minimizing the external influences of evaporation and dilution.

Factor 2. This group contains the higher valence state and intermediate ionic-size metals — Ti, Mo, Fe, Al, Ni and Pb. With the exception of Mo, they are frequently associated and considered to be moderately to very immobile elements (Hawkes and Webb, 1962, p. 17). Even Mo becomes immobile in an iron-rich environment. Although Mo is often thought of as a "pathfinder" for U in many ore deposits, it does not show a significant correlation with U in water. This is probably primarily a function of the different soluble

TABLE III

Selected factor groups resulting from *R*-mode factor analysis of the water data matrix (parentheses denote secondary relationships). Values are factor loadings which indicate the strength of the element to the grouping. A minus sign in front of the element indicates a negative relation to the group

Group 1		2		3		4		5		6	
(a) *Unfiltered-unacidified samples*											
As	1.00	Mn	1.00	Ti	1.00	Ba	1.00				
Li	0.96	P	0.94	Hg	0.92	Co	0.73				
B	0.93	Be	0.94	Zr	0.92	Fe	0.72				
K	0.80	Cr	0.90	(Pb)	0.60	U	0.68				
Na	0.70	V	0.89	(Fe)	0.49	Mg	0.59				
(Co)	0.46	Ni	0.87			Ca	0.51				
(Ca)	0.42	Al	0.85			-Zn	-0.90				
		Th	0.84			(K)	0.65				
		Mo	0.83			(Na)	0.57				
		Pb	0.73								
		Cu	0.67								
		(U)	0.55								
		(Mg)	0.44								
		(Ca)	0.46								
(b) *Unfiltered-acidified samples*											
Mo	1.00	U	1.00	Hg	1.00	B	1.00	Ba	1.00	Zn	1.00
Pb	0.94	Li	0.78	Al	0.76	K	0.65	(Al)	0.71	As	0.82
Ti	0.76	Na	0.69	(V)	0.67	(As)	0.55			Ni	0.65
Co	0.73	(K)	0.50	(Ni)	0.58					Cu	0.62
V	0.68	(Ni)	0.62	(Cu)	0.59					Be	0.62
Zr	0.63	(Cu)	0.59	(Be)	0.59					Cr	0.62
(P)	0.49	(Be)	0.59	(Cr)	0.59					P	0.62
(Cu)	0.49	(Cr)	0.59	(P)	0.59					Ca	0.61
(Be)	0.49	(P)	0.59	(Mn)	0.46					Mn	0.55
(Cr)	0.49	(Ca)	0.57							Mg	0.54
(Mn)	0.48	(Mg)	0.46								
(c) *Filtered-unacidified samples*											
U	1.00	Ti	1.00	Hg	1.00						
Mg	1.00	Mo	0.96	Co	0.89						
Na	1.00	Fe	0.95	V	0.66						
K	0.98	Al	0.94	Ra	0.62						
Li	0.96	Ni	0.77								
Ca	0.95	Pb	0.53								
Ba	0.87										
As	0.74										
B	0.73										
Co	0.49										
-Zn	-0.62										
-Mn	-0.68										
(d) *Filtered-acidified samples*											
Li	1.00	Ca	1.00	Fe	1.00	Zn	1.00				
As	0.95	Ni	0.88	Ba	0.63	Mn	0.91				
K	0.92	Mg	0.61	-Hg	-0.79	Co	0.86				
Na	0.83	Al	0.49	(Mn)	0.69	-Ti	-0.96				
B	0.66	-Cu	-0.76	(Cu)	0.66	(Ni)	0.68				
U	0.65	(Cd)	0.54			(Cd)	0.67				
-Cd	-0.92	(U)	0.57								
(Mg)	0.49	(B)	0.49								

complexes formed by these two elements in water, but may also be a result of the mixing of uranium-bearing and non-uranium-bearing terranes.

Factor 3. This group contains Hg, Ra and V. Hg and Ra are similar in charge and considerably larger than the factor group 2 metals and would be expected to associate with each other. V would more likely have been expected to occur in the second group. Unfortunately, element size and charge have minimal importance in determining correlations in water as compared to solubility, which is difficult to evaluate without better knowledge of the ionic species with which the element is complexed.

Stream sediments

Variations with grain size

Stream sediment samples, taken in conjunction with the water samples, were separated into the seven size fractions described above and analyzed for

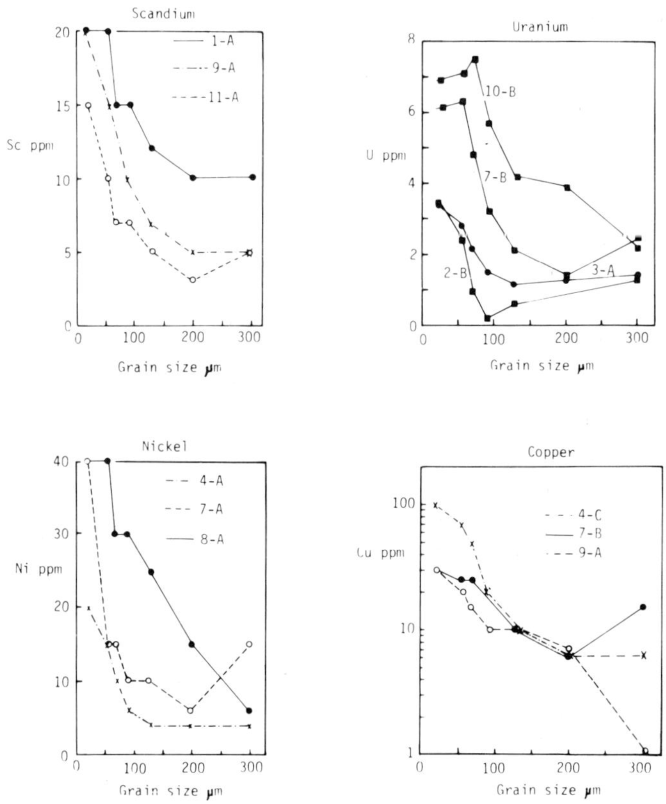

Fig. 6. Element versus grain size for U, Cu, Ni and Sc. The samples came from the following localities in the United States: *1-A* = San Juan River, Pagosa Springs, Colo.; *2-B*, *3-A* = Chinle Wash, Mexican Water, Ariz.; *4-A*, *4-C* = Paria River, Lees Ferry, Ariz.; *7-A*, *7-B* = Rio Puerco, Rio Puerco, N.M.; *8-A* = Jemez River, Jemez, N.M.; *9-A* = Cherry Creek, Franktown, Colo.; *10-B* = Huerfano River, Redwing, Colo.; *11-A* = Coyote Creek, Golondrinas, N.M.

various elements to determine if there is a significant correlation between these elements and grain size, which might influence the type of sediment sample needed for exploration. The average correlation coefficient between the log of the grain size and U concentration for all sites (195 samples) is −0.70 (Fig. 6). This highly significant negative correlation coefficient indicates that if anomalous concentrations of U are to be contrasted with background values, the sample must not be diluted with the coarse, low-U fraction; only grain size fractions less than 90 μm (170 mesh — U.S. sieve series) should be analyzed for U. An average of 73% of the total U concentration of these samples is contained within the less than 90-μm size fraction. Detailed discussion can be found in Wenrich-Verbeek (1976).

TABLE IV

Correlation coefficients for element concentration versus grain size

Element	Correlation coefficient	Number of samples	Level of significance
Th	−0.553	189	x*
U	−0.700	195	x*
SiO_2	0.486	42	x*
Mn	−0.35	210	x*
Ag	−0.63	10	x
B	−0.60	56	x*
Ba	−0.10	188	—
Be	−0.335	107	x*
Co	−0.514	132	x*
Cr	−0.676	205	x*
Cu	−0.605	210	x*
La	−0.586	61	x*
Mo	−0.536	62	x*
Nb	−0.247	134	x*
Ni	−0.473	205	x*
Pb	−0.525	180	x*
Sc	−0.650	127	x*
Sn	−0.552	17	x
Sr	−0.338	107	x*
V	−0.558	210	x*
Y	−0.600	135	x*
Zr	−0.598	211	x*
P	−0.332	35	—
Ce	−0.600	21	x*
Ga	−0.438	100	x*
Yb	−0.566	101	x*
Nd	−0.573	30	x*
As	−0.187	99	—

x* = significant at the 99% confidence level; x = significant at the 95% confidence level; — = not significant.

Table IV shows similar highly significant negative correlations between most other elements and grain size. Silica is the only phase which has a significant positive correlation with grain size. Some elements, including U (Fig. 6), occasionally show a slight increase in concentration for the coarsest fraction (greater than 250 μm — 60 mesh). In the case of U this is consistent with the presence of some detrital U minerals, notably brannerite and carnotite (E.H. Hornbrook, personal communication, 1976), sometimes occurring among the coarser grain sizes.

Correlations between the various elements

Correlations and significance levels for U and coexisting elements are shown in Table V. Twenty-one elements show a significant correlation with U, although only nine — Th, Nb, Y, P, Ce, Yb, Li, Se and Mg — show strong positive correlations with U, whereas SiO_2 shows a strong negative correlation. Other elements reflecting weak positive, though significant, correlations are Ag, Be, Cr, Ni, Pb, Sc, Zr, Ga, Zn, Ca and Ti. The weaker correlations are a function of the correlation coefficient calculations combining sampling sites that show no correlation between U and these elements with sites showing very strong correlations. This, of course, is a result of the varied minerals incorporating U, and the fact that these less frequently correlating elements do not occur as commonly in the uranium-bearing phases as do the above strongly correlating elements. Many of these strongly associated elements may be used not only as indicators of the U host, but also as U "pathfinders" in situations where U may not have been leached or transported as detrital material in anomalous concentrations while its "pathfinders" were.

Chemical analysis of major and trace elements along with a petrographic analysis is a useful indicator of the possible sources of the U — i.e., the host material. For instance, a large Zr enrichment with the U is indicative of a concentration of zircon in the sediments but not particularly indicative of U deposits. On the other hand, an increase in U with an increase in clay content (Al content) is suggestive of U leached from a moderately concentrated U source (large enough to produce a stream sediment anomaly) and adsorbed onto the clays.

Multivariate *R*-mode factor analysis made on 30 of the elements (Table VI) shows four distinct factor groups; the factor loading indicates the strength of their association. U forms a very distinct group with Th, Zr, As, Ca, Cu, Nb, Yb and Y — elements (other than As and Cu) for which U may readily substitute in the crystal lattice because of the similar ionic size and/or charge. Anion complexes of As form bonds with U. Perhaps the association of U with Cu is due to similar adsorption properties. Group 4 contains elements which are not compatible with the other 27 elements either because of a very large ionic radius (K, Rb, Be) or because of a very small ionic radius (Be — low factor loading to the other three elements in the group). The elements in groups 1 and 3 are very similar geochemically, and it is unclear why they are separate. All of these elements are of medium ionic size and have low valences (generally 2+). Thus, the factor groups generally segregate elements with similar crystal chemical behaviour.

TABLE V

Correlation coefficients, number of sample pairs, and significance levels for U and coexisting elements

Element	Correlation coefficient	Number of sample pairs	Level of significance
Ag	0.540	16	×
Al	0.156	102	—
As	0.174	94	—
B	0.038	80	—
Ba	-0.012	196	—
Be	0.201	119	×
Ca	0.300	102	×*
Ce	0.548	39	×*
Co	-0.105	146	—
Cr	0.304	197	×*
Cu	-0.004	203	—
Fe	0.156	103	—
Ga	0.375	114	×*
Hg	0.137	61	—
K	-0.041	102	—
La	-0.075	75	—
Li	0.554	96	×*
Mg	0.427	103	×*
Mn	0.037	203	—
Na	0.051	68	—
Nb	0.452	133	×*
Nd	0.012	50	—
Ni	0.277	194	×*
P	0.908	36	×*
Pb	0.270	189	×*
PO_4	0.179	27	—
Rb	0.107	63	—
Sc	0.366	131	×*
Se	0.443	70	×*
SiO_2	-0.686	57	×*
Sn	-0.041	32	—
Sr	-0.146	115	—
Ti	0.266	69	×
Th	0.994	199	×*
V	-0.080	202	—
Y	0.638	146	×*
Yb	0.602	104	×*
Zn	0.296	95	×*
Zr	0.192	204	×*

×* = significant at the 99% confidence level; × = significant at the 95% confidence level; — = not significant.

TABLE VI

Selected factor groups resulting from multivariate *R*-mode factor analysis of the data matrix. Values are factor loadings, indicating the strength of the element to the grouping of the elements in stream sediments

Group	1	2		3		4	
Al	1.00	U	1.00	V	1.00	K	1.00
Pb	0.95	Th	0.97	Mo	0.86	Rb	0.85
Ni	0.93	Zr	0.89	Mn	0.81	Ba	0.60
Na	0.86	As	0.87	Mg	0.65	Be	0.001
Fe	0.86	Ca	0.86				
Sn	0.85	Cu	0.75				
Co	0.82	Y	0.65				
Ti	0.80	Nb	0.65				
Se	0.80	Yb	0.61				
Li	0.78						
Hg	0.75						
Cr	0.64						
Zn	0.63						

Comparison of water and stream sediment sampling

The correlation of U and coexisting elements between water and stream sediments is not as good as might be expected. The only elements that appear to show a correlation are the major elements Al, Fe, Ca and Si. This is because:

(1) their concentrations are sufficiently high so that substantial amounts from the sediment can be dissolved into the water. The amount of element concentrated in the water is, of course, dependent on the element's solubility which adds a large additional geochemical variable to any correlation between element content in water and that in the stream sediments.

(2) these elements are also present in some of the suspended material that passes through the 0.45-μm filters into the "dissolved fraction" of water (Jones et al., 1974). This material, notably kaolinite, silica and amorphous $Fe(OH)_3$, rapidly equilibrates with the water (Kennedy, 1971; Jones et al., 1974). Thus, these minerals in the bottom sediment are also in the suspended load, accounting for the correlation of Al, Fe and Si between the two media.

In general, the U concentration, as well as that of most other elements, in stream sediments varies less between sample sites than the U concentration in water samples. In some cases, a variation of two orders of magnitude for water accompanies no difference in the stream sediments, while in other instances anomalous stream sediment U concentrations occur with only a correspondingly moderate U concentration in water. These inconsistencies in part reflect (1) the inhomogeneities in stream sediments, and the consequent sampling problem, as demonstrated by some of the replicate samples in this study that were removed from specific, but different, locations within the stream channel, and (2) in part due to the dependence of U on its solubility in water. Consequently, it is recommended that both water and the associated sediment be sampled in a geochemical exploration program.

CONCLUSIONS

Some optimum sampling methods for U exploration, utilizing surface waters and stream sediments have been established. These can also be applied to other elements shown by this study to behave similarly in both media. These methods are:

(1) For water sampling, field-filtration and acidification are recommended. Filtration overcomes the problem of changing concentrations in solution due to leaching of elements from the suspended fractions during storage, while acidification minimizes ion loss onto the container. Although these effects are minimal for clear water, they are not for turbid water where filtration becomes essential for reproducible U determinations.

(2) Because of the highly significant correlations between U and the alkalies and alkaline earths in water, conductivity measurements can be effectively and easily used to normalize U values in water samples, so that effects of evaporation and dilution, caused by increased stream discharge, can be minimized.

(3) The highly significant negative correlation coefficient between U and grain size indicates that if anomalous concentrations of U are not to be overlooked, the sample must not be diluted with the coarse, low-U fraction; only grain size fractions less than 90 μm (170 mesh) should be analyzed for U.

(4) The correlation of U and coexisting elements between water and stream sediments is not as good as might be expected. This study has shown that anomalies found in one of these two media are frequently not reflected in the other, and hence sampling of both surface water and its coexisting sediment is strongly recommended.

ACKNOWLEDGEMENTS

The author is grateful to D.B. Collins, K.V. Noyce and S.S. Burnside for patient assistance in the field, and to D.B. Collins, S.S. Burnside and J.F. Mascarenas for data compilation and assistance in the preparation of this manuscript. Appreciation is also expressed to Drs. M.B. Goldhaber, J.S. Leventhal, and G.M. Reimer for stimulating discussions. Thanks also goes to personnel of the U.S. ERDA Oak Ridge Gaseous Diffusion Plant, Oak Ridge, Tenn. (ORL), particularly to Gordon Cagle for assistance in obtaining fluorimetric determinations of uranium in water, and other elements in both water and stream sediments.

REFERENCES

Barker, F.B., Johnson, J.O., Edwards, K.W. and Robinson, B.P., 1965. Determination of uranium in natural waters. U.S. Geol. Surv. Water-Supply Paper 1696-C: 25 pp.

Brown, E., Skougstad, M.W. and Fishman, M.J., 1970. Methods for collection and analysis of water samples for dissolved minerals and gases. In: Techniques of Water-Resources Investigations of the United States Geological Survey. U.S. Geological Survey, U.S. Government Printing Office, Washington, D.C., Book 5, Chapter A1.

Hawkes, H.E. and Webb, J.S., 1962. Geochemistry in Mineral Exploration. Harper and Row, New York, N.Y., 415 pp.

Jones, B.F., Kennedy, V.C. and Zellweger, G.W., 1974. Comparison of observed and calculated concentrations of dissolved Al and Fe in stream water. Water Resour. Res., 10: 791—793.

Kennedy, V.C., 1971. Silica variation in stream water with time and discharge. In: Non-equilibrium Systems in Natural Water Chemistry. Advances in Chemistry Series, No. 106. American Chemical Society, Washington, D.C., pp. 94—130.

Wenrich-Verbeek, K.J., 1976. Water and stream-sediment sampling techniques for use in uranium exploration. U.S. Geol. Surv. Open-File Rep. 76—77: 30 pp.

Wenrich-Verbeek, K.J., Cadigan, R.A., Felmlee, J.K., Reimer, G.M. and Spirakis, C.S., 1976. Recent developments and evaluation of selected geochemical techniques applied to uranium exploration. Proc. 1976 IAEA Symp. on Exploration of Uranium Ore Deposits (in press).

Journal of Geochemical Exploration, 8(1977)357—364 357

A STREAM SEDIMENT ORIENTATION PROGRAMME FOR URANIUM IN THE ALLIGATOR RIVER PROVINCE, NORTHERN TERRITORY, AUSTRALIA

M.F. FOY and J.E. GINGRICH

Consulting Geologist, 11 Dudley Street, Randwick, N.S.W. (Australia)
Terradex Corporation, Walnut Creek, Calif. 94596 (U.S.A.)

(Revised version accepted February 23, 1977)

ABSTRACT

Foy, M.F. and Gingrich, J.E., 1977. A stream sediment orientation programme for uranium in the Alligator River province, Northern Territory, Australia. J. Geochem. Explor., 8: 357—364.

Sediment samples were collected from streams draining the Koongarra uranium deposit and the small uranium mines in the South Alligator Valley. Determinations for U, Cu and Pb on various size fractions, taken from each of these samples, indicated that the best results were obtained for U from the minus 200-mesh fraction, but the train from the Koongarra ore deposit was very short. Cu and Pb were not found to be very useful as indicator elements for U. Alpha-track films were used to determine the Rn content of each sample and the ratio of alpha-track film reading to U content was found to define anomalous drainage areas around the mineralization in the Koongarra area. The areas so defined were of sufficient magnitude to be defined in a reconnaissance stream sediment programme.

INTRODUCTION

Following the discovery of the Koongarra U orebody in 1970 (Foy and Pedersen, 1975) and the recognition of the Alligator River Uranium Province, various attempts were made to apply stream sediment techniques to U exploration in the Koongarra area. None of these orientation surveys yielded results which could be interpreted as defining the Koongarra orebody nor the U mineralization at anomalies A and X (see Fig. 3). This paper describes a later survey, which in addition to conventional determinations of U, Cu and Pb on various size fractions, also attempted to indicate the relative quantity of Rn present in the sample. In the Koongarra area, samples were collected from closer intervals than in previous surveys; comparative samples were collected from the South Alligator Valley, where U was known in the drainage system. The areas are 80 km apart and located approximately 220 km east and southeast of Darwin (see Figs. 2 and 3). The climate is monsoonal.

REGIONAL SETTING

The Koongarra U deposit is situated on gently sloping ground,comprising steep-dipping Lower Proterozoic schists below a rugged, faulted escarpment of unconformable, gentle-dipping Middle Proterozoic sandstone rising 200—300 m above the plain. U mineralization is confined to the Lower Proterozoic schists. The South Alligator area is a valley, 3 km in width, of steep-dipping Lower Proterozoic siltstone between escarpments of unconformable gentle-dipping Middle Proterozoic sandstone and volcanics. Numerous small uranium mines occur up to 150 m above the valley floor. U mineralization occurs both in Lower and Middle Proterozoic rocks. The geology of the Koongarra area is described more fully by Foy and Pedersen (1975) and that of the South Alligator Valley by Foy (1975).

The Koongarra orebody shows downslope dispersion of oxidized U which appears to be limited to approximately 100 m from the primary ore zone. Minerals identified from the oxidized zone include sklodowskite, kasolite, renardite, metatorbernite, saleeite and curite while the primary zone contains dominantly uraninite with traces of galena and chalcopyrite and minor sklodowskite and kasolite along fractures. Considerably lower spectrometer readings for the oxidized ore compared with the primary ore, where U is in equilibrium with its daughters, indicated leaching of U daughter products from this secondary ore zone. Speculation that U daughter elements could be concentrated in stream sediments had been raised by the occurrence of the numerous "black swamp" radiometric anomalies in the area. The difficulty of measuring minute concentrations of U daughter elements anticipated in active stream channels was solved by exposing alpha-track films to sediment samples thus obtaining a relative measure of Rn gas.

SAMPLE TREATMENT AND ANALYSIS

Large samples in excess of 1 kg were collected from the active stream channels. Nineteen samples were collected from the South Alligator Valley and twenty-five from the Koongarra area. After drying, mixing and splitting a 100-g sample was placed in an evaporating dish and a track etch cup placed over the evaporating dish, the size of which was selected to be a tight fit for the cups (Fig. 1). Alpha-sensitive films contained in the track etch cups detect Rn gas and the relative Rn concentrations are expressed in tracks per square millimetre (t/mm^2), as described by Gingrich and Fisher (1976). A sixty-day exposure period was used to obtain maximum effect on the alpha-track films and four blank samples were used to monitor contamination from atmospheric Rn. Such contamination was found to be very low, the four blanks returning a mean of 0.4 t/mm^2.

From the remainder of the sample an assay split was prepared, and after sizing, assay splits were also taken of the plus 10-mesh, the 10—40-mesh, the

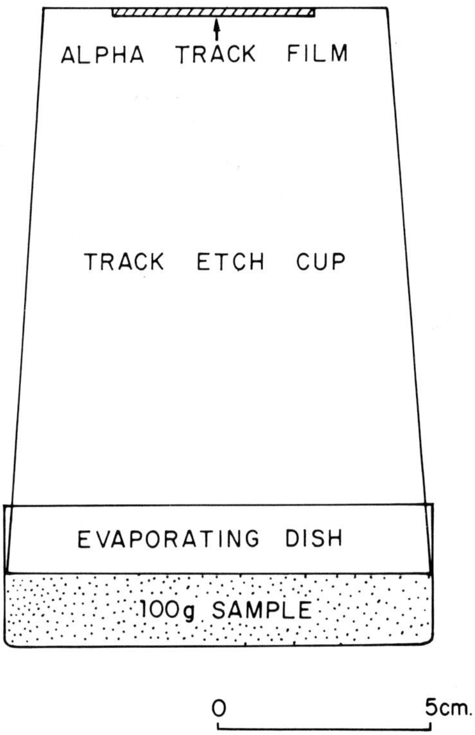

Fig. 1. Track etch cup in position over stream sediment sample.

40—80-mesh, the 80—200-mesh and the minus 200-mesh fractions. Determinations were carried out on the above assay splits for Cu and Pb by atomic absorption spectrometry and for uranium oxide by fluorometric analysis at Geochemical and Mineralogical Laboraties, Darwin.

RESULTS

Data treatment. From the basic data frequency-distribution graphs for each area, element and size fraction were plotted. From these graphs, any obviously anomalous samples were defined and the remaining samples used for statistical treatment. The background means and standard deviations were calculated and thresholds (mean + 2 standard deviation) were then determined. Weak anomalies were defined as being greater than the threshold value but less than the mean + 2.5 standard deviation. Strong anomalies were defined as being greater than the mean + 2.5 standard deviation.

Base metals. The results of the determinations for Cu are shown in Table I and Pb in Table II.

TABLE I

Background values and ranges of values for Cu (all results in ppm)

Mesh size fraction	Koongarra area				South Alligator Valley			
	N	mean	s.d.	range	*N*	mean	s.d.	range
plus 10*	18	9.3	7.2	2—28	19	25.2	10.2	8—44
10—40	25	5.0	3.6	2—16	19	15.7	9.7	4—36
40—80	25	3.7	2.5	2—10	19	16.8	9.2	4—36
80—200	25	7.0	4.2	4—18	19	25.3	11.9	10—46
minus 200	25	10.4	6.4	2—26	19	28.2	12.5	14—56
Total samples	25	5.4	3.8	2—18	19	19.6	8.9	8—34

N = number of samples; s.d. = standard deviation.

* In the Koongarra, samples of the plus 10-mesh fraction sufficient material for determination was only extracted from 18 samples.

TABLE II

Background values and ranges of values for Pb (all results in ppm)

Mesh size fraction	Koongarra area				South Alligator Valley			
	N	mean	s.d.	range	*N*	mean	s.d.	range
plus 10	18	11.4	9.1	4—36	19	20.1	5.6	12—30
10—40	25	3.4	2.6	2—12	19	12.3	6.3	4—30
40—80	25	3.7	2.6	2—10	19	13.7	7.3	4—36
80—200	25	6.1	5.1	2—24	19	17.7	7.8	6—44
minus 200	25	17.4	4.5	8—28	19	25.2	8.5	14—56
Total sample	25	6.8	3.5	4—16	19	17.9	4.2	10—26

N = number of samples; s.d. = standard deviation.

Only a few of the determinations could be regarded as anomalous and these are not regarded as being sufficiently definitive. Generally the backgrounds and absolute values are enhanced in the plus 10- and minus 200-mesh fractions. The mean values of Cu and Pb in the South Alligator Valley for all fractions are also much higher (by a factor of 3—4) than those of the Koongarra area.

Uranium. The results of determinations for U are shown in Table III and additional data shown in Table IV. While the calculated backgrounds for

TABLE III

Background values (extreme highs excluded) and ranges of values for uranium oxide, total population (all results in ppm, × indicates less than 0.5 ppm)

Mesh size fraction	Koongarra area				South Alligator Valley			
	N	mean	s.d.	range	*N*	mean	s.d.	range
plus 10	17	2.2	2.4	×—95	16	8.9	5.8	3—110
10—40	23	1.0	1.4	×—12	16	4.8	3.7	1— 90
40—80	23	0.9	1.0	×—17	16	5.9	4.5	1—100
80—200	23	2.0	1.8	×—36	16	8.5	4.2	3—130
minus 200	23	8.3	5.1	4—70	16	11.2	4.4	6—190
Total sample	23	1.7	1.3	1—21	16	6.6	3.4	2— 66

N = number of samples; s.d. = standard deviation.

Koongarra are extremely low, these were not inconsistent either with background rock values established from diamond drilling confirmed by other analytical techniques or backgrounds from other stream sediment surveys in the area. The mean values of U in the South Alligator Valley are much higher than those of the Koongarra area.

Four anomalous U samples (⩾18.5 ppm minus 200-mesh) were found at Koongarra. In reconnaissance stream sediment sampling, it is doubtful if samples in such closely spaced positions would be taken (Fig. 3). As expected several anomalous samples (⩾20 ppm minus 200-mesh) were found in the South Alligator Valley (Fig. 2). Generally the contrasts between anomalies and background appeared best in either the plus 10-mesh or the minus 200-mesh fraction and were noticeably depressed in the minus 10- to plus 80-mesh fractions.

Radon. The Rn results are shown in Table IV. In the Koongarra area, the mean film reading was 11.2 t/mm^2 with a standard deviation of 13.4 and a range of 3—72 t/mm^2, while in the South Alligator Valley the mean was 12.5 t/mm^2 with a standard deviation of 12 and a range of 3—52 t/mm^2. These results are very similar but it should be remembered that the Rn concentration depends on both the quantity of U present and its state of equilibrium. Complete determination of the quantity and state of daughter products present in each sample was beyond the scope of this survey, but it was found that some approximation could be made by dividing the relative Rn value from the alpha-track film by the uranium oxide (in ppm) determined from the original sample and arriving at a dimensionless ratio.

When these derived ratios (Table IV) were examined Koongarra was found to have a ratio background mean of 4.7 with a standard deviation of 1.9 and a range from 0.7 to 18.0, while the South Alligator area had a ratio back-

TABLE IV

Basic data for uranium oxide and Rn (for sample locations see Figs. 2 and 3)

Sample No.	U_3O_8 (ppm) plus 10-mesh fraction	U_3O_8 (ppm) minus 200-mesh fraction	U_3O_8 (ppm) total sample	Rn (t/mm^2)	Rn/U_3O_8 (total sample)
1	4	8	4	8.6	2.1
2	5	9	5	22.7	4.5
3	22	190	41	18.8	0.5
4	5	10	2	4.3	2.1
5	7	10	6	8.6	1.4
6	8	13	4	6.8	1.7
7	14	12	9	3.8	0.4
8	11	7	6	3.4	0.6
9	19	20	14	6.8	0.5
10	6	13	8	23.1	2.9
11	16	12	12	8.1	0.7
12	23	22	12	9.0	0.7
13	5	8	6	6.4	1.1
14	5	14	3	2.6	0.9
15	38	52	49	52.2	1.1
16	7	8	7	17.1	2.4
17	5	7	4	12.8	3.2
18	3	6	4	10.3	2.6
19	110	100	66	18.9	0.3
20	0.5	4	1	6.4	6.4
21	—	12	5	24.4	4.9
22	3	8	2	8.6	4.3
23	2	18	2	4.3	2.1
24	4	6	3	19.7	6.6
25	4	5	2	12.8	6.4
26	>0.5	5	1	7.7	7.7
27	0.5	7	1	3.0	3.0
28	3	20	2	7.7	3.8
29	—	4	0.5	7.3	14.5
30	0.5	4	0.5	5.6	11.1
31	2	8	2	4.7	2.4
32	6	9	3	9.0	3.0
33	—	6	1	12.4	12.4
34	—	5	0.5	4.3	8.6
35	—	6	0.5	6.0	12.0
36	9	23	2	4.7	2.4
37	95	70	21	14.5	0.7
38	2	4	0.5	9.0	18.0
39	—	16	4	71.8	18.0
40	—	7	4	15.4	3.8
41	>0.5	6	1	5.6	5.6
42	0.5	4	1	6.4	6.4
43	>0.5	6	0.5	3.8	7.9
44	>0.5	5	0.5	6.0	12.0

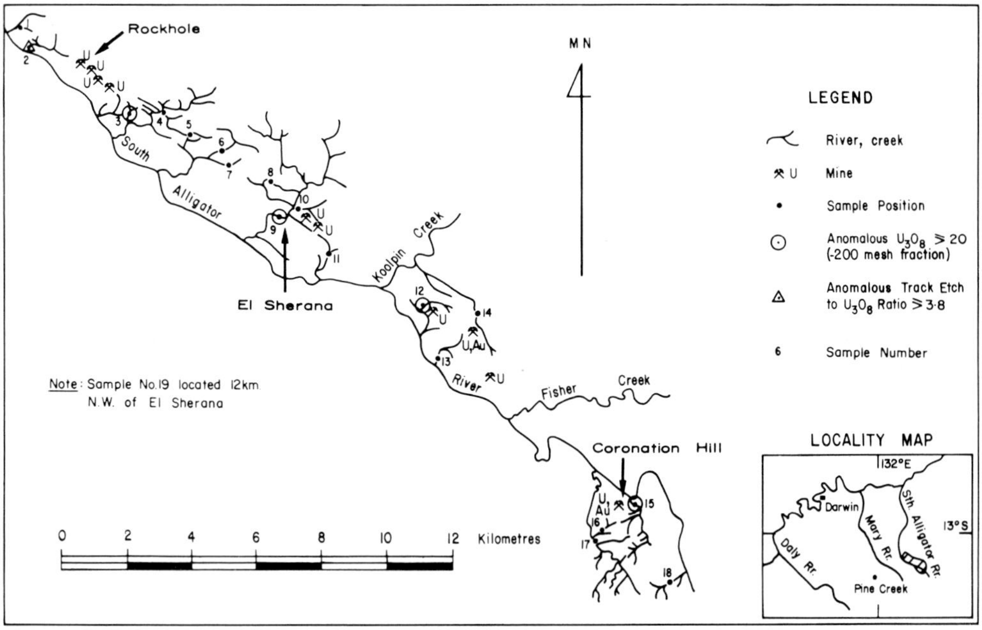

Fig. 2. Anomaly and sample locations, South Alligator Valley.

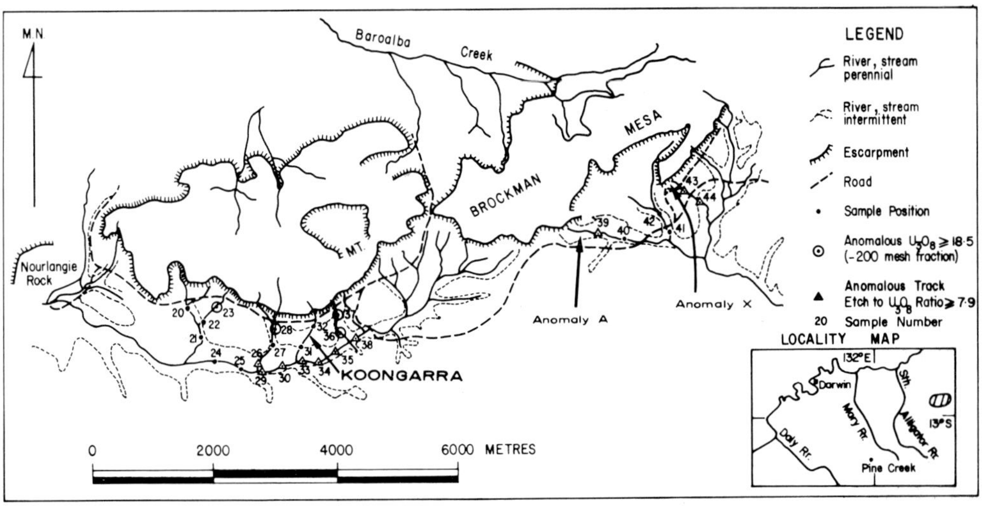

Fig. 3. Anomaly and sample locations, Koongarra area.

ground mean of 1.6 with a standard deviation of 1.1 and a range from 0.3 to 4.5. Only one of the South Alligator samples could be considered as weakly anomalous (Fig. 2) whereas five of the Koongarra samples could be considered strongly anomalous and three weakly anomalous (Fig. 3).

These anomalous samples in the Koongarra area defined a zone around and down drainage from the orebody, showing a train in the major drainage of sufficient size to be recognizable in a reconnaissance programme. The known U mineralization at anomalies A and X was also defined by these anomalous ratios.

CONCLUSIONS

Cu and Pb were found to be poor indicators for U in the Koongarra and South Alligator Valley. Uranium oxide in both the plus 10- and minus 200-mesh fractions was found to be a useful indicator but only showed a very short train. The derived ratio of Rn to uranium oxide was found to be the best indicator of an orebody similar to Koongarra. While this method of stream sediment geochemistry appears to show some promise, it requires testing over a much wider area.

ACKNOWLEDGEMENTS

The authors thank the management of Noranda Australia Limited for permission to publish this paper and Terradex Corporation for supplying the track etch cups.

REFERENCES

Foy, M.F., 1975. South Alligator Valley uranium deposits. In: C.L. Knight (Editor), Economic Geology of Australia and Papua New Guinea, 1. Metals. Australasian Institute of Mining and Metallurgy, Melbourne, Vic., pp. 301—303.

Foy, M.F. and Pedersen, C.P., 1975. Koongarra uranium deposits. In: C.L. Knight (Editor), Economic Geology of Australia and Papua New Guinea, 1. Metals. Australasian Institute of Mining and Metallurgy, Melbourne, Vic., pp. 317—321.

Gingrich, J.E. and Fisher, J.C., 1976. Uranium exploration using the track etch method. Proc. IAEA Int. Symp. on Exploration of Uranium Ore Deposits, Vienna, 29 March—2 April, 1976. International Atomic Energy Agency, Vienna.

Journal of Geochemical Exploration, 8 (1977) 365—379

URANIUM ISOTOPES IN GROUNDWATER: THEIR USE IN PROSPECTING FOR SANDSTONE-TYPE URANIUM DEPOSITS

J.B. COWART and J.K. OSMOND

Department of Geology, Florida State University, Tallahassee, Fla. (U.S.A.)

(Revised version accepted December 30, 1976)

ABSTRACT

Cowart, J.B. and Osmond, J.K., 1977. Uranium isotopes in groundwater: their use in prospecting for sandstone-type uranium deposits. J. Geochem. Explor., 8: 365—379.

The relative abundances of dissolved ^{238}U and its daughter ^{234}U appear to be greatly affected as the uranium is transported downdip in sandstone aquifers. In an actively forming uranium accumulation at a reducing barrier, an input of ^{234}U occurs in proximity to the isotopically non-selective precipitation of uranium from the water. The result is a downdip water much lower in uranium concentration but relatively enriched in ^{234}U. The measurement of isotopic, as well as concentration, changes may increase the effectiveness of hydrogeochemical exploration for uranium.

The investigation includes the uranium isotopic patterns in aquifers associated with known uranium orebodies in the Powder River and Shirley Basins, Wyoming, and Karnes County, Texas, U.S.A. In addition, the Carrizo sandstone aquifer of Texas was studied in detail, and the presence of an uranium accumulation inferred.

INTRODUCTION

The concentration of uranium in water has been used extensively for the exploration of uranium orebodies; however, reliable interpretation of these data results only from their integration with other measured chemical parameters. A promising variation of this approach is analysis of the relationship of the two naturally occurring isotopes, ^{238}U and ^{234}U, although it has been attempted only infrequently for uranium exploration and seldom if ever for hydrogeochemical exploration.

The first part of the 4n + 2 decay series is shown in Fig. 1. As can be observed, ^{234}U is a daughter of ^{238}U. The other naturally occurring uranium isotope, ^{235}U, is the parent nuclide of a separate (4n + 3) decay series. The ratio, $^{235}U/^{238}U$, is essentially constant throughout nature with the exception of the Oklo "fossil reactor" (Lancelot et al., 1975). In contrast, the relative abundances of ^{234}U and ^{238}U vary over a wide range within those environments at the surface of the earth which are subject to weathering processes. Systems which are not open to weathering are found to have the

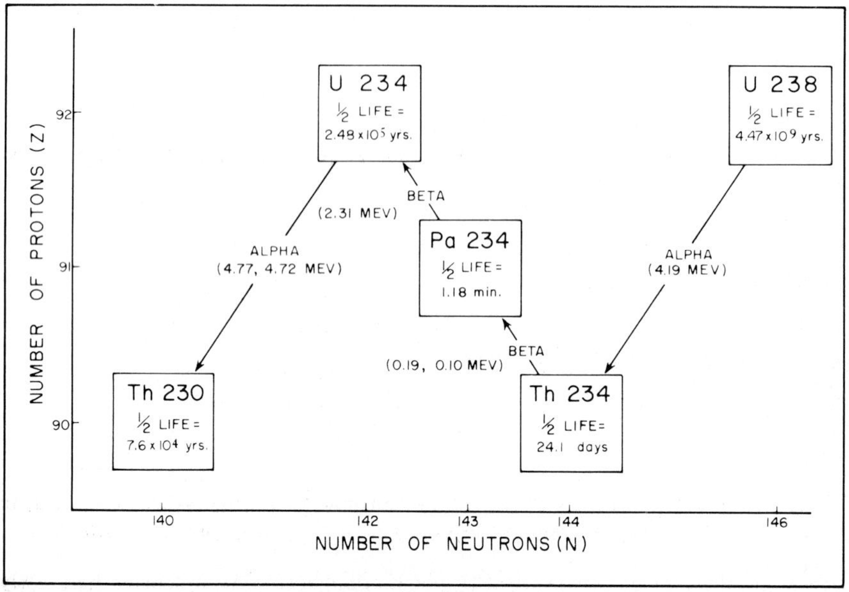

Fig. 1. The first part of the uranium (4n+2) decay series. The combined half-life of the nuclides intervening between ^{238}U and ^{234}U is less than 25 days.

equilibrium relationship between ^{234}U and ^{238}U. Because the equilibrium amount of ^{234}U is very small, 0.0055 wt. %, the relationship is usually discussed in terms of the alpha activity ratio $^{234}U/^{238}U$, or "A.R." At secular equilibrium the ratio is 1.00. It is apparent that the radiogenic origin of ^{234}U, abetted by the open nature of the weathering environment, must be the condition permitting the isotope fractionation. Published values of A.R.'s range from approximately 0.5 to 15, with both solids and liquids attaining these values (Cherdyntsev et al., 1967; Kaufman et al., 1969; Sakanoue and Komura, 1971; Osmond and Cowart, 1976). In comparison, stable isotope fractionation, which results from the difference in mass of two isotopes, is measured in parts per thousand. Other heavy radioactive isotopes which fractionate, such as ^{232}Th and ^{228}Th, have relatively long-lived intervening nuclides which are elementally different and thereby promote fractionation by chemical processes. The long-lived ^{234}U is separated from parent ^{238}U by ^{234}Th and ^{234}Pa which combined have a half life of only about 25 days and an average lifetime ($1/\lambda$) of about 36 days.

The separation of the uranium isotopes can be considered to be the result of (1) selective leaching of the ^{234}U atoms, (2) the direct alpha recoil of ^{234}Th across a boundary, or (3) a combination of the two processes.

For selective leaching to occur, ^{234}U atoms must be more vulnerable to extraction from a lattice than is ^{238}U. Events or conditions which would promote such selectivity include lattice damage induced by radioactive decay, location of daughter atoms in weakly bonded or interstitial sites and oxidation of the daughter to the +6 state as a result of the recoil within the lattice (Rosholt et al., 1963, 1964; Chalov and Merkulova, 1966; Cherdyntsev, 1971). Direct alpha recoil of ^{234}Th from a solid into a surrounding liquid has been observed in the laboratory (Kigoshi, 1971) and is considered by several investigators to be an important fractionation mechanism in the natural environment (Cowart and Osmond, 1974; Kronfeld, 1974; Kronfeld et al., 1975; J.N. Rosholt, personal communication, 1976).

Uranium in the +6 oxidation state may be very mobile whereas the +4 state is quite immobile. Various studies have indicated that in permeable media, uranium mobility is not only a function of redox conditions but also pH, carbonate and sulphate availability (Lisitsin, 1962; Hostetler and Garrels, 1962). Sandstone-type uranium accumulations occur at location of redox or sorption barriers, with infiltrating downdip-flowing water providing the transport medium. Given the proper chemical composition and long-term stable conditions, waters bearing only a modest amount of uranium could cause a significant amount of accumulation at a barrier.

It has been demonstrated that sandstone-type deposits are not usually in isotopic equilibrium. The unaltered interior of the orebodies is usually deficient in ^{234}U (A.R.<1) whereas peripheral altered ores often contain excess ^{234}U (Rosholt et al., 1963, 1964, 1965; Smith and Jackson, 1969).

K.E. Ivanov and R.G. Kudryashova in 1963 (reported in Cherdyntsev, 1971, pp. 124—125) noted moderately high uranium concentrations and A.R.'s in waters associated with several uranium deposits in the Soviet Union.

METHODS

Samples of groundwater from wells in the various areas were collected. Sample size ranged from 1 to 54 litres. Nitric acid was added to obtain pH = 1. A carrier in the form of ferric nitrate and a known activity of ^{232}U were added to each sample. The sample was heated to near boiling and the pH was adjusted to pH = 11 by adding ammonium hydroxide, in order to coprecipitate the radioelements with ferric hydroxide floc.

Purification procedures included decantation and centrifugation, ether extraction and anion exchange.

The resulting purified uranium was electroplated on a stainless steel planchet and was analyzed by means of alpha pulse height spectroscopy. Isotope dilution calculations provide yield and concentration values.

Detailed description of procedures can be found in Cowart (1974).

LOCATION AND GEOLOGIC SETTING

Carrizo aquifer area

The portion of the Carrizo which was investigated is located in central south Texas, U.S.A. The Carrizo Sandstone is a member of the Claiborne Group of Eocene age. It is thought that the sands were once part of an extensive dune field (Lonsdale, 1935). The Carrizo is a prolific source of irrigation water for this region. Almost the entire amount of water in the aquifer originates by infiltration of meteoric water in the outcrop area (Alexander and White, 1966); therefore, the uranium in the groundwater must have its source in the Carrizo itself. It should be noted that there is no known occurrence of uranium in the Carrizo other than that inferred from the data presented in this paper. The aquifer is well confined and there is little leakage in or out of the sand.

The Carrizo consists mostly of quartz sand with smaller amounts of bentonitic clay, shale and lignite. Downdip from the outcrop region, isolated unoxidized crystals of pyrite are present (Lonsdale, 1935). The thickness of the sand varies from about 180 m near the outcrop to approximately 400 m some 90 km downdip from the outcrop. The Carrizo dips to the south or southeast at 20—25 m per kilometre (Alexander and White, 1966).

Highland area

The Highland uranium deposits are located in the southern Powder River Basin in east central Wyoming, U.S.A. As this basin filled with fluvial sediments, the north-south-trending axis of the syncline shifted somewhat east of the axis of the underlying Precambrian basement. Cainozoic sedimentary rocks comprise most of the near surface and outcropping rocks, but around the edge of the basin uplifted Palaeozoic and Mesozoic age rocks crop out (Hagmaier, 1971; Dahl and Hagmaier, 1974).

During the early Tertiary, north-flowing streams in a region of subdued relief deposited clastic sediments such as are found in the Fort Union Formation of Paleocene time. Sandstones of the Fort Union Formation are the host rocks for the Highland uranium deposits which are in the southern part of the Powder River Basin. These are poorly consolidated arkosic sandstones which contain organic and tuffaceous material in abundance and have high permeability. The uranium deposits are found in discrete sandstone tongues which extend from the main sandstone unit (Dahl and Hagmaier, 1974).

According to Dahl and Hagmaier (1974) "The Highland uranium deposits exhibit many of the geologic characteristics of the sandstone roll-type that are considered epigenetic in origin. . . In this area, there is both a physical and chemical relationship between uranium occurrence and sandstone boundary. Physically, the finer-grained units near the edge of the sandstone bodies tended to direct mineralizing groundwater into the boundary area where the higher organic content was instrumental in producing a reducing environment necessary for mineral deposition."

"The ore mineral is primarily coffinite with lesser uraninite. Coffinite occurs in the form of thin sooty layers and irregular spherical or botryoidal masses. . . on individual detrital mineral grains in the sandstone. . . Pyrite is widely distributed within the ore zones and unaltered protore."

"A comparison of chemical analyses of cores and equivalent uranium calculated from natural gamma-ray logs of generalized zones indicated that the deposits are essentially in radioactive equilibrium. For grades less than 0.08 percent U_3O_8, the U/*e*U ratio is slightly less than 1, whereas grades greater than about 0.12 percent U_3O_8 have a U/*e*U slightly greater than 1."

Shirley Basin area

The Shirley Basin mining district is about 80 km south-southwest of the Highland area. The uranium deposits are found in sandstones of the Wind River Formation of early Eocene age. The Tertiary clastic sediments of the Basin are derived from the nearby mountains which were uplifted during the Laramide orogeny (Harshman, 1972). As in the Highland area, the ore-bearing sandstones of the Shirley Basin area are of arkosic composition and the unoxidized zones contain abundant pyrite; however, the only ore mineral is uraninite.

"The sequence of events that gave rise to the Shirley Basin ore deposits is believed to have been as follows: (1) downward percolation of vadose waters that leached uranium from weathered granite in nearby highlands and/or from the tuffaceous constituents of rocks overlying the ore-bearing intervals, (2) lateral movement of this water through the most permeable beds of the Wind River Formation, (3) reaction of the oxidizing ore-bearing solutions with the rocks through which they moved, causing a lowering of the Eh and pH of the solutions and deposition of the ore and associated minerals, and (4) passage of the spent ore-bearing solutions beyond the zone of deposition and out of the basin" (Harshman, 1972).

Karnes County area

The Karnes County uranium district is located about 75 km southeast of San Antonio, Texas, and about 55 km east-southeast of the Carrizo Sandstone study area.

The most important host rock for the uranium deposits in this area is the Deweesville Sandstone member of the Whitsett Formation (Upper Eocene). The deposits are located "within or just below the zone of oxidation" (Bunker and MacKallor, 1973).

RESULTS

In the Highland area and the Shirley Basin, which are mining areas containing ores situated in reducing environments, and in the Carrizo Sandstone, large variations of uranium concentration and $^{234}U/^{238}U$ activity ratio occur across known or inferred redox barriers. Such variations, if present at all, are much subdued for the Karnes County deposits which are, at least in part, located in a shallow oxidizing environment.

Table I lists the uranium concentration and A.R. values for all the sample locations discussed in this paper as well as available data on well depths and depth to aquifer.

Carrizo area. In the Carrizo area, the greatest uranium concentrations are invaribly from those wells located on the outcrop and wells located within a distance of 5 km of the outcrop (Fig. 2, Table I). In moving downdip from this "outcrop regime", the uranium concentration of the water decreases abruptly, by at least an order of magnitude, along a zone sub-parallel to the outcrop. The A.R. of the outcrop regime is usually between 0.50 and 1.00 although in one area the A.R.'s are as high as 1.16.

Isotopic, as well as concentration, changes occur as the waters move downdip from the outcrop regime. The western samples flow first through what we term a transition zone with low uranium concentration and rather low A.R.'s (0.65–1.08) and then flow into a "downdip regime" which has low concentrations and high A.R.'s. Profiles *A-A'*, *B-B'* and *C-C'* (Fig. 3) illus-

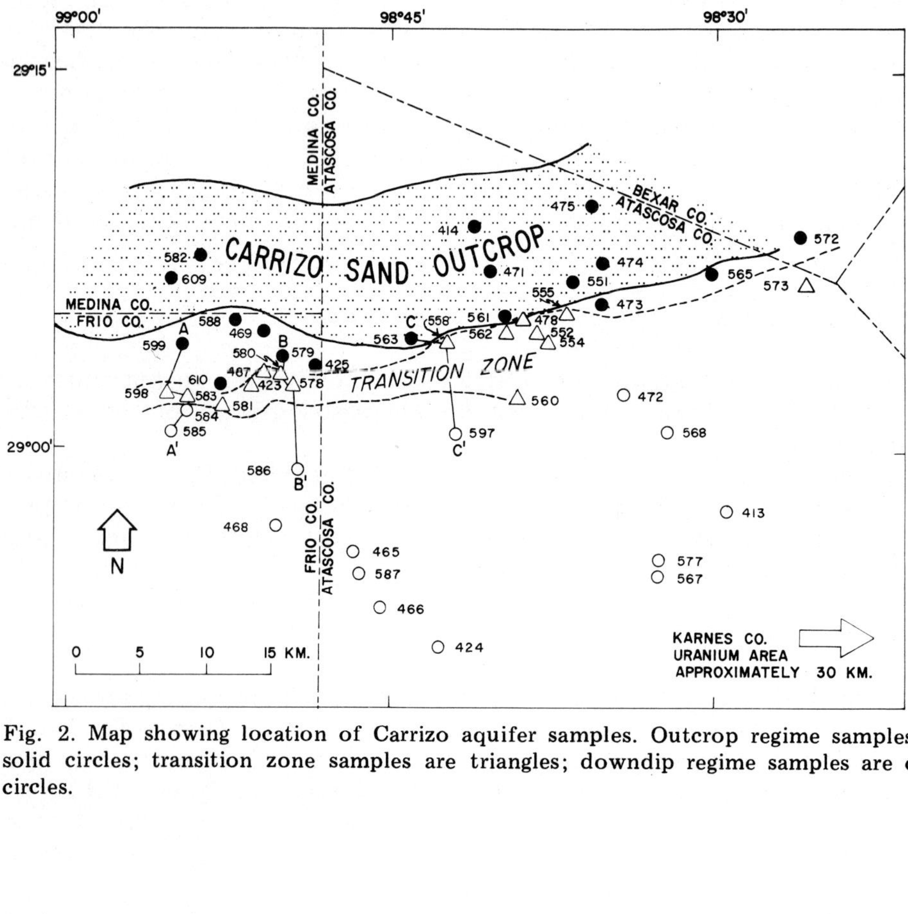

Fig. 2. Map showing location of Carrizo aquifer samples. Outcrop regime samples are solid circles; transition zone samples are triangles; downdip regime samples are open circles.

TABLE I

Uranium isotope and concentration data

Sample No.	Activity ratio $^{234}U/^{238}U$*	Concentration (μg/litre)*	Top of Carrizo Sandstone (m below land surface)**	Production zone of well (m below land surface)***
Carrizo Aquifer samples				
413	1.41 ± 0.27	0.03 ± 0.004	397	___ — 519
414	0.90 ± 0.04	0.20 ± 0.006	outcrop	___ — 51
423	0.69 ± 0.03	0.20 ± 0.007	130	141 — 181
424	6.68 ± 0.85	0.006 ± 0.001	473	519 — 570
425	0.52 ± 0.01	1.49 ± 0.03	76	76 — 139
465	9.02 ± 1.03	0.003 ± 0.001	328	297 — 377
466	6.36 ± 1.68	0.004 ± 0.001	374	___ — 488
467	1.09 ± 0.16	0.03 ± 0.003	122	146 — 207
468	5.48 ± 1.63	0.004 ± 0.001	290	___ — 354
469	1.03 ± 0.03	0.63 ± 0.02	61	66 — 73
471	0.77 ± 0.03	2.56 ± 0.06	outcrop	___ — 96
472	1.77 ± 0.30	0.008 ± 0.001	160	228 — 289
473	0.81 ± 0.04	1.70 ± 0.03	38	___ — 134
474	0.87 ± 0.05	0.21 ± 0.01	outcrop	___ — 37
475	0.58 ± 0.02	1.20 ± 0.03	outcrop	___ — 53
478	1.75 ± 0.23	0.02 ± 0.002	46	
551	0.84 ± 0.04	0.51 ± 0.04	outcrop	82 — 110
552	1.29 ± 0.21	0.009 ± 0.002	53	___ — 92
554	0.97 ± 0.05	0.04 ± 0.005	61	97 — 135
555	1.00 ± 0.08	0.04 ± 0.008	46	118 — 173
558	1.37 ± 0.22	0.006 ± 0.002	23	75 — 88
560	1.04 ± 0.11	0.02 ± 0.006	153	
561	0.82 ± 0.02	0.80 ± 0.03	15	61 — 106
562	0.82 ± 0.08	0.03 ± 0.003	53	___ — 113
563	0.84 ± 0.04	0.59 ± 0.05	outcrop	
565	0.67 ± 0.02	0.32 ± 0.03	53	73 — 92
567	5.18 ± 0.74	0.01 ± 0.003	412	489 — 499
568	1.71 ± 0.39	0.004 ± 0.001	252	259 — 342
572	0.79 ± 0.03	0.51 ± 0.04	outcrop	87 — 142
573	1.63 ± 0.25	0.003 ± 0.001	122	
577	5.14 ± 0.86	0.003 ± 0.001	404	___ — 610
578	1.00 ± 0.04	0.12 ± 0.01	130	___ — 187
579	0.71 ± 0.02	3.30 ± 0.15	76	80 — 122
580	0.56 ± 0.02	0.25 ± 0.03	131	131 — 194
581	1.08 ± 0.12	0.02 ± 0.007	137	191 — 221
582	1.16 ± 0.08	3.53 ± 0.27	outcrop	
583	1.08 ± 0.11	0.01 ± 0.008	114	___ — 183
584	4.63 ± 0.32	0.05 ± 0.01	130	192 — 200
585	4.44 ± 0.47	0.007 ± 0.002	160	___ — 261
586	8.70 ± 0.91	0.008 ± 0.001	252	___ — 275
587	5.18 ± 1.08	0.004 ± 0.001	366	___ — 458

TABLE I (continued)

Sample No.	Activity ratio $^{234}U/^{238}U$*	Concentration (μg/litre)*	Top of Carrizo Sandstone (m below land surface)**	Production zone of well (m below land surface)***
588	1.16 ± 0.09	0.30 ± 0.02	23	—122
597	2.86 ± 0.43	0.005 ± 0.001	175	166 —226
598	1.12 ± 0.15	0.006 ± 0.001	92	—
599	0.71 ± 0.01	1.64 ± 0.27	43	51 — 85
609	1.06 ± 0.09	1.14 ± 0.09	outcrop	
610	0.79 ± 0.02	0.44 ± 0.03	99	122 —183
Highland area samples				
797	4.70 ± 0.13	5.73 ± 0.23		
798	4.88 ± 0.32	6.58 ± 0.49		
799	3.34 ± 0.49	2.70 ± 0.22		
800	4.43 ± 0.36	12.48 ± 1.07		
801	4.38 ± 0.07	33.60 ± 0.77		
803	5.73 ± 0.95	0.12 ± 0.02		
1057	1.36 ± 0.05	36.85 ± 0.62		
Shirley Basin samples				
632	5.04 ± 0.85	0.11 ± 0.02		
636	1.37 ± 0.05	1.33 ± 0.06		
Karnes County samples				
600	3.31 ± 0.14	13.80 ± 1.61		
602	2.25 ± 0.12	19.60 ± 3.70		
605	1.88 ± 0.08	5.79 ± 1.03		
606	2.60 ± 0.11	7.83 ± 1.21		

*Uncertainty ranges based on standard counting statistics only.
**Values for depth of the Carrizo Sandstone below land surface is derived from Alexander and White (1966, fig. 29).
***Each well in this table is reported to produce water from the Carrizo Sandstone although the production zone within the aquifer is not known in all cases.

trate these changes. Where the well spacing is close enough, it appears that the boundary between the transition and the downdip zones is sharply delineated.

The eastern Carrizo samples do not display the high A.R.'s found to the west and so no distinct boundary between the transition and downdip zones is observed.

Fig. 3. Three profiles showing the progressive concentration and isotope changes of dissolved uranium in Carrizo aquifer water. Closed circles indicate activity ratio values, × 's indicate concentration values. Broken and dotted lines indicate inferred abrupt changes. Location of profiles is shown in Fig. 2.

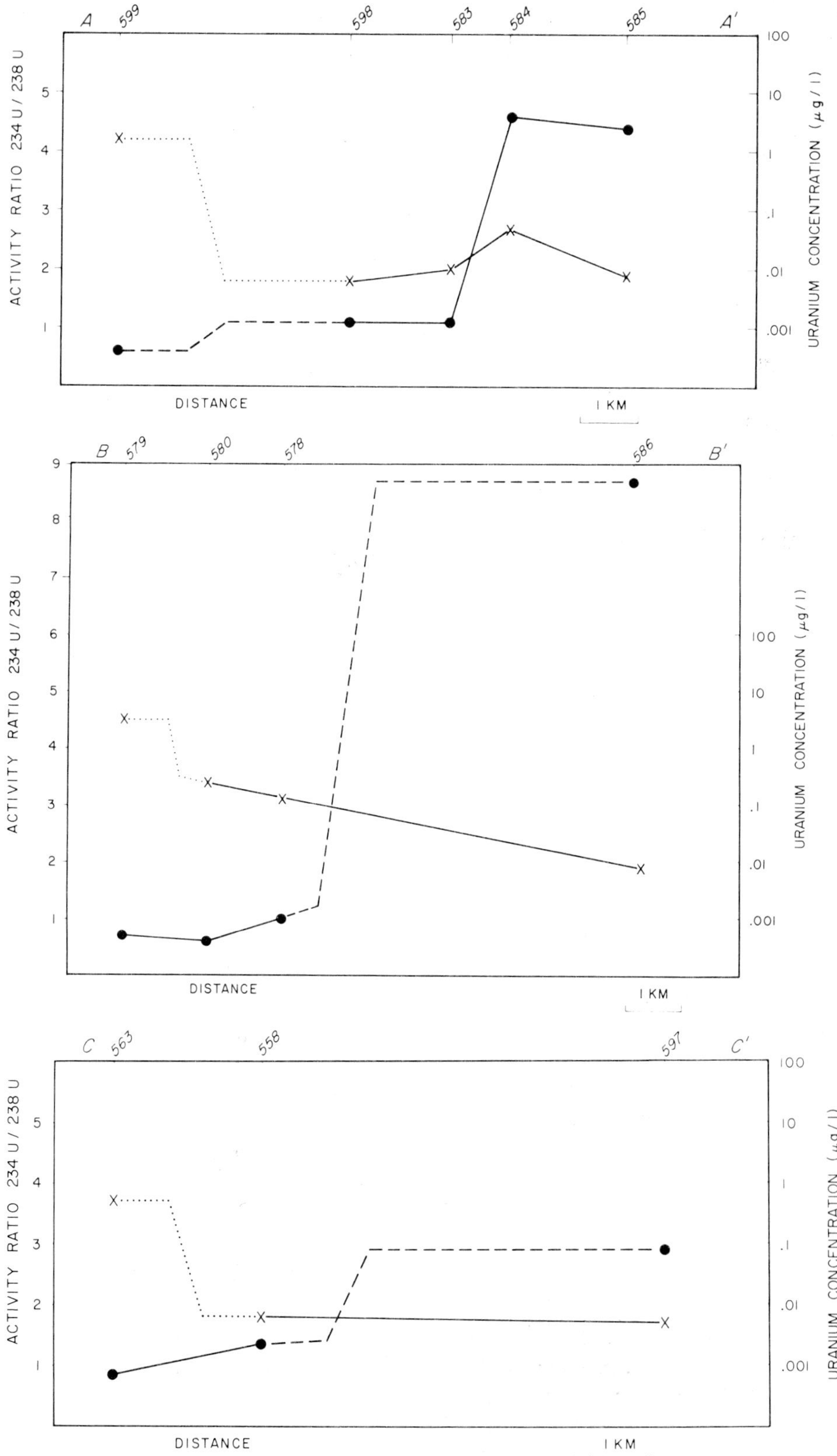
A
599
598
583
584
585
A'
ACTIVITY RATIO 234 U / 238 U
URANIUM CONCENTRATION (μg / l)
100
10
1
.1
.01
.001
DISTANCE
1 KM
B
579
580
578
586
B'
ACTIVITY RATIO 234 U / 238 U
URANIUM CONCENTRATION (μg / l)
DISTANCE
1 KM
C
563
558
597
C'
ACTIVITY RATIO 234 U / 238 U
URANIUM CONCENTRATION (μg / l)
DISTANCE
1 KM

The direction of groundwater flow is southward in the western half of the area shown in Fig. 2 and changes to southeastward in the eastern part. Thus, samples 567 and 577 are related to the western samples, not the eastern group.

Highland area. The results from the Highland Mine area are shown in profile *D-D'* (Fig. 4) and the location in Fig. 5. Dissolved uranium in five wells in the immediate vicinity of the orebody showed A.R.'s greater than that observed for the single well sampled updip from the mine (Table I); the uranium concentrations of water near the orebody are in general lower than the updip sample. A sample located approximately 5 km downdip from the orebody had by far the least concentration of any sample in the system but had the greatest A.R. Although this sample lies beyond the limits of the Highland sands, the well produces water which is though to have traversed the orebody zone (Dahl and Hagmaier, 1974).

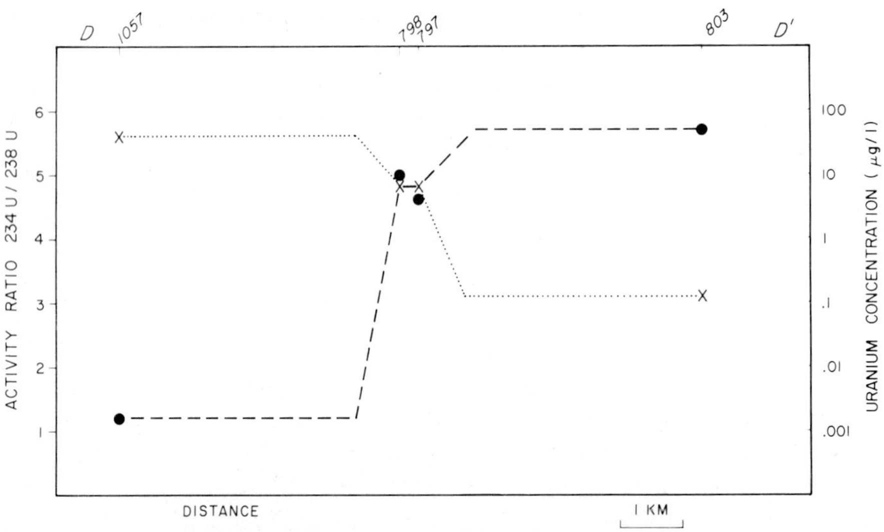

Fig. 4. Profile showing concentration and isotope changes of dissolved uranium in the vicinity of the Highland deposit. Closed circles indicate activity ratio values, X 's indicate concentration values. Broken and dotted lines indicate inferred abrupt changes. Location of profile is shown in Fig. 5.

Shirley Basin. A sample taken at the Shirley Basin townsite located about 3 km from the mine, had a greater concentration and a much lower A.R. than a sample taken somewhat below the ore zone at the mine.

Karnes County. All samples taken in the Karnes County area had concentrations of about the same magnitude as those from the ore deposit zone at the Highland mine and had A.R.'s that ranged from 1.9 to 3.3. No downdip, low concentration, high A.R. samples were found.

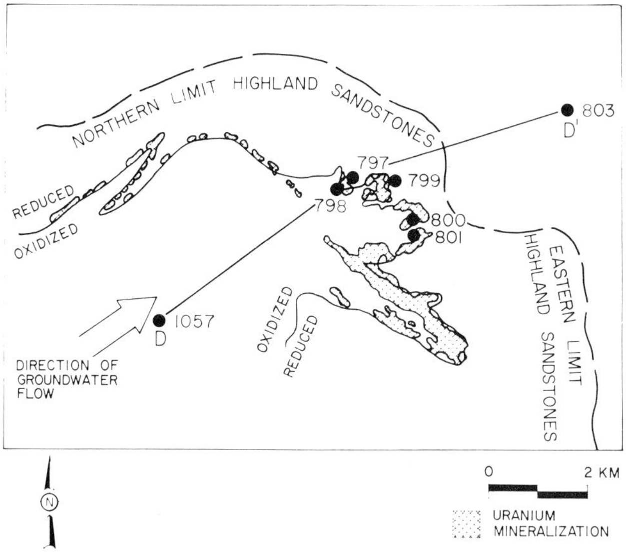

Fig. 5. Map showing location of samples in the Highland area. Although sample 803 is not located within the Highland sands, it is likely that the water from this well traversed the zone of uranium accumulation. (Modified from Dahl and Hagmaier, 1974.)

DISCUSSION

The three types of situations depicted in Fig. 6 are useful in explaining both concentration and isotope variations in the vicinity of redox barriers. If the present-day barrier exists up-flow from an accumulation of uranium, the alpha recoil input of ^{234}Th (hence ^{234}U) into water of low uranium concentration results in a high A.R. for the water (Type I). For Type II, the uranium accumulation and redox barrier are coincident. In this case, some of the alpha recoil atoms are introduced into relatively high concentration waters which are subject to subsequent loss of uranium at the barrier. Such precipitation of uranium is isotopically non-selective. The water emerging from such a system will have a low uranium concentration (as did Type I) but the A.R. will not be as great as found associated with Type I. Type III represents a system wherein a uranium accumulation is bathed by oxidizing waters which promote the destruction of the accumulation. The uranium concentration will increase as it travels through the orebody. Preferential leaching of the ^{234}U as well as alpha recoil ^{234}Th (= ^{234}U) may cause an elevation of the A.R. during the early stages of destruction. Assuming the destruction is completed within 100,000 years or so, the early loss of ^{234}U ensures that, ultimately, uranium deficient in ^{234}U will be mobilized from the remaining accumulation.

The three types are idealized systems which are unlikely to exist in such perfect form in nature. Elements of two and possibly three of the types may exist together in a given uranium accumulation although one type may be dominant for that time and place.

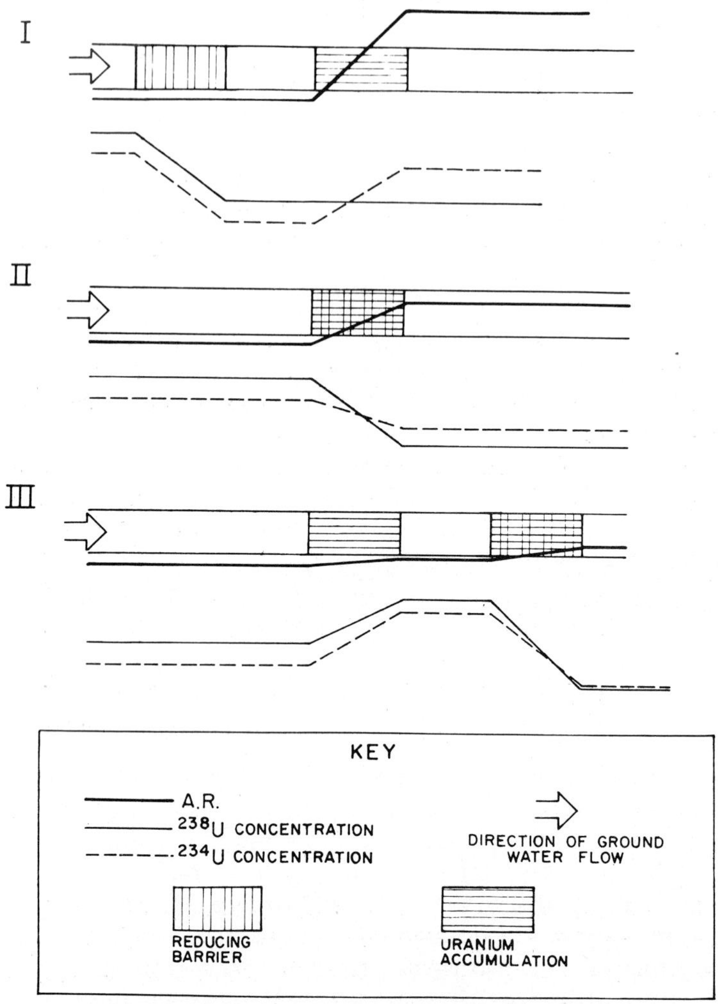

Fig. 6. Three configurations of a uranium precipitation —alpha recoil system in a confined aquifer. The relative positions of ^{234}U and ^{238}U concentrations and the concentration changes shown are schematic. Actual concentration, at equilibrium, of ^{234}U is about 5 orders of magnitude less than ^{238}U.

I. Reducing barrier updip from main uranium accumulation. It is assumed that the updip migration was recent and, as yet, only a negligibly small quantity of uranium has precipitated at that location.

II. Reducing barrier coincident with uranium accumulation. Precipitation of uranium isotopes is not preferential; input by alpha recoil from previously precipitated uranium is the cause of increase of ^{234}U concentration relative to ^{238}U concentration.

III. Reducing barrier downdip from previous uranium accumulation which is now in an oxidizing environment. It is, therefore, being dissolved, transported to the reducing barrier, and reprecipitated. Because the uranium concentrations in water are greater than in I and II above, the change in relative concentration between ^{234}U and ^{238}U is significantly less. It is assumed that the quantity of uranium in both accumulation zones of III is similar. (From Cowart, 1974).

The pattern of uranium isotope variation in the western portion of the Carrizo study area can be explained on the basis of precipitation of uranium at a barrier and the input of ^{234}Th (with subsequent decay to ^{234}U) at an accumulation of uranium which is located further downdip. This downdip accumulation of uranium is the former site of a barrier which has since moved somewhat updip to become the present day barrier. Conditions dictating such an updip migration could include climatic changes and/or hydrologic changes such as a decrease in permeability at the site of accumulation.

In the eastern area of the Carrizo, the absence of high A.R.'s downdip from the barrier suggests that there is no downdip accumulation separate from the present-day barrier to serve as the source for ^{234}U. With reference to Fig. 6, the western Carrizo would typify Type I whereas the eastern Carrizo would be similar to Type II.

The changes in uranium concentration and A.R. observed at the Highland area (Fig. 4) seem to resemble those of Carrizo, at least for the samples obtained (profiles A-A',B-B', C-C' in Fig. 3). In the case of the Carrizo, the existence of an accumulation of uranium which serves as a source of ^{234}U is inferred. In the Highland mine area, a known orebody seems to be the source of high A.R.'s. In both cases the high A.R. signal generated is observable for at least several kilometres downdip from the source. Because of the paucity of water wells in the vicinity of the Highland mine (only one was available updip from the mine) the abruptness of uranium concentration and A.R. changes can only be inferred.

Because the deposit is probably of the roll type (Dahl and Hagmaier, 1974), the oxidizing environment must extend to just up-flow from the uranium deposit. Such a set of conditions coupled with the uranium data suggest a system similar to that shown in Type II (Fig. 6).

If accumulations of uranium serve as a source which generates high A.R.'s in through-flowing water and if these waters retain this isotopic character for some distance in the direction of flow, then it seems likely that this phenomenon may be of use in uranium exploration. We do not wish to convey the notion that high A.R.'s found in groundwaters can only result from association with sizeable accumulations of uranium. Most high A.R.'s found are probably explicable on the basis of rapid selective leaching in the soil for shallow samples or, for deep samples, by alpha recoil input into the water over a long period of time. It is an abrupt increase in the uranium isotope ratio coupled with a decrease in concentration which may be the significant phenomenon for uranium ore exploration.

Water samples taken from the vicinity of two other uranium deposits are consistent with the above ideas but are too few to confirm them. In the Shirley Basin mining district, the townsite well sample, probably similar to the source waters of the orebody, has an A.R. similar to that of the Highland updip sample; water associated with the orebody is similar in A.R. to the Highland orebody waters. The samples from Karnes County, Texas, all have somewhat elevated A.R.'s but neither "outcrop" nor "downdip" locations have been located. The more oxidized nature of these deposits may contribute to this "halo" effect.

CONCLUSIONS

Sandstone-type orebodies accumulate at a reducing barrier within the groundwater circulation system. Where regional uplift has brought such orebodies into a more oxidizing regime, the circulating groundwater dissolves and disperses the ore, causing a concentration halo (Type III, Fig. 6). This model is the basis for standard hydrogeochemical exploration for uranium.

However, where conditions have remained more stable, and the accumulation process continues, no such concentration halo is produced. In this case, the presence of the mineralization is indicated by a decrease in uranium concentration in the downdip water, and more particularly, by a significant, sometimes dramatic increase in the $^{234}U/^{238}U$ activity ratio, similar to that at the Highland Mine. In effect, it appears that accumulations of uranium in sandstones generate an isotopic signal which is discernable for a number of kilometres in the direction of flow. Comparison of these data with water samples in the area of recharge may prove to be a useful indicator for the locating of such accumulations.

ACKNOWLEDGEMENTS

We are grateful to the following people for assistance in sample collection: J.L. Hagmaier, J. Marshall, M. Worley, Exxon Minerals, Casper, Wyoming; S. Trudeau and W.F. Baumann, Continental Oil Company, Falls City, Texas; E. Cooley and J. Whitman, Petrotomics Inc., Casper Wyoming. The authors were assisted in the preparation of figures by Majorie Knapp, Alan Brown, and Dennis Cassidy. Financial support for portions of this study were provided by the U.S. National Science Foundation, Grant Number GA-41474.

REFERENCES

Alexander, W.H. and White, D.E., 1966. Ground-water resources of Atascosa and Frio Counties, Texas. Texas Water Dev. Board, Rep. 32, 211 pp.

Bunker, C.M. and MacKallor, J.A., 1973. Geology of the oxidized uranium ore deposits of the Tordilla Hill—Deweesville area, Karnes County, Texas; a study of a district before mining. U.S. Geol. Surv., Prof. Paper, 765: 37 pp.

Chalov, P.I. and Merkulova, K.I., 1966. Comparative oxidation rates of U-234 and U-238 atoms in certain minerals. Dokl. Akad. Nauk. SSSR, 167: 146—148 (English translation).

Cherdyntsev, V.V., 1971. Uranium-234. Israel Program for Scientific Translations, Jerusalem, 234 pp. (English translation of "Uran-234", Atomizdat, Moskva, 1969).

Cherdyntsev, V.V., Kislitsina, G.I. and Zverev, V.L., 1967. Isotopic composition of uranium and thorium in rocks and products of active volcanism. Dokl. Akad. Nauk SSSR. 172: 177—179.

Cowart, J.B., 1974. ^{234}U and ^{238}U in the Carrizo sandstone aquifer of south Texas. Ph.D. Dissertation, Florida State University, Tallahassee, Fla., 80 pp. (unpublished).

Cowart, J.B. and Osmond, J.K., 1974. ^{234}U and ^{238}U in the Carrizo sandstone aquifer of south Texas. In: Isotope Techniqes in Ground-Water Hydrology, 2. Proceedings of a Symposium Held in Vienna, 1974. International Atomic Energy Agency, Vienna, pp. 131—149.

Dahl, A.R. and Hagmaier, J.L., 1974. Genesis and characteristics of the southern Powder River Basin uranium deposits, Wyoming, U.S.A. In: Formation of Uranium Ore Deposits. Proceedings of a Symposium Held in Athens, 1974. International Atomic Energy Agency, Vienna, pp. 201—218.

Hagmaier, J.L., 1971. Groundwater flow, hydrogeochemistry, and uranium deposition in the Powder River Basin, Wyoming. Ph.D. Dissertation, University of North Dakota, Grand Forks, N.D., 166 pp. (unpublished).

Harshman, E.N., 1972. Geology and uranium deposits, Shirley Basin area, Wyoming, U.S.A., U.S. Geol. Surv. Prof. Paper, 745: 82 pp.

Hostetler, P.B. and Garrels, R.M., 1962. Transportation and precipitation of uranium and vanadium at low temperatures, with special reference to sandstone-type uranium deposits. Econ. Geol. 57: 137—167.

Ivanov, K.E. and Kudryashova, R.G., 1963. Voprosy Prikladnoi Radiogeologii (D. Ya. Surazhskii, Editor). Gosatomizdat, Moskva, 278 pp.

Kaufman, M.I., Rydell, H.S. and Osmond, J.K., 1969. $^{234}U/^{238}U$ disequilibrium as an aid to hydrologic study of the Floridan aquifer. J. Hydrol., 9: 374—386.

Kigoshi, K., 1971. Alpha-recoil thorium-234: dissolution into water and the uranium-234/uranium-238 disequilibrium in nature. Science, 173: 47—48.

Kronfeld, J., 1974. Uranium deposition and Th-234 alpha-recoil: an explanation for extreme U-234/U-238 fractionation within the Trinity aquifer, Earth Planet. Sci. Lett., 21: 327—330.

Kronfeld, J., Gradsztan, E., Muller, H.W., Radin, J., Yaniv, A. and Zach, R., 1975. Excess ^{234}U: an aging effect in confined water. Earth Planet. Sci. Lett., 27: 342—345.

Lancelot, J.R., Vitrac, A. and Allegre, C.J., 1975. The Oklo natural reactor: age and evolution studies by U-Pb and Rb-Sr systematics. Earth Planet. Sci. Lett., 25: 189—196.

Lisitsin, A.K., 1962. Form of occurrence of uranium in groundwaters and conditions of its precipitation as UO_2, Geochemistry, 1962: 876—884.

Lonsdale, J.T., 1935. Geology and ground-water of Atascosa and Frio Counties, Texas. U.S. Geol. Surv. Water-Supply Paper, 676: 90 pp.

Osmond, J.K. and Cowart, J.B., 1976. The theory and uses of natural uranium isotopic variations in hydrology. At. Energy Rev., 14: 621—679.

Rosholt, J.N., Shields, W.R. and Garner, E.L., 1963. Isotopic fractionation of uranium in sandstone. Science, 139: 224—226.

Rosholt, J.N., Garner, E.L. and Shields, W.R., 1964. Fractionation of uranium isotopes and daughter products in weathered granite and uranium-bearing sandstone, Wind River Basin region, Wyoming. U.S. Geol. Surv. Prof. Paper, 501-B: 84—87.

Rosholt, J.N., Butler, A.P., Garner, E.L. and Shields, W.R., 1965. Isotopic fractionation of uranium in sandstone, Powder River Basin, Wyoming, and Slick Rock District, Colorado. Econ. Geol. 60: 199—213.

Sakanoue, M. and Komura, K., 1971. Anomalously high $^{234}U/^{238}U$ ratio in a natural sample. Nat. Phys. Sci., 233: 80—81.

Smith, R.F. and Jackson, J.M., 1969. Variations in ^{234}U concentration in natural uranium. U.S. At. Energy Comm., Rep. Ky-581, 9 pp.

Journal of Geochemical Exploration, 8(1977)381—395 381

RADIOACTIVE SPRINGS GEOCHEMICAL DATA RELATED TO URANIUM EXPLORATION

ROBERT A. CADIGAN and J. KAREN FELMLEE

U.S. Geological Survey, Denver Federal Center, Denver, Colo. 80225 (U.S.A.)

(Revised version accepted February 23, 1977)

ABSTRACT

Cadigan, R.A. and Felmlee, J.K., 1977. Radioactive springs geochemical data related to uranium exploration. J. Geochem. Explor., 8: 381—395.

Radioactive mineral springs and wells at 33 localities in the States of Colorado, Utah, Arizona and New Mexico in the United States were sampled and studied to obtain geochemical data which might be used for U exploration.

The major source of radioactivity at mineral spring sites is ^{226}Ra. Minor amounts of ^{228}Ra, ^{238}U and ^{232}Th are also present. Ra is presumed to have been selectively removed from possibly quite deep uranium-mineralized rock by hydrothermal solutions and is either precipitated at the surface or added to fresh surface water.

In this way, the source rocks influence the geochemistry of the spring waters and precipitates. Characteristics of the spring waters at or near the surface are also affected by variations in total dissolved solids, alkalinity, temperature and co-precipitation. Spring precipitates, both hard and soft, consist of four major types: (1) calcite travertine; (2) iron- and arsenic-rich precipitates; (3) manganese- and barium-rich precipitates; and (4) barite, in some instances accompanied by S. Ra and U, if present in the spring water, are co-precipitated with the barite, Mn-Ba and Fe-As precipitates.

Using parameters based on U and Ra concentrations in waters and precipitates, spring-site areas are tentatively rated for favourability as potential uraniferous areas.

INTRODUCTION

The purpose of this study is to determine whether geochemical and radiochemical data obtained from radioactive mineral springs can be used in the exploration for deeply buried U deposits. This is a progress report on a still expanding geochemical study.

Our investigation began in 1974. Approximately 70 sites of flowing springs have been visited. We have been concerned not only with the spring waters but also with their various types of precipitates. Data have been accumulated for 33 of these sites on the elemental composition of the waters and precipitates and on the mineral and radionuclide composition of the radioactive precipitates. The geochemistry of some springs suggests that they are related to nearby (20 km distant) significant U or Th deposits.

The term "mineral springs" is used in this paper in a general sense. Most mineral springs contain at least 1000 mg/l dissolved solids, but some of the springs and wells discussed contain less than that. Many are considered to be surface manifestations of mineralizing hydrothermal systems with different geochemistry but resembling the Delta County, Colorado, spring system described by Cadigan et al. (1976).

AREA OF STUDY

Data available for this report are confined to samples from the radioactive mineral springs in the southern Rocky Mountains of Colorado and New Mexico and the Basin and Range Province of New Mexico, Arizona and Utah. The Colorado Plateau region, which occupies the centre of this area, contains only a few mineral or thermal springs, and these are located near the margins of the plateau. Sample localities and kinds of analytical data available from each site are shown in Fig. 1.

PREVIOUS WORK

Much work has been done on the measurement of U and Ra in springs and wells. Public health agencies world-wide have been concerned with quality of water for human use. Reports such as those by O'Connell and Kaufmann (1976), Scott and Barker (1958, 1962), and Andrews and Wood (1973) were concerned with an accumulation of data on radioelements in natural water in the human environment and their potential effects on health. Reports on radioactive springs by Wollenberg (1975) and by us, primarily consider the possibility of using quantitative data on radioactive mineral springs as a tool in geochemical exploration for deeply buried U ore deposits. Scott and Barker (1962) also considered the metallogenic implications of Ra and U in springs and wells. They concluded that ratios of Ra to total dissolved solids (Ra/TDS) and U to total dissolved solids (U/TDS) as well as concentrations of Ra and U compared with the norm were the best parameters to use to rate different geotectonic regions or different localities within regions in terms of uranium enrichment. Earlier reports (circa 1900) viewed mineral springs, both radioactive and non-radioactive, as an economic resource based on their principal use as health baths or mineral tonics.

SAMPLING AND ANALYTICAL METHODS

At each locality, temperature, pH and specific conductance are measured on water of one or more springs in order to obtain an estimate of total dissolved solids (TDS) and to determine within-site variation. Specific ion electrodes are used to measure selected major ions and estimate water type in

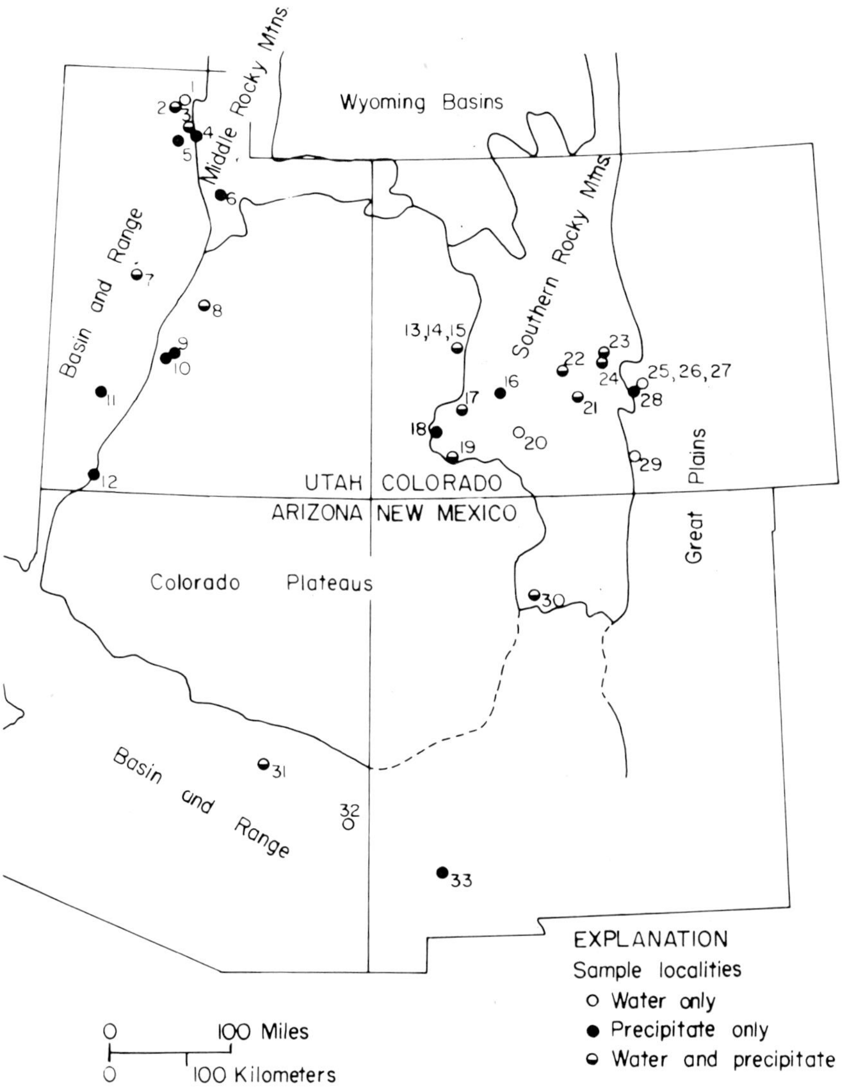

Fig. 1. Sample localities of selected springs and flowing or pumped wells. The symbol indicates the kind of samples for which data are available. 1 = Crystal Hot Springs, 2 = Stinking Hot Springs, 3 = Utah Hot Springs, 4 = Ogden Hot Springs, 5 = Hooper Hot Springs, 6 = Midway Hot Springs, 7 = Baker Hot Springs, 8 = Cold Stinking Springs, 9 = Redhill Hot Springs, 10 = Joseph Hot Springs, 11 = Thermo Hot Springs, 12 = Pah Tempe Hot Springs, 13 = Col. Chinn flowing well, 14 = Doughty Springs, 15 = Alum Gulch, 16 = Cebolla Hot Springs, 17 = Box Canyon Hot Springs, 18 = Dunton Hot Springs, 19 = Baker's Bridge Hot Springs, 20 = Wagon Wheel Gap Hot Springs, 21 = Mineral Hot Springs, 22 = Poncha Hot Springs, 23 = Guffey Soda Springs, 24 = Taylor Soda Springs, 25, 26, 27 = pumped well, 28 = Lower Red Creek Soda Springs, 29 = La Veta Sulfur Springs, 30 = Soda Dam Hot Springs, 31 = Salt Banks Springs, 32 = Clifton Hot Springs, 33 = Faywood Hot Springs.

the field. A hand-carried (Mt. Sopris Instruments Company) scintillometer is used to make a radiometric survey. If TDS content is low and there is no evidence of above-background radiation, no further measurements and no samples are taken. If there is significant radioactivity, a 2-litre sample of raw water is collected for Ra and U analyses. If TDS content is high, the following samples are collected: a 4-litre sample of filtered water acidified with 6 ml of 6*N* HNO_3 for analysis of trace metals; a 1-litre sample of filtered water acidified with 3 ml of HNO_3, for analysis of major cations; a 250-ml filtered sample, for major anions; and a 250-ml raw sample, for alkalinity and miscellaneous parameters. Analysis for trace metals is done by emission spectrography on evaporated residue. Ra is determined by co-precipitation with barium sulphate, partial equilibration with Ra subsequently produced, and counting of alpha radiation. Analysis for U is done using a fluorophotometric method (Thatcher and Barker, 1957; Barker et al., 1965).

Samples of travertine or other hard crystalline precipitates and soft precipitates are collected for analysis by emission spectrographic and X-ray diffractometer methods. Precipitate samples are collected from the areas of highest radioactivity (total count) at the site. Travertine or soft precipitates with measured radioactivity 50 times background or higher are commonly sampled for analysis by gamma-ray spectrometer and a multichannel analyzer system to determine radionuclide content and disequilibrium relationships. These samples are also analyzed for U and Th by delayed neutron activation (DNA).

Using a four-channel gamma-ray spectrometer, samples are also collected from precipitates showing the highest radiation at the energy level of U (^{214}Bi) and higher instead of highest total count. We have deliberately biased our sampling to concentrate on springs showing anomalously high field radioactivity, especially those with radioactive precipitates. We have done this for two reasons:

(1) The radioactive precipitates contain radioactive decay products of U and Th in large enough concentrations that isotope studies can be done to determine the importance of the U relative to the Th decay series in the source rocks.

(2) Springs showing high field radioactivity are more likely to contain high concentrations of Ra in the water relative to regional background levels as determined by Scott and Barker (1962). Determination of local anomalies relative to local background would then have to be substantiated by more sampling in smaller areas. The purpose of this report is to delineate radioactive areas on a reconnaissance level which need to be examined further in terms of local radioactivity patterns, groundwater flow and geologic structure before exploratory activities such as drilling can be instigated.

GEOCHEMICAL ENVIRONMENTS

Accumulated evidence suggests that the following represent extreme radioactive mineral spring environments which invite further investigation:

(1) Springs with considerable amounts of precipitates, hard or soft, and local total radiation of $\geqslant$10,000 cps on a portable scintillometer.

(2) Springs with low TDS (<1000 mg/l), few precipitates, but above-background radiation (400—1000 cps) and significantly high amounts of dissolved Ra (>5 pg/l) and U (>5 μg/l) in the water.

(3) Springs with mineable deposits of S, Fe minerals, Mn minerals or barite, parts of which are radioactive (1000—4000 cps).

DATA

Geochemical data for the spring water and precipitate samples show that both physical properties and geochemistry vary over a wide range. Table I shows the range of values and median value for each variable. Data on precipitates are entirely from our own samples. Data on water were compiled from analyses of our own samples as well as analyses published in other reports since 1961 and those available from unpublished sources of the U.S. Geological Survey. The complete tabulated data for this report are contained in a U.S.G.S. open-file report by J.K. Felmlee and R.A. Cadigan (in preparation).

MINERALOGY

The term "travertine" has been used to describe the hard crystalline precipitates associated with many mineral springs. Travertine may consist of almost pure calcite or a mixture of minerals that can be differentiated by X-ray diffraction. Calcite most commonly comprises 10% or more of each sample of precipitate. Other minerals that comprise 10% or more of one or more samples are barite, manganite, haematite, goethite, gypsum, fluorite, romanechite [$BaMn^{2+}Mn_8{}^{4+}O_{16}(OH)$] and pyrolusite. Quartz, feldspar, dolomite and mica are among the minerals identified that represent detritus in the samples. X-ray diffractometer traces do not account for the presence of As or W, which suggests that they are present in amorphous or non-crystalline forms. Some precipitates contain native S in minor to major amounts. At some sites iron- and manganese-rich precipitates as well as S have been mined as ores. One "travertine" deposit was found to be composed almost entirely of barite; part of it has been mined because of its high radioactivity, at first as a possible U ore, then later as a soil conditioner.

FACTORS CONTROLLING URANIUM AND RADIUM IN WATER AND PRECIPITATES

The hydrogeochemical environment of a radioactive mineral spring is comprised of many complex interacting materials and geochemical factors or processes. One can surmise that some of the factors operating in the environments are variations in the following: lithology of the rock units containing

TABLE I

Summary of analyses for samples from localities shown in Fig. 1 (27 water samples from 21 sites; 54 precipitate samples from 25 sites)

Parameter		Water			Parameter		Precipitate		
		minimum	maximum	median			minimum	maximum	median
Temperature (°C)		7	78	30	Ca	(ppm)	10,000	>100,000	>100,000
pH		2.9	9.6	7.1	Mg	(ppm)	200	20,000	5000
Alkalinity*		0	1610	—	Fe	(ppm)	300	>100,000	20,000
Specific conductance (μmhos/cm at 25°C)		170	52,000	3200	Mn	(ppm)	30	>100,000	700
					Si	(ppm)	700	>100,000	20,000
					Al	(ppm)	200	>100,000	5000
TDS	(mg/l)	87	43,500	2120	Na	(ppm)	500	>100,000	7000
SiO_2	(mg/l)	0.7	86	36	K	(ppm)	<7000	100,000	<7000
Ca	(mg/l)	3.9	930	110	Ti	(ppm)	1	5000	150
Mg	(mg/l)	0.4	330	24	As	(ppm)	<1000	>100,000	<1000
Na	(mg/l)	15	15,000	560	B	(ppm)	<20	200	<20
K	(mg/l)	1.6	870	36	Ba	(ppm)	3	>100,000	1000
HCO_3	(mg/l)	0	4330	323	Be	(ppm)	<1.5	500	10
CO_3	(mg/l)	0	13	0	Co	(ppm)	<5	200	<5
SO_4	(mg/l)	1.9	1500	160	Cr	(ppm)	<1	50	2
Cl	(mg/l)	3.4	26,000	320	Cu	(ppm)	<1	150	5
F	(mg/l)	0.3	12	2.4	Mo	(ppm)	<3	50	<3
NO_3	(mg/l	<0.1	0.48	<0.1	Nb	(ppm)	<10	20	<10
PO_4	(mg/l)	<0.01	0.77	0.03	Ni	(ppm)	<5	50	<5
Fe	(μg/l)	<10	47,000	120	Pb	(ppm)	<10	1500	<10
Mn	(μg/l)	<1	2200	130	Sb	(ppm)	<200	700	<200
Ba	(μg/l)	6	7600	100	Sc	(ppm)	<5	100	<5
Sr	(μg/l)	24	38,000	2500	Sr	(ppm)	200	15,000	5000
Mo	(μg/l)	<0.5	760	18	V	(ppm)	<7	150	<7
Li	(μg/l)	7	14,000	1500	W	(ppm)	<100	10,000	<100
B	(μg/l)	4	23,000	780	Y	(ppm)	<10	500	15
Al	(μg/l)	10	10,000	50	Zr	(ppm)	<10	500	20
Ra	(pg/l)	0.52	420	9.1	Ga	(ppm)	<5	150	<5
U	(μg/l)	<0.4	95	<0.4	Ge	(ppm)	<10	100	<10
					Li	(ppm)	<100	300	<100
					eU	(ppm)	<10	6010	250
					RaeU	(ppm)	435	8500	—**
					U	(ppm)	<0.20	38.32	—**
					Th	(ppm)	<4	11.61	—**

*Only 16 samples.
**Only 17 of the most radioactive samples.

the hydrothermal plumbing systems; rates of flow of water from depth; composition and pressure of dissolved gases; distances to related mineralizing geothermal systems associated with hot igneous intrusive complexes; and geochemical environments of such complexes.

The objective identification of effects of such factors, and an indication of their relative importance, can be obtained through statistical multivariate analysis of the variations measured in selected parameters. The particular factors identified in the following section are limited to those that strongly affect groups of the selected parameters. Furthermore, it should be noted that not all variation in the selected parameters is accounted for by the identified factors.

Factor analysis

Any number of factors can be computed, from two up to the number of variables. The interested reader is referred to the comprehensive text of Harman (1967) and detailed papers by Imbrie (1963), Steiner (1965) and Cadigan (1972). The results of the factor analysis of water samples are shown in Table II. Factor matrices for 2—10 rotated factors were studied, and the four-factor matrix was selected as most adequately establishing the order of importance and identifying the major geochemical factors or processes affecting the measured variables. These four factors account for 77% of the matrix variance.

We identified the first four factors as the most important. They are interpreted to be: total dissolved solids, alkalinity, temperature and co-precipi-

TABLE II

Reordered varimax factor matrix for water samples resulting from statistical multivariate analysis of samples from 21 of the 33 selected sample localities. A data matrix of 27 samples and 23 variables was used

Variable	TDS Factor 1 loading	Alkalinity Factor 2 loading	Temperature Factor 3 loading	Co-precipitation Factor 4 loading	Communality
Cl	*0.975*	0.100	0.077	-0.027	0.97
Na	*0.956*	-0.038	0.090	-0.039	0.93
TDS	*0.953*	-0.185	0.108	-0.117	0.97
K	*0.923*	-0.080	0.102	-0.185	0.90
Li	*0.904*	0.012	0.155	-0.117	0.86
Sr	*0.897*	-0.180	0.249	0.051	0.90
B	*0.877*	-0.055	0.091	-0.090	0.79
Ca	*0.798*	(-0.394)	0.165	0.014	0.82
Mg	*0.699*	(-0.276)	-0.241	-0.361	0.75
Ra	*0.596*	0.121	(-0.469)	-0.289	0.67
Mo	*0.511*	0.068	0.125	0.012	0.28
pH	-0.023	*0.837*	-0.038	(0.308)	0.80
$HCO_3 + CO_3$	(*0.406*)	*0.796*	-0.045	0.031	0.80
NO_3	0.171	*-0.463*	-0.072	-0.029	0.25
Mn	0.018	*-0.518*	(0.384)	(0.279)	0.49
Al	(*0.428*)	*-0.804*	-0.109	-0.120	0.86
F	0.117	0.261	*0.889*	0.057	0.88
SiO_2	0.216	-0.124	*0.810*	-0.311	0.82
Temperature	0.222	0.034	*0.686*	(0.371)	0.66
SO_4	0.234	(-0.407)	*0.667*	-0.440	0.86
Ba	(*0.600*)	-0.099	-0.069	*0.645*	0.79
U	(*0.458*)	-0.048	-0.149	*-0.756*	0.81
Fe	0.194	(-0.046)	0.110	*-0.804*	0.86

Italicized factor loadings are those for variables most affected by the factor under which they are listed. If a variable is significantly affected by a second or third factor, its loading for that factor is shown in parentheses. Communalities are the sums of the squares of the factor loadings for each variable. They represent the proportion of variance in each variable accounted for by the four factors; thus, four factors account for 67% and 81%, respectively, of the variance in Ra and U concentrations.

tation. From Table II, it is apparent that as total dissolved solids content increases, most variables are positively affected. Alkalinity increases are accompanied by an increase in the definitive HCO_3 plus CO_3 ions, an increase in pH, and a decrease in Al and other ions shown. Temperature increases are characterized by increases in F, SiO_2 and SO_4 and by a decrease in Ra. The effects of the co-precipitation factor are more subtle and are keyed to concentrations of Ba. There is a tendency for barium-rich water to have higher values for temperature, pH and Mn and lower values for Fe, U, SO_4 and other ions shown. This suggests that the Ba tends to remove the SO_4 from solution by forming $BaSO_4$ which precipitates at the surface and co-precipitates U and Ra. The accompanying increase in pH would also cause precipitation of Mg, Fe and SiO_2.

Multivariate *R*-mode factor analysis was also done on data for precipitates from 24 localities. The first five factors were selected as the most important affecting the measured elements in the precipitates. Five factors explain 69% of the correlation matrix variance.

Factor groups of variables are summarized in Table III. An attempt is made to identify the factors as geochemical processes characterizing the springs and controlling the distribution of elements in the precipitates. Factor groups are taken from the reordered varimax factor matrix.

TABLE III

Factor groups of variables for precipitate samples. Factor analysis of precipitates was based on a data matrix of 54 samples and 32 variables

Detrital contamination		Mn precipitation	Fe and As precipitation	Heavy metals precipitation	Ba precipitation
Si	(V)	W	As	Ni	Ba
Al	(Mg)	Nb	Be	Co	eU
K	(Na)	Mn	Fe	V	Sr
Cr	(Y)	Mo	Ge	Sc	(Sc)
Zr	(Ni)	Sb	Na	Pb	(Zr)
Cu	(Ba)	(Ga)	Y	(Cu)	
Ga		(Ba)	(eU)	(Mo)	
Ti		(Co)	(B)	(Zr)	
B		(eU)	(Pb)	(Cr)	
Li				(Al)	
Negatively affected					
(Ca)		Mg	—	(Sr)	La
		Ca			
		(Na)			
Percent of total variance of sample composition accounted for by each factor					
30		16	9	8	5

The first factor is contamination of spring-related chemical precipitates by locally derived clay, silt, sand and evaporite minerals transported by wind or water. This process is negatively related to the precipitation of Ca (mostly $CaCO_3$). These represent two major opposing processes occurring in the deposition of mineral spring precipitates. The second factor, deposition of black and reddish-black manganese-rich precipitates, characteristically includes deposition of W, Mo, Sb and Nb, and tends to be mutually exclusive of the deposition of Ca, Mg and Na minerals. The third factor, deposition of reddish-brown Fe and As precipitates, characteristically includes deposition of Be, Ge, Na and Y. Ra, represented by eU (a total radiation count), As and related elements are co-precipitated by Fe. The fourth factor, precipitation of the heavy metals Co, Ni, V and Pb, a classically hydrothermal suite of metals, is possibly related to igneous source rocks. The fifth factor is precipitation of Ba as barite and co-precipitation of Ra (eU) and Sr.

Factor 5 confirms the strong association between Ba and Ra previously noted. Most sites where manganese- and iron-rich precipitates are being deposited are typically and, in part conspicuously, radioactive. These factor groups are compatible with field observations and support the validity of our sampling programme. In other words, results of the statistical analysis of the sample data supports what we think we see in the field.

RADIOACTIVITY AND RADIOISOTOPE GEOCHEMISTRY

Radioactivity in mineral springs and in the precipitates associated with them is believed to represent, in effect, an attenuated halo related to unoxidized U deposits. The fact that Ra is present in the spring water in quantities far out of equilibrium with accompanying U suggests that Ra is more mobile than U. The lack of U in turn suggests that most of it is in an unoxidized state at the source. Groundwater which has been in contact with near-surface U deposits in which the U is mostly oxidized contains Ra and U in near-equilibrium proportions or contains an excess of U. Various amounts of U and its daughter products are transported in solution to the surface, where they may or may not be precipitated depending largely on the kind and amount of other constituents in the water. Graphic comparisons between Ra, U and the Ra enrichment factor (REF) in the water and in the precipitates were made using analytical data on 16 precipitate and 9 water samples from 9 localities (Table IV) and are shown in Fig. 2.

The amount of Ra in the spring precipitates does not correlate significantly with the amount of Ra in the spring water. On the other hand, the amount of U in the precipitates shows a tendency to increase as the amount in the water increases.

The REF, or the number of times the amount of Ra in the sample exceeds the Ra value for equilibrium with the U in sample, ranges from 10 to 2000 in water and from about 50 to 50,000 in the precipitates. At every spring for which data are available, the computed REF is greater in the precipitates

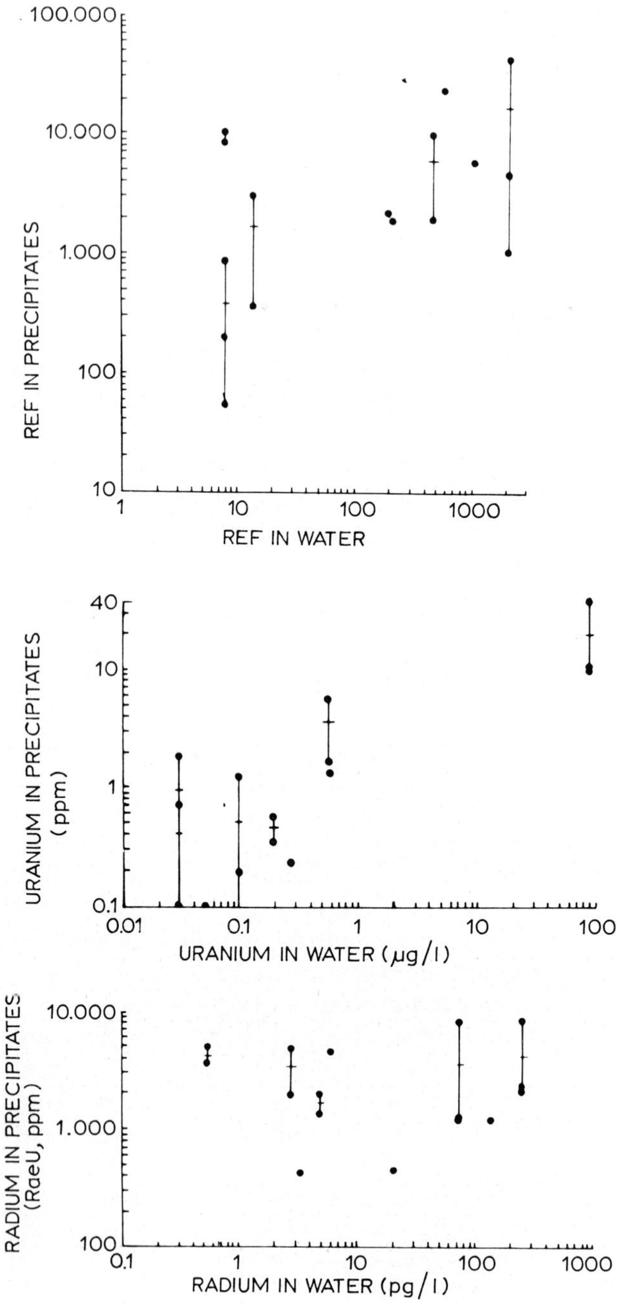

Fig. 2. Relationships between Ra, U and the Ra enrichment factor (REF) in water and precipitates. Data are listed in Table IV. Only values for the most radioactive samples at each locality are shown. Where there is more than one precipitate per locality the mean value is shown by a horizontal bar. REF is defined in Table IV.

TABLE IV

Ra-U equilibrium relationships in spring water and precipitates

Location	Precipitates			Water		
	RaeU* (ppm)	U (ppm)	RaeU/U or REF**	Ra (pg/l)	U (μg/l)	2.9 × Ra/U or REF**
2	8500	<0.20	⩾43,000	76	0.10	2200
	1300	1.24	1000	—	—	—
	1250	0.20	6400	—	—	—
4	460	0.24	1900	21	0.28	220
7	435	<0.20	⩾2200	3.5	0.05	200
9	4600	<0.20	⩾23,000	6.3	0.03	610
14	4900	1.65	3000	2.9	0.59	14
	2000	5.53	360	—	—	—
17	1975	<0.20	⩾9900	5.1	0.03	490
	1400	0.72	1900	—	—	—
22	4900	0.57	8600	0.55	<0.4	⩾4
	3700	0.36	10,000	—	—	—
24	8500	9.75	870	260	95	8
	2300	11.49	200	—	—	—
	2100	38.32	55	—	—	—
30	6500	1.36	4800	230	0.6	1100

*RaeU, radium equivalent uranium, or the amount of U that would be required to support the amount of Ra in the sample.

**REF, radium enrichment factor, or the number of times the amount of Ra in the sample exceeds the equilibrium amount of Ra for the U present in the sample.

than it is in the water, and the greater the enrichment in the water, the greater tends to be the enrichment in the precipitates. Ra enrichment in the water indicates that U is being left behind in the source rocks causing disequilibrium between Ra and U in the water. Recent work by O'Connell and Kaufmann (1976) indicates that ^{238}U, ^{234}U and ^{230}Th — the precursors of ^{226}Ra — are in approximate equilibrium; the major point of disequilibrium in the chain is between ^{230}Th and ^{226}Ra. It is ^{226}Ra, therefore, not its parents, that is mobile and is transported from the parent ^{238}U source rocks; much ^{238}U is unaccounted for and must be present either in, or at the source end of, the hydrothermal pipeline.

The highly soluble ^{226}Ra is transported in bicarbonate and chloride waters and is co-precipitated on degassing, oxidation or cooling of the hydrothermal solutions. The amount of Ra concentrated in the precipitates is quite variable at a given locality and depends largely on the degree of co-precipitation of Ra with $BaSO_4$, arsenic-bearing iron oxide or hydroxide, and manganese oxide or hydroxide. Co-precipitation of Ra with $CaCO_3$ seems to be minimal, and those precipitate samples not containing significant Ba, Fe, As or Mn also do not show anomalous radioactivity, indicating only background ^{226}Ra.

Radioactive decay of ^{226}Ra in the precipitates and water yields ^{222}Rn, which in turn yields ^{214}Bi, the source of most of the gamma-ray energy level radiation attributed to ^{238}U in a series equilibrium situation. Enough ^{228}Ra, a daughter product of ^{232}Th, is also present in most radioactive precipitates to yield a measurable amount of ^{208}Tl, source of gamma-ray energy level radiation assigned to ^{232}Th in a series equilibrium situation. As a result, most of the total count as well as K, U and Th gamma-ray spectrometer readings come from ^{226}Ra and ^{228}Ra and their daughter products. $^{228}Ra/^{226}Ra$ ratios exceed 1:1 in at least two spring precipitates, as revealed by isotopic analysis and equilibrium studies by J.N. Rosholt, J.R. Dooley and A.N. Kahn of the U.S. Geological Survey (written communication, 1976) based on Ra data from C.M. Bunker. Ratios of 2:1 and 6:1 suggest that the source rocks for radioisotopes in water at these springs contain more ^{232}Th than ^{238}U.

DISCUSSION OF RESULTS

Geochemical factors

We have not reached the point in the study of radioactive springs where we can say "This is the way to use the geochemistry of mineralized springs to find uranium deposits." We have, however, accumulated much basic geochemical data. To obtain a geochemical profile of a radioactive mineral spring, we needed precipitate samples as well as water samples to determine the distribution of U, Ra and various other elements and to identify the mineral phases present. Mineral deposits observed as part of modern spring precipitate accumulations are evidence of the springs' capability for transporting and depositing large amounts of metals in high concentrations and suggests that the sources of the metals for some springs may be of significant size.

The evidence suggests that the hydrothermal solutions are leaching relatively more soluble Ra from parent U deposits. The form of the uranium deposits is not known with certainty, but the presence of Th daughters suggests vein-type deposits in igneous or metamorphic rock because sedimentary-type deposits have virtually no Th. In some instances, the isotopic disequilibrium suggests rapid movement of the metal-rich hydrothermal solutions to the surface (Cadigan et al., 1976). The springs are mostly vents for carbon dioxide- and steam-propelled waters. They do not appear to be artesian or gravity flow springs. Sulphur gases are fairly common; the related surface precipitation suggests that S minerals are the products of near-surface oxidation of the S ions (HS^-?) carried in the solutions and are not transported from depth.

Factor analysis supplies evidence of geochemical processes operating in the spring waters and at the sites of deposition of precipitates. Major factors affecting the chemistry of the waters appear to be, in order of importance, variations in the quantity of total dissolved solids, alkalinity, temperature and oxidation. Alkalinity changes as gas pressure in the solutions change and carbon dioxide is released as solutions approach the surface. Some solutions literally effervesce. Temperatures decrease steadily as the spring waters rise from depth; oxidation of ions in solution begins where oxygen is first encountered and continues to the surface.

Factors operating at the sites of deposition of precipitates are, in order of importance, contamination by detrital material and evaporites; Mn precipitation, possibly due to oxidation; an increase in alkalinity and decrease in temperature; Fe and As precipitation, probably an effect of oxidation; heavy metals precipitation, possibly a co-precipitation process with Mn; and barite precipitation, which co-precipitates U and Ra with it. The precipitation of barite appears to depend on the late formation of SO_4^{2-} — a product of oxidation of S ions carried in solution.

Classification of springs

The mineral springs may be classified into four types: (1) those producing only calcareous travertine; (2) those also producing hard or soft iron- and arsenic-rich precipitates with co-precipitated Ra; (3) those also producing manganese- and barium-rich precipitates with co-precipitated Ra; and (4) those also producing $BaSO_4$ in hard and soft precipitates with co-precipitated Ra and, locally, native S. All spring precipitates are subject to varying degrees of contamination by detritus and evaporites. The latter three types of springs are of most interest in uranium exploration.

Tentative use of spring data for evaluation of economic potential

A reconnaissance-type evaluation of the relative U potential of the 33 studied localities, based on the use of spring radioactivity, was made using the following uranium- and radium-associated parameters:

(1) eU (in precipitates), the radiometric equivalent U.

(2) U (in water), the amount of U in solution.

(3) Ra (in water), the amount of Ra in solution.

(4) U/TDS (in water), the ratio of U in solution to total dissolved solids (Scott and Barker, 1962).

(5) Ra/TDS (in water), the ratio of Ra in solution to total dissolved solids (Scott and Barker, 1962).

(6) REF (in water), the Ra enrichment factor, or the number of times the quantity of Ra present exceeds the amount equivalent to the U present in a series equilibrium situation.

The 33 localities were ranked according to values of the six parameters and then a composite listing was made. The eight top-ranking localities are shown in Fig. 3. The grouping of localities is based on their occupying the same geologic environment and their being possibly related to the same hydrothermal system.

The Stinking Hot Springs area (sites 1, 2, 3 and 5) includes Crystal Hot

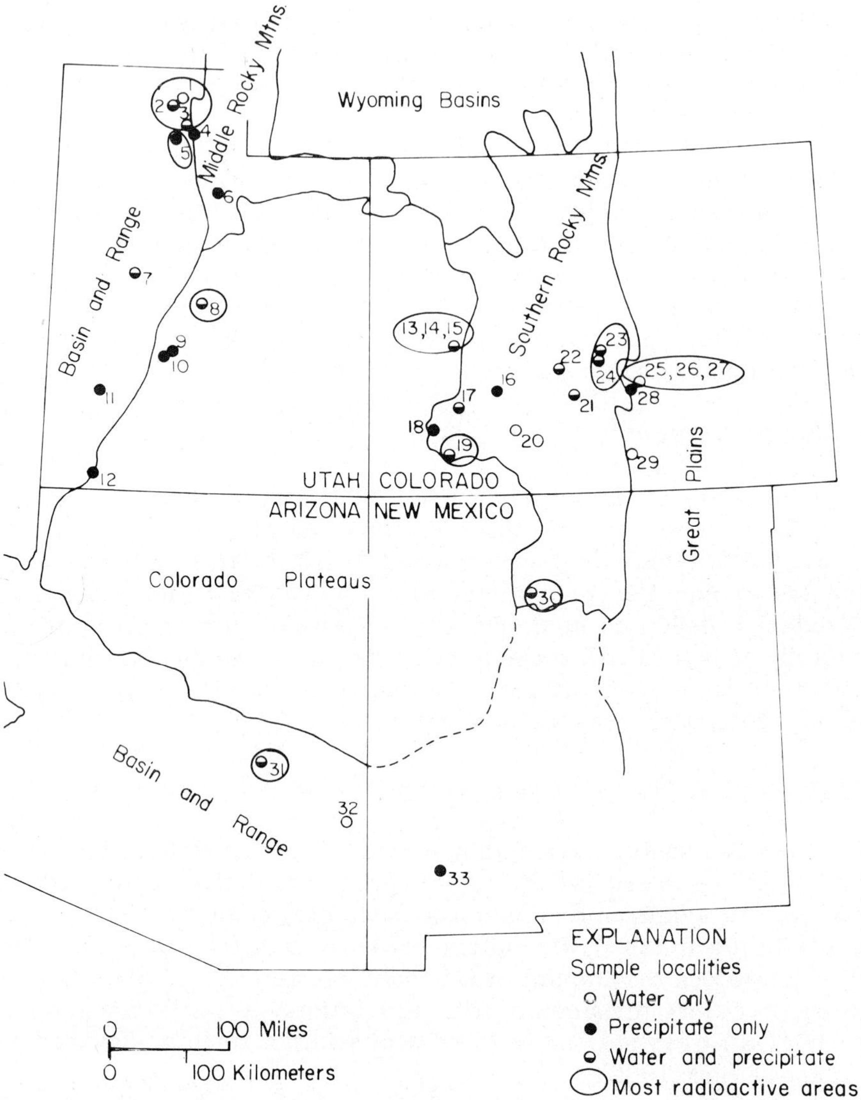

Fig. 3. Localities of the eight most radioactive areas of those sampled. For key to numbered spring sites, see Fig. 1.

Springs, Utah Hot Springs and Hooper Hot Springs and lies near the western front of the Wasatch Mountains northeast and north of the Great Salt Lake. The Stinking Hot Springs site is the most radioactive and most interesting of any visited.

The Taylor Soda Springs area (sites 23, 24, 25, 26 and 27) includes the region of pumped wells in Pueblo County because they may be related to the same mineralizing system. Observations made during a one-week visit to the area, a study of Pueblo County Health Department records, and a report by Scott and Voegeli (1961) support this possibility.

Other areas of similar potential are present in the western United States but have not yet been studied. We hope that our efforts will encourage others to take a careful look at radioactive springs in the course of evaluating the U ore potential of unexplored regions.

REFERENCES

Andrews, J.N. and Wood, D.F., 1973. Radium-226, radon-222, and lead-210 in Bath Thermal Springs compared with some environmental waters. Health Phys., 27: 307—310.

Barker, F.B., Johnson, J.O., Edwards, K.W. and Robinson, B.P., 1965. Determination of uranium in natural waters. U.S. Geol. Surv. Water-Supply Paper, 1696-C: 25 pp.

Cadigan, R.A., 1972. Geochemical anomalies and alteration in the Moenkopi Formation, Skull Creek, Moffat County, Colorado. U.S. Geol. Surv., Prof. Paper, 761: 21 pp.

Cadigan, R.A. and Felmlee, J.K., 1975. Use of the geochemistry of radioactive springs in uranium exploration (abstract). In: Abstracts of the 1975 Uranium and Thorium Research and Resources Conference. U.S. Geol. Surv., Open-file Rep., 75—595: 8.

Cadigan, R.A., Felmlee, J.K. and Rosholt, J.N., 1976. Radioactive mineral springs in Delta County, Colorado. U.S. Geol. Surv. Open-file Rep., 76-223: 39 pp.

Harman, H.H., 1967. Modern Factor Analysis. Chicago University Press, Chicago, Ill., 474 pp.

Imbrie, J., 1963. Factor and vector analysis programs for analyzing geological data. U.S. Office Naval Research, Tech. Rep., 6, ONR Task No. 389-135, Contract 1228 (26): 65 pp.

O'Connell, M.F. and Kaufmann, R.F., 1976. Radioactivity associated with geothermal waters in the western United States — basic data. U.S. Environmental Protection Agency, Tech. Note ORP/LV-75-8A: 25 pp.

Scott, R.C. and Barker, F.B., 1958. Radium and uranium in ground water in the United States. In: Proceedings of the Second United Nations International Conference on the Peaceful Uses of Atomic Energy, United Nations, New York, N.Y., 2: 153—157.

Scott, R.C. and Barker, F.B., 1962. Data on uranium and radium in ground water in the United States, 1954 to 1967. U.S. Geol. Surv., Prof. Paper 426: 115 pp.

Scott, R.C. and Voegeli, P.T., Sr., 1961. Radiochemical analyses of ground and surface water in Colorado, 1954—1961. Colo. Water Conserv. Board, Basic Data Rep., 7: 27 pp.

Steiner, D., 1965. A multivariate statistical approach to climatic regionalization and classification. Tijdschr. K. Nederl. Aardrijkskundig Genootschap, 82: 329—347.

Thatcher, L.L. and Barker, F.B., 1957. Determination of uranium in natural waters. Anal. Chem., 29: 1575—1578.

Wollenberg, H.A., 1975. Radioactivity of Nevada hot-spring systems. Geophys. Res. Lett., 1: 359—362.

Journal of Geochemical Exploration, 8(1977)397—414

SOIL HYDROCARBON GEOCHEMISTRY, A POTENTIAL PETROLEUM EXPLORATION TOOL IN THE COOPER BASIN, AUSTRALIA

S.B. DEVINE and H.W. SEARS

Geological Survey of South Australia, Eastwood, S.A. (Australia)
Amdel Laboratories, Frewville, S.A. (Australia)

(Revised version accepted March 9, 1977)

ABSTRACT

Devine, S.B. and Sears, H.W., 1977. Soil hydrocarbon geochemistry, a potential petroleum exploration tool in the Cooper Basin, Australia. J. Geochem. Explor., 8: 397—414.

The ethane, propane, butanes and pentanes concentrations of shallow soil samples collected in 1975 from 1070 sites over the Permian Cooper Basin have been analyzed. The samples were taken at a spacing of 300 m along two regional traverses; on a widely spaced irregular grid from above the known gas fields of Daralingie and Della; and above potential gas field areas. Geochemical anomalies appear to be associated with all the known fields traversed and with structures having petroleum potential either as anomalies directly above the gas fields or as halo anomalies.

The most useful variables mapped are: (1) the sum of soil concentration of the four hydrocarbons (expressed in ppb of soil by weight), and (2) that sum divided by the percentage of acid-soluble material. Average and highest hydrocarbon concentrations measured respectively are ethane 0.59, 18.9; propane 0.34, 10.9; butanes 0.21, 4.6; pentanes 0.20, 3.8. The acid-soluble material averaged 4.2% with the highest value at 62.8% and moisture content at 60° C averaged 5.7% with a highest value of 25.7%.

Some apparent anomalies are probably spurious. The soil geochemical prospecting method has a potential in petroleum exploration only if interpreted in conjunction with good structural and stratigraphic information.

INTRODUCTION

The petroleum exploration industry has shown spasmodic interest in so-called direct hydrocarbon-finding methods since the 1930's. Interest has been reactivated recently by the advent of the gas chromatograph and computer processing which make analyses of soil hydrocarbons simple and cheap, and by the increased need in the minds of many geologists to search for stratigraphic traps.

An initial experimental soil hydrocarbon survey in 1973 covered the area surrounding the Della gas field in the Cooper Basin (Devine and Sears, 1975). The survey of 379 samples established that measurable quantities of ethane, propane, butane and pentane occur in the soils and that their concentrations show mappable variations. The sum of the four hydrocarbons and that sum divided by the amount of acid-soluble material proved to be the most useful variables to map. Anomalies were mapped over the Della gas field and an undrilled nearby anticline.

This paper reports the results of a second survey in which samples were taken from 1070 sites during July to November, 1975. The survey consisted of two regional lines, a widely spaced grid over the Daralingie gas field, and some further samples from the Della gas field area (Fig. 1). One regional line runs from south of the limits of the Nappamerrie Formation subcrop to the centre of the basin and the second ties the Daralingie and Della gas fields. The purpose of these regional lines was to assess any regional gradient which might be related to depth of burial of the Permian gas-bearing sediments and the thickness and presence of the overlying Nappamerrie Formation.

The sampling over the Daralingie Field was designed to provide evidence of any empirical relationship between soil hydrocarbons and the gas field after the encouraging results from the Della field area. Daralingie is on the opposite site of the basin from Della and is a wet gas field, whereas Della contains dry gas. A further difference is that the reservoir sandstones in the Della field are in the upper part of the Permian Gidgealpa Group (Toolachee Formation) whereas in the Daralingie field they are in the lower part (Patchawarra Formation). Further sampling was done near the Della field to compare results from the new sampling technique and to get a better definition of the anomaly over the nearby undrilled structure.

The Cooper Basin is a good place for experimentation for several reasons. The topography is flat and the near-surface strata are flat-lying so that the area does not have the potential complication of different strata with varying abilities to carry hydrocarbon gases subcropping beneath the soil. The surface detrital influx is quartzose and clayey; there is little or no detrital carbonate from any other known petroleum province which might cause contamination of soil samples (McCrossan et al., 1972).

The Cooper Basin contains established oil and gas fields which can be used to assess empirically their relationship to soil hydrocarbon content. At present, it has sixteen commercially viable gas fields and two small oil fields. Only three of the gas fields are currently producing. Further petroleum discoveries are possible but many of the larger easily defined structural targets have already been drilled. Consequently, if geochemical prospecting can be shown to be effective, it has the potential to supplement the seismic information and increase drilling success.

METHODS

Sampling

Sample lines were located according to accessibility. Sample points are 300 m apart except in the southernmost part of the regional north-south line where the spacing is 500 m. In general, the lines over the Daralingie and

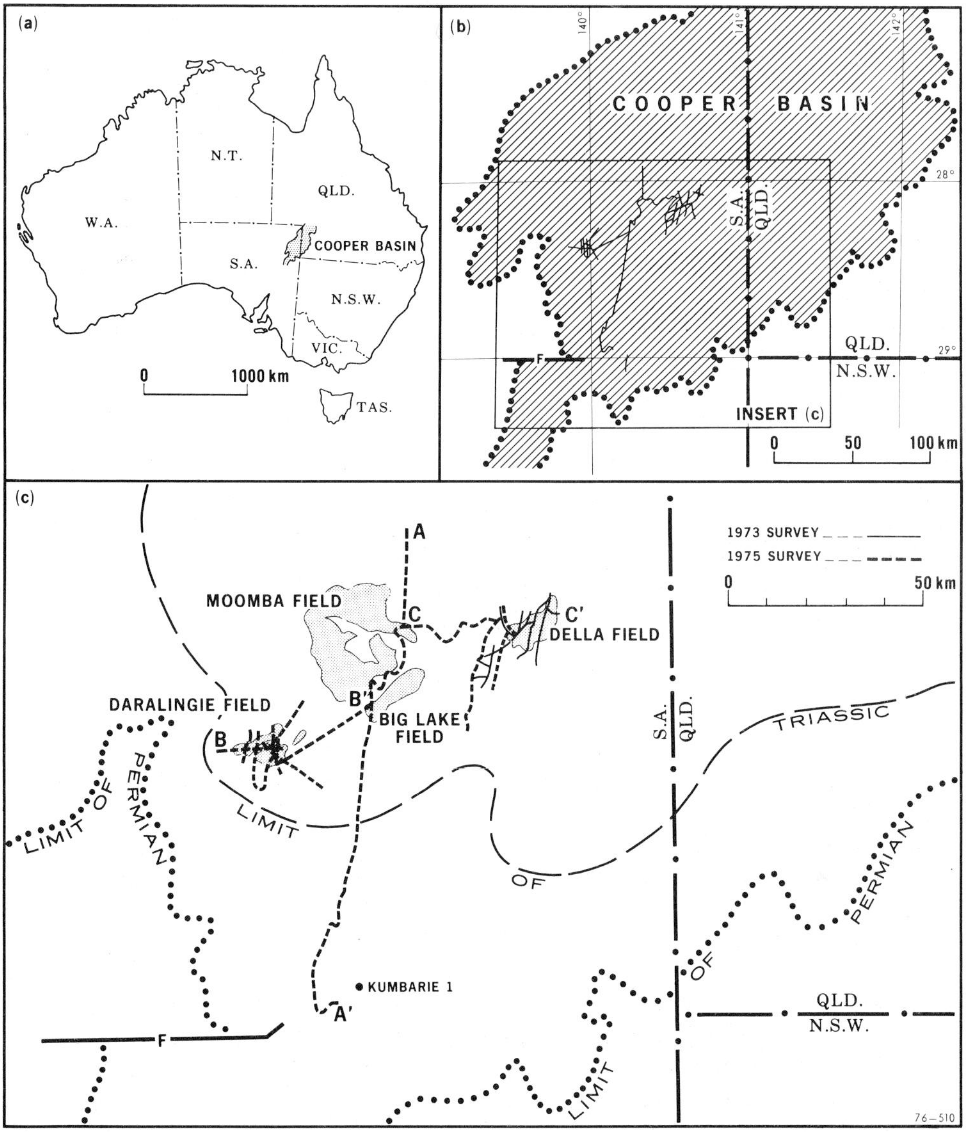

Fig. 1 (a) Locality map. (b) Cooper Basin, extent of Permian sequence (Gidgealpa Group). (c) Geochemical sampling traverses.

Della fields are spaced farther apart than is desirable but the expense of establishing a closer grid was not appropriate at this experimental stage. Complete confidence cannot be placed in the isoconcentration lines mapped (Figs. 2 and 3) because, with the large distance between most lines, the data can be contoured in alternative ways in some places. Nevertheless, we consider that the isoconcentration lines, as with most contouring, are a legitimate method of interpreting and presenting the data. The precision of the contours improves as the sample lines are approached.

A sample spacing of no more than 300 m along each line is justified because many of the anomalies tend to be confined to narrow zones as Debnam (1964) also found over some Canadian fields. A "smoothing function" was applied to all data to assist in the presentation of the contours: if sample point B is flanked by sample points A and C with total hydrocarbon concentrations of b, a and c respectively then the smoothed concentration at B is $(a + 2b + c)/4$.

The samples were taken between 2.8 and 3.1 m below surface using a mechanical drilling unit powered by an air compressor (300 l/minute; 260 cu.ft/minute) with pressure rating up to 7000 g/cm^2 (100 psi). Holes were rotary drilled to 2.8 m and then a bottom-hole core sample was obtained by hammering in a modified cylinder (Stace and Palm, 1962) with the drill's downhole hammer. Cylinders with diameters of either 5 or 7.5 cm were used. After discarding the friable top, which could be contaminated with cavings, a core of about 15 cm was heat-sealed into a polythene bag with a 0.15 mm wall thickness. The samples were air-freighted 800 km to Adelaide for analysis.

Hydrocarbon analyses

The analytical procedures used are essentially those of Debnam (1964) with minor modifications, including those discussed previously (Devine and Sears, 1975).

For hydrocarbon extraction, a representative sample was leached with 2N hydrochloric acid. A sample portion of 50 g was used rather than a 20-g portion used in the first survey. This increase in sample weight together with a twofold increase in sensitivity, achieved by very rigid reagent blank control, enabled a detection level of 0.2 μg of total C_2—C_5 hydrocarbons per kilogram (ppb) to be attained. Exhaust gases from the vacuum pump can contribute noticeably to background unless vented from the laboratory.

The samples were given minimum preparation prior to analysis as it was considered likely that some, if not most, of the samples could lose much of their hydrocarbon gases on exposure to air (Horvitz, 1972). This loss would be further aggravated by any sample grinding.

The samples comprised free-flowing, well-sorted quartzose sands, poorly sorted clayey sands with large gypsum crystals and calcrete nodules and damp, sticky clays. Gravel including large gypsum crystals and calcrete nodules larger than about 10 mm were handpicked and rejected before weighing and splitting. The moisture content was determined gravimetrically on a 100-g portion after drying for 12 hours at 60°C. The clays had a moisture content of up to 26% so that representative sampling was difficult. They were first broken by hand, then further divided with a spatula.

With samples containing more than a few percent carbonate, blockages in the gas delivery lines to the potassium hydroxide scrubbers occurred unless these scrubbers had been refilled with fresh cold solution. For these problem samples, a cold water bath was also applied to the scrubbers and the acid was added very slowly.

After aliquots of all samples had been analyzed using hydrochloric acid extraction, a suite of 36 samples was selected for an alternative digestion technique using 50% phosphoric acid in place of the 2*N* hydrochloric acid. This was done to check if the digestion technique of Horvitz (1972) would give a higher recovery of hydrocarbons.

The technique was modified so that the sample under vacuum was heated in a boiling water bath under reflux for 30 minutes. To achieve this, the apparatus was fully evacuated and the valve to the hydrocarbon trap was shut to give a completely closed system. The acid was then added and the boiling water bath was raised to immerse the flask completely. The liberation of gases apart from carbon dioxide and water vapour allowed the temperature of the digest to rise to near boiling water temperature and this was held for 30 minutes when the water bath was removed and the apparatus filled with displacement water.

The hydrocarbon trap was immersed in liquid nitrogen and after a minute, the tap between this trap and the apparatus was slowly opened and the gaseous contents of the train evacuated through the trap. Then the train was purged with outside air and the rest of the determination was made as before.

The phosphoric acid treatment did not significantly improve the recovery of hydrocarbons compared with the 2*N* hydrochloric acid treatment but it did dissolve much of the clay and more of the gypsum than had the hydrochloric acid. As it takes twice the time for digestion, its use in preference to hydrochloric acid could be of no advantage.

REGIONAL GEOLOGY

The petroleum in the Cooper Basin is generated and trapped in Permian sediments which consist of sandstone, siltstone, shale and coal. No outcrop occurs. The top of the Permian rocks varies from 600 to 3300 m below the surface and the sequence varies from zero to 2000 m in thickness. The Permian rocks are conformably overlain by the Triassic Nappamerrie Formation, which ranges up to 400 m in thickness. The Nappamerrie consists predominantly of shale and siltstone and acts as a seal on the Permian sediments, except in the southern part of the basin (Youngs, 1975). It separates the Permian rocks from the overlying Mesozoic sediments of the Great Australian Basin which suboutcrop beneath a Tertiary cover.

RESULTS

The results of the geochemical survey are presented as isoconcentration maps (Figs. 2 and 3) and sections (Fig. 4). The two variables mapped are: (1)

"*first variable*" the sum of the contents (in ppb) of the four hydrocarbons analyzed, and (2) "*second variable*" that sum divided by the percentage of acid-soluble material.

Regional variations

The regional north-south line (*AA*′, Figs. 1, 2 and 3) runs from near the southern margin of the Permian sediments across the southern limit of the Nappamerrie Formation to near the deepest part of the basin. This line was sampled to test the hypothesis that, if a flux of hydrocarbon gases permeates upwards from sediments bearing hydrocarbons, then the flux measured in the soil at the surface ought to be highest where the hydrocarbon-bearing sediments are shallowest and the relatively impermeable cap rock is absent or thinnest. The results confirm the hypothesis. The average sum of hydrocarbons south of the Nappamerrie Formation limit is 2.3 ppb from 131 samples versus 1.2 ppb from 922 samples north of the limit. The trend is evident to the eye in the top profile of Fig. 4. The contrast also holds for the second variable — the sum of hydrocarbons divided by percentage of acid-soluble material — for which the average values are 1.9 ppb/% south of the Nappamerrie limit versus 0.74 ppb/% north of the limit.

A further contrast between regions is shown by comparing the Della and Daralingie areas: both variables are considerably higher in the Della area than in the Daralingie area. An explanation may lie in the fact that the gas pay-sands at Daralingie are 200 m deeper than at Della, a difference of about 10%. The Nappamerrie Formation is thicker at Della but the pay-sands at Della are at the very top of the Toolachee Formation whereas at Daralingie the pay-sands are stratigraphically lower in the Patchawarra Formation. This means that the Daralingie pay-sands are overlain by 230 m of Permian and Triassic sediments, whereas Della has a thinner cap of 90 m of Triassic sediments. Perhaps of even more significance, at Daralingie, which is close to the southern limit of the Nappamerrie Formation, the Toolachee Formation has been flushed by invading water from the overlying Great Australian Basin sediments (Youngs, 1975).

The contrast in overall higher hydrocarbon levels at Della as against Daralingie implies that a single threshold will not apply throughout the basin. Areas within the basin will need to be interpreted independently of other areas taking into account all relevant geophysical and geological data.

First variable anomalies

In the Della area the appropriate empirical threshold value for the first variable (total hydrocarbon content) appears to be 5 ppb, whereas 3 ppb appears more appropriate to the Daralingie area (Fig. 2). Anomalies so defined in the Della area are nicely placed over the proven-probable limits of the Della field and the undrilled structures to the northeast and southwest of

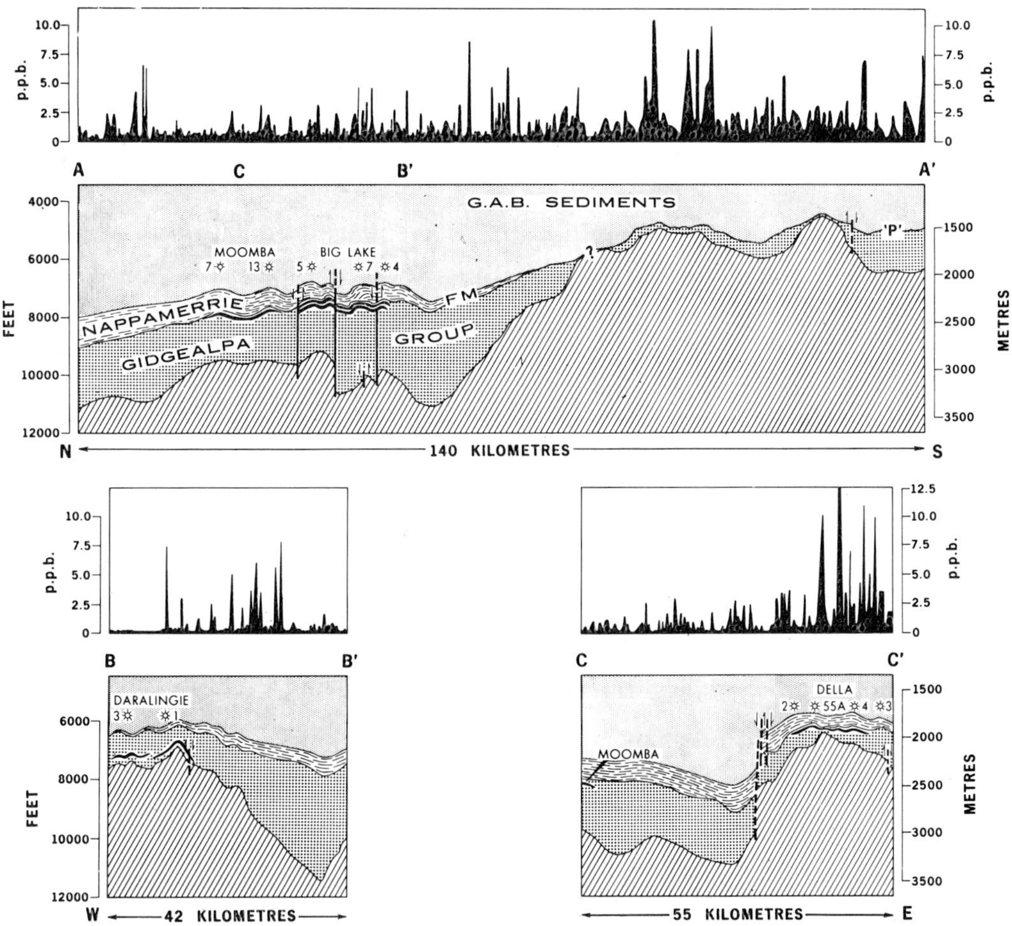

Fig. 4. Diagrammatic stratigraphic sections and soil hydrocarbon concentration profiles (sum of ethane, propane, butanes and pentanes).

the Della field. The two northernmost anomalies, outside possible field limits, are apparently spurious.

The southernmost anomaly in the Della area lies at the head of a "structural valley" defined by seismic structure contour maps. In the Daralingie area, the easternmost and southernmost anomalies are also at the head of "structural valleys". All three of these anomalies are on single lines and require confirmation by further sampling. The southernmost Della anomaly disappears when the hydrocarbon content is divided by the percent acid-soluble material (the second variable) but the Daralingie anomalies remain (Fig. 3).

The possibility that such apparently spurious anomalies rise from gas pools in sandstones which "pinch out" to the head of "structural valleys" to form hydrocarbon traps cannot be discounted. That type of trap has not yet been tested in the Cooper Basin. It is in prospecting for these more obscure traps that soil geochemistry could have its most valuable application in the Cooper Basin.

In the Daralingie area, the other first variable anomalies appear to be associated with gas pools. Two strong anomalies are displaced southwards from the large anticlinal area of proven-probable reserves and the northern one of the two anomalies is localized by the large northeast-trending fault. The undrilled anticlinal area of possible gas reserves to the north also has two associated anomalies, at least one of which lies above a fault. Neither the small western area of proven-probable reserves nor the area of possible reserves farther west in Daralingie show any associated first variable anomaly but a second variable anomaly is present over both areas.

The southwesterly displacement relative to the defined gas reserves of these anomalies in the Daralingie area may be attributed to the southwesterly hydrodynamic gradient in the overlying Great Australian Basin sediments. Displacement of anomalies at Della is not so obvious, but they may have been displaced relative to the point of origin of the hydrocarbon flux without transgressing the limits of the Della field. The need to interpret geochemical anomalies in relation to seismically defined structural information is obvious.

Second variable anomalies

By dividing all the values of the sum of hydrocarbons by the percentage of acid-soluble material to produce the second variable mapped, some advantages as well as ambiguities to interpretation are obtained (Fig. 3). Values of the second variable are higher at Della, on average, than at Daralingie as is the case with the first variable. Empirically interpreted anomalies in both the Della and Daralingie areas are more restricted than for the first variable. Anomalies in the Della area probably need to be restricted to zones containing more than 2 ppb/% to bear a meaningful relationship to the gas field outline and the undrilled structures of possible gas reserves. On this basis, two areally small anomalies occur above the proven-probable reserves of the Della field. Three anomalies lie around the margin of the undrilled culmination to the southwest but none occurs over the undrilled culmination to the northeast.

In the Daralingie area, values of the second variable as low as 1.0 ppb/% or even perhaps 0.5 ppb/% can be considered anomalous by their relationship to the field outlines. Taking the higher value as threshold, two single-point anomalies, and one double-point anomaly are placed over field limits but the two southernmost and the easternmost are located at the heads of "structural valleys" (Fig. 3).

A feature of the isoconcentration patterns in the Daralingie area for both the first and second variables is an apparent "halo" pattern. The concentration zones appear to form rings centred on the No. 2 well. In the second variable plot (Fig. 3), a second halo within the larger one surrounds the area of proven-probable reserves around the No. 3 well. The halo pattern has been suggested by some workers as the normal hydrocarbon anomaly to be sought around drilling prospects (e.g. Horvitz, 1939). It is difficult to delineate a halo pattern around the Della field although this may be because sampling may not extend far enough to the east and south. On the western and northern sides of the Della field and nearby undrilled anticlinal structures, the belt of values in the 1.0—2.0 ppb/% range could be part of an encompassing halo anomaly. With further sampling, we may find that anomalies above the fields and halo anomalies surrounding the fields are both legitimate anomalies. Thus, some of the anomalies which otherwise appear spurious may in fact be part of a halo.

Criteria for anomalies

The identification of anomalies among geochemical data is no less a problem with hydrocarbons than with trace metals. The regional effects discussed above in relation to Della and Daralingie suggest that a universal threshold cannot be applied throughout the Cooper Basin. In essence the value(s) to be chosen as threshold(s) has to be chosen on the basis of "whatever works". Our experimentation in the Cooper Basin to this point has been concerned with analysis and interpretations to determine what might work to assist petroleum exploration.

Soil hydrocarbon concentrations

The basic statistics of the hydrocarbon analyses for the 1070 sites sampled are shown in Table I. The range for the soil hydrocarbon analyses is wide and can be described as "extremely noisy". It should be noted here that values of analyses for our earlier 1973 survey (Devine and Sears, 1975) were erroneously reported as ppm instead of ppb. Reproducibility of analyses of aliquots from the same sample is good (correlation coefficients for the individual hydrocarbons lie between 0.82 and 0.90 for 21 samples, significant at the 0.01 level).

Cross plots between duplicate samples show a high degree of scatter although correlations are significant. For example, the correlation coefficient for the sum of hydrocarbons between 104 duplicate samples, each taken at 3 m depth from holes 1 m apart, is 0.618 (significant at the 0.01 level). The differences in the results for duplicate samples appear to be due to differences in the soil itself and not to the sampling technique used. Samples taken 1 m apart can be seen to be mineralogically different in the hand specimen. It is doubtful that the sampling technique needs to be improved over the coring technique used in this survey.

TABLE I

Summary of basic statistical values for analytical results from 1070 samples

	$C_2+C_3+C_4+C_5$ (ppb)	C_2 (ppb)	C_3 (ppb)	C_4 (ppb)	C_5 (ppb)	Acid-soluble material (%)	Moisture (%)	Days from sampling to analysis
Arithmetic mean	1.33	0.59	0.34	0.21	0.20	4.19	5.68	43
Geometric mean,								
loge value	−0.36	1.37	−1.79	−2.10	−2.14	0.72		
(numeric value)	(0.70)	(0.26)	(0.17)	(0.12)	(0.12)	(2.05)		
Standard deviation,								
loge value	1.02	1.20	1.06	0.90	0.89	1.24		
(numeric value)	(2.77)	(3.32)	(2.90)	(2.46)	(2.44)	(3.45)		
Highest value	37.0	18.9	10.9	4.6	3.8	62.8	25.70	56
Lowest value	<0.20	<0.05	<0.05	<0.05	<0.05	0.10	0.10	31
Correlation coefficients								
$C_2+C_3+C_4+C_5$		0.97	0.97	0.94	0.89	0.38		
C_2			0.93	0.87	0.79	0.37		
C_3				0.92	0.83	0.38		
C_4					0.88	0.36		
C_5						0.33		

The hydrocarbon gases in the soil tend to maintain a proportional relationship among themselves, i.e., a soil sample with a high ethane concentration tends to have a correspondingly high propane concentration, etc. The high correlation coefficients (Table I) attest to the relationship.

The hydrocarbon concentrations show an association with the acid-soluble content (correlation coefficients shown in Table I are statistically significant at 0.01 level). If this statistical association is symptomatic of a causative relationship, two possible causes may be advanced. The first hypothesis is that the acid-soluble material, mainly calcite and to a lesser extent gypsum, exerts some control on the hydrocarbon concentration of the soil. These are pedogenic minerals and it may be hypothesised that hydrocarbon gases escaping to the surface in a flux over a long period might become trapped in minerals forming the soil. This would mean that a fossil hydrocarbon flux which occurred at the time of the pedogenic mineral formation is being measured.

In our opinion, an equally likely explanation is that the gases are absorbed to the soil particles and/or occur in the soil air, with or without bias to certain minerals in the soil, and that the more vigorous and prolonged extraction reactions for samples with the larger contents of acid-soluble minerals more effectively scour the hydrocarbons from the soil.

The concentrations of hydrocarbons appear to decrease with storage time before analysis. The correlation coefficients between concentrations of the hydrocarbons and the number of days from sampling to analysis, taking all samples into consideration, are between −0.28 and −0.32. Although these correlations are low, they are statistically significant and suggest that in future surveys the time between sampling and analysis should be minimized and as far as possible standardized.

The concentrations of hydrocarbons do not increase with an increase in depth of the samples from 1.5 to 3 m. Although the mean concentration of the sum of hydrocarbons is greater at 3 m (0.96 ppb) than at 1.5 m (0.61 ppb) measured from 106 sample sites, the variability of the data is too great (standard deviation 1.45 and 0.52 respectively) to say that the concentrations are significantly higher in the deeper samples. The concentrations of the sum of hydrocarbons in the 1.5 and 3 m samples from the 106 sites show a positive correlation. The correlation coefficient of 0.38 is statistically significant (to 0.01 level). A corollary from these data appears to be that 1.5-m samples may suffice in place of 3-m samples.

Inorganic chemical analyses by hydrofluoric acid dissolution of 38 selected soil samples confirmed that those samples which contain higher proportions of carbonate also give off higher amounts of hydrocarbons (Table II).

The samples were selected to provide a full range of hydrocarbon values above the detection limit. The correlation coefficients of Table II imply that the main contribution to the acid-soluble content is calcium carbonate which contains minor amounts of Mg and trace quantities of P. The moisture content is an indirect and imprecise measure of the clay content of the soils. It

TABLE II

Correlation coefficients for soil hydrocarbon content and inorganic chemical analyses[1] of 38 selected samples

	$C_2+C_3+C_4+C_5$	Acid-soluble material	CaO	$CaCO_3$ [2]	$CaSO_4$ [3]	MgO	P_2O_5	SiO_2	Moisture
Arithmetic mean	7.0	15.7	10.4	15.5	4.9	0.7	0.07	64.9	7.5
Acid-soluble material	0.70								
CaO	0.66	0.88							
$CaCO_3$ [2]	0.70	0.87	0.96						
$CaSO_4$ [3]	(−0.18)	(0.01)	(0.12)	(−0.18)					
MgO	0.52	0.45	0.47	0.43	(0.13)				
P_2O_5	0.50	0.80	0.74	0.73	(0.01)	0.49			
SiO_2	−0.49	−0.75	−0.86	−0.72	−0.45	−0.65	−0.72		
Moisture	(0.32)	(0.26)	(0.28)	(0.17)	(0.33)	0.65	(0.26)	−5.7	

[1] Other analyses, none of which have a significant correlation with hydrocarbon content include: Al_2O_3, Fe_2O_3, FeO, K_2O, MnO, Na_2O, TiO_2, Ba, Cr, Cu, Pb, Rb, Sr, V, Zn.
[2] Evolved CO_2 calculated to $CaCO_3$.
[3] SO_4^{2-} calculated to $CaSO_4$.
[4] The threshold values for significance of the correlation coefficients where n = 38 are 0.413 at 0.01 level and 0.514 at 0.001 level. Correlation coefficients shown in brackets are not significant at 0.01 level.

shows no correlation with the hydrocarbon content of the soils. Al_2O_3 is another indirect index of clay content which also shows no correlation with hydrocarbons. The SiO_2 content confirms the low hydrocarbon yield from quartz-rich samples.

While these correlations appear to indicate that calcium carbonate is the source of hydrocarbons in the soils, the evidence is not conclusive. The acid treatment in the hydrocarbon extraction procedure dissolves all the carbonate but little of the gypsum, iron oxides, clay and quartz from the soil samples. The possible scouring effect from the CO_2 evolved from samples rich in carbonate has been discussed above.

CONCLUSIONS

Concentrations of soil hydrocarbons in the Cooper Basin have been observed in the range from below detection limit (0.2 ppb) to 37 ppb for the combined total value of the ethane, propane, butanes and pentanes contents. The values tend to be highest where the hydrocarbon-bearing sediments are shallowest and the cap rock is thinnest or absent. Thus, the soil hydrocarbon concentrations vary in a regional sense from one area of the Cooper Basin to another, e.g. from the Della to Daralingie area, suggesting that threshold limits for the interpretation of anomalies will vary between zones within the basin.

The range of soil hydrocarbon content is wide and makes interpretation of statistical results difficult. The variability in duplicate samples 1 m apart appears to be due to differences in soil mineralogy.

The most useful hydrocarbon variables to map are firstly, the sum of the concentrations of the four hydrocarbons and secondly, that sum divided by the percentage acid-soluble fraction. Empirically determined anomalies of both variables occur above the Della and Daralingie fields and over nearby anticlines which are likely to contain gas. Halo anomalies may also be interpreted around these fields.

Regional single-line anomalies occur above all fields traversed providing both variables are considered. This probably means that two or three lines over potential fields should show some indications of soil hydrocarbon anomalies, implying that a relatively widely spaced grid could find anomalies associated with gas pools in structural and stratigraphic traps. Such anomalies could then be confirmed by closely gridded sampling. It may be useful to follow-up closely gridded sampling with several sample holes of 200—300 m depth in search of a hydrocarbon gradient with depth before proceeding to expensive hydrocarbon test wells. This hypothesis remains to be tested.

Geochemical anomalies can only be useful when interpreted in conjunction with good-quality structural and stratigraphic information from seismic surveys. Spurious anomalies are likely to occur but should be recognizable in the context of the structural and stratigraphic information available.

ACKNOWLEDGEMENTS

It is a pleasure to acknowledge the assistance of R.C.N. Thornton in field supervision, D.R. Vinall in field supervision and office compilation, B. Frost and A. Francis in drafting and Dr. B. Leach in computer processing and statistics. Delhi International Oil Corporation and Santos Ltd. assisted with field logistics. This paper is published with the approval of the Director of Mines, South Australia. Responsibility for errors and shortcomings remains ours.

REFERENCES

Debnam, A.H., 1964. Determination of hydrocarbons in soils by gas chromatography. Geol. Surv. Can., Paper 64-15: 17 pp.

Devine, S.B. and Sears, H.W., 1975. An experiment in soil geochemical prospecting for petroleum, Della Gas Field, Cooper Basin. Aust. Pet. Explor. Assoc. J., 15: 103—110.

Horvitz, L., 1939. On geochemical prospecting, I. Geophysics, 14: 210—228.

Horvitz, L., 1972. Vegetation and geochemical prospecting for petroleum. Am. Assoc. Pet. Geol. Bull., 56: 925—940.

McCrossan, R.G., Ball, N.L. and Snowdon, L.R., 1972. An evaluation of surface geochemical prospecting for petroleum, Olds-Caroline area, Alberta. Geol. Surv. Can., Paper 71-31: 101 pp.

Stace, H.C.T. and Palm, A.W., 1962. A thin walled tube for core sampling of soils. Aust. J. Exp. Agric. Anim. Husb., 2: 238—241.

Youngs, B.C., 1975. The hydrology of the Gidgealpa Formation of the western and central Cooper Basin. Geol. Surv. S. Aust, Rep. Inv., 43: 35 pp.

Journal of Geochemical Exploration, 8(1977)415—429

INTEGRATED GEOCHEMICAL EXPLORATION FOR DEEP-SEATED SOLID AND GASEOUS MINERAL RESOURCES

H.J. RÖSLER, P. BEUGE, J. PILOT and G. TISCHENDORF

Section of Geosciences, Department of Geochemistry and Mineralogy, Bergakademie Freiberg, Freiberg (German Democratic Republic)

(Revised version accepted April 5, 1977)

ABSTRACT

Rösler, H.J., Beuge, P., Pilot, J. and Tischendorf, G., 1977. Integrated geochemical exploration for deep-seated solid and gaseous mineral resources. J. Geochem. Explor., 8: 415—429.

The location and qualitative evaluation of deep-seated deposits requires the application of various geochemical methods at different stages. Together with geological-tectonic analysis, the first step should be the lithogeochemical characterization of the geological environment (and of geologically comparable units) to obtain a good understanding of the genesis of the rocks and mineral deposits. In this way, one is able to estimate the ore-bearing potential of the region. During prospecting, a great variety of geochemical methods should be used. The analysis of mobile elements and isotopes (mercury, noble gases, hydrocarbons, radon) in conjunction with geophysical, especially geothermal, methods is needed. The application of a carefully chosen system of geochemical and geophysical methods is an essential factor for successful exploration, especially for hidden deposits.

INTRODUCTION

The German Democratic Republic is situated in the Variscan consolidated block of Central Europe. The area has a geologically complicated structure which has been studied closely for a long time. Consequently, most outcropping or thinly covered deposits have been found and some are already worked out. The present aim of geological reconnaissance work is to discover deep-seated deposits. Research is centred on two areas: (1) in the south, for minerals in the metamorphic Proterozoic and Lower Palaeozoic basement, and (2) in the north, for hydrocarbon deposits in thick Upper Palaeozoic and Mesozoic sediments.

Because of the complexity of the geology, it is necessary to emphasize genetic aspects of the rocks and minerals during exploration. In general, geochemical exploration can only be successful when based on new genetic models; consequently, detailed genetic studies have to be carried out.

Geochemical investigations are considered to be of major importance for

the first stage of exploration, particularly in previously unstudied areas. The investigations are of two types, with different aims:
(1) *Regional geochemical investigations* to determine the primary distribution of elements and the effects of geochemical specialization (analysis of a state).
(2) *Detailed investigations* to determine the systematics of element migration in specific geological units (analysis of a process).

ANALYTICAL TECHNIQUES

The various analytical techniques used in the course of the investigations described in this paper are summarized in Table I.

REGIONAL GEOCHEMICAL INVESTIGATIONS

This work is concerned with determining average geochemical values for comparable, large-sized geological units, for example within granitoids and metamorphic rocks, and with establishing standard geochemical sections in sediments. By means of such geochemical inventories, it should be possible to give certain rock complexes and regional units a so-called "geochemical passport". This provides an indication of the ore-bearing potential of the rock units and their suitability for geochemical prospecting. The granitoids of the G.D.R., for example, are characterized by a definite set of elements (Fig. 1) which provide a basis for subdivision into prospective and non-prospective for deposits of those elements genetically related to the intrusion process.

The geochemical specialization of the individual granitoid formations is Thus, the lineament-bound granitoids are enriched by magmatic supplies from the lower crust (areas *IV*, *V*) with the elements Sr, Ba, Cu, Ca and Pb. In the post-kinematic granites formed palingenetically in the highly metalliferous Erzgebirge anticline (area *VI*), Sn, Rb, F and Li were accumulated in an extraordinary way.

DETAILED INVESTIGATIONS

A number of investigations have been made to determine the systematics of element migration in certain geological formations and the behaviour of mobile elements along faults. Examples are given below.

Primary dispersion in igneous rocks

In the Saxonian Ore Mountains, the relationships between metasomatic Sn deposits of the greisen type and metallogenetically specialized granite bodies and their country rock are being studied. Since the country rocks are acid magmatites and metamorphites (porphyries, gneisses, mica schists) there

TABLE I

Analytical methods, summarized from Adamski (1976), Beuge (1976), Mayer (1977) and Pilot (1970)

	Mercury	Nitrogen	Fluorine	Inert gases
Sample material	soil, rocks	rocks	rocks	soil, rocks, soil air
Disintegration:				
soil	extraction with $H_2SO_4 + H_2O_2$	—	—	degassing in vacuum at 80° C
rock	thermal distillation at 900° C	melting at 1600° C in high vacuum (without flux)		melting at 1600° C in high vacuum (without flux)
Separation	aqueous solution or gold trap	eliminate: H_2O, CO_2 by cooling trap (−194° C); O_2 by passing through heated Cu + CuO + Ti-turnings	distillation with HCl in a water vapour stream	as for N, but without Ti-turnings
Concentration	as amalgam in gold trap		aqueous solution	Ar adsorption on activated charcoal
Determination	flameless AAS	mass spectrometer	colorometrically with eriochrom-cyamin-R	mass spectrometer

are few differences in petrophysical qualities — therefore, most geophysical exploration methods are not successful. For this reason, primary dispersion haloes around orebodies become particularly important in prospecting. As a start, their structure and extent in the vicinity of well-known deposits has been determined. Using methods of correlation, a certain sequence of elements was obtained (Fig. 2; Dahm, 1972). Each element takes a distinct space-time position within the temperature field of the country rock. This position is governed by the type and degree of migration of each element in the rocks. It is obvious that the position of the elements is normally not the result of a *unique* process of deposition, but the result of a *multistage* mineralization process. Nevertheless, knowledge of the laws of distribution is an important prerequisite for the interpretation of endogenous primary haloes.

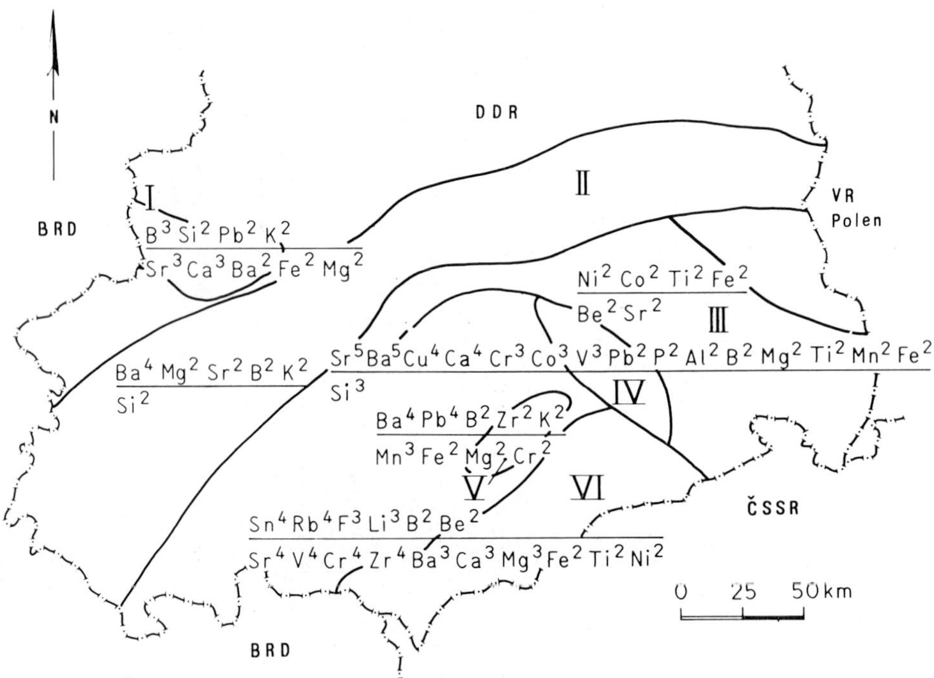

Fig. 1. Geochemical characteristics of granitoids in the southern part of the G.D.R. according to Tischendorf et al. (1972, 1978).

$\frac{Sn^4 \quad Rb^3}{Sr^5 \quad Ba^2}$ means: $\frac{\text{enriched elements in relation to the planetary granite average}}{\text{impoverished elements in relation to the planetary granite average}}$

The index is a measure of enrichment or impoverishment. *I*=Harz, syenogranites; *II*= Mitteldeutsche Schwelle, granodiorites; *III*=Lausitz, granodiorites; *IV*= Meissener Massiv, monzonites to monzogranites; *V*=Granulitgebirge, monzogranites; *VI*=Erzgebirge, metallogenetically specialized monzo- and syenogranites.

Metamorphic dispersion processes

Until recently, acid igneous rocks have been regarded as the source of the elements in epigenetic deposits, but now it is widely believed that they can also originate from the upper mantle or be mobilized by metamorphic processes. Despite many studies of metamorphic rocks, there are, so far, only few quantitative data on mobilization by metamorphism, with respect either to time or quantity of element release. In any case, many of the studies that have been done do not provide useful information since the original materials were of different lithologies. For this reason, we have studied an "artificial" metamorphic sequence consisting of eight metamorphic stages. Each metamorphic stage was represented by a composite consisting of five to ten random sub-samples derived from different localities (mainly in the G.D.R.) and ages but having the same diagenetic or metamorphic grade. Gyttja (organic-rich, anaerobic freshwater mud) was chosen as the facies of the original material, since it is the most frequent sediment found in a clay environment. The samples have been examined mineralogically and geochemically. The results for five elements in which mobilization occurs at different grades are shown in Fig. 3.

The elements gradually leave the rocks when subjected to metamorphic

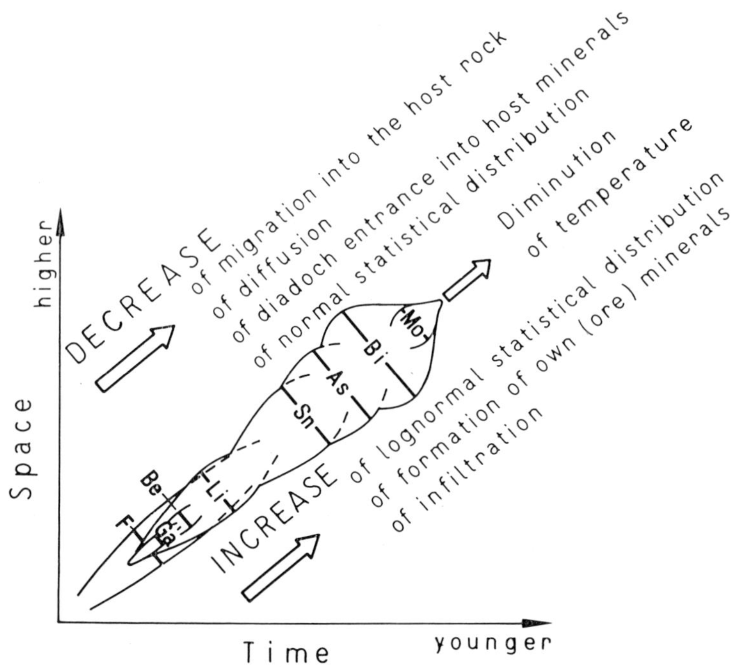

Fig. 2. Diagram to show the evolution of some elements in time and space in Sn greisen deposits of Saxonian Ore Mountains and their primary haloes. (After Dahm, 1972.)

stress. They are released either continuously (e.g. water) or more or less discontinuously (e.g. Hg). The manner in which the elements are held in rocks of different metamorphic grades varies considerably and breaking these bonds prior to mobilization will therefore be a very complex process. Nevertheless, given the same original material and the same metamorphic grade, a quantitative statement of the overall effect can be made, thus indicating at what stage and in what quantity an element may or must migrate from a rock. By means of such investigations, it is possible to fix certain geological fields as barren or prospectively deposit-bearing with respect to a definite element. Thus, as a sedimentary rock is only able to release Hg in large quantities by metamorphism up to the phyllite stage, this is a precondition for the formation of Hg deposits. The problem of the source of N in natural gases can be partially solved using similar considerations. It is possible that gas deposits in sediments that are underlain by high-grade metamorphic rocks have obtained their N primarily from these rocks. Rocks at a lower stage of metamorphism, such as phyllites, are unable to produce the essential quantitiy of N.

Dispersion in sedimentary rocks

Geochemical investigations on sedimentary rocks have shown some correlations between certain elements and the natural hydrocarbons (Rösler et al., 1977b).

Using Hg as an example, it is shown that the sapropelitic facies is the most active geochemical barrier in the Hg cycle and a strict correlation exists in the sediments between Hg and organic carbon (Fig. 4). Also, if compared with other elements such as V, Cr, Ni, Pb and Cu, Hg shows — relative to its

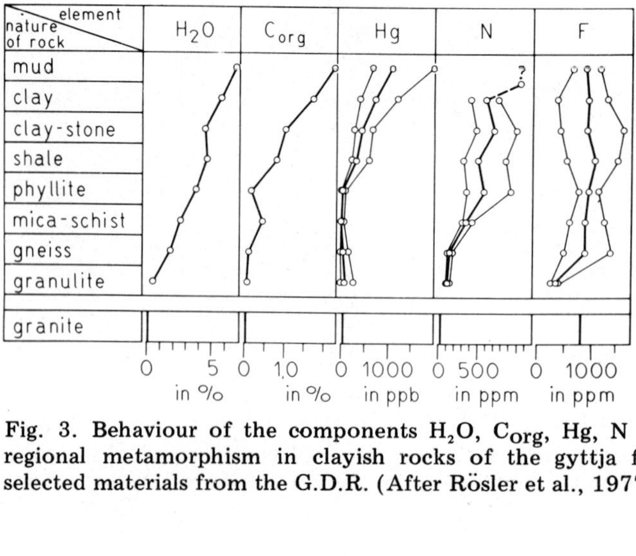

Fig. 3. Behaviour of the components H_2O, C_{org}, Hg, N and F during the progressive regional metamorphism in clayish rocks of the gyttja facies. Trends determined on selected materials from the G.D.R. (After Rösler et al., 1977a.)

clarke of 53 ppb (Beuge, 1976) — the strongest enrichment in organic substances (e.g. plankton, mineral oil, brown coal).

In the physicochemical conditions of coalification, the organometallic and adsorptive bonds of Hg are partly destroyed. Under these conditions, elementary Hg and hydrocarbons can migrate in a similar manner; they can be concentrated together in geological traps and in this way Hg becomes a good indicator of deposits of hydrocarbons, especially of natural gases. These findings are confirmed by numerous practical results.

Dispersion related to structural features (geochemistry of faults)

Migration from deep-seated deposits is only possible for a few mobile elements. The necessary prerequisites are distinct geothermal anomalies and a relatively high permeability in the rocks covering the deposits. Essentially, these elements move along tectonic inhomogeneities, so that geochemical analysis of these can therefore be of great importance. Hg, noble gases (He, Ar, Ne), Rn and, sometimes, hydrocarbons and CO_2, are known as good indicators of faults and lineaments. The migration behaviour of elements along faults is difficult to explain fully, because it depends on many factors, the essential ones being:

(1) The degree of openness or the effective permeability of the fault zone for diffusion, i.e. the relative rates of diffusion and free flow.

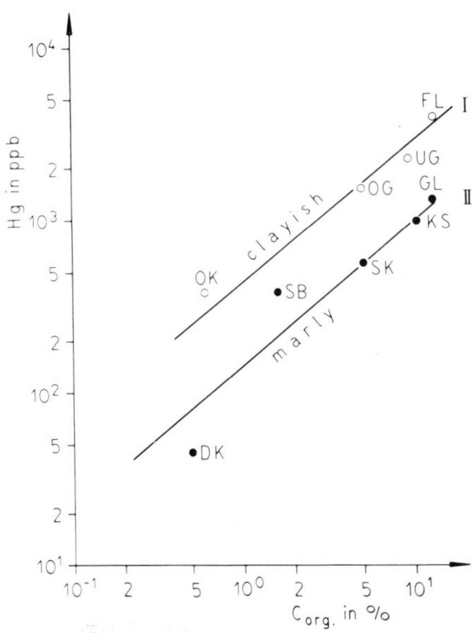

Fig. 4. Correlation of Hg with C_{org} in clay-rich (*I*) and marly (*II*) black shales, Thuringian Forest, G.D.R. Formation of copper shale, Lower Zechstein: DK = Dachklotz, SB = Schwarze Berge, SK = Schieferkopf, KS = Kammschale, GL = Grobe Lette, FL = Feine Lette; Formation of the Silurian graptolite shale: OG = Oberer Graptolithenschiefer, OK = Ockerkalk, UG = Unterer Graptolithenschiefer.

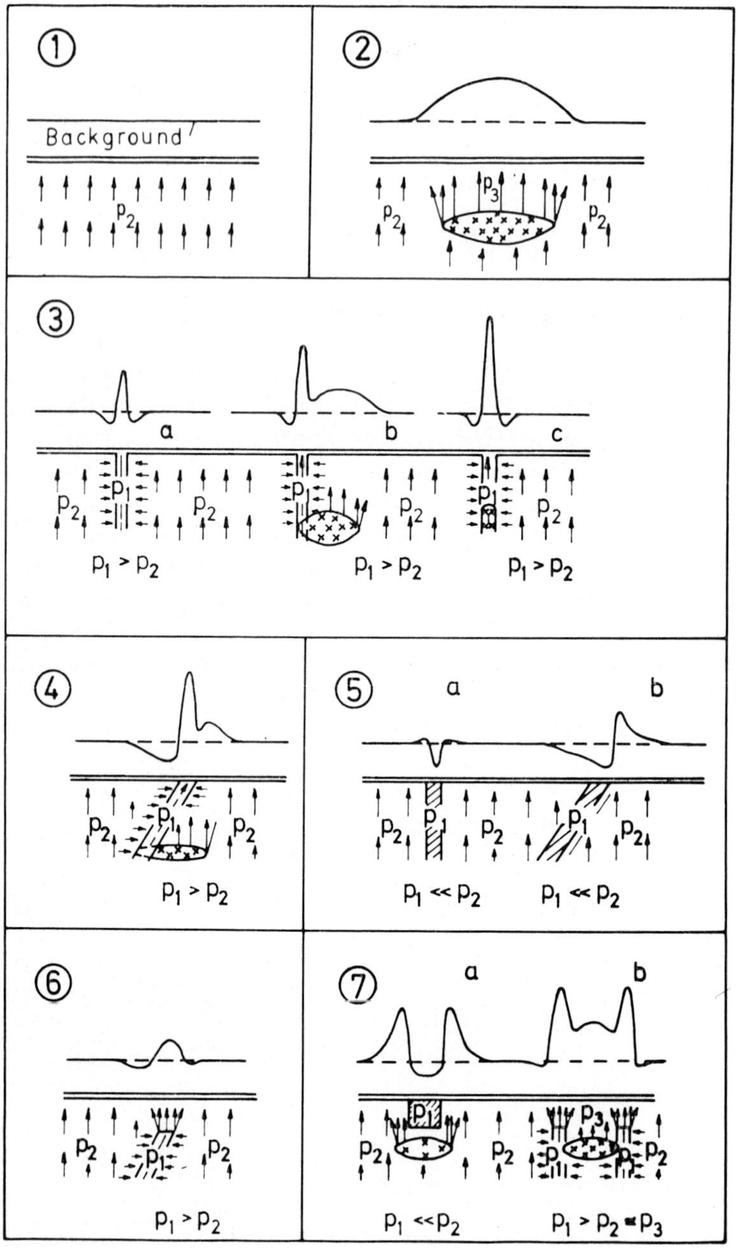

Fig. 5. Schematic presentation of the types of anomalies of mobile components in different geological conditions considering different diffusion permeabilities between faults and the country rock according to Mayer (1977). Diffusion permeabilities: p_1 = within the fault; p_2 = within country rock uninfluenced by a gas deposit; p_3 = within country rock influenced by a gas deposit.

(2) The difference in permeability between the fault zone and the country rock.
(3) The position of the element source to the fault zone.
(4) The geometrical-geological location of the fault zone.
(5) The geochemical character of the migrating elements.

The possible routes of migration are summarized in Fig. 5. Knowledge of the behaviour of elements in mineralized and non-mineralized faults, as well as in the country rock, is essential for the evaluation of anomalies. As an example the behaviour of Hg near a non-mineralized fault (A) and at a thermal contact (B) are compared in Fig. 6. The similarity in behaviour in both cases is remarkable and is probably due to similar thermal conditions (Oelsner et al., 1973).

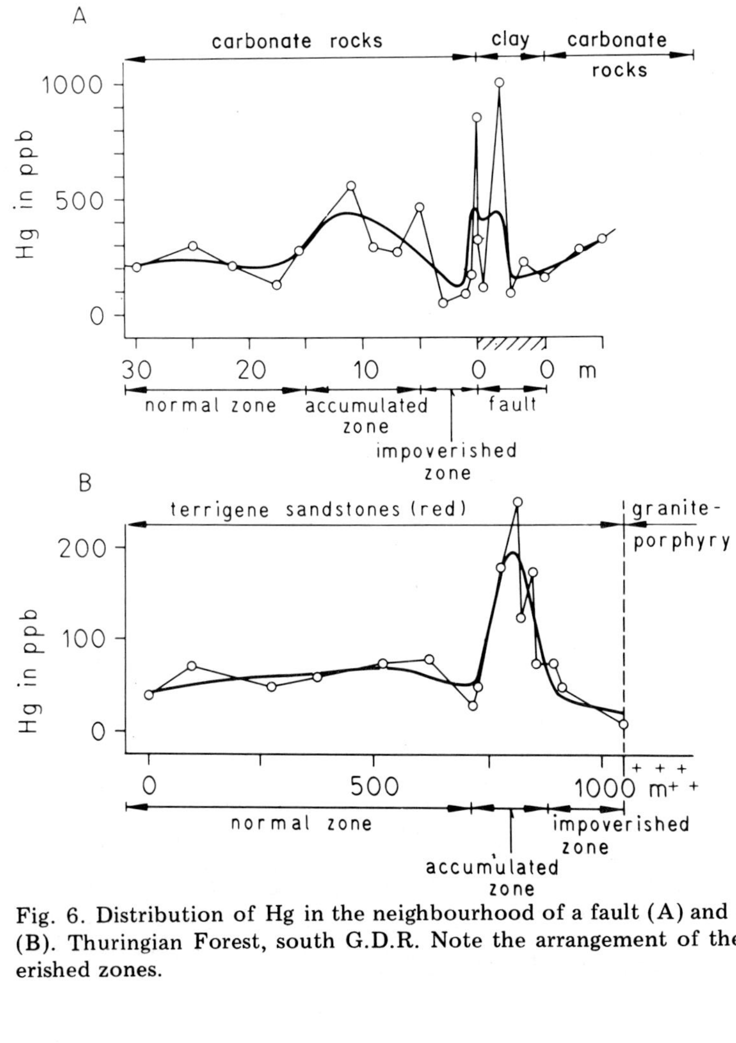

Fig. 6. Distribution of Hg in the neighbourhood of a fault (A) and of a magmatic contact (B). Thuringian Forest, south G.D.R. Note the arrangement of the enriched and impoverished zones.

APPLICATIONS TO PROSPECTING

As already indicated, one can distinguish two types of element migration when prospecting deep-seated deposits, namely: (1) elements moving mainly in a gaseous state (noble gases, N_2, Hg, Rn), or (2) elements moving mainly in uprising solutions (F, Cl, also Cu, Zn, etc.).

In both cases, the elements are extremely mobile, i.e., volatile or easily soluble. The recognition of the predominant form of migration is, however, difficult. Deposits are rarely indicated by only one element. Thus, in a given situation, it may be necessary to use different properties of several elements

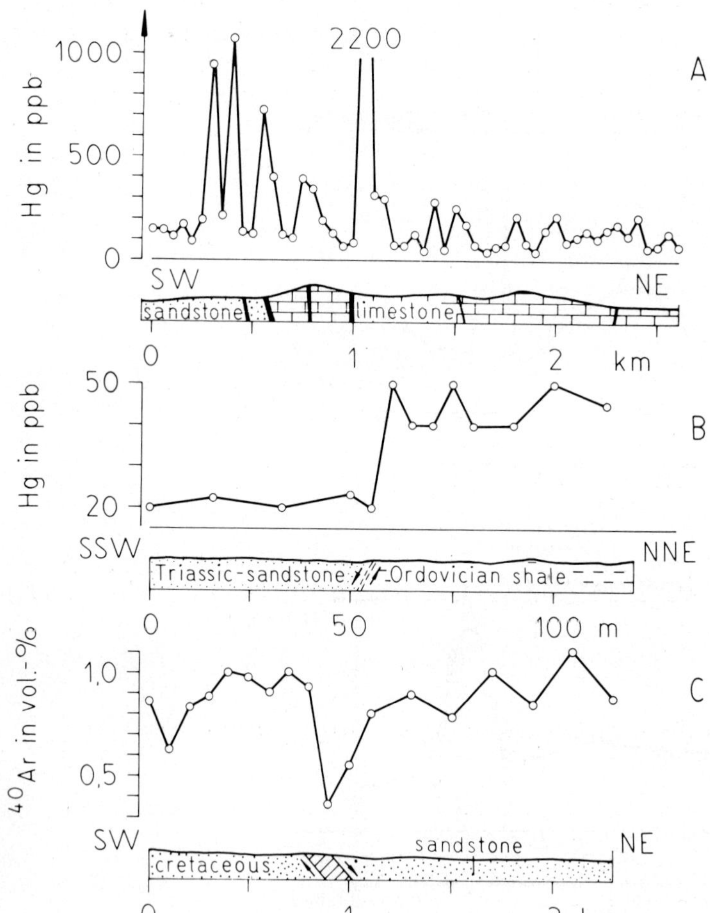

Fig. 7. Types of fault indications by means of Hg and noble gases. Thuringian Forest, south G.D.R. A. Positive anomaly in comparison with the regional clarke, Saxon Vogtland. B. Anomalies originated by different country rocks, Thuringian, Gera. C. Appearance of a minimum caused by an inclined fault, Magdeburg, G.D.R.

and of their isotopes, perhaps in combination with certain geophysical methods. The applicability of this approach is shown by the following examples.

Indication of faults

Faults may be located by Hg and noble gases, although the nature of the response may vary (Fig. 7). The absolute abundance of the respective elements indicating faults depends to a large extent on the regional geochemical situation. As shown in Table II, Hg-values above faults within and outside hydrocarbon-bearing deposits are very different.

The behaviour of noble gas isotopes (Ar, He, Ne) on expected faults in the basement are demonstrated in Fig. 8A,B. The isotopes of an element behave rather similarly. Isotope abundance ratios are suitable for the indication of faults by either positive or negative anomalies. It is remarkable that anomalies in the ratios do not indicate the fault itself but the more permeable adjacent country rock. Anomalies at faults within thick Mesozoic and part Upper Palaeozoic sediment series are even stronger because of the greater quantities of the elements released during diagenesis and low-grade metamorphism.

Fault systems in a salt diapir, identified geophysically, are indicated by both absolute isotope contents and isotope ratios of the noble gases (Fig. 9).

The migration behaviour of noble gas isotopes above faults in the basement and in sedimentary rocks is presented in a very generalized form in Fig. 10. The differences in the intensity of their migration are remarkable. Increased ^{40}Ar contents in sedimentary series are the consequence of present salt deposition.

Use of pathfinders

Additional information about type of mineralization, depth of formation, mineral paragenesis, etc., can be determined using elements with different

TABLE II

Comparison of Hg anomalies (in ppb) above fault zones inside and outside hydrocarbon-bearing deposits in European Lower Permian beds

	Position to the hydrocarbon-bearing field	
	inside	outside
Regional clarke	75	45
Average anomalous content	1800	380
Maximum anomalous content	4000	500
Relative to the regional clarke	25—55	8—12

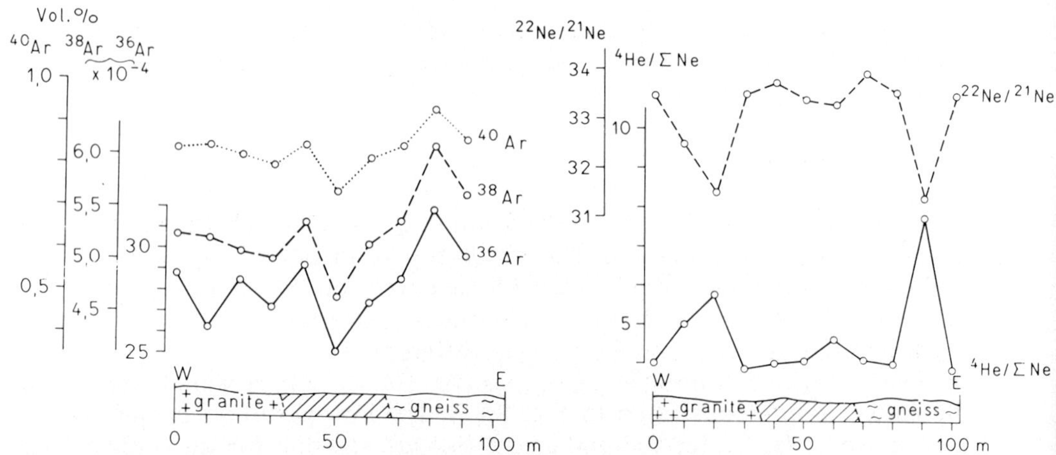

Fig. 8. Indication of a fault by the ratios of noble gas isotopes in the basement, Freiberg, Saxony. A. Ar isotopes B. He and Ne isotopes.

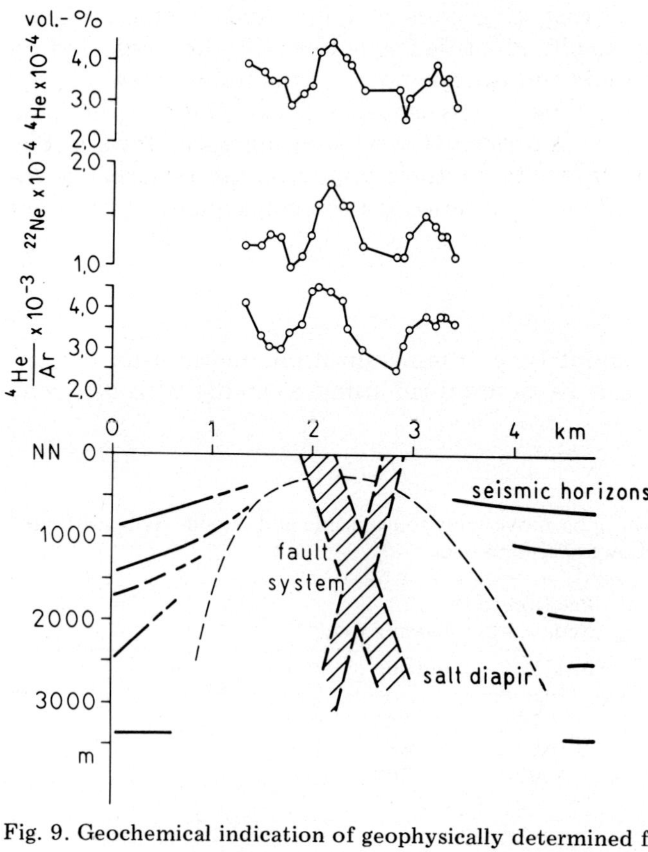

Fig. 9. Geochemical indication of geophysically determined faults in a salt diapir.

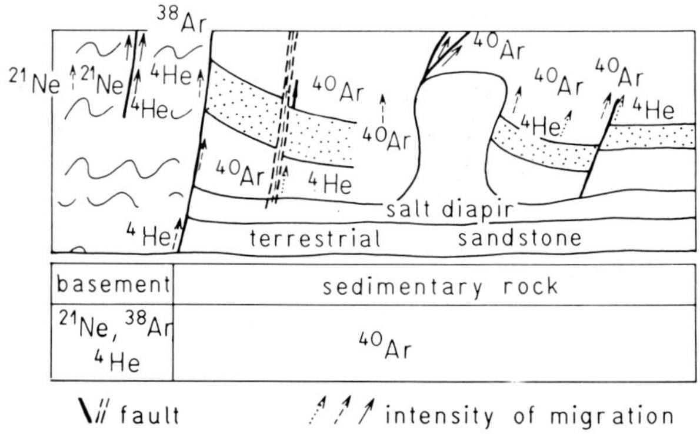

Fig. 10. Abundance of noble gas isotopes with different sources indicating faults.

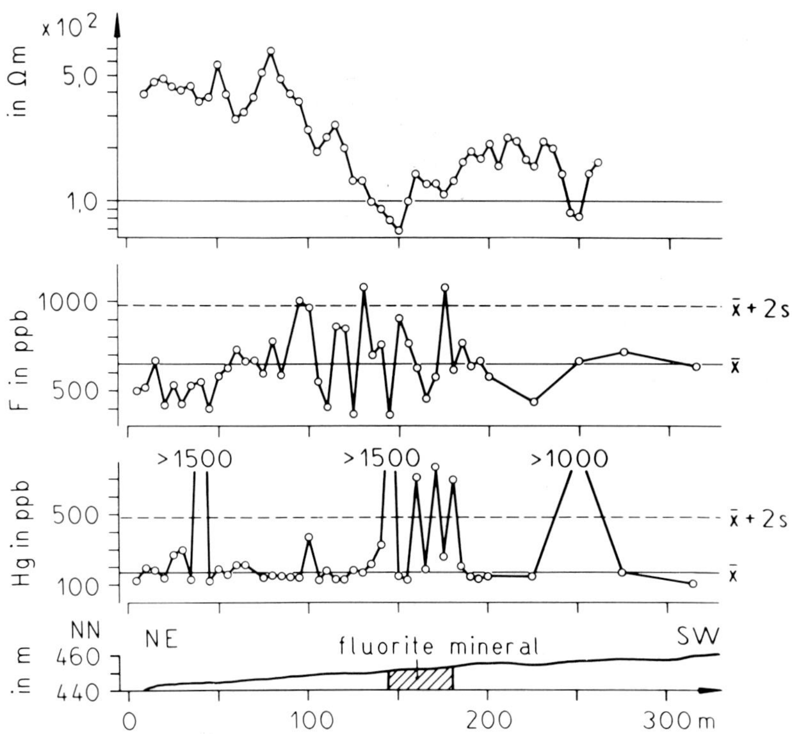

Fig. 11. Distribution of F (ppm) and Hg (ppb) in the soil above a fluorite deposit in combination with geophysical measurements of the resistivity, Vogtland, Saxony.

geochemical behaviour. During exploration for fluorite vein deposits, the results shown in Fig. 11 were obtained. Whereas the Hg and resistivity anomalies fix directly the position of the mineralized faults, increased F contents are present only in the soil downslope.

Relation between geochemical and geophysical features

The integration of geochemical and geophysical methods is very effective. Thus, prospecting activities for ore deposits should be planned principally as a combination of geological, geochemical and geophysical (especially radiometric and electrical) methods.

During exploration, the only indications of deep-seated deposits that can be expected at the surface are from very mobile elements and isotopes. These are accumulated under favourable geothermal conditions. Theoreti-

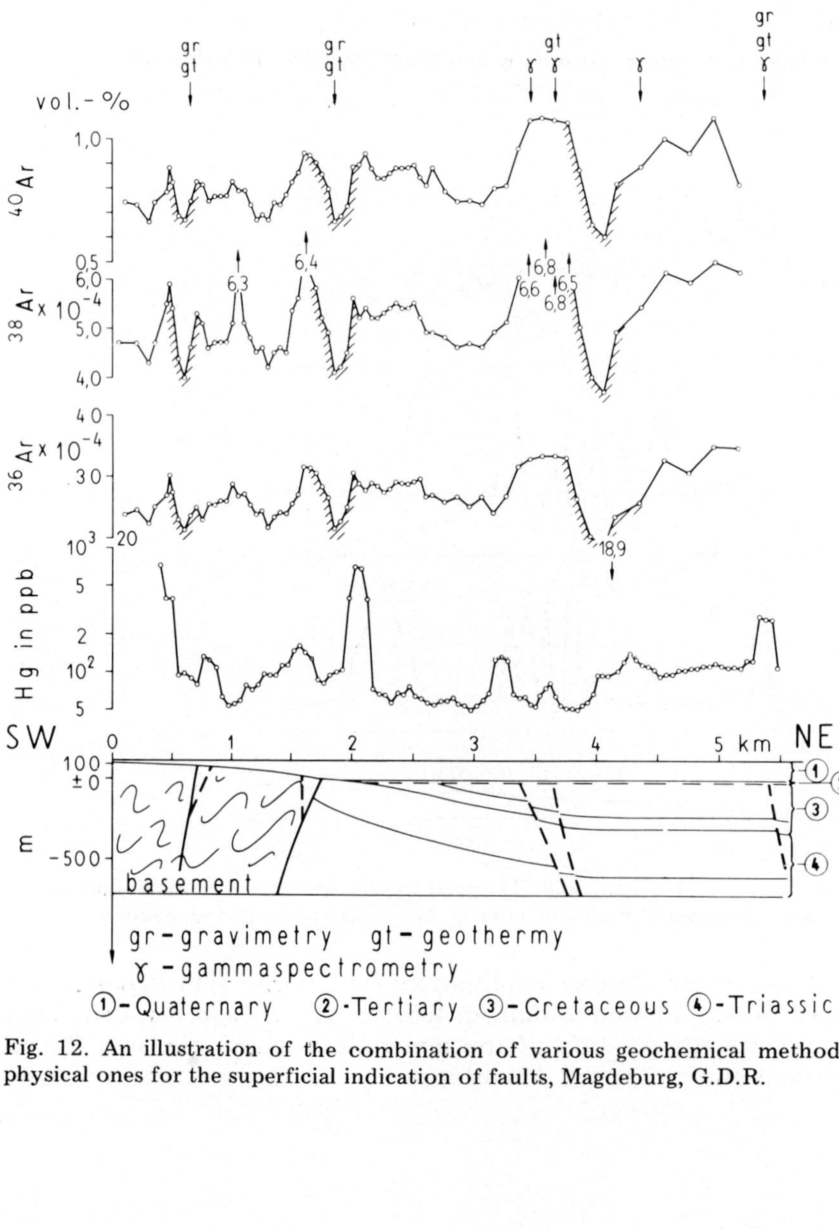

Fig. 12. An illustration of the combination of various geochemical methods with geophysical ones for the superficial indication of faults, Magdeburg, G.D.R.

cally, it may be expected that geothermal anomalies coincide with geochemical ones. The gaseous disintegration products of U and Th will migrate similarly. Therefore gamma spectrometric measurements provide equivalent indications. A summary of such combined prospecting activities for locating faults and finding deep-seated deposits is given in Fig. 12.

REFERENCES

Adamski, B., 1976. Ein Beitrag zur Geochemie des Stickstoffs. Dissertation, Bergakademie Freiberg, Freiberg (unpublished).

Beuge, P., 1974. Beitrag zur Geochemie des Quecksilbers unter besonderer Berücksichtigung des Territoriums der DDR. Dissertation, Bergakademie Freiberg, Freiberg (unpublished).

Beuge, P., 1976. Zur Geochemie des Quecksilbers in Magmatiten und Einzelmineralen. Freiberg Forschungsh. C 313.

Dahm, K.P., 1972. Gesetzmässigkeiten in der Ausbildung der primären Dispersionshöfe von Zinnlagerstätten im Erzgebirge. Geologie, 21: 521—535.

Mayer, K., 1977. Beiträge zur Geochemie der Edelgase, 1. Entstehung, Eigenschaften und Verteilung der Edelgase und ihrer Isotope. Z. Geol. Wiss. (in press).

Oelsner, Ch., Beuge, P., Mayer, K. and Pilot, J., 1973. Komplexer Störungsnachweis mit geochemischen und geophysikalischen Methoden. Neue Bergbautech., 3: 482—489.

Pilot, J., 1970. Isotopengeochemie. Situation, Konzeption, Entwicklung, Möglichkeiten. Freiberg Forschungsh. C 255.

Rösler, H.J., Beuge, P. and Adamski, B., 1977a. Das Verhalten chemischer Elemente bei der Diagenese und Metamorphose. Z. Angew. Geol. 23: 53—56.

Rösler, H.J., Beuge, P., Adamski, B. and Pilot, J., 1977b. Suche und Erkundung von Erdgas-Lagerstätten mit Hilfe nichtorganischer flüchtiger Elemente und Isotope. Z. Angew. Geol., 23: 261—267.

Tischendorf, G., Pälchen W. and Lange, H., 1972. Zur geochemischen Charakterisierung granitoider Gesteine. Geologie, 21: 554—569.

Tischendorf, G., Pälchen, W., Röllig, G., Lächelt, S. and Lange, H., 1978. Die Granitoidformationen im Südteil der DDR. Jahrb. Geol., 11/12 (in press).

Journal of Geochemical Exploration, 8(1977)431—456 431

CONTRASTING GEOCHEMICAL EXPRESSIONS OF COPPER MINERALIZATION AT NAMOSI, FIJI

M.D. LEGGO

AMAX Exploration (Australia), Inc., Suva (Fiji)

(Revised version accepted March 1, 1977)

ABSTRACT

Leggo, M.D., 1977. Contrasting geochemical expressions of copper mineralization at Namosi, Fiji. J. Geochem. Explor., 8: 431—456.

Three areas of Cu mineralization have been extensively diamond drilled at Namosi on the island of Viti Levu, Fiji. Two of these areas, Waisoi and Waivaka, were initially located by stream sediment geochemistry whilst the third, Wainabama, provided very little stream sediment geochemical response because of dilution by externally derived material. One-kilometre sample spacing at the first two prospects would produce two or three locations anomalous in either total Cu or cold-extractable Cu.

Despite similar topographic, climatic and geological settings, detailed soil and pit geochemistry was useful in siting drill holes at Waivaka, moderately useful at Wainabama, but of little value at the principal prospect of Waisoi. Two main factors are involved. Firstly, recent superficial cover of both alluvial and "mud avalanche" origin has obscured the surface geochemistry in places. Secondly, the geochemical expression varies as a function of pyrite content, its geometry relative to the Cu mineralization and the relationship of the sulphide mineralization to weathering and drainage. Cu, Pb, Zn, Mo, Au and pH determinations were used to reach these conclusions.

Magnetics and induced polarization showed a good correlation with hydrothermal magnetite and sulphides respectively at Waivaka and, to a lesser extent, at Wainabama, and were used to define and seek extensions of mineralization. A combination of these two geophysical techniques at Waisoi was useful in understanding the generalized geology, but was not directly applicable in the detailed definition of drill targets. Quartz veining density was the other principal quantitative approach used at the three prospects and this proved to be of considerable value.

INTRODUCTION

A large, low-grade copper sulphide system with many features typical of southwestern Pacific porphyry Cu deposits is currently being investigated in Fiji. The geology of this system is briefly outlined, the geochemistry as investigated so far is described, and other exploration techniques employed in the programme are mentioned.

Special Prospecting Licence 1014 was granted to Central Mining Finance (a subsidiary of Australian Anglo American Limited, hereafter referred to as Anglo) in 1968 and originally covered an area of 694 km^2. This has subsequently been reduced to an area of 194 km^2. In 1972 R.S.T. (Fiji) Inc. (with AMAX Exploration (Australia) Inc., hereafter referred to as AMAX, acting as its agent) entered into a joint venture agreement with Anglo and are the current managers of the exploration programme.

DESCRIPTION OF AREA

Location and climate

The prospecting area is located in the headwaters of the Waidina River in the province of Namosi, about 30 km WNW of Suva on the island of Viti Levu, Fiji (Fig. 1). Except for the principal river valley, the area is uninhabited, undeveloped and covered by dense tropical rainforest. Elevations range from less than 100 m above sea level to approximately 1200 m. Annual rainfall exceeds 500 cm. Average monthly maximum and minimum temperatures range from 26 to 33°C and from 18 to 22°C respectively. The area has been without vehicle access, but a road is now being established.

Regional geology

In the Namosi area, Eocene to Lower Miocene basic meta-agglomerates and volcanic flows of the Wainimala Group are unconformably overlain by Upper Miocene-Pliocene andesites, agglomerates and conglomerates of the Namosi Andesite Formation (Rodda and Band, 1967; Gill, 1970). The unconformity is marked by a basal conglomerate 15—60 m thick. The area as a whole represents one of a number of Upper Miocene centres of andesitic volcanicity (Branch, 1972). The Wainimala rocks have been affected by deformation and greenschist facies regional metamorphism. Both the Wainimala and the Namosi rocks have been contact metamorphosed and hydrothermally altered in the vicinity of later intrusives. Quartz diorite porphyries have intruded the Wainimala and Namosi volcanic and volcaniclastic rocks in the Namosi area and these often have associated copper sulphide mineralization. To date, three of the prospects located, Waisoi, Waivaka and Wainabama (Fig. 1), have been diamond drilled.

Waisoi deposit

Geology. The Waisoi deposit, located on Waisoi Creek, is in plan a large oval-shaped system of overlapping zones of intrusion, alteration and mineralization (Fig. 2). So far, eighty-one diamond drill holes have been completed on this prospect revealing two contiguous centres of mineralization, Waisoi West and Waisoi East, separated by a low-grade zone which appears to be structurally controlled by a major SW-NE-trending fault system.

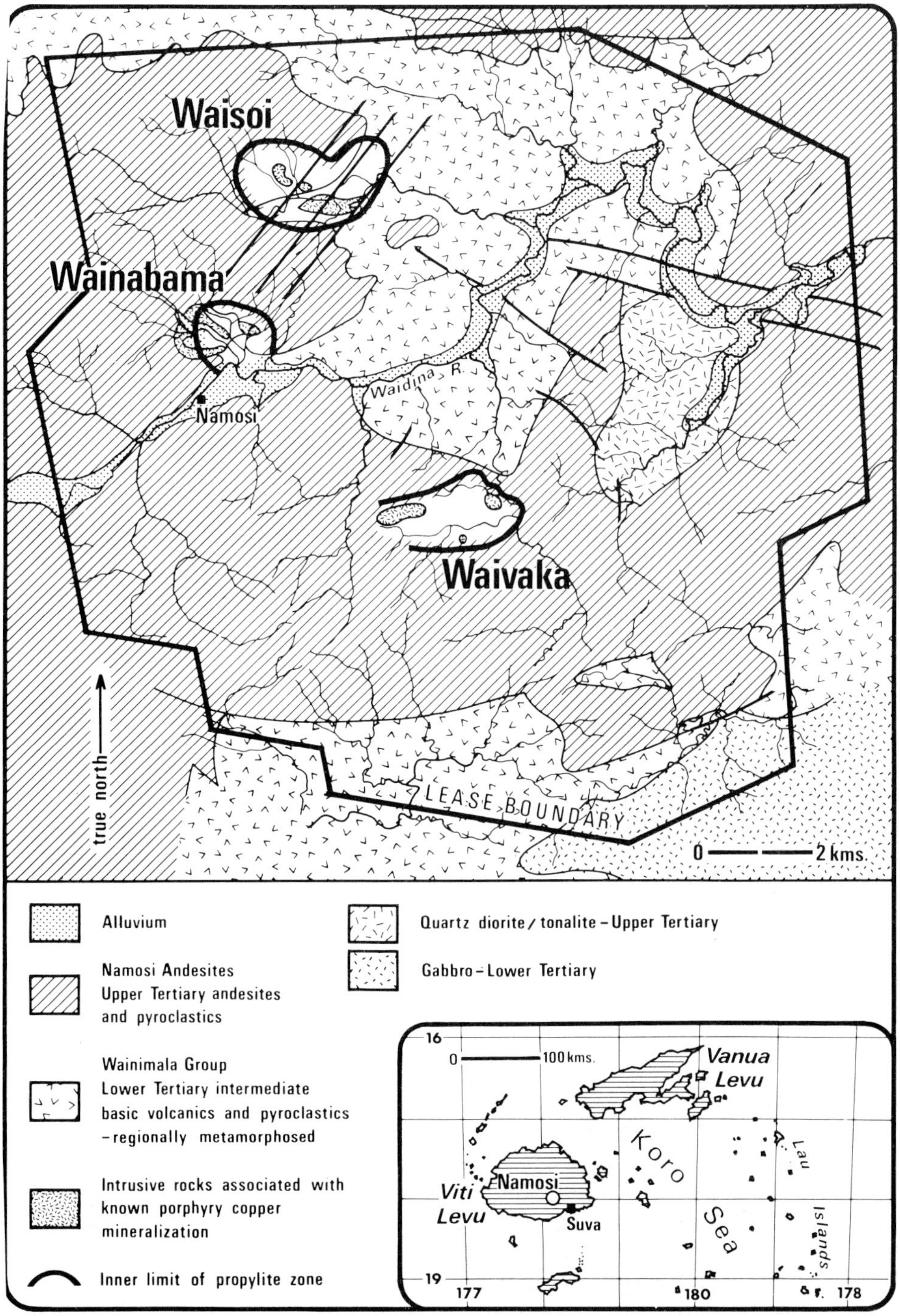

Fig. 1. Location map and generalized geology, S.P.L. 1014, Namosi, Fiji.

Three distinct intrusive types are currently recognized in the drilled area at Waisoi. These are known as the P1 and P2 phases of quartz diorite porphyry, and a quartz porphyry (quartz diorite—diorite porphyry, QP). The P1 phase of the quartz diorite porphyry does not outcrop. Hammond and McKay (1976) record that a number of distinct hydrothermal alteration assemblages occur in the intrusives and wall rocks at Waisoi. Pervasive alteration affects the intrusives in mineralized areas and chlorite-clay-sericite assemblages are developed; in addition, the central portion of the quartz porphyry in Waisoi East has been altered to a quartz-sericite-clay assemblage with associated pyrite. In the wall rocks, the outer alteration at Waisoi is represented by a well-developed and distinctive propylite zone. This is defined by the presence of large amounts of disseminated and fracture-controlled epidote with associated chlorite, calcite and pyrite. A chlorite-sericite alteration assemblage is well developed in wall rock inside the propylite zone and is usually seen in the form of ragged blotches or clots or, less commonly, as narrow selvages adjacent to fractures carrying chalcopyrite-pyrite mineralization. Towards porphyry contacts (P1 and QP), the chlorite-sericite assemblage grades laterally into a biotite-chlorite-sericite assemblage. Chalcopyrite and bornite are the most common sulphides associated with this type of alteration. A strongly foliated biotite-quartz-sericite assemblage has also been noted immediately adjacent to the quartz porphyry. Chalcopyrite, pyrite and bornite in about equal proportions, plus lesser molybdenite, are the sulphide minerals associated with this type of alteration. Superimposed on all alteration types is a clay-chlorite-calcite assemblage associated with steep fault zones, shears and fractures; pyrite is usually abundant, sphalerite and less commonly galena occur, but there is little or no copper sulphide content.

Strong lateral and vertical zonation of primary sulphides is evident at Waisoi. Passing inwards from the strongly pyritic (over 3% by volume) propylitic alteration halo, chalcopyrite becomes dominant over pyrite, traces of bornite appear and the total sulphide content decreases to 1—2% by volume. The inner mineralization on the western side of the Waisoi prospect is marked in places by dominant bornite, in which case pyrite is rarely present. The copper sulphides and pyrite occur as disseminations in intrusives and wallrocks, as dry fracture coatings and in quartz veins.

Oxidation and weathering. Depth of complete oxidation at Waisoi varies from surface to a usual maximum of about 30—80 m below crests of spurs. Partial oxidation can occur to considerable depths (300 m) in deeply fractured rocks. Drill hole data suggest that, on the north side of Waisoi Creek, the base of oxidation reflects an old planar weathering surface dipping fairly gently to the southeast. Oxidation and weathering have been so thorough that the present topography appears to have little effect on the morphology of the base of oxidation. Cu minerals in the oxidized zone include neotocite, cuprite, malachite, chrysocolla and native Cu. Supergene Cu minerals are chalcocite and trace covellite; these occur most commonly in Waisoi East.

Waivaka deposit

The Waivaka deposit, on Waivaka Creek, has an alteration/sulphide system comparable in areal extent with Waisoi (Figs. 2 and 3). However, within this zone, six small concentrations of Cu occur closely associated with quartz veining and in a close spatial relationship with small, discrete, quartz diorite porphyry intrusions. Quartz veining and biotite—K-feldspar alteration are locally intense, but decline very rapidly, resulting in several small, relatively high-grade deposits. The Cu mineralization and zones of high-grade hydrothermal alteration are contained within an indistinct propylite halo and the Waivaka system is best regarded at the present level of erosion as separate, telescoped, partially overlapping, mineralized and altered centres. Nineteen holes have been diamond drilled at Waivaka.

D.P. Taylor (1975) has recorded that mineralization consists of chalcopyrite, bornite, pyrite, magnetite and rare traces of molybdenite, sphalerite and haematite. Chalcopyrite is consistently the most abundant mineral, after magnetite, in the areas of economic interest. Bornite occurs erratically; generally, it is most common in areas of the most intense veining. Cu mineralization is found almost entirely in quartz veins, on fractures and as disseminations from veins and fractures into both porphyry and volcanics. In areas of intense veining and mineralization, magnetite is abundant along vein margins and in the veins themselves. Pyrite is rare in mineralized areas, being either a low percentage vein or fracture mineral, or absent.

Wainabama deposit

At Wainabama (Faga, 1976), where fifteen diamond drill holes have been bored, a larger multi-phase intrusion exhibits pervasively disseminated biotite-chlorite and minor localized K-feldspar alteration with low-grade associated Cu mineralization; local higher-grade Cu sulphides are associated with increased quartz vein density and these are related to a particular intrusive porphyry phase (Fig. 12). Chalcopyrite, bornite and pyrite are the dominant sulphides observed although some minor molybdenite is seen in places. Sulphides occur as vein fillings, fracture coatings and disseminations. Pyrite occurrence is mainly restricted to fault structures. Magnetite is well developed along quartz vein margins and also as thin monomineralic stringers.

Superficial cover

Significant superficial cover of recent origin occurs at both the Waisoi prospect (where it is termed "boulder terrace") and the Waivaka prospect ("landslip deposits") where stabilized "mud avalanche" deposits of andesitic material occur. These rock units are composed of epiclastic/pyroclastic rocks shed from the retreating inner escarpment of the remnant volcanic apron deposits located to the north of Waisoi (Korobasabasaga Range) and the south

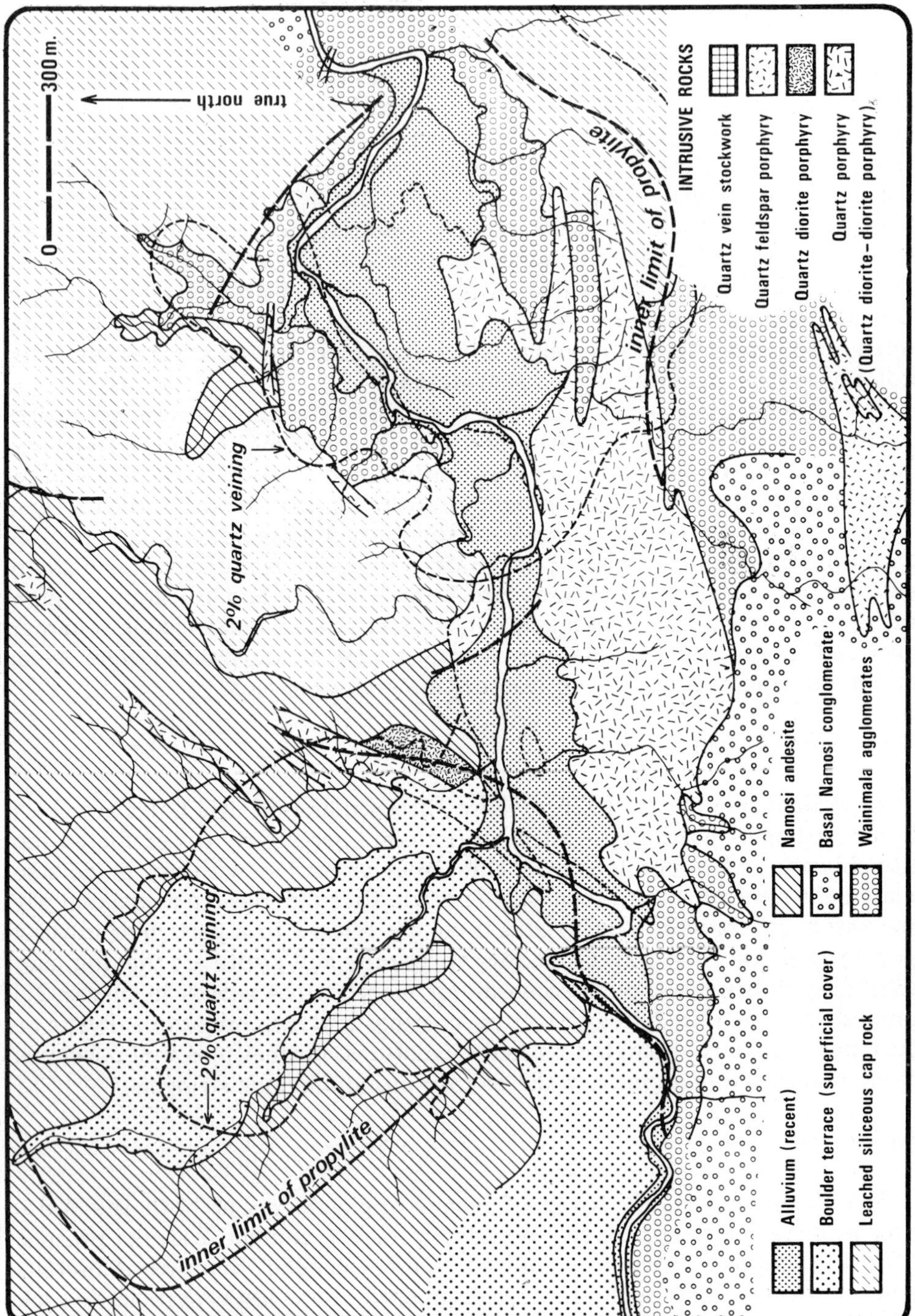

Fig. 2. Geological map and quartz veining, Waisoi prospect.

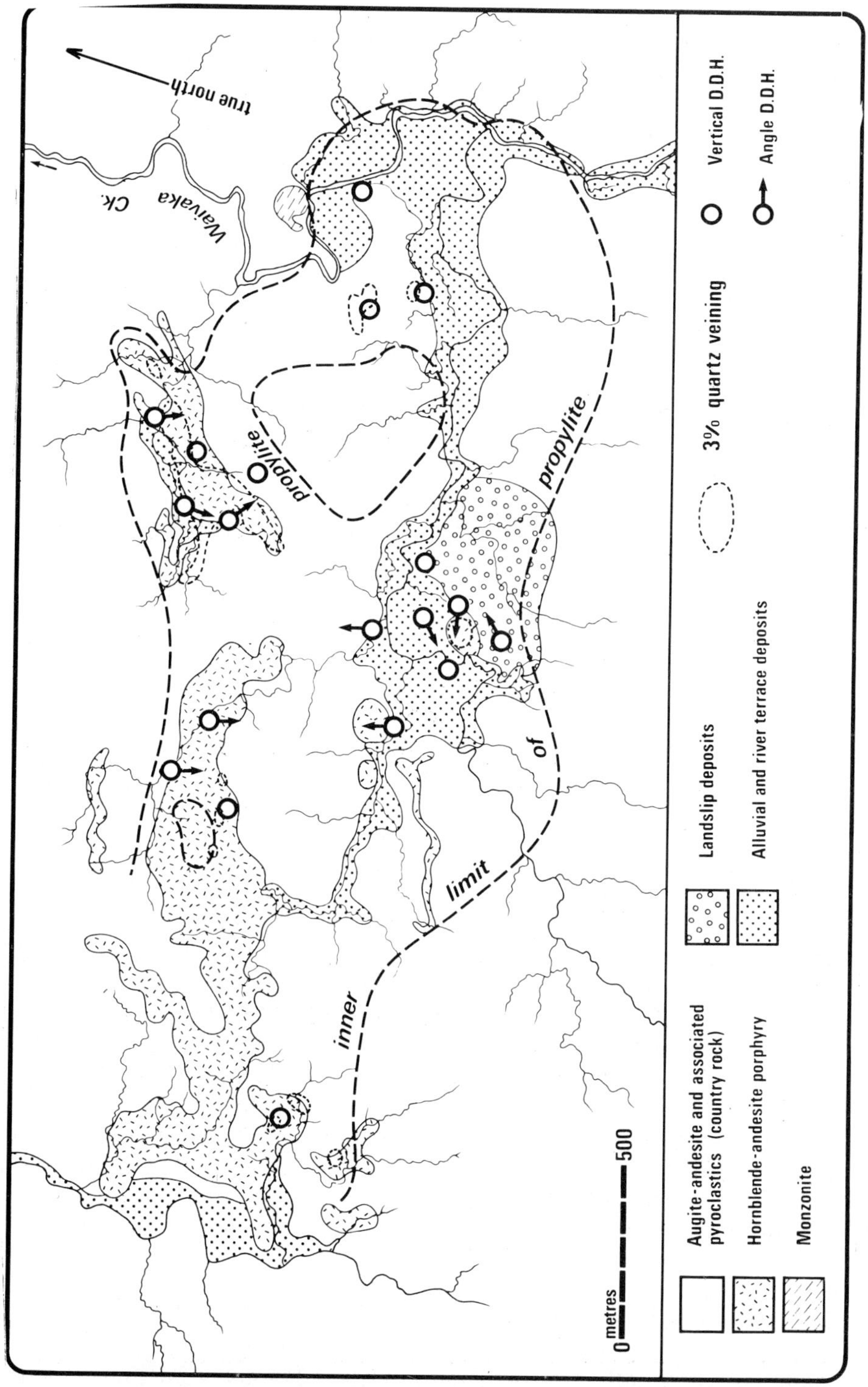

Fig. 3. Geological map and quartz veining, Waivaka prospect.

of Waivaka. Other superficial deposits include alluvial cover (e.g. at Wainabama) and minor landslip cover.

Soils

The soils of the Namosi area, using local terminology (Twyford and Wright, 1965), are mainly steepland soils formed under a tropical climate with no dry season. Two main soil types are developed: firstly, humic latosols derived from parent material of intermediate composition, usually stoney and bouldery clays; secondly, red yellow podzolic soils derived from parent materials of acidic composition and varying from bouldery clays to stoney sandy clays to stoney silty clay loams. An anomalous profile occurs around Waisoi East consisting of pale, severely leached, crumbly to porcellanous cap rock. Hammond and McKay (1976) have shown that this capping corresponds with present topographic highs marginal to the core of copper sulphide mineralization and it is interpreted as being derived from strongly pyritic wall rocks. Limonitic gossans are common and original rock textures are rarely preserved in this profile.

SAMPLING AND ANALYTICAL METHODS

Stream sediments

The early stream sediment work (1968-69) by Anglo had a sample density of 2.4 samples/km^2. Significant tributaries to main drainages were sampled, resulting in approximate sampling intervals of 150 m along the main streams. The minus 0.2-mm (80-mesh) material was analyzed for Cu, Co, Ni, Pb, Zn and cold-extractable Cu (cxCu). Follow-up stream sediment sampling and analysis for Cu, Co, Ni, Pb and Zn but not cxCu, was also performed by Anglo in selected areas using a sampling density of approximately 9/km^2. Corresponding base of slope samples were taken at a depth of about 50—100 cm. Analytical methods used were perchloric acid/atomic absorption spectrometry (AAS) for Cu, Co, Ni, Pb and Zn and citric acid/dithizone papers for cxCu. Re-sampling and analysis were undertaken by AMAX in 1972 at the Waisoi prospect. Initially, nine streams were re-examined and the minus 0.2-mm fraction was analyzed as follows; Cu, Pb, Zn by $HClO_4$/AAS; cxCu using a thick slurry with 0.5*M* HCl; Mo by fusion/AAS; pH on a thixotropic soil mixture prepared with boiled de-ionized water. Recently Mo distribution at Waisoi has been re-studied and stream sediment Au contents have been determined. Mo was analyzed by $HClO_4$/HNO_3/AAS; Au by aqua regia/MIBK/AAS.

Ridge and spur and grid soil sampling

Soil samples were collected by Anglo along ridge and spur lines cut upslope to a major control ridge line. Sample interval was approximately 62 m along lines and sample depth ranged from 20 to 40 cm.

Detailed soil sampling was undertaken by Anglo at a number of anomalous areas, the two most significant being Waisoi and Waivaka. This was carried out along chain and compass cut lines from a surveyed base line on a 400 ft × 50 ft (123 m × 15 m) grid closing down to 100 ft × 50 ft (31 m × 15 m) in places. The minus 0.2-mm soil fraction was analyzed for Cu only by $HClO_4$/AAS.

Analyses of core composites from early diamond drilling at Waisoi led AMAX to use Au and Mo in addition to Cu, Pb and Zn in a second soil sampling programme (46 cm depth) in an attempt to refine potential drilling targets. Grid spacing employed was 123 m × 31 m.

The close spatial association of the andesine porphyry intrusives with mineralization at Waivaka led to the idea of distinguishing between the porphyries and andesites in the often deeply weathered soil profiles at Waivaka by using the Ni-Co fractionation index (S.R. Taylor, 1966). Analyses of appropriate core samples showed, however, that a distinction could not be made on this basis in view of the low levels of Ni and Co and limited variation in concentration. Core composites and some pits from Waivaka have low Mo contents (mostly <15 ppm; maximum for a 15-m interval, 58 ppm) and this suggested that analyses for Mo in surface samples would have been of little assistance.

Pits

Pits were dug at both the Waivaka prospect, where over 200 pits were channel sampled, and Waisoi prospect, where 100 pits were channel sampled and analyzed. Samples were collected over a combination of set intervals and intervals related to soil horizons. Maximum depth rarely exceeded 10 m.

ADDITIONAL EXPLORATION METHODS EMPLOYED

Helicopter magnetometer (Barringer M104 proton precession magnetometer), ground magnetometer (M-700 McPhar fluxgate magnetometer) and induced polarization/resistivity surveys (time domain equipment, pole-dipole array) have been conducted over portions of S.P.L. 1014 and covered the Waisoi, Waivaka and Wainabama prospects in some detail.

In addition, following the work of Fountain (1972), quartz veining density expressed as vol.% was estimated semi-quantitatively in outcrops and pits wherever possible and these data were delineated using 2%, 5% and 10% contours.

RESULTS AND INTERPRETATION

Stream sediments

The Cu threshold for the entire area was estimated statistically as 180 ppm using a probability graph and histograms, but the value of 230 ppm was used as the threshold owing to uncertainty about the nature of the distribution of Cu values. This is approximately lognormal although there may be, in fact, four overlapping populations. S.R. Yardley (written communication, 1969) using histograms, estimated background values for Cu and Zn for the six main rock units within the original prospect area (Table I).

Values up to 1400 ppm Cu and >100 ppm cxCu were recorded at Waisoi and 1300 ppm and >100 ppm respectively at Waivaka. Background for cxCu is ⩽10 ppm and the threshold is approximately 20 ppm. Using a cumulative frequency plot, the threshold value for reconnaissance Zn values was established as 350 ppm. According to S.R. Yardley (written communication, 1969), the regional values for Ni, Co and Pb remained constantly low and fluctuated little.

Reconnaissance and follow-up stream sediment geochemistry by Anglo successfully located and largely defined four prospects, Waisoi, Waivaka, Waisomo and Wainimolaunia. The latter two prospects did not reach the diamond drilling stage and are not discussed further.

Values of 250 ppm total Cu and 50 ppm cxCu were recorded for a location in Waisoi Creek at the junction with the Waidina River, 3 km below the Waisoi prospect. One-kilometre sample spacing using either tributaries or the main stream (Waisoi Creek) would have produced two to three anomalous locations (Fig. 4). Sampling of tributaries to Waisoi Creek at 300-m intervals would have clearly indicated the anomalous zone. This is in contrast to the lack of stream sediment geochemical expression at Wainabama (Fig. 4).

Stream sediments from Waisoi may contain up to 0.35 ppm Au, but the lack of precision of the data, as established by duplicate analyses, precludes its usefulness other than as a general indicator element for this area.

TABLE I

Background stream sediment values for Cu and Zn for various rock units in the original licence area

		Cu (ppm)	Zn (ppm)
Wainimala Group		55	100
Mendrausuthu Group	Namosi Andesites	80	120
(Suva Series)	Sediments	60	100
Tonalite/granodiorite plutonics		15	50
Vango volcanics		40	150
Gabbro intrusive		75	100

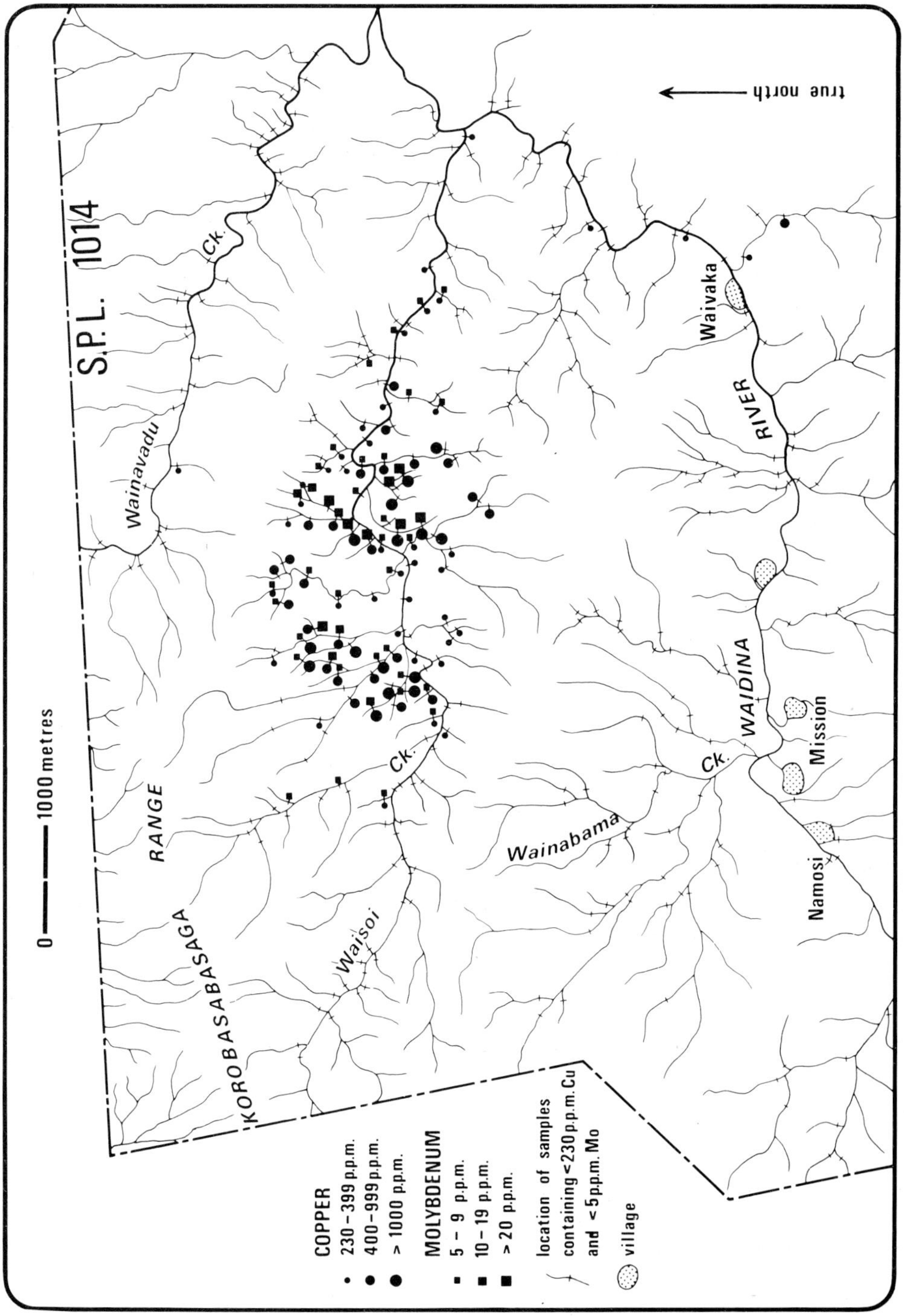

Fig. 4. Stream sediment Cu and Mo values for the Waisoi and Wainabama prospects.

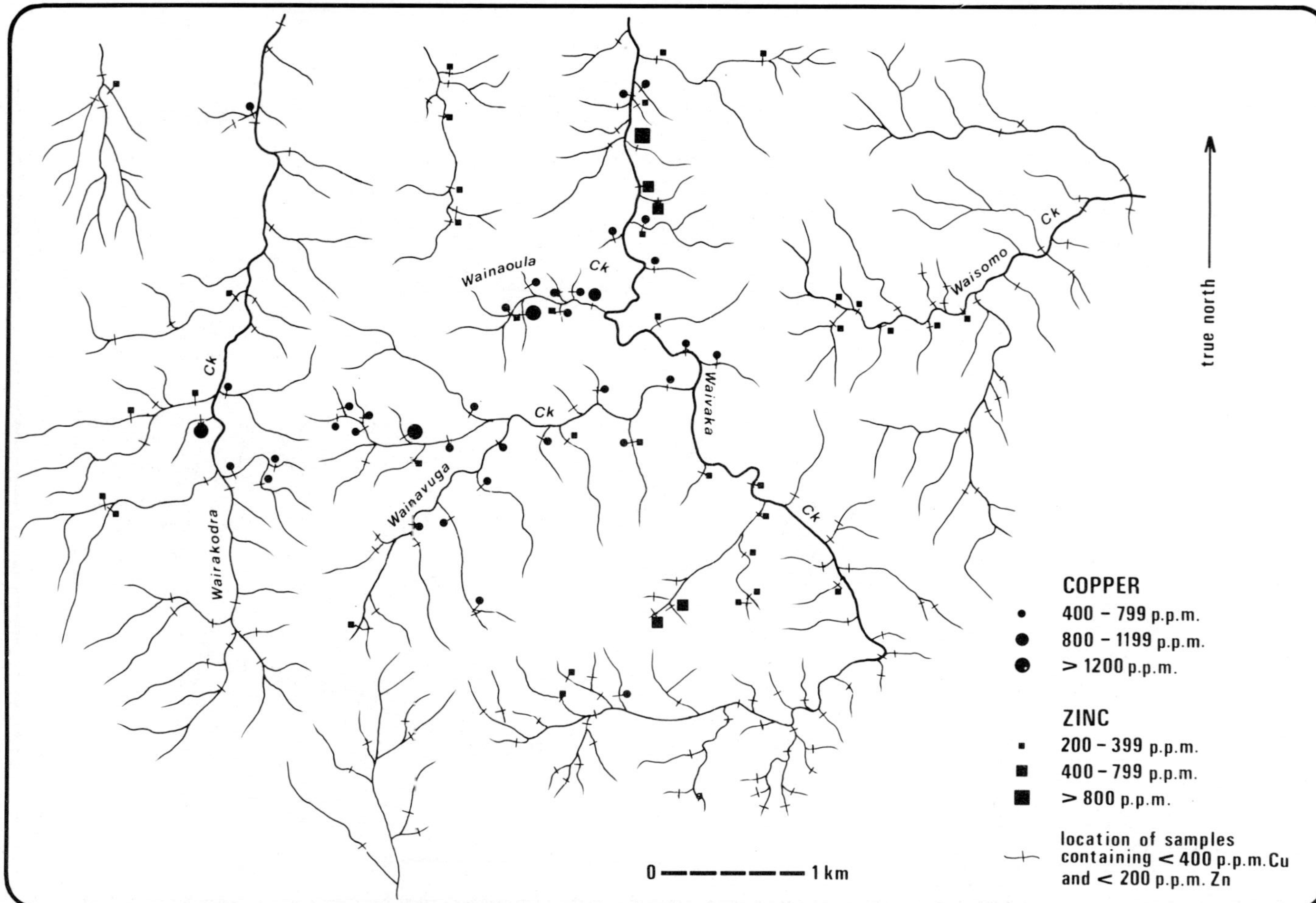

Fig. 5. Cu and Zn values in stream sediments from the Waivaka prospect.

The Mo contrast between the Waisoi and Wainabama areas is fairly striking (Fig. 4). Neither Wainabama nor Wainavadu Creeks have stream sediments containing more than 4 ppm Mo, whereas in Waisoi Creek they can contain up to 34 ppm and a widespread anomalous zone is developed. Mo is a good positive indicator of a porphyry environment, but its absence does not necessarily indicate a lack of Cu mineralization.

The Wainabama prospect was not detected in the regional geochemical stream sediment survey, the maximum Cu value recorded being 180 ppm. Quartz veining in a weathered outcrop was observed by a geologist (R.J. Fountain in 1973) and this led to further examination. The very limited stream sediment geochemical response to the Wainabama mineralization is a function of its position relative to the Waidina alluvial flats — it has been largely swamped by externally derived background material from upstream catchment areas (Fig. 4).

The approximate size of the Waivaka stream sediment Cu anomaly is 4 × 1½ km (Fig. 5). Sampling at 1-km intervals would have successfully located the prospect.

The anomalous stream sediment Cu zone at Waivaka is encircled by a zone of anomalous Zn values. Although the areas of mineralization at Waivaka appear to be discrete, at the spacing employed for stream sediment sampling, the Cu and Zn anomalies are regional in scale. Closer-spaced sampling may reveal smaller discrete patterns. A low-level aeromagnetic survey suggests that the intrusives currently exposed at Waivaka are apophyses to a large intrusive at depth and it is probable that any smaller discrete patterns would be superimposed on the regional scale anomaly.

Systematic stream sediment Mo analysis was not undertaken at Waivaka. The maximum value recorded in some reconnaissance samples was 7 ppm, all other results being 4 ppm or less. No attempt was made to analyze for Au. The Au content at Waivaka, established by drilling, rarely exceeds 0.15 ppm over a 15-m interval. The maximum value for a 15-m intersection was 0.5 ppm Au, with a Cu content of 4%.

Soil and pit sampling, Waisoi

Soil geochemistry. The results of Anglo's base of slope and ridge and spur Cu values at Waisoi are shown in Fig. 6. Although the contour intervals are not strictly comparable, this should be compared with Fig. 7 in which the Cu results of the AMAX soil survey are presented. It can be seen that the refinement provided by the latter probably did not warrant the effort.

During the main soil survey, 1620 samples were collected and analyzed. Au and Mo, in addition to Cu, Pb and Zn, were determined, firstly, because of the probable direct relationship between Cu and Au (as suggested by analyses of drill core composites from the first four holes) and the likelihood that the mobility of Au in the weathering environment is not pH dependent (Learned and Boissen, 1973); secondly, because of the possibility of an in-

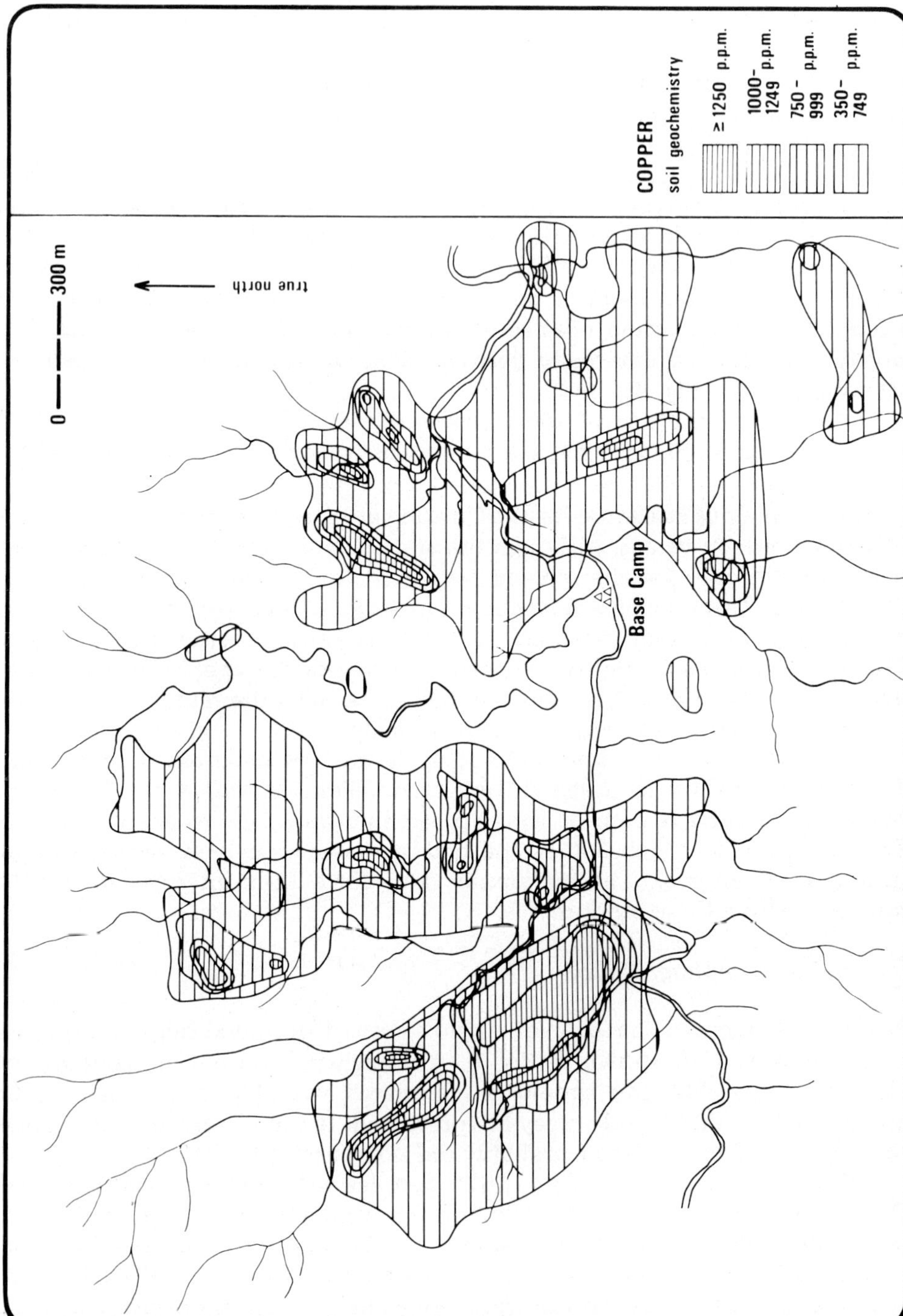

Fig. 6. Ridge and spur soil Cu values, Waisoi prospect.

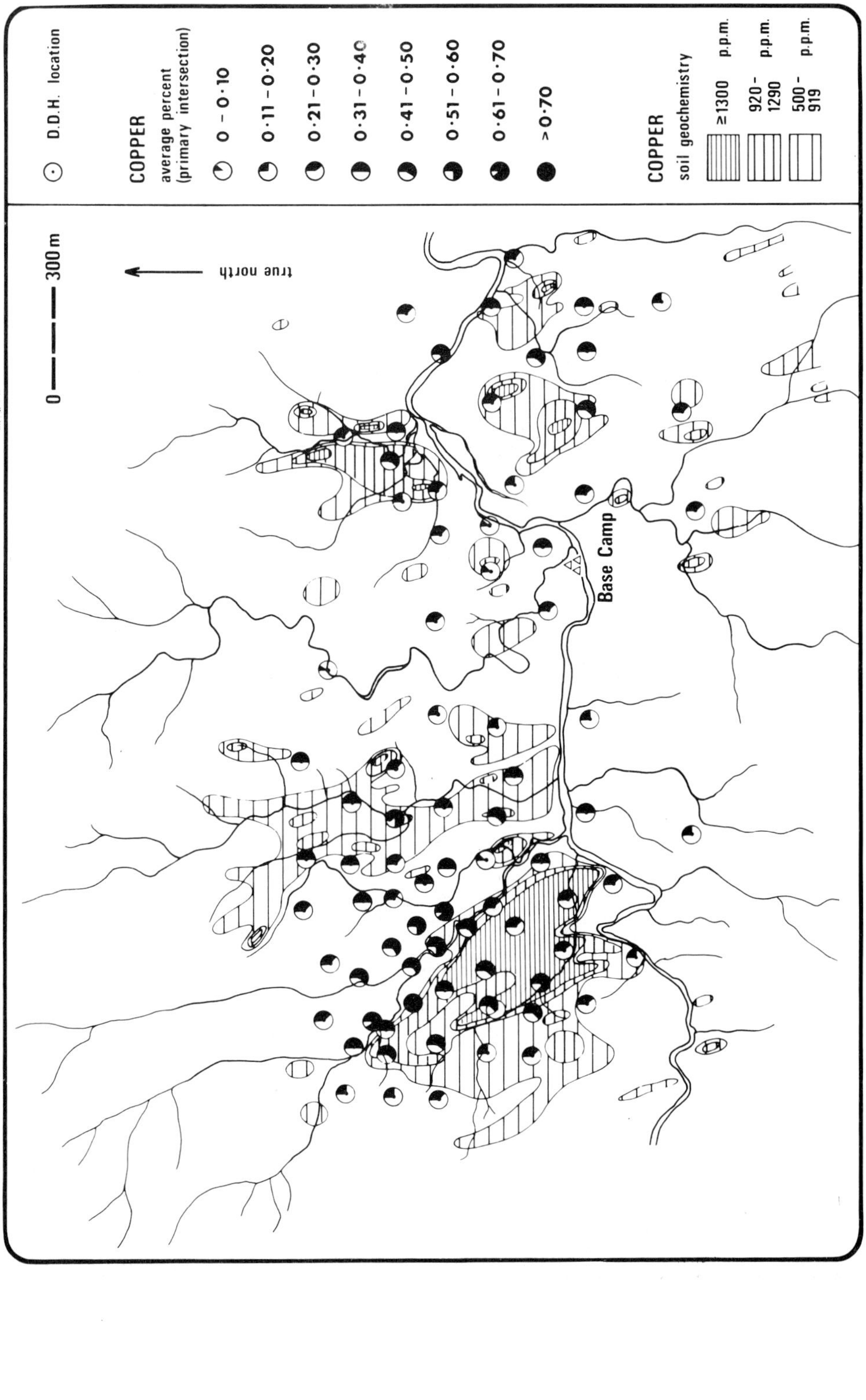

Fig. 7. Grid soil and average diamond drill hole Cu values, Waisoi prospect.

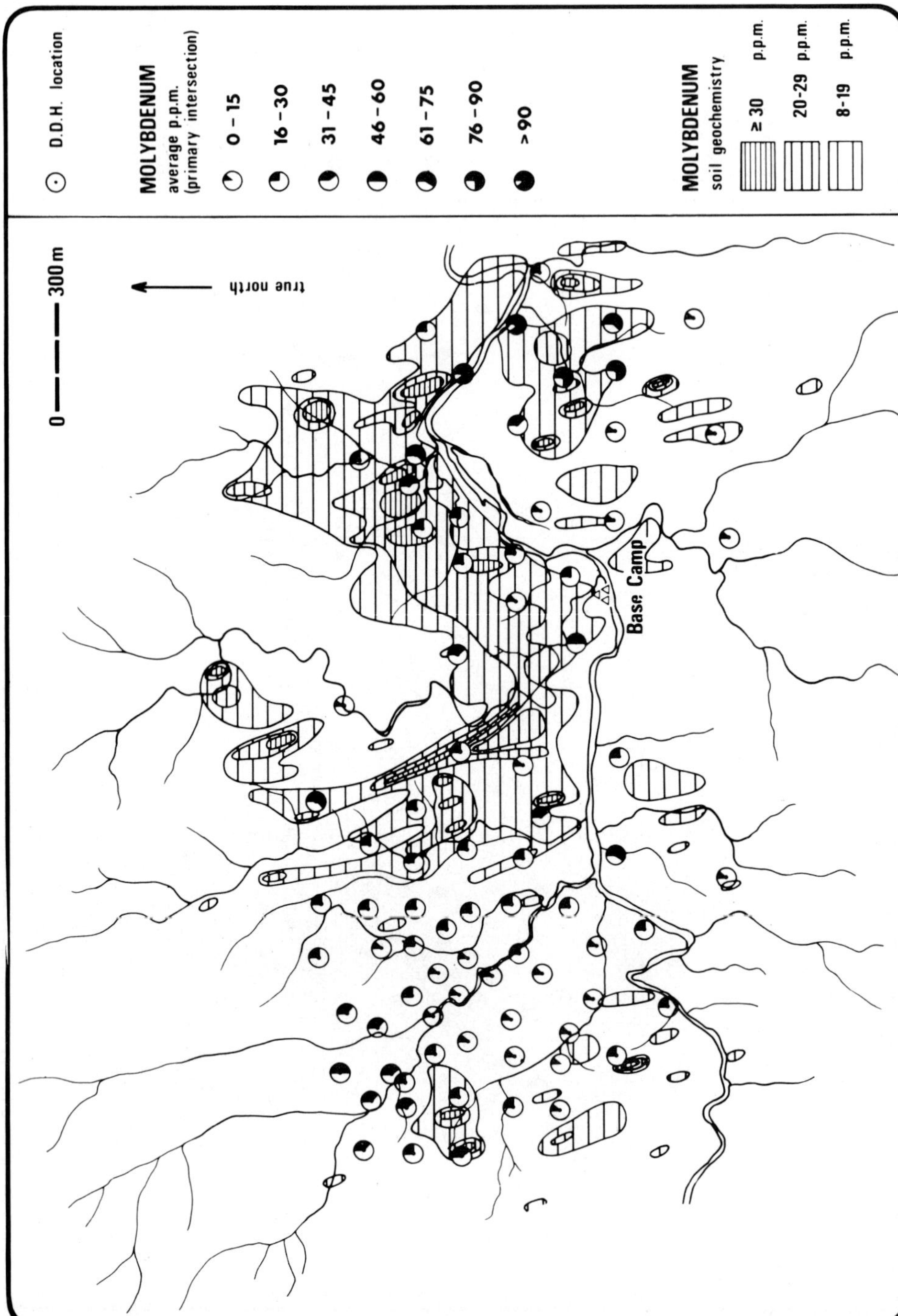

Fig. 8. Soil and average diamond drill hole Mo values, Waisoi prospect.

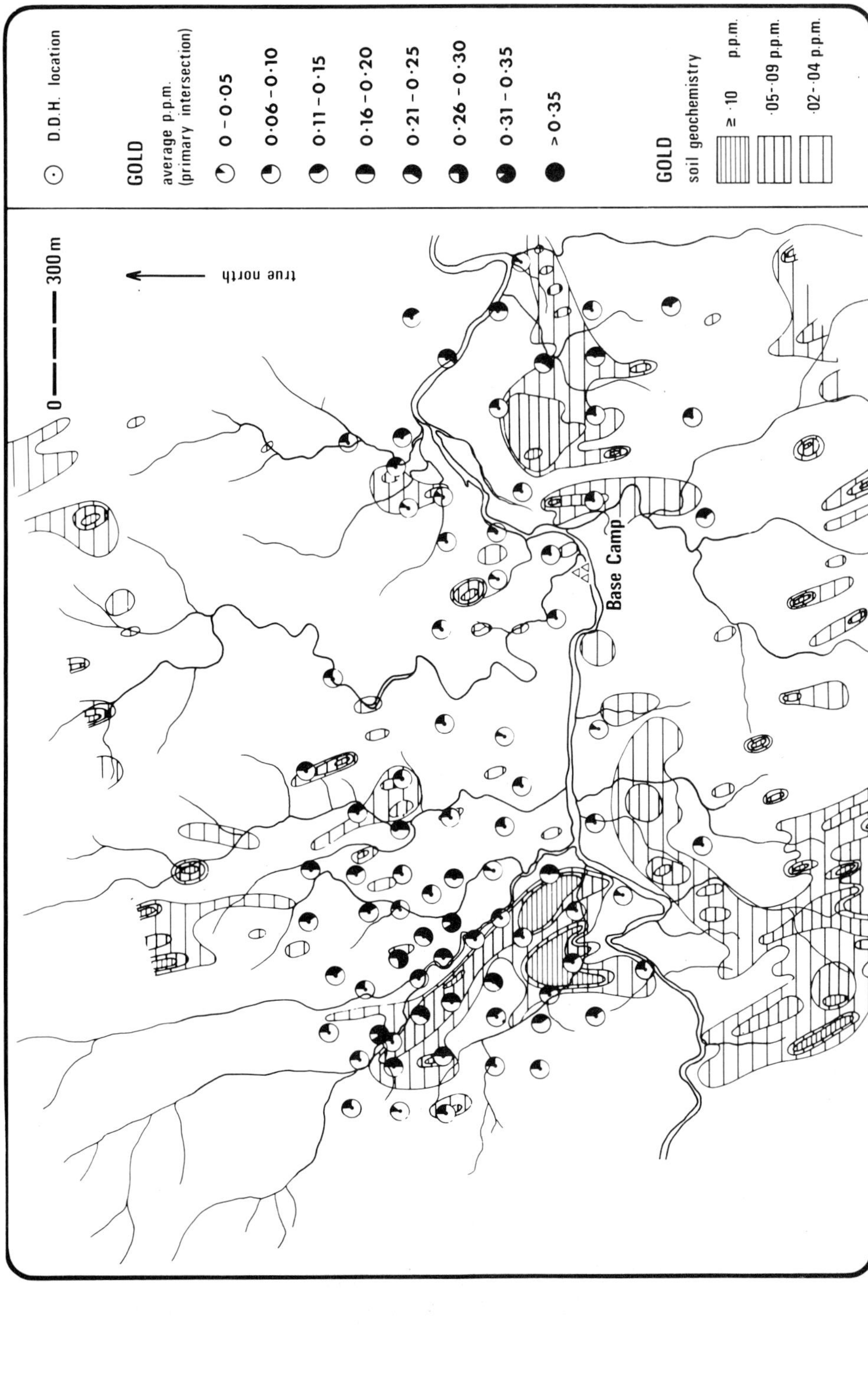

Fig. 9. Soil and average diamond drill hole Au values, Waisoi prospect.

verse relationship between Mo and Cu in soils, again as a function of pH (Learned and Boissen, 1973).

Using a 500-ppm soil copper contour, four principal anomalous areas, totalling approximately 2 km × 1 km, are defined. These are probably in fact two main areas, the one to the west of the base camp being divided by a wedge of superficial post-mineralization boulder cover, that to the east being separated by Waisoi Creek alluvium. The maximum soil Cu value recorded was 2800 ppm.

Using a threshold of 8 ppm, the largest and strongest soil Mo anomaly (maximum 232 ppm) lies to the east of the base camp on the north side of Waisoi Creek (Fig. 8). Other anomalous areas are located to the south, on the opposite side of Waisoi Creek, and northwest of the base camp. All three anomalies are only partly coincidental with the Cu anomalies; the strongest soil Cu anomaly largely lacks anomalous Mo values.

A much better coincidence occurs between anomalous Au areas (threshold 0.02 ppm, Fig. 9) and the Cu anomalies, but this is complicated by the occurrence of anomalous Au values in the propylite zone in the agglomerates. In soil, the maximum value was 0.50 ppm Au.

Two types of Zn anomaly (Fig. 10) are found, one located in propylite bounding the Cu anomalies, and the other clearly related to drainage, with anomalies following creeks and associated alluvium. In part, this results from drainage intersecting zinc-rich propylite zones, but it may also be a function of the correlation between Zn content with faulting, and faulting with drainage. Some support for the latter is found in the occasional diversion of stream courses away from linear Zn anomalies.

Pb values are low and distributed rather erratically although tending to be higher in association with Zn in the propylite areas.

Statistical analysis. Prior to the main AMAX soil survey, five east-west lines across the prospect area were re-sampled at 30.5-m centres. These samples, collected by augering to 46 cm depth, were analyzed for Cu, cxCu, Mo and soil pH. The product-moment correlation coefficient matrix for these variables in 164 samples is given in Table II.

The only correlation found was cxCu with total Cu. No apparent relationships exist between the other parameters. The geometric mean for cxCu is 7 ppm, (range 1—140 ppm). The arithmetic mean for pH is 5.08 (standard deviation 0.31; range 4.05—5.86).

An *R*-mode factor analysis was performed on the soil data from the main survey, eigenvalues were extracted and the cumulative proportion of total variance found. Three factors were chosen which account for 72% of the total variance. This is fairly satisfactory, considering both the large volume and the inhomogeneity of the data. The unrotated factor matrix did not display any particular correlation between factors and elements. However, on rotation to a simple oblique structure (oblimin rotation) the three factors display quite clear associations with the elements. The rotated factor matrix is recorded in Table III.

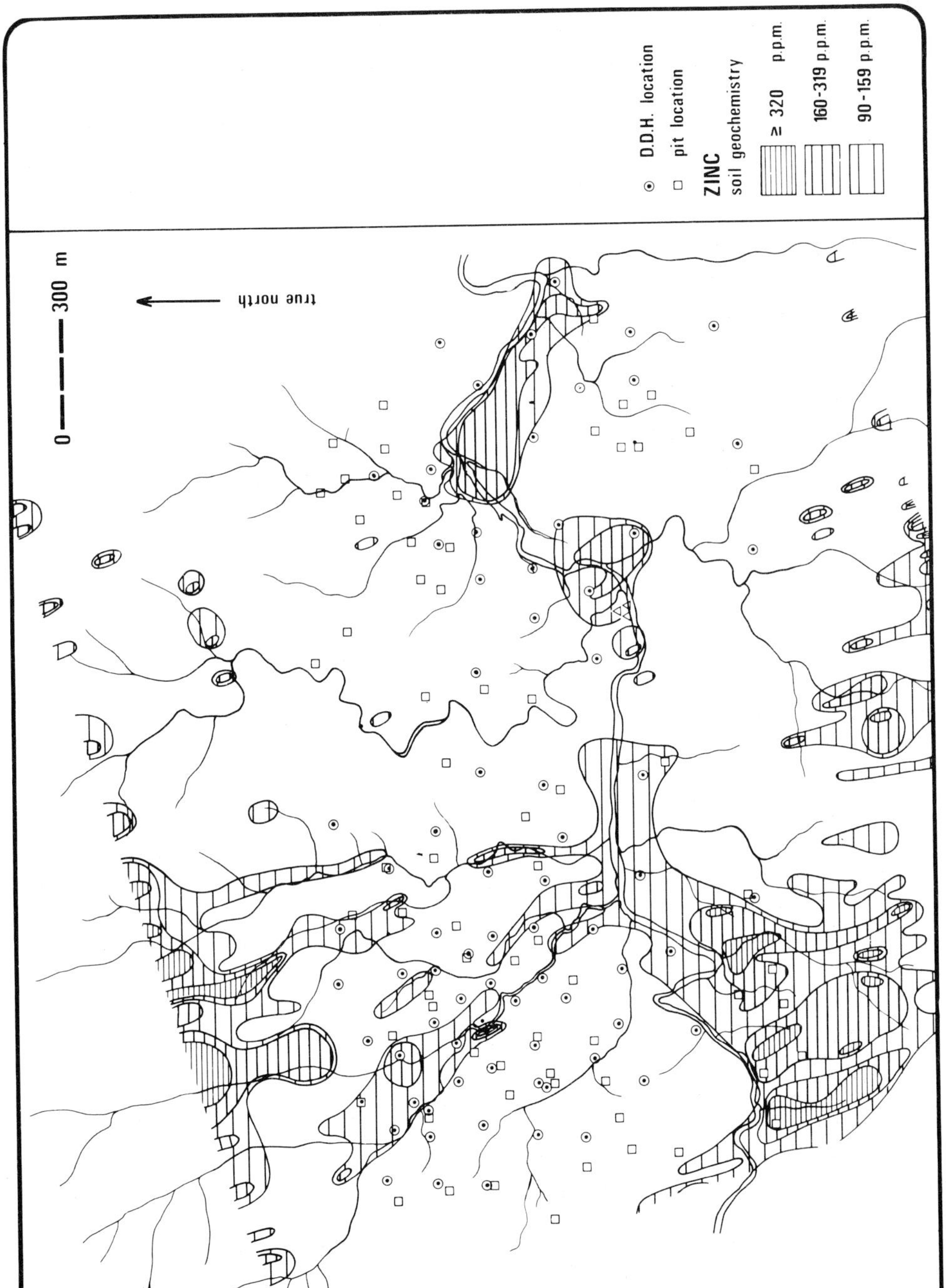
D.D.H. location
pit location
ZINC
soil geochemistry
≥ 320 p.p.m.
160-319 p.p.m.
90-159 p.p.m.
0
300 m
true north

Cu shows an exclusive dependence on factor 3. Pb has a predominant dependence on factor 1, with a small negative dependence on factor 2. Zn shows a strong dependence on factor 1 also, with a small positive dependence on factor 2. The Au values are spread over the three factors. Mo displays a prime dependence on factor 2. From the above it can be seen that factor 1 is principally a Pb-Zn factor with some Au dependence, factor 2 is principally an Mo factor with minor Zn and a greater Au association than factor 1. Factor 3 is distinctly a Cu factor with moderate influence by Au. Computerized grey-level mapping treatment (Howarth, 1971) of both element and factor data facilitated the interpretation of results.

The above factors make some spatial sense, factor 1 being related to the propylite, factor 2 in part to the quartz porphyry intrusive and factor 3 to the Cu mineralization. Factor 2, however, is not simply a function of the presence or proximity of the quartz porphyry, since a zone of relatively high Mo values is found in soils and, more particularly, in drill core on the northwest side of Waisoi West where there is no known quartz porphyry intrusive.

Soil pH. Studies of pH through soil horizons into weathered bedrock show that the A_0 horizon when present is more acid than the A horizon (arithmetic and geometric mean of 14 pits: 4.70, arithmetic standard deviation: 0.31), which in turn is usually slightly more acid than the B and C horizons (arithmetic mean for both equals 5.01, arithmetic standard deviation 0.48 and 0.25, respectively; the geometric mean for both equals 5.00). Variations in pH found during early trial soil traverse lines were due in part to different soil horizons being sampled at the arbitrary 46 cm depth. The effect of original rock mineralogy, deep leaching, high rainfall and, possibly to a minor extent, superimposed superficial pH caused by humic acid content, has obscured the pH generated by the oxidation and weathering of the sulphide mineralization and thus precluded the establishment of any relationship between Cu, Mo and pH. In fact, pit (C horizon) and drilling data fail to reveal any relationship between Cu and Mo and thus Mo is eliminated as a directly useful element in looking for concentrations of copper sulphides.

Pit geochemistry. Although Cu, Au and Mo values in pits show about a 66—70% agreement with soils in the broad sense of defining anomalous areas (using similar concentration limits for pits as for soils), the agreement is poor when detailed absolute values are compared. It can be concluded that, except where pits penetrated the boulder terrace unit at Waisoi West, pit geochemical data added little useful information to the soil data. In turn, the detailed soil geochemistry was of little assistance in locating mineralization, except to confirm the distribution of the anomalous Cu areas already established by the Anglo ridge, spur and base of slope sampling.

This is in contrast to the Waivaka and Wainabama prospects and is the result of strong leaching at Waisoi. The deep leaching is a consequence of (1) basin-like drainage into the principal Cu mineralized areas from the high pyrite, propylitic rims; (2) considerable fracturing, and (3) a higher pyrite to

TABLE II

The product-moment correlation coefficient matrix for the variables cxCu, Cu, Mo and pH in soils from Waisoi

	cxCu	Cu	Mo	pH
cxCu	1.0000	0.5252	0.0312	0.1169
Cu	0.5252	1.0000	0.1171	0.0917
Mo	0.0312	0.1171	1.0000	−0.0050
pH	0.1169	0.0917	−0.0050	1.0000

TABLE III

The rotated *R*-mode factor matrix for soil data from Waisoi

Variable	Factor 1	Factor 2	Factor 3
Cu	−0.08662	0.07423	0.94182
Pb	0.82215	−0.16380	−0.03330
Zn	0.77180	0.29762	0.00546
Au	0.21917	−0.44624	0.48487
Mo	−0.11386	−0.89628	−0.03713

total sulphide ratio than at Waivaka or Wainabama. This has caused markedly reduced soil and pit Cu values relative to underlying primary rock contents (Fig. 7). In addition, the most anomalous soil Cu area is significantly displaced from the higher-grade Cu zone. Depletion of Au at surface may have occurred to a minor extent (less than a factor of 2) although downhole variance makes this difficult to establish with confidence. Surface and near-surface Au values cannot be used to predict underlying Cu grades closely, despite the broad linear correlation between the two in the diamond drill core. Except for the northwest corner of Waisoi West, Mo in soils and pits reflects fairly closely the underlying primary rock content (Fig. 8).

Soil and pit sampling, Waivaka

There is very good agreement between anomalous areas defined by soil sampling, pit sampling and ridge, spur and base of slope sampling. There is no consistent trend in Cu values down the pit profile with 30% of the highest values occurring as bottom intervals (average depth 4—5 m), 17% as top intervals (approximately 0—1 m). Leaching of Cu has not occurred and the surface values provide a reliable reflection of underlying primary Cu grades (Fig. 11). Superficial "mud avalanche" or landslip material has obscured the soil geochemistry on the south side of one anomalous area, but "windows", pits and creek exposures enabled some bedrock to be sampled.

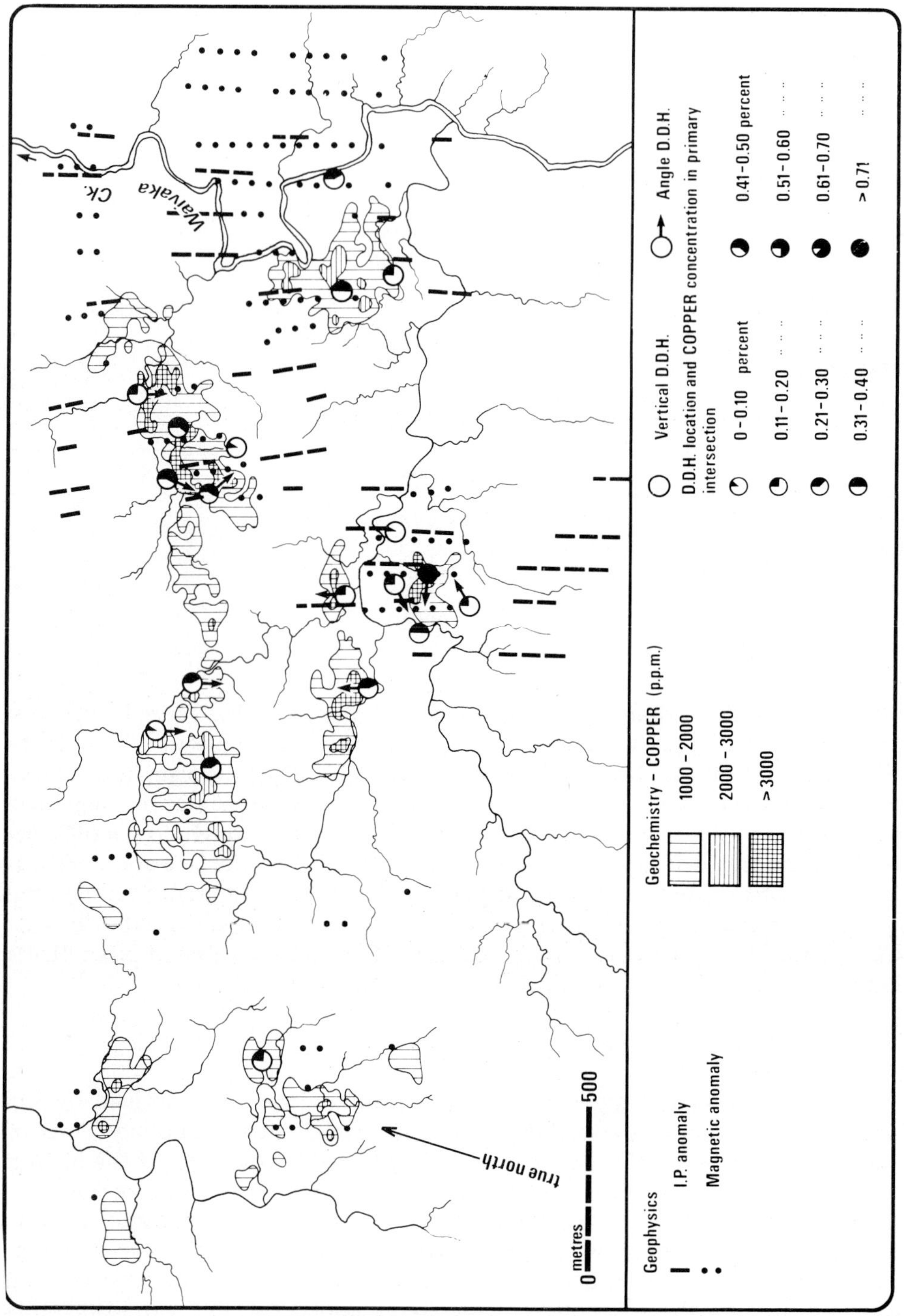

Fig. 11. Soil and average diamond drill hole Cu values, ground magnetic and I.P. anomalies, Waivaka prospect.

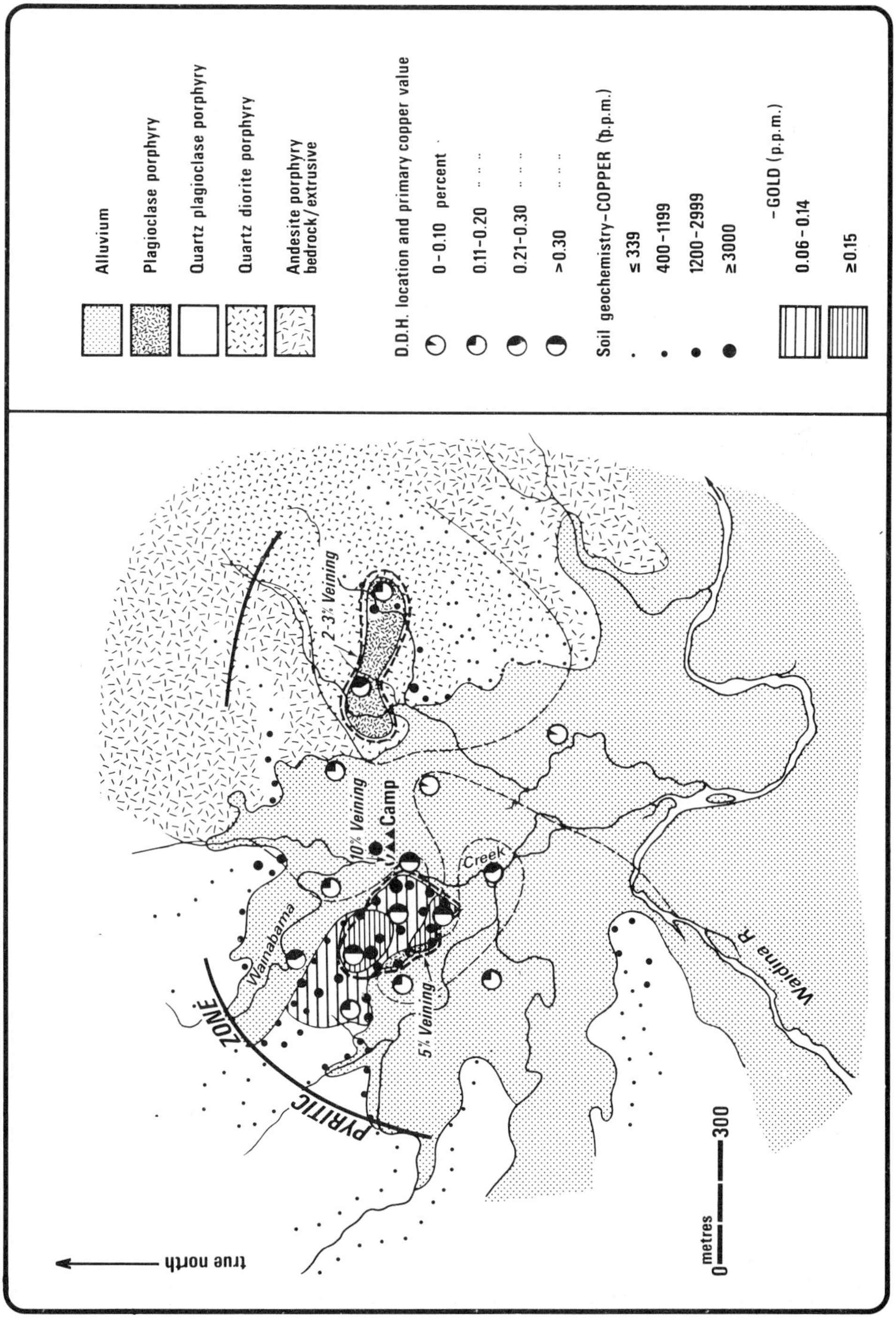

Fig. 12. Soil and average diamond drill hole Cu values, geology and quartz veining, Wainabama prospect.

Soil and pit sampling, Wainabama

The soil Cu values at Wainabama provide an excellent indication of Cu content in the underlying fresh rock (Fig. 12). There is very little pyrite present with the Cu mineralization and no significant leaching has occurred. Surface Au values also correspond well with the Au tenor of the underlying rock. As at Waisoi, the Au variance in core makes it difficult to establish whether minor depletion has occurred at the surface. The Au content of the mineralization at Wainabama is about 1½—2 times as high as that at Waisoi, relative to Cu content. Mo content exceeds an average of 13 ppm in only two diamond drill holes (average 25 ppm in each, maximum value over 15 m: 44 ppm). Both the latter were collared in alluvium. In view of the low, primary rock Mo content, it is not surprising that Mo is largely absent in the soils (detection limit 1 ppm), only six soils recording greater than 4 ppm Mo.

Additional exploration methods

Quartz veining density. This was established as a very useful semi-quantitative measure of Cu grade, particularly at Waivaka and Wainabama. At Waivaka, 70% of pits that contain quartz veining or stockworks have Cu values greater than 2000 ppm. At Waisoi, the relationship between veining and tenor of Cu mineralization is complicated by the presence of at least three generations of quartz veining. Even so, except for the eastern side of Waisoi West, the principal Cu-mineralized zone is located inside the 2% veining contour. There is no obvious correlation of surface and near-surface Mo values with the 2 and 5% quartz veining contours. A possible correlation of Au with higher quartz veining density occurs in the western soil anomaly, except where blanketing by superficial post-mineralization colluvial cover occurs. Au values in soils and pits on the south side of Waisoi Creek, on the extension of the western soil Cu anomaly, are associated with Zn not Cu and, despite a veining density of 2—3%, significant Cu mineralization does not extend very far south of Waisoi Creek. In contrast, despite the sparse quartz veining at the surface in the central soil Cu anomaly, the presence of some Au values suggested that there was the possibility of significant Cu mineralization and this was borne out in the drilling. Quartz veining density was likewise useful at Wainabama (Fig. 12) even though three generations of veining were recognized.

Ground magnetometer survey. This was of most value at Waivaka (Fig. 11) where the mineralized quartz veining carries considerable magnetite. Not only could the favourable rock type (porphyries) be mapped against the andesites (lower magnetic susceptibility), but the ground magnetics provided a basis for discrimination of non-prospective ground. At Waisoi, the situation is more complex. Magnetic relief is small and, although the porphyries generally form "lows", the base of the Namosi andesite forms local "highs" and the agglomerate and andesites produce moderate magnetic response (in contrast to Waivaka), detailed correlations and predictions are difficult. A number of magnetically inferred linears support the presence of a series of prominent NE- and NW-trending faults. At Wainabama, the area of high magnetic susceptibility coincides closely with the principal mineralized area and is very restricted in extent. This is related to hydrothermal magnetite content and compares closely with the Waivaka situation.

Induced polarization surveys. At Waivaka, induced polarization (I.P.) surveys, like the ground magnetics, provided a valuable detailed guide to locating and defining possible extensions to known areas of mineralization (Fig. 11). Other I.P. anomalies, particularly those without coincident magnetic highs and lying outside soil geochemical anomalies and areas of quartz veining, have been interpreted as due to the effect of increased pyrite. At Waisoi, only a gross picture of sulphide zoning was obtained; the I.P. measurements and a complementary resistivity pattern suggest two chargeable rings surrounding lower sulphide zones, with evidence for a pyritized zone between the two. Mapping and drilling confirmed this picture of sulphide distribution. In Waisoi West, the lower sulphide zone carries Cu and virtually no pyrite; in Waisoi East, the chargeable ring is Cu-mineralized and has associated pyrite, whilst the inner core is very poorly mineralized. A related feature is the more widespread occurrence of supergene enrichment in Waisoi East than in Waisoi West. At Wainabama, a crudely horseshoe-shaped I.P. anomaly is centred on the main Cu-mineralized body. The two extending limbs have been shown to be related to pyrite content. A pyritic zone boundary is shown in Fig. 12 rather than a limit of propylite, since the propylite, although partially coincident with the pyritic zone, is harder to define at Wainabama. Surface effects (alluvial cover) eliminated the usefulness of the resistivity values as a guide to rock type or mineralization.

CONCLUSIONS

Within a relatively restricted topographic, climatic and geological environment considerable variation was found in the usefulness of the various exploration techniques and parameters employed in searching for porphyry Cu mineralization.

Cu and cxCu in stream sediments clearly located and outlined the two largest areas of Cu mineralization, but did not locate the third centre because of dilution by background material. Anomalous Mo associated with Cu is a favourable pointer to a porphyry environment. Au content can be ambiguous. Zn is anomalous in a zone peripheral to one porphyry Cu system. Increased care must be taken in assessing major alluvial-filled drainage systems where the stream sediment geochemistry can be obscured.

Ridge and spur soil geochemistry was effective in outlining the mineralized zones at all prospects except where there was landslip or "mud avalanche" cover. The latter feature is not unusual in young volcanic terrains in tropical environments. The efficacy of using detailed soil and pit geochemistry varied as a function of original sulphide content (particularly pyrite), individual elemental concentrations and correlations and weathering history.

As a direct pathfinder for Cu mineralization, geochemical Au determinations are preferred to those for Mo. Depending on the degree of fracturing, the original rock mineralogy and the weathering and leaching history, strongly acid soils can be developed independently of the original sulphide content; pH determinations may therefore be of little value.

I.P. and ground magnetics (rather than aeromagnetics) were valuable in helping to elucidate the geology. When significant amounts of sulphides and hydrothermal vein magnetite are present, geophysical methods can be applied to defining the extent of the mineralization. Quartz veining density is also recommended as a parameter with the qualification that a number of generations of veining, not all sulphide mineralized, may be present.

ACKNOWLEDGEMENTS

This is a sketchy compilation of much hard work undertaken by past and present Anglo and AMAX staff. In particular I would like to mention S.R. Yardley, S. Harrison, R.J. Fountain, J.M. Hammond, D.F. McKay, G. Faga, D.P. Taylor, B.K. McKnight, E.A. Bowen, T. Mattiske, C.D.S. Bates and A.G.L. Paine. J.E. Thompson and W.W. Shaw provided the concepts and enthusiasm to keep the project going during the lean times.

REFERENCES

Branch, C.D., 1972. A vulcanological interpretation of Viti Levu and Vanua Levu, Fiji, with comments on the porphyry copper potential of the area. Barringer (Fiji) Limited, Rep., 18 pp. (unpublished).

Faga, G., 1976. Completion report on the Wainabama prospect. AMAX Exploration (Australia) Inc., Rep., 7 pp. (unpublished).

Fountain, R.J., 1972. Geological relationships in the Panguna porphyry copper deposit, Bougainville Island, New Guinea. Econ. Geol., 67: 1049—1064.

Gill, J.B., 1970. Geochemistry of Viti Levu, Fiji and its evolution as an island arc. Contrib. Mineral. Petrol., 27: 179—203.

Hammond, J.M. and McKay, D.F., 1976. Geology of the Waisoi porphyry copper prospect, Namosi Province, Fiji. AMAX Exploration (Australia) Inc., Rep., 31 pp. (unpublished).

Howarth, R.J., 1971. Fortran IV program for grey-level mapping of spatial data. Math. Geol., 3: 95—121.

Learned, R.E. and Boissen, R., 1973. Gold — a useful pathfinder element in the search for porphyry copper deposits in Puerto Rico. In: I.L. Elliott and W.K. Fletcher (Editors), Geochemical Exploration 1974. Elsevier, Amsterdam, pp. 93—103.

Rodda, P. and Band, R.B., 1967. Geology of Viti Levu. Geol. Surv. Fiji Rep., 1966: 8—16.

Taylor, D.P., 1975. Report on exploration activities in the Waivaka area, S.P.L. 1014, Namosi, Fiji. AMAX Exploration (Australia) Inc., Rep., 14 pp. (unpublished).

Taylor, S.R., 1966. The application of trace element data to problems in petrology. Phys. Chem. Earth, 6: 171—175.

Twyford, I.T. and Wright, A.C.S., 1965. The Soil Resources of the Fiji Islands. Government Printer, Suva, Fiji, 570 pp.

Journal of Geochemical Exploration, 8(1977)457—471

DETECTION OF NATURALLY HEAVY-METAL-POISONED AREAS BY LANDSAT-1 DIGITAL DATA

B. BØLVIKEN[1], F. HONEY[2], S.R. LEVINE[3], R.J.P. LYON[3] and A. PRELAT[4]

[1] *Geological Survey of Norway, N-7001 Trondheim (Norway)*
[2] *Commonwealth Scientific and Industrial Research Organization, Wembley, W.A. (Australia)*
[3] *Stanford Remote Sensing Laboratory, Department of Applied Earth Sciences, Stanford University, Stanford, Calif. 94305 (U.S.A.)*

(Revised version accepted March 1, 1977)

ABSTRACT

Bølviken, B., Honey, F., Levine, S.R., Lyon, R.J.P. and Prelat, A., 1977. Detection of naturally heavy-metal-poisoned areas by LANDSAT-1 digital data. J. Geochem. Explor., 8: 457—471.

Natural poisoning of soil and vegetation due to heavy metals originating from sulphide deposits in the bedrock has been demonstrated at several locations in Norway. At some of these, the features of poisoning are rather widespread, suggesting that the recognition of such naturally poisoned areas by means of satellite imagery and satellite digital tapes might be possible.

An occurrence of natural Cu poisoning near Karasjok, northern Norway, was selected as a training area. LANDSAT-1 computer compatible tapes were used to prepare enhanced colour ratio images and to calculate ratios of calibrated *channel* reflectances. *Ground* reflectance measurements were also carried out using radiometers with band passes equivalent to those on the LANDSAT multispectral scanner system.

Compared with the surroundings the LANDSAT data over the training area showed unique spectral characteristics. These are also indicated on the enhanced ratio images, and the digital data can be utilized in supervised discriminant analysis and unsupervised "cluster" analysis. In particular, the area of the natural Cu poisoning shows low values in the band 7/band 5 reflectance ratio. At present, a detailed, precise numerical correlation between the ground and satellite reflectance data is not possible due to differences in resolution, although spatial correlation of spectral trends can be established.

It appears that features of natural heavy metal poisoning can be located with the 0.45-hectare resolution of the LANDSAT multispectral scanner system. The search for such features might therefore be used as a tool in mineral exploration. The current resolution limitation can be overcome by utilizing scanners at aircraft altitudes.

INTRODUCTION

In Norway, several occurrences of natural heavy-metal-poisoned soil and vegetation have been found (Låg et al., 1970; Låg and Bølviken, 1974). It seems that such natural poisoning is a fairly common feature at locations where sulphide mineralization occurs in bedrock. This type of poisoning is not caused by man; it is the result of natural processes by which heavy metals are extracted from crystalline mineralization in bedrock and overburden, transported downslope in groundwater solution, and reprecipitated in humus-rich soils where the heavy-metal-containing solutions emerge at the surface. Some of the heavy metals are strongly bound to humus, and humus-rich soils may, by this process, accumulate heavy metals from the moving groundwater. The soil gradually becomes restrictive for the natural plant growth. As a consequence, the rate of humus production is reduced. If the breakdown of humus occurs faster than the supply of new organic matter, the soil will become increasingly rich in heavy metals, and in some cases eventually become toxic to plants.
in some cases eventually become toxic to plants.
and stunted growth within areas of normal growth. At more advanced stages of poisoning, the vegetation may be dying, deficient or absent, so that patches of barren soil appear. The sizes of patches of poisoning may range from 10^{-2} to 10^{3} m^2. Several patches may together form irregular patterns over quite large areas. The dry matter of samples of the topsoil of such patches might contain thousands of parts per million of heavy metal, sometimes as much as 10% Pb and 5% Cu.

The heavy metals causing the poisoning originate some distance uphill from the poisoned areas. This distance might vary from a few up to hundreds of metres. Having originally established the indications of heavy metal poisoning, the phenomenon may be easily recognized in the field. Since the metal source might be sulphides or other heavy metal mineralization in the bedrock, the search for areas of natural heavy metal poisoning can be used as a prospecting method. Obviously, in areas of glacial terrain, looking for patches of natural heavy metal poisoning might sometimes be more efficient than hunting for mineralized boulders, which normally are of much smaller size than poisoned patches. Moreover, patches of natural heavy metal poisoning might occur at the surface even though no mineralized outcrops or boulders are visible. Observations of poisoning as a prospecting method has already been used with some success in northern Norway (A. Bjørlykke, personal communication).

The question arises: could areas of natural heavy metal poisoning be detected by reflectance measurements obtained from satellites or other remote platforms? If so, one would have a potential tool for regional mineral exploration. A cooperative research programme between the Geological Survey of Norway and the Stanford Remote Sensing Laboratory has been started, which aims at answering this question. Initially, we have studied the possibilities for using LANDSAT-1 images and digital data to detect areas naturally poisoned by heavy metals. Currently, the use of data from digitized multi-channel air photographs are being investigated. We anticipate that future use of airborne scanners for detecting areas of poisoning is a realistic possibility.

The present paper is a report on the first results of this research programme. LANDSAT-1 data from two test areas with known occurrences of natural heavy metal poisoning were investigated to see if the reflectances of the poisoned patches were different from those of the surrounding normal areas. The two test areas are (1) the Snertingdal area, southern Norway (Pb mineralization), and (2) the Karasjok area, northern Norway (Cu mineralization) (Fig. 1).

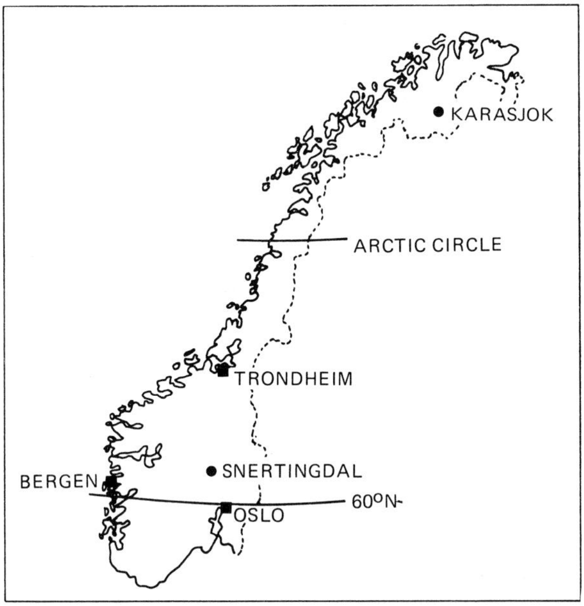

Fig. 1. Locations of selected occurrences of naturally heavy-metal-poisoned areas in Norway. Snertingdal: Pb poisoning; Karasjok: Cu poisoning.

Although occurring over a fairly large area (approximately 10 hectares), the individual patches of Pb poisoning at Snertingdal are rather small, in most cases of the order of 10 m^2 or less (Låg and Bølviken, 1974). It soon became clear, therefore, that the resolution of the LANDSAT-1 data (unit picture element or pixel approximately 60 m × 80 m) is not adequate for the detection of this type of poisoning. Consequently, the Snertingdal area is not considered here, but we hope to analyze the data from this locality later after having done more detailed aircraft and ground reflection measurements.

At Karasjok, the poisoning is more widespread. In the following an account is given about the Karasjok area. After a short description of the locality, the main satellite data are presented followed by some conclusions which can be drawn from these data. The investigation was done in collaboration with the Sydvaranger Mining Company which holds the mining rights in the area.

DESCRIPTION OF KARASJOK SURVEY AREA

The area is south of the village of Karasjok (Fig. 1) about 300 m above sea level in a dry and rather cool region (annual precipitation ca. 300 mm, mean temperature -2.6°C). The mineralization consists of disseminated chalcopyrite and pyrite in a nearly flat-lying muscovite- and amphibole-bearing gneiss. Above the gneiss, there is a concordant black schist rich in carbon and pyrrhotite. Both the gneiss and the schist are weathered down to approximately 0.5 m (B. Rosholt, personal communication).

The superficial deposits generally consist of till, mostly fine sand, a few metres thick. In the mineralized area, however, a thickness of 0.5—1 m is common. The predominant tree species are birch (*Betula pubescens*) and some pine (*Pinus sylvestris*). The ground-cover vegetation is comprised mainly of bilberry (*Vaccinium myrtillus*) and cowberry (*V. vitis-idaea*). Låg and Bølviken (1974) observed patterns of natural poisoning in the field and these could be verified by high Cu contents in soil and vegetation samples (Table I). The poisoned areas may be recognized (see Figs. 2 and 3) as:

(1) Open areas in the otherwise dense birch forest. These vary in size and form from small patches to areas several thousands of square metres in extent, often with long axes normal to the map contours.

TABLE I

Ash, heavy metals and pH in dry matter (105°C) of soil; ash and heavy metals in dry matter of vegetation from poisoned areas at a Cu occurrence Karasjok, northern Norway. Each sample for analysis is a composite of at least 10 subsamples spread evenly at the type of area being sampled (after Låg and Bølviken, 1974)

	N	Ash (%)	Cd (ppm)	Cu (ppm)	Ni (ppm)	Pb (ppm)	Zn (ppm)	pH
Soil samples, depth 2—5 cm	9	85.1	0.6	7400	32	18	36	5.0
Soil samples, depth 20—25 cm	5	96.6	0.9	3000	38	20	58	5.1
Betula pubescens, leaves	4	1.9	trace	7	6	4	102	
Betula, first-year twig	4	1.2	trace	5	1	2	116	
Betula, second-year twig	4	1.5	trace	12	4	3	178	
Deschampsia flexuosa	2	5.8	0.2	271	7	3	42	
Festuca ovina	3	3.9	trace	334	5	4	97	
Juncus trifidus	1	3.5	0.5	69	9	3	112	
Viscaria alpina	1	5.7	0.3	895	10	3	63	

Figures indicate arithmetic means of *N* samples.

Fig. 2. Naturally copper-poisoned area in birch forest (*Betula pubescens* with some *Pinus sylvestris*), Karasjok, northern Norway. Dark area in the middle right-hand side of the picture is barren soil. Ground-cover vegetation on the left hand side consists mainly of *Juncus trifidus, Festuca ovina, Deschampsia flexuosa* and *Viscaria alpina*.

Fig. 3. Naturally copper-poisoned area appearing as open space in the otherwise dense birch forest (*Betula pubescens* with some *Pinus sylvestris*), Karasjok, northern Norway. The normal ground-cover vegetation in the forest is dominated by *Vaccinium myrtillus* and *V. vitis-idaea* while *Juncus trifidus, Festuca ovina* and *Deschampsia flexuosa* prevail in the open space where Cu concentrations in the soil is high (see Table I). This area can be recognized on the aerial photograph (Fig. 4).

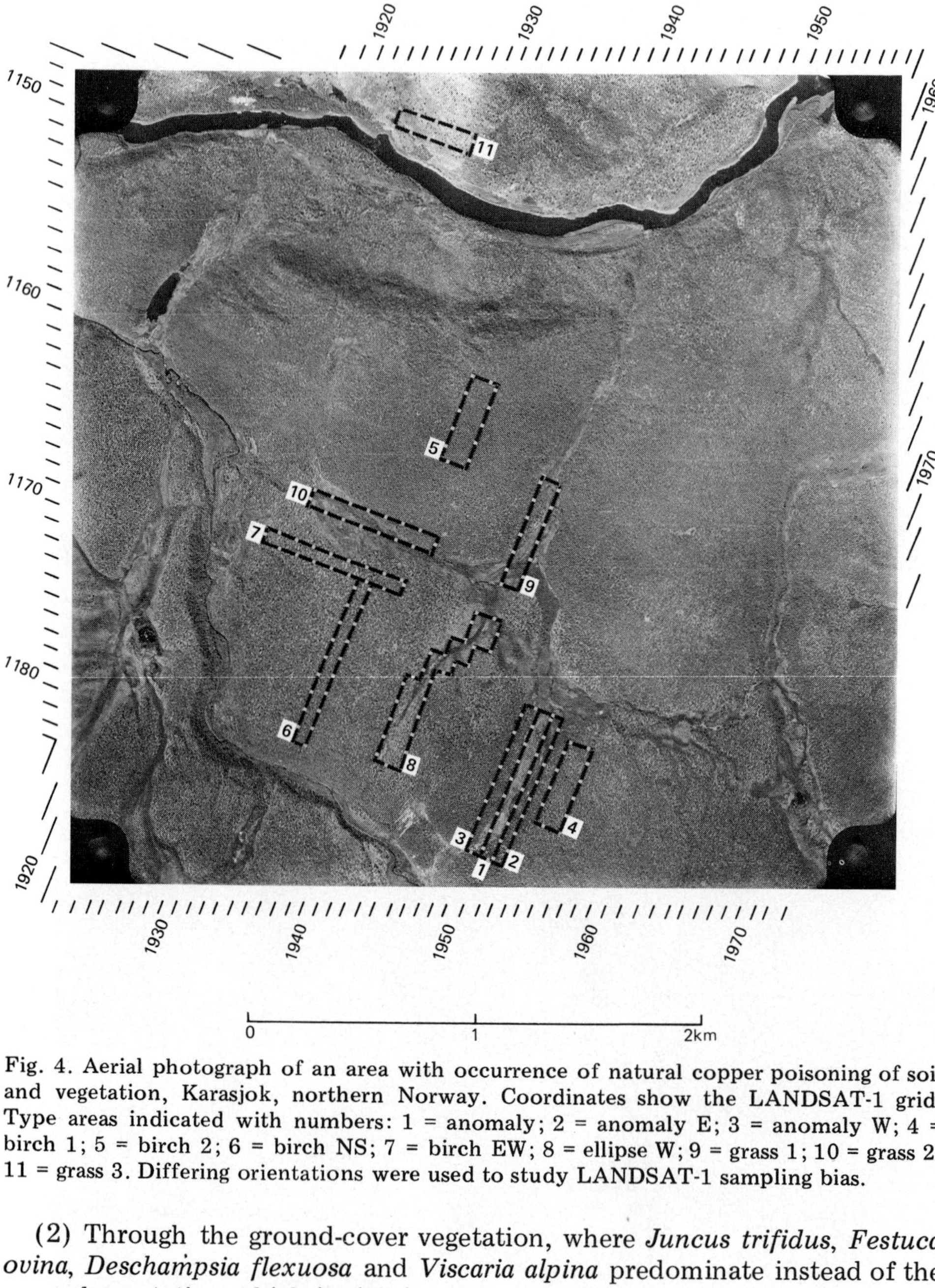

Fig. 4. Aerial photograph of an area with occurrence of natural copper poisoning of soil and vegetation, Karasjok, northern Norway. Coordinates show the LANDSAT-1 grid. Type areas indicated with numbers: 1 = anomaly; 2 = anomaly E; 3 = anomaly W; 4 = birch 1; 5 = birch 2; 6 = birch NS; 7 = birch EW; 8 = ellipse W; 9 = grass 1; 10 = grass 2; 11 = grass 3. Differing orientations were used to study LANDSAT-1 sampling bias.

(2) Through the ground-cover vegetation, where *Juncus trifidus*, *Festuca ovina*, *Deschampsia flexuosa* and *Viscaria alpina* predominate instead of the normal vegetation which is rich in various herb species and *Vaccinium myrtillus* and *V. vitis-idaea.*

(3) As barren areas up to several tens of square metres in extent at places where the water issues at the surface.

Some of these features can also be seen on the black and white aerial photographs. However, the main area of Cu poisoning is hard to distinguish from the normal waterlogged bogs in the area (Fig. 4).

Measurements of the reflectances of the vegetation and soils using 4-channel EXOTECH radiometers (with LANDSAT equivalent bandpasses) confirmed that sufficiently large differences existed for discrimination of the poisoned areas using the LANDSAT digital imagery data if the areas were greater than 0.45 hectare.

COMPUTER PROCESSING OF SATELLITE DATA

Two numerical classification techniques were applied to the LANDSAT-1 reflectance data from the survey area to evaluate the results and use them as a tool for future detection of poisoned areas. These methods were:

(1) *Stepwise discriminant analysis.* The BMD07M stepwise discriminant computer program which is part of a series of bio-medical statistical programs compiled by the UCLA Health Service Computing Facility, was used to calculate discriminant functions among the different type areas and to plot the pixels of each type area as a function of canonical variables. For a description of the computational procedure to obtain the discriminant functions and canonical variables see Dixon (1970).

(2) *Pattern recognition.* The cluster analysis program used for the analysis of the LANDSAT data is part of the Stanford Remote Sensing Laboratories STANSORT package, running on a DEC-10 system. The STANSORT system has been described in detail previously (Honey, 1975).

Some of the features of these programs used in this analysis were: (a) density slicing — selection of portion of the digital range of any of the four LANDSAT channels to enable enhancement of specific target materials; (b) ratioing — inter-channel ratios, followed by density slicing for selective enhancement; (c) conversion of data to relative reflectance values, using predetermined atmospheric contribution parameters (Lyon et al., 1975); and (d) cluster analysis.

The cluster analysis algorithm provides the most powerful technique in searching for anomalous areas. The sequence of operations is:

(1) Normalization of the data to reduce the total radiance in the four channels to a constant value. This has the effect of reducing the variability due to topography.

(2) Cycling through the data, generating new "standard" signature ranges when the current pixel does not fit into the standards previously generated, or classifying it with the appropriate symbol when it does fit one of those generated earlier. The limit of classes in this system is twenty-six — if this number is exceeded the tolerance (width of the standard signature "gate") is increased and the data recycled.

(3) Presentation of the classification on a screen and/or line printer.

The algorithm uses a rapid, non-statistical approach to the classification, merely comparing shapes of signatures. It can also be termed a parallelepiped classifier, but it is not true "clustering".

SATELLITE IMAGERY AND DIGITAL DATA ANALYSIS

Examples of black and white satellite images of the Karasjok survey area are shown in Fig. 5 (channel 5) and Fig. 6 (ratio channel 7/5). Location of the satellite grid on the ground was achieved by the use of photographic enlargements of channel 5 and 6 images. By identifying pixel-size lakes on these images it was possible to superimpose the satellite grid on the aerial photograph (Fig. 4).

Based on previous knowledge of the area (Låg and Bølviken, 1974), interpretation of aerial photography, helicopter reconnaissance and ground reflectance measurements, eleven type areas were selected for further study. These areas are indicated (areas 1—11) in Fig. 4.

Area 1 is known to be anomalous, with Cu concentrations of up to 0.5% in the soil. On the aerial photograph this area appears to be a bog, but in fact it is an open space in the birch forest which is partly wet and partly dry. In some places the higher plants of the ground cover vegetation are absent

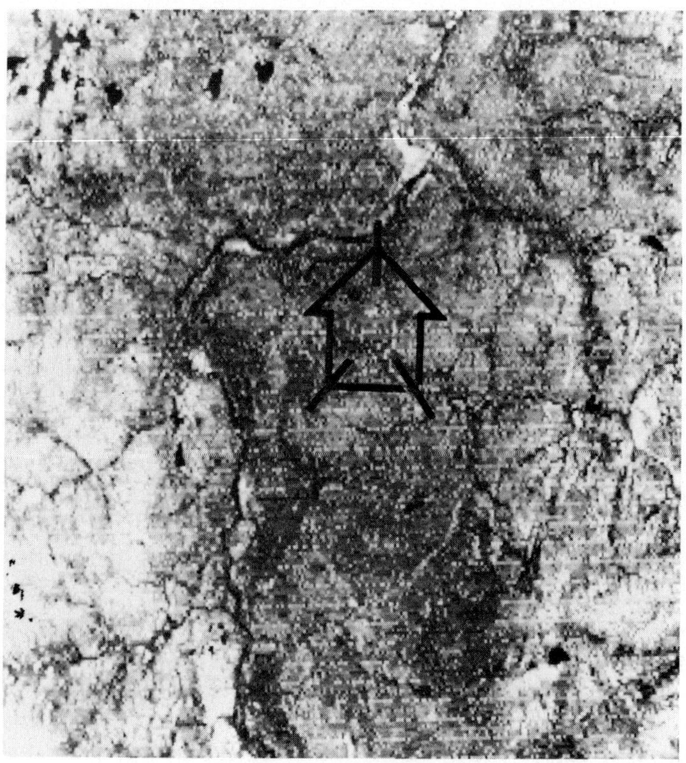

Fig. 5. Channel 5 LANDSAT-1 image from an area with natural Cu poisoning at Karasjok, northern Norway. The poisoned area (in the middle of the north arrow) cannot be identified in the image. Characteristic elliptical features seem to be apparent, which may possibly have some connection with mineralization (see also Fig. 6). Satellite prints were obtained before field checking commenced.

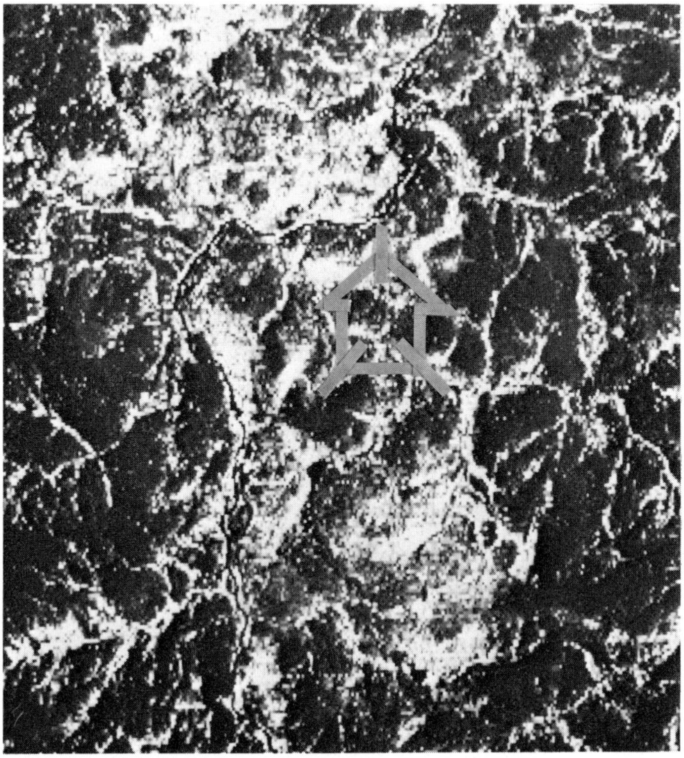

Fig. 6. Black and white image reflectance ratio channel 7/5 of LANDSAT-1 digital data from an area with natural Cu poisoning at Karasjok, northern Norway. The poisoned area (in the middle of the arrow) cannot be identified, but characteristic elliptical features apparently have some connection with mineralization (see also Fig. 5). Satellite prints were obtained before field checking commenced.

and the soil appears barren. Where higher plants are present, *Juncus trifidus*, *Festuca ovina*, *Deschampsia flexuosa* and *Viscaria alpina* predominate at the expense of *Vaccinium myrtillus*, *V. vitis-idaea* and the normal herb species.

Areas 2 and 3 are adjacent to area 1; area 2 on the eastern side, and area 3 on the western side. Area 2 and 3 have fairly normal ground-cover vegetation (*Vaccinium* sp. with herbs) and some birch.

Areas 4—7 are typical for the normal forest of the area consisting of birch (*Betula pubescens*) with scattered pine (*Pinus sylvestris*). The ground-cover vegetation is dominated by *Vaccinium myrtillus* and *V. vitis-idaea* and various herbs.

Area 8 is chosen because it forms a part of an elliptical structure on the satellite images.

Areas 9—11 are rather wet open spaces in the otherwise dense birch forest, with normal ground cover vegetation of various grasses. These areas can be easily confused on the aerial photograph with areas of natural Cu poisoning.

The means and standard deviations of the reflectances and reflectance ratios were computed for each type area, and the results are shown in Table III and plotted in Fig. 7. Some observations on these results are:

(1) The copper-poisoned areas cannot be recognized visually on the channel 5 satellite images but were easily seen using channel 7/channel 5 (Figs. 5 and 6). However, various curvilinear and linear features seem to be present on the images. At present, it is not known if these features may have any connection with the Cu mineralization, but experience from other places (Saunders et al., 1973) indicates that this possibly could be the case.

(2) The reflectance *ratios* seem to be appropriate parameters for distinguishing areas of natural Cu poisoning from other type areas. The 7/5 ratio is lower for area 1 than for any other type area studied. It is interesting to note that the grass areas (9—11) (which on the aerial photograph can easily be misinterpreted as areas of natural Cu poisoning) show very high values for the 7/5 reflectance ratio. Moreover, grass areas are unique in showing higher values for the 6/5 ratio than for the 7/4 ratio. There is no evidence to indi-

TABLE II

Values used to convert LANDSAT-1 brightness numerics to reflectance data, Karasjok, northern Norway. Calculations at dark target reflectance are based on own measurements and data taken from Romanova (1964). The bright target reflectance was obtained at Karasjok. (Methodology of Lyon et al., 1975)

	N	CH4	CH5	CH6	CH7
Dark target radiance	(18)	16.24±0.66	9.57±0.72	6.76±0.83	0.88±0.03
Bright target radiance	(14)	31.79±5.78	31.71±7.64	30.14±7.36	12.85±3.87
Dark target reflectance (Romanova, 1964)		6.0	3.0	1.5	1.0
Bright target reflectance (measured at Karasjok)	(6)	22.6±0.68	26.0±0.52	31.5±0.63	32.8±1.31

N = number of measurements; CH = channel numbers.

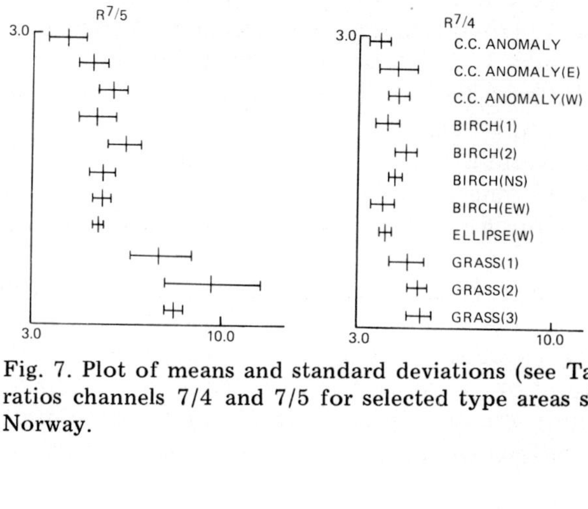

Fig. 7. Plot of means and standard deviations (see Table III) of LANDSAT-1 reflectance ratios channels 7/4 and 7/5 for selected type areas shown in Fig. 4, Karasjok, northern Norway.

TABLE III

Reflectance and reflectance ratios. LANDSAT-1 data from an area of natural Cu poisoning near Karasjok, northern Norway. For locations of the various type areas, see Fig. 4. The 7/4 and 7/5 values are plotted in Fig. 7

Type area		Reflectance				Reflectance ratios					
		4*	5	6	7	7/6	7/5	7/4	6/5	6/4	5/4
1. Anomaly	$\bar{X}$	10.97	9.27	30.82	40.17	1.30	4.33	3.69	3.36	2.83	0.89
$N = 9$	S	1.26	0.90	1.32	2.65	0.07	0.66	0.42	0.34	0.21	0.15
	V	0.11	0.10	0.04	0.07	0.05	0.15	0.11	0.10	0.07	0.17
2. Anomaly E	$\bar{X}$	10.79	8.13	30.64	40.64	1.33	5.22	4.54	3.94	2.93	0.77
$N = 9$	S	1.05	0.79	1.62	1.82	0.09	0.53	0.82	0.51	0.32	0.11
	V	0.10	0.10	0.05	0.04	0.07	0.10	0.18	0.13	0.11	0.14
3. Anomaly W	$\bar{X}$	9.67	7.70	31.4	41.0	1.34	5.90	4.43	4.41	3.31	0.76
$N = 9$	S	0.50	0.55	1.93	1.89	0.05	0.56	0.32	0.35	0.25	0.05
	V	0.05	0.07	0.06	0 05	0.04	0.09	0.07	0.08	0.08	0.07
4. Birch 1	$\bar{X}$	9.91	7.67	28.93	39.69	1.39	5.26	4.13	3.84	2.98	0.82
$N = 10$	S	0.59	0.86	2.16	1.86	0.14	0.69	0.41	0.62	0.27	0.10
	V	0.06	0.11	0.07	0.05	0.10	0.13	0.10	0.16	0.09	0.12
5. Birch 2	$\bar{X}$	9.40	6.92	32.39	45.26	1.40	6.47	4.86	4.68	3.47	0.77
$N = 10$	S	0.52	0.71	1.77	2.86	0.09	0.71	0.47	0.48	0.28	0.09
	V	0.06	0.10	0.05	0.06	0.06	0.11	0.10	0.10	0.08	0.12
6. Birch NS	$\bar{X}$	9.40	7.40	28.93	38.90	1.36	5.57	4.16	4.15	3.09	0.77
$N = 10$	S	0.52	0.57	1.67	2.08	0.08	0.58	0.22	0.41	0.23	0.09
	V	0.06	0.08	0.06	0.05	0.06	0.10	0.05	0.10	0.07	0.12
7. Birch EW	$\bar{X}$	9.60	6.93	27.63	38.12	1.38	5.50	3.97	3.99	2.88	0.72
$N = 10$	S	0.52	0.33	2.41	1.51	0.10	0.33	0.20	0.36	0.22	0.05
	V	0.05	0.08	0.09	0.04	0.07	0.06	0.05	0.09	0.07	0.07
8. Ellipse W	$\bar{X}$	10.24	7.78	30.44	41.73	1.37	5.36	4.08	3.91	2.97	0.76
$N = 19$	S	0.58	0.77	1.43	2.14	0.07	0.16	0.20	0.12	0.15	0.05
	V	0.06	0.10	0.05	0.05	0.05	0.03	0.05	0.03	0.05	0.08
9. Grass 1	$\bar{X}$	10.02	6.63	38.47	50.73	1.31	7.80	5.06	5.91	3.85	0.70
$N = 10$	S	0.66	1.13	4.23	6.42	0.08	1.24	0.73	0.89	0.53	0.17
	V	0.07	0.17	0.11	0.13	0.06	0.16	0.14	0.15	0.14	0.24
10. Grass 2	$\bar{X}$	9.11	5.27	35.36	48.61	1.37	9.66	5.40	6.94	3.91	0.79
$N = 10$	S	0.33	1.00	2.07	2.83	0.05	1.87	0.32	1.26	0.27	0.12
	V	0.04	0.19	0.06	0.06	0.04	0.19	0.06	0.18	0.07	0.15
11. Grass 3	$\bar{X}$	9.83	6.60	44.23	55.15	1.23	8.33	5.60	6.70	4.47	0.70
$N = 6$	S	0.41	0.00	2.28	2.16	0.05	0.33	0.41	0.36	0.37	0.00
	V	0.04	0.00	0.05	0.04	0.04	0.04	0.07	0.05	0.08	0.00

N = number of pixels; $\bar{X}$ = arithmetic mean; S = standard deviation of individual observations; V = coefficient of variation

*Channel numbers

cate that factors other than the copper poisoning cause the low 7/5 reflection ratio (elsewhere typically indicative of a low biomass). Waterlogging could not be the only reason because (a) the poisoned areas are not especially wet, and (b) bog areas have other reflectance characteristics than the poisoned area.

(3) The resolution of the LANDSAT-1 system is restricted by the pixel size of approximately 60 m × 80 m, therefore only patterns with this minimum size can be distinguished on the ground. Area 1 consists of 9 pixels, the majority of which represent copper-poisoned areas observed during earlier ground surveys (Låg and Bølviken, 1974). Possible poisoned areas within the other type areas are too small to be detected from space.

The results of the previously described classification and clustering procedures were:

(1) *Stepwise discriminant analysis.* Five of the eleven type areas outlined above (see Table III and Fig. 4) were selected to test if the anomaly area 1 could be separated from the four others; grass (type area 9), grass (type area 10), birch (type area 2) and an area adjacent to the anomaly (type area 3). Fig. 8 shows a plot of the results. The coordinates are canonical variables which represent numerical transformation of the LANDSAT-1 reflectance data given in Table II. In the plot, the area 1 anomaly is separated as a differ-

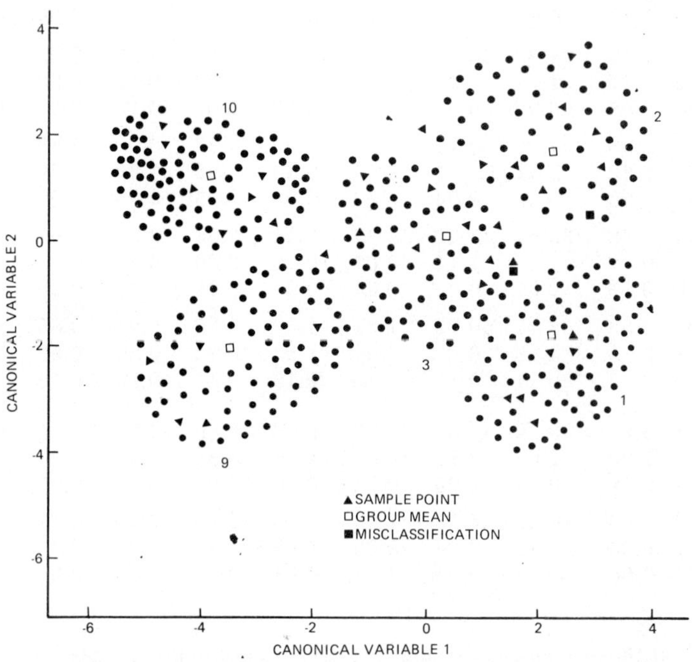

Fig. 8. Plot of five different types of control areas (1, 2, 3, 9, 10, shown in Fig. 4) as a function of the two best canonical variables calculated by stepwise discriminant analysis. The values of the canonical variables represent numerical transformations of the LANDSAT-1 reflectance data given in Table II. Solid dots show the classification domains and are not data points ($\Sigma N = 47$).

Fig. 9. Results of unsupervised pattern recognition of LANDSAT-1 reflectance data from an area with natural Cu poisoning; Karasjok, northern Norway. Type area (1) anomaly is used as a training set (see Fig. 4). Symbol C = poisoned, Cu.

ent group except for one pixel classified as belonging to the area 3, and one classified as birch (area 2). These two pixels are located near the centre of the area 1 anomaly (see Fig. 4). The amount of birch present on them, apparently, is sufficient to alter the spectral reflectance in a way that they seem to be more similar to the area 3 and area 2 groups than to the area 1 anomaly group.

(2) *Pattern recognition.* The pixels located in area 1 (Fig. 4) were selected as a training group and the STANSORT pattern recognition technique was used to search for similar areas. Fig. 9 shows the results of the pattern recognition of the survey area (800 hectares, 2000 pixels). The pixels with the "C"s are classified as potential poisoned areas. The most promising area is located southeast of area 1. The area will be investigated on subsequent ground surveys. The "C"s north and west of area 1 are probably errors caused by the classification procedure employed.

CONCLUSIONS

An area of natural Cu poisoning associated with a Cu deposit at Karasjok, northern Norway, can be characterized in satellite data primarily by the *low* values of the 7/5 channels reflectance ratio. Normal grass areas, which on the aerial photographs appear to be similar to the poisoned areas, show *high* values for the 7/5 channels reflectance ratio. Heavy metal poisoning seems to be a common feature near Norwegian sulphide deposits. Recording of multispectral reflectance from remote platforms, therefore, appears to be a promising regional exploration method for disseminated, shallow Cu deposits in this part of Norway.

Numerical classification techniques, such as pattern recognition and stepwise discriminant analysis proved to be adequate tools for distinguishing a copper-poisoned area from normal areas. These techniques could be applied to satellite data for routine searching for new poisoned areas.

Characteristic elliptical features appear in the channel 5 and 6 images and are more accentuated in the reflectance ratio colour images (in particular 6/5). These elliptical features might have some relation to the mineralization, a possibility which should be further investigated.

The spatial resolution of the LANDSAT-1 data is approximately 0.45 hectare which is not sufficient to identify all known patches of natural Cu poisoning at Karasjok or the occurrences of natural Pb poisoning found at other locations in Norway. This resolution problem may be overcome by using reflectance data obtained from digitized multispectral airphotography or airborne scanners.

Multispectral aerial photography has been obtained for these areas and is being digitized and analyzed by computer in an attempt to overcome the resolution limitations of LANDSAT-1.

ACKNOWLEDGEMENTS

The authors wish to express their gratitude to the Sydvaranger Mining company for the information and assistance offered to perform this study. Special thanks are due to Mr. Bernt Røsholt, exploration geologist of the Sydvaranger Mining Company, who provided the geological information. Mr. Røsholt found the mineralization at Karasjok on 1968 during the follow up of a geochemical anomaly obtained in a stream sediment survey suggested by Professor J.A.W. Bugge and conducted by the Geological Survey of Norway 1967 as contract work for the Sydvaranger Mining Company.

REFERENCES

Dixon, W.J. (Editor), 1970. Biomedical Computer Programs. Publications in Automatic Computations No. 2. University of California Press, Los Angeles, Calif., 773 pp.

Honey, F., 1975. STANSORT: a fully interactive tape reading program for ERTS II, philosophy and operational usage (with users manual). Stanford Remote Sensing Lab. Rep., No. 75-13, 30 pp.

Honey, F.R., Prelat, A.E. and Lyon, R.J.P., 1974. STANSORT: Stanford remote sensing laboratory pattern recognition and classification system. Proc. 9th Int. Symp. on Remote Sensing of Environment, Ann Arbor, Mich., 9 pp.

Låg, J. and Bølviken, B., 1974. Some naturally heavy-metal poisoned areas of interest in prospecting, soil chemistry and geomedicine. Nor. Geol. Unders., 304: 73—96.

Låg, J., Hvatum, O.Ø. and Bølviken, B., 1970. An occurrence of naturally lead poisoned soil at Kastat near Gjovik, Norway. Nor. Geol. Unders., 266: 141—159.

Lyon, R.J.P., Honey, F.R. and Ballew, G., 1975. A comparison of observed and model-predicted atmospheric perturbations on target radiances measured by ERTS. Proc. Inst. Electr. Electron. Eng. Symp. on Application of Remote Sensing and Digital Imagery to Mineral Exploration, Houston, Texas, 6 pp.

Levine, S.R., 1975. An interactive program for producing computer-enhanced ERTS images (IMAGE), II. Methodology and development. Stanford Remote Sensing Lab. Rep., No. 75-4.

Romanova, M.A., 1964. Air survey of sand deposits by spectral luminance. Laboratory of Aeromethods of the Academy of Sciences of the USSR, State Scientific and Technical Press for Literature on Petroleum and Solid Fuels, Leningrad.

Saunders, D.F., Thomas, G.E., Kinsman, F.E. and Beatty, D.F., 1973. ERTS-1 imagery use in reconnaissance prospecting-evaluation of the commercial utility of ERTS-1 imagery in structural reconnaissance for minerals and petroleum. Type III Final Report to NASA. U.S. Department of Commerce, Natl. Tech. Inf. Serv., E74-10345.

Journal of Geochemical Exploration, 8(1977)473—481

RECOGNITION OF MINERALIZED AREAS BY A REGIONAL GEOCHEMICAL SURVEY OF THE TILL-BLANKET IN NORTHERN FINLAND

HERMAN STIGZELIUS

Geological Survey of Finland, Espoo (Finland)

(Received October 19, 1976)

ABSTRACT

Stigzelius, H., 1977. Recognition of mineralized areas by a regional geochemical survey of the till-blanket in northern Finland. J. Geochem. Explor., 8: 473—481.

A regional geochemical sampling programme of glacial drift was initiated by the Geological Survey of Finland in 1971. By the end of 1975 a total of 100,000 samples of till had been collected in northern Finland, where glacial drift is of rather local origin. The till is sampled at a depth of 1.5 m at 100-m intervals along lines about 1 km apart. The minus 0.06-mm fraction is analyzed by emission spectroscopy and the results are processed by computer. Maps showing the distribution of the different elements are drawn by a Calcomp plotter on a 1:20,000 scale.

Areas in which the till has a uniform chemical composition are delineated as lithological moraine units. These are mostly compatible with local bedrock units and geophysical features. The survey is successful in recognizing mineralized areas.

INTRODUCTION

Finland is part of the Fennoscandian (Baltic) Shield, which was glaciated during the Pleistocene. The crystalline bedrock is covered by tills and associated sands, gravels and clays. This glacial terrain is rather flat and covered by lakes and ponds (10% of the land area) and peat lands (30%). Outcropping bedrock is sparse (3%) and outcrops are unevenly distributed among the different rocks, metamorphic and plutonic.

The use of conventional mineral exploration techniques is hampered by these terrain conditions. Airborne magnetic, electromagnetic and radiometric surveys have been used extensively, but geophysical information is not adequate for localizing mineral exploration targets. Furthermore, the results of geochemical stream sediment surveys do not produce sufficient information in a terrain characterized by a peat land drainage pattern. Till, being far the most widespread element of the glacial suite, is a suitable sampling medium for geochemical survey work in Finland.

In 1971, the Geological Survey started an extensive programme of geochemical sampling of till in northern Finland. The programme aims at finding new natural resources and, thus, new opportunities for all-year employment in this remote and economically undeveloped part of the country.

THE SURVEY AREA

The location of the survey area is indicated in Fig. 1. The average thickness of the Pleistocene deposits in northern Finland is about 6 m (Mäkinen, 1975). Five till beds have been identified. The three upper beds are considered to be Weichselian in age and the two lower beds Saalian or older (Hirvas et al., 1976). Up to 80% of the pebble material in the Weichselian tills are of local origin (Kujansuu, 1976). Lithologically, the tills reflect the composition of the underlying bedrock rather well.

The survey area covers a belt of greenstones and other supracrustal metamorphic rocks surrounded and intruded by plutonic rocks. The belt is considered to have a rather higher mineral potential than the surrounding granites, gneisses and granulites.

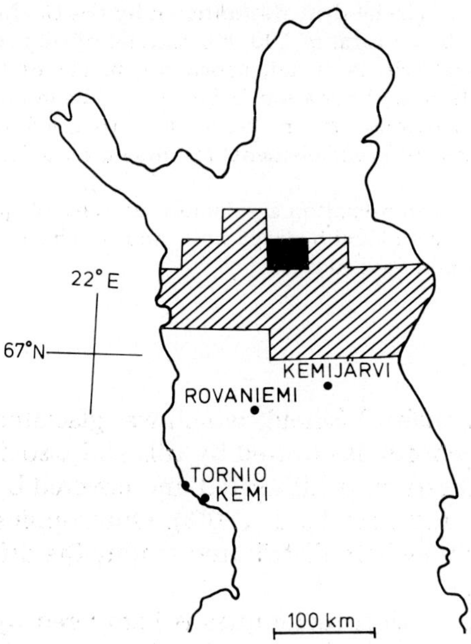

Fig. 1. Location of the survey area within the borders of Finland. The location of map sheet 3723 shown in black.

METHODS

The till samples are taken at a depth of about 1.5 m at 100-m intervals along lines about 1 km apart. The samples are collected using light-weight, petrol-operated percussion drills fitted with through-flow shell bits 25 and 35 mm in diameter. Much of the sampling is carried out in winter, when the movement of the sampling teams in the trackless terrain is made easy by use of skis and motor sledges.

The samples are dried and screened. The minus 0.06-mm fraction is analyzed by emission spectroscopy using an ARL 31000 Quantometer and the results are processed on an HP 3000 computer. At present, seventeen elements are analyzed: Si, Al, Fe, Mg, Ca, Na, K, Ti, V, Cr, Mn, Co, Ni, Cu, Zn, Pb and Ag. Complementary analytical services are provided by atomic absorption spectrophotometry, energy-dispersive X-ray fluorescence and delayed neutron methods.

Maps indicating the distribution of different elements are drawn by the computer on the scale 1:20,000. Each map covers an area of 10 km × 10 km and twelve such maps together make a 1:100,000 map sheet covering 30 km × 40 km. In such an area, 8000—12,000 samples are collected, which corresponds to a frequency of 7—10 samples/km^2.

At the end of 1975, a total of 100,000 till samples had been collected from the survey area. One map sheet (No. 3723), covering 1200 km^2, had been completed, with maps and listed analytical data published on microfiche cards (Geological Survey of Finland, 1975). A further description of sampling, analytical and data processing techniques applied in this survey has recently been presented by Kauranne (1975).

RESULTS

The results are shown as symbol maps drawn by a Calcomp plotter on the scale 1:20,000, one map for each element to be studied. The symbols on the maps, as shown in the legend of Fig. 2, indicate the grade of each sample classified as follows:

a — low	20% of the population
b — normal	60% of the population
c — high	10% of the population
d — very high	5% of the population
e — extremely high	2.5% of the population
f — top anomaly	2.5% of the population
	100%

This is the classification normally used in the survey, but it is easily changed by the computer in order to show special characteristics.

The maps are carefully studied in order to find anomalous metal contents. Correlations with geological and geophysical features are noted in order to delineate lithological moraine units with uniform chemical composition. The composite map analysis is carried out on the scales 1:50,000 and 1:100,000 on maps photographically compiled from the 1:20,000 maps. In the area

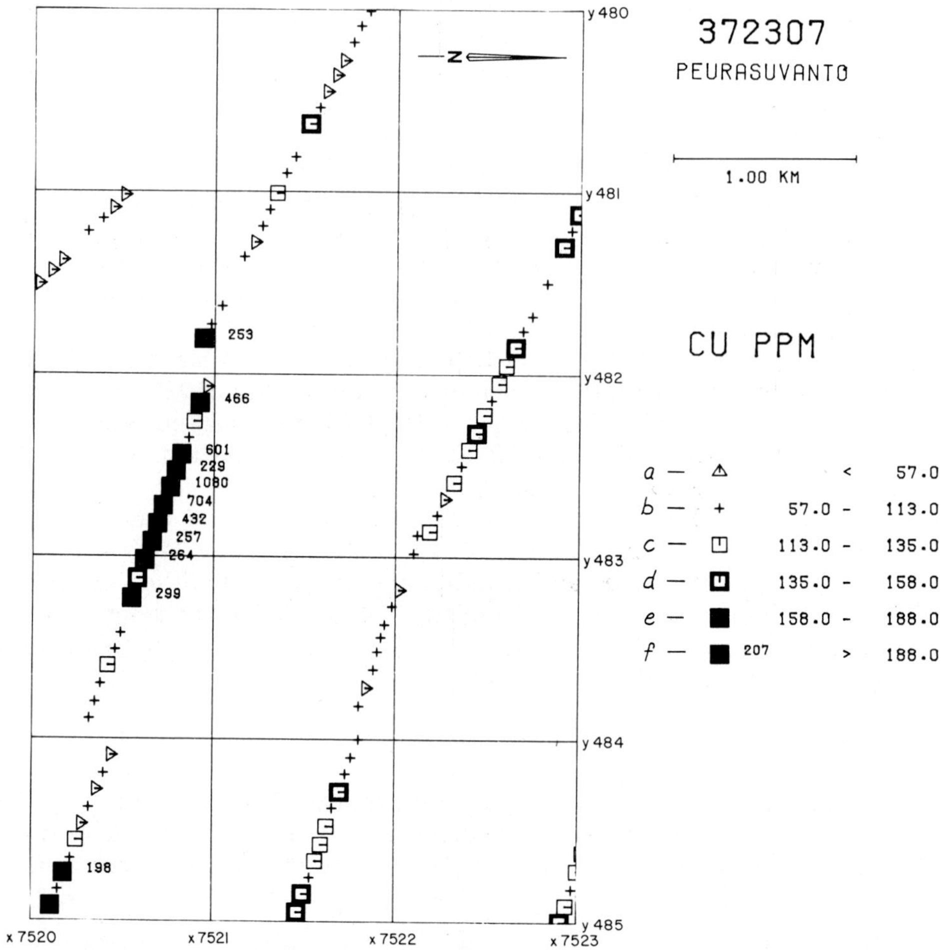

Fig. 2. The Cu anomaly in the Rumanilmankuusikko area. Abstract from map sheet 3723.

first studied (map sheet 3723), the Cu distribution map is of most interest from an economic point of view. Two examples of high Cu values are given here.

Rumanilmankuusikko. At this locality (x = 7521, y = 483; Fig. 2), an anomaly with a top value of 1080 ppm Cu is noted in an area completely covered by bogs and moraine hills. In a preliminary follow-up investigation, 3300 ppm Cu was encountered in one moraine sample and some mineralized boulders were found with Cu contents up to 0.4%. The exploration of the target continues.

Rookkiaapa. This area (x = 7529, y = 495; Fig. 3), with an anomalous Cu

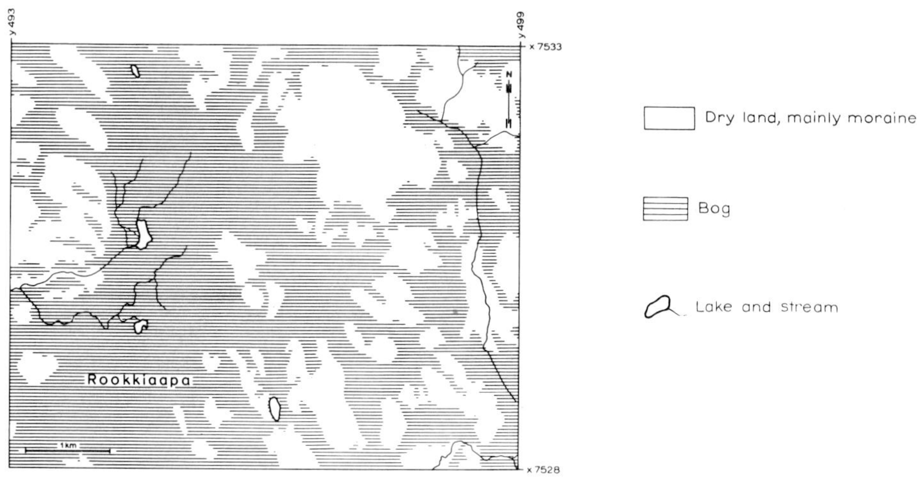

Fig. 3. Simplified field map of the Rookkiaapa area. Abstract from map sheet 3723. No houses, roads or tracks exist within the area.

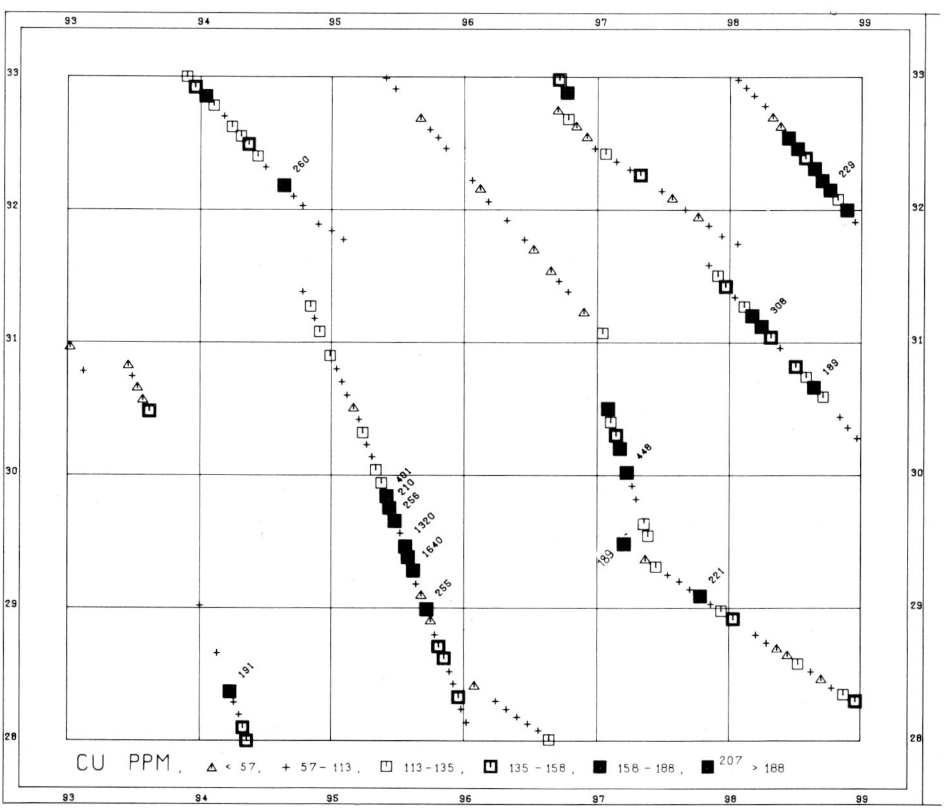

Fig. 4. The Cu anomaly in the Rookkiaapa area. Symbols as in Fig. 2. The map covers the area shown in Fig. 3.

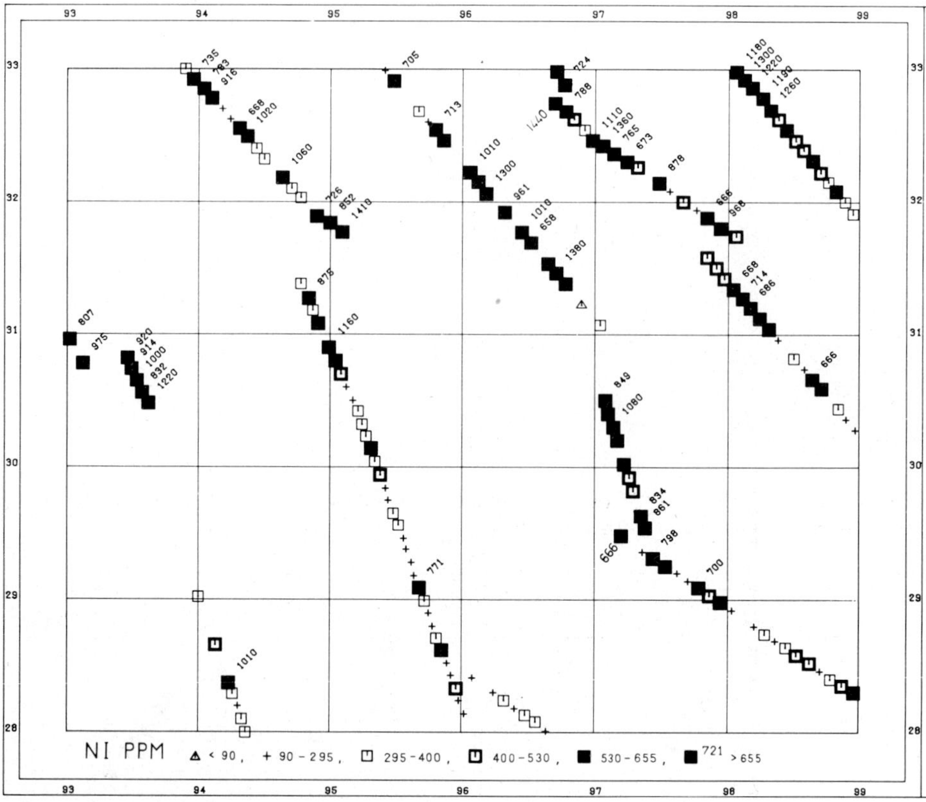

Fig. 5. The distribution of Ni in the Rookkiaapa area. The map covers the area shown in Fig. 3.

content in the till, is completely covered by a swamp inaccessible in summer (the Finnish word *aapa* indicates a wet bog). No roads or tracks exist within the area. The Cu distribution map (Fig. 4) indicates a strong uniform anomaly trending NE-SW, with top values of 1640 and 1320 ppm Cu with some other high Cu values further northeast. In comparison, the distribution maps for Ni (Fig. 5) and Cr (Fig. 6) show high values within a moraine unit in the northern part of the map; they are apparently associated with local chlorite-amphibole schists.

The Cu/Ni ratio was plotted to elucidate the relationship between the Cu anomaly and the Ni-Cr-Mg-rich formation (Fig. 7). The map shows that the lowest ratios occur where Ni, Cr and Mg have high background values, whereas high ratios coincide with the highest Cu anomaly outside the chlorite-am-

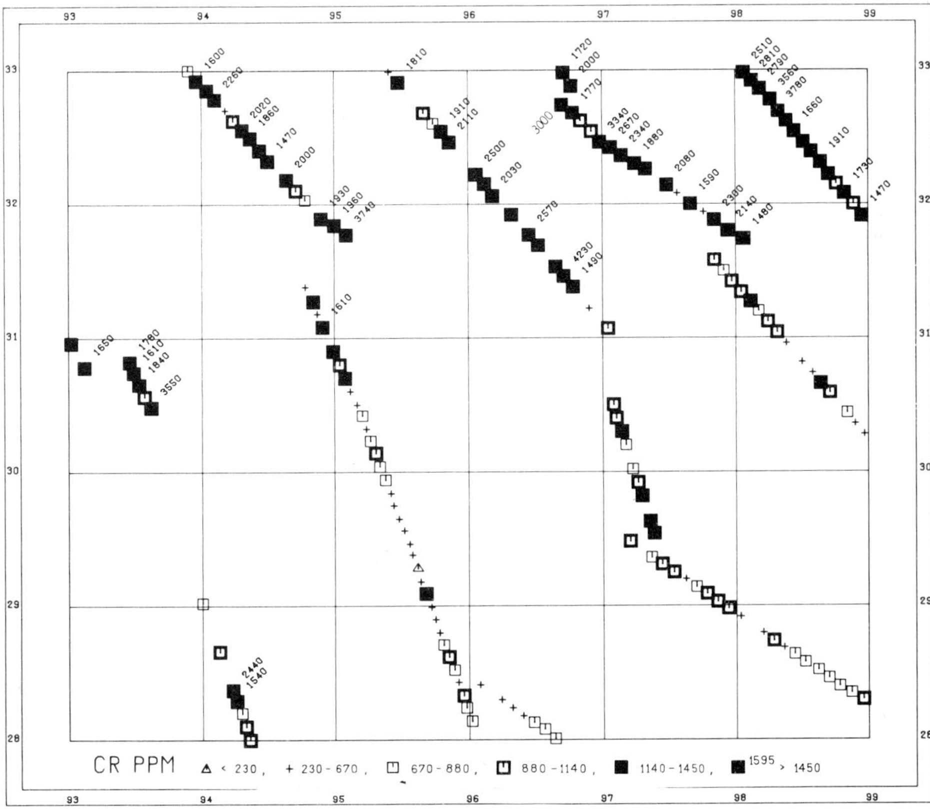

Fig. 6. The distribution of Cr in the Rookkiaapa area. The map covers the area shown in Fig. 3.

phibole schist area. The Cu anomaly is obviously related to a separate moraine unit.

The zone of high Ni background coincides well with the aeromagnetic anomaly recorded in the low-altitude survey carried out in the same area by the Geological Survey (Fig. 8). In contrast, the magnetic map contains no anomaly on the location of the highest Cu anomaly. Similarly, neither the electromagnetic nor the gamma-radiometric maps give any conclusive information about the nature of the Cu anomaly. The correlation between some geochemical and geophysical features in the Rookkiaapa area indicates that the main part of the till has not been transported more than a few hundred metres and that the moraine units apparently consist mainly of local weathered rock material. The Rookkiaapa terrain will be further explored.

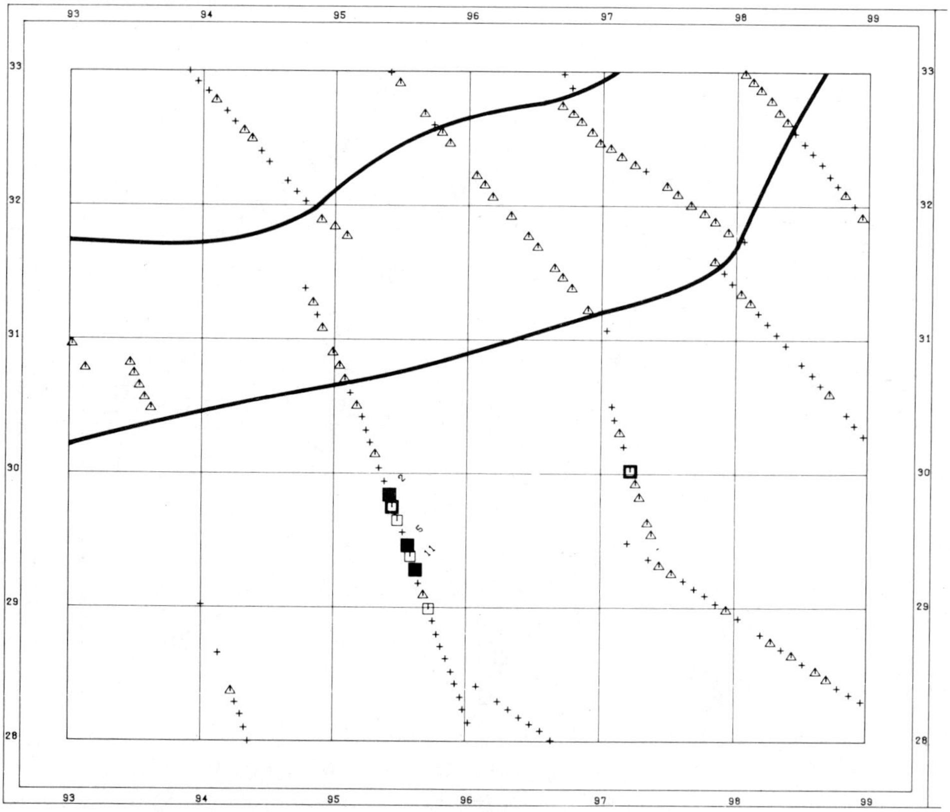

Fig. 7. Cu/Ni ratios in the Rookkiaapa area. Black squares indicate samples with very high Cu/Ni ratios (0.9—11); bold line delimits an area dominated by low values (ratios < 0.2). The map covers the area shown in Fig. 3.

CONCLUSIONS

The results hitherto obtained indicate that in the north of Finland, the chemical composition of the till blanket and the underlying bedrock are quite similar and that a regional geochemical survey of the till-blanket, in combination with a composite map analysis, can be successfully used to define mineralized areas. It seems that a new dimension in the exploration of glaciated terrain is growing from this composite approach.

ACKNOWLEDGEMENTS

The author is indebted to the staff members of the Geological Survey, especially Dr. L.K. Kauranne, Mr. Nils Gustavsson and Dr. Marjatta Okko.

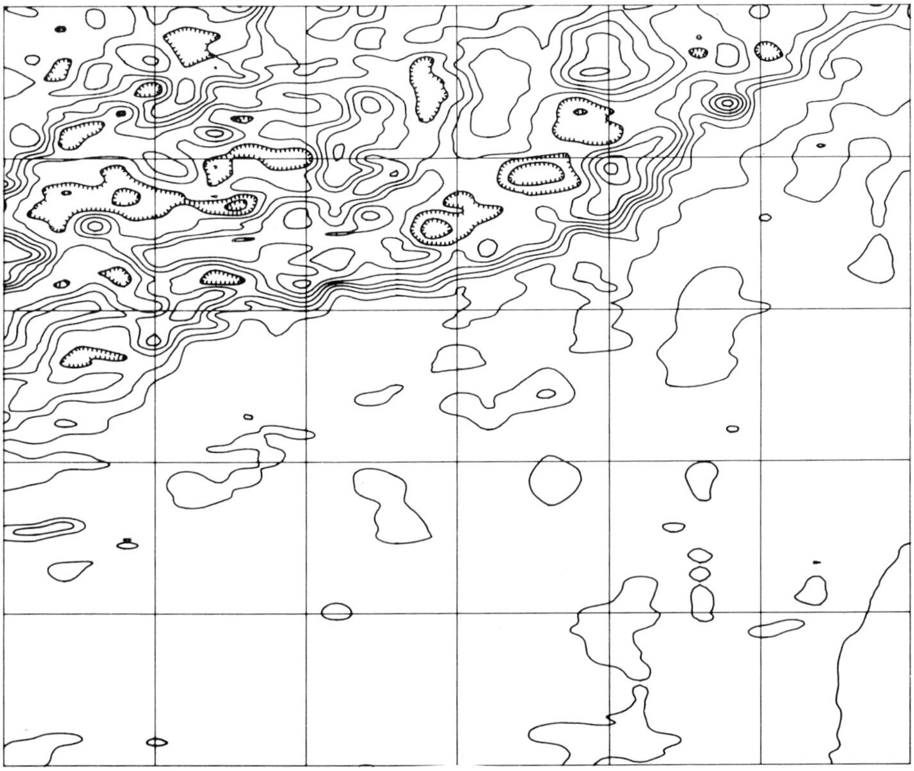

Fig. 8. Aeromagnetic map of the Rookkiaapa area. Contour interval is 500 nT. The map covers the area shown in Fig. 3.

REFERENCES

Geological Survey of Finland, 1975. Geochemical Data, Map Sheet Peurasuvanto 3723. Archival Record, Geological Survey of Finland, Espoo (available in microformat).

Hirvas, H., Kujansuu, R. and Tynni, R., 1976. Till stratigraphy in northern Finland. In: Quaternary Glaciations in the Northern Hemisphere. I.G.C.P. Proj. 73/1/24, Rep. No. 3. Billingham, Prague, pp. 256—273.

Kauranne, L.K., 1975. Regional geochemical mapping in Finland. In: M.J. Jones (Editor), Prospecting in Areas of Glacial Terrain. Institution of Mining and Metallurgy, London, pp. 71—81.

Kujansuu, R., 1976. Glaciogeological surveys for ore-prospecting purposes in northern Finland. In: R.F. Legget (Editor), Glacial Till. Royal Society of Canada, Ottawa, Ont., Spec. Publ. No. 12. pp. 225—239.

Mäkinen, K., 1975. Maanpeitteen paksuudesta Keski-Lapissa (manuscript). Turun Yliopisto, Turku, Finland.

Journal of Geochemical Exploration, 8(1977)483—494

SEQUENTIAL SOIL ANALYSIS IN EXPLORATION GEOCHEMISTRY

S. GATEHOUSE[1], D.W. RUSSELL[2] and J.C. VAN MOORT[2]

[1] *Geopeko Limited, Sydney, N.S.W. (Australia)*
[2] *Geology Department, University of Tasmania, Hobart, Tasmania (Australia)*

(Revised version accepted February 10, 1977)

ABSTRACT

Gatehouse, S., Russell, D.W. and Van Moort, J.C., 1977. Sequential soil analysis in exploration geochemistry. J. Geochem. Explor., 8: 483—494.

A system of progressive leaching of soil samples was developed by which it is possible to locate the distribution of metals over the various soil phases. The sequence includes a new method of reduction of iron oxides by hydrazine chloride, which is described in detail. The system enables one to determine on a single 2-g sample the elemental concentrations in the following phases: water-extractable, ion-exchangeable, reducible Mn, organic material, reducible Fe, clay and residue.

Application of the sequential analysis to eight Tasmanian soil profiles indicates a much stronger scavenging effect by iron oxides than is generally recognized. The quantitative expression of the scavenging effects aids in discriminating between true and false "anomalies". No geochemical soil sampling programmes should be undertaken without analysis of manganese and iron oxide fractions.

INTRODUCTION

Some soil components are enriched in several elements. For example, humus is reputedly enriched in Cu, U and, possibly, in Ni and Zn. Hydrous manganese and iron oxides adsorb Co, Ni, Cu and Zn and many cases of adsorption of elements (e.g. Cu and Ni), on clays are known. However, the relative strength of these sorbing and complexing properties are not known. In order to solve this problem, a method has been developed by which the relative amounts of major and trace elements in the different soil phases can be evaluated.

The soils studied come from the western part of Tasmania, an area with an annual precipitation of more than 250 cm. The soils are rather acid at pH 4.5.

SEQUENTIAL SOIL ANALYSIS

The principle of sequential soil analysis was described by Jackson (1956) as part of his procedures for the dispersion of soil minerals. To avoid the complications of his methods, a sequence was developed in which the samples were kept throughout the procedures in large glass centrifuge tubes. In the following order the water-soluble components are removed by a distilled water extraction, easily exchangeable cations are extracted in ammonium acetate, manganese oxides are reduced by hydroxylamine hydrochloride, organic material is destroyed by hydrogen peroxide, iron oxides are reduced by hydrazine chloride, whilst clay and silt residues are separated by centrifugation and subsequently leached with perchloric acid. During the initial procedures the pH is kept at 4.5. Carbon is determined on a separate sample.

The order of the sequence must be observed as the later steps are the more aggressive ones. For example, reduction of iron oxides will also reduce manganese oxides and early removal of clay will be incomplete.

The distilled water extraction (soil/water ratio 1:5) and the extraction with 1*M* ammonium acetate (buffered at soil pH) are classical procedures. The former extraction tends to be insignificant in wet climates. The ammonium acetate extraction exchanges cations in general, including exchangeable manganese.

The reduction of Mn with hydroxylamine hydrochloride (Chao, 1972) was modified in so far as the extraction was performed with buffered ammonium acetate instead of nitric acid. This method extracts most of the manganese and some iron oxides. Hydroxylamine hydrochloride is a stronger reducing agent than the more commonly used hydroquinone (Leeper, 1947) which is incapable of reducing bulk MnO_2.

After the removal of Mn, the hydrogen peroxide treatment extracts the metals complexed to the organic material, which is subsequently destroyed. The metals released are again prevented from sorption on clays by the presence of ammonium acetate. However, it should be noted that the hydrogen peroxide treatment will oxidize sulphides if they are present (Lynch, 1971). It was found that monomineralic powders of chalcopyrite, pyrite and sphalerite, and to a lesser extent galena, reacted strongly with the peroxide.

The free Fe in soils is usually reduced by sodium dithionite (Galabutskaya and Govarova, 1934), but the high concentrations of the salt used exclude further analysis by atomic absorption spectrophotometry. Other methods available for the reduction of iron oxides are equally unsuitable (Asami and Kumada, 1960). Consequently, reduction by hydrazine chloride has been used (Gatehouse, 1973). The reducing action of hydrazine in acid solution is attributed to the following half reactions:

$$N_2H_5^+ \rightleftharpoons N_2 + 5H^+ + 4e^- \qquad E^\circ = -0.23 \text{ volt}$$

$$N_2H_5^+ \rightleftharpoons N_2H_3 + 2H^+ + e^- \qquad E^\circ = -0.088 \text{ volt}$$

(N_2H_3 decays spontaneously according to:

$$2H^+ + 2N_2H_3 \rightleftharpoons N_2 + 2NH_4^+ \quad \therefore \quad \Delta G < 0 \text{ kcal/mole})$$

The second reaction is considered by Latimer (1964) to be applicable to one electron oxidizing agents, e.g.:

$$Fe^{3+} + e^- \rightleftharpoons Fe^{2+}$$

The reaction between hydrazine and soil iron oxides (represented by goethite) should be as follows:

$$FeOOH + N_2H_5^+ + H^+ \rightleftharpoons N_2H_3 + Fe^{2+} + 2H_2O \qquad \Delta G = -23.3 \text{ kcal/mole}$$

As can be seen from the free energy change the forward reaction is energetically favoured. Soil iron (Fe^{III}) oxides are reduced to Fe^{II} ions by acidified hydrazine solutions.

Precipitation of lead chloride ($PbCl_2$), the most insoluble of possible species in solution, may occur during the extraction if the Pb content of the sample exceeds 3.5%. During subsequent extracting with ammonium acetate the precipitated lead chloride, however, will be dissolved again (cf. Stecher, 1968).

Optical examination of silicate minerals before and after extraction revealed no obvious corrosion due to the hydrazine treatment. Colorimetric analyses for silica were performed on three of the extracted solutions by the method of Rainwater and Thatcher (1960). The amount of silica extracted from the soils was in all cases less than 50 ppm. The total silica contents of the soils were not measured, but was probably not less than 50% of the total soil. On this basis the amount of silica extracted by the hydrazine was less than 0.005% of the total soil silica.

The final determination of the metal content in the clay fraction does not reflect metals adsorbed on it, as these have already been removed.

The hot perchloric acid digestion was chosen to enable comparison with company data. The reliability of this method is not very good (Gedeon et al., 1977), although probably more than 80% of the total amounts of Fe, Mn, Zn, Pb and Cu would have been leached.

It should be stressed that all techniques outlined above do not rigidly discriminate between metal concentrations present in the various soil phases. However, they do give a general indication of the sites occupied and provide a reproducible system of analysis, provided the procedures outlined below are observed.

DETAILED ANALYTICAL PROCEDURES

(1) *Water extraction.* Weigh 2 g of air-dried soils of appropriate grain size, transfer to 50-ml glass centrifuge tubes, add 10 ml distilled water, soak for 1 hour, fit tubes tightly with stopper, shake vigorously for 20 minutes on

mechanical shaker, add 1 drop of 0.1% polyacrylamide flocculant solution and centrifuge till clear. Transfer the supernatant solutions with care and filter into a 25 ml volumetric flask, add 1 ml 1 : 4 HNO_3 and bring to volume with distilled water. Keep the residues in the centrifuge tubes.

(2) *Ammonium acetate extraction.* Add 10 ml of 1*M* ammonium acetate solution (buffered with acetic acid to soil pH 4.5) to the residues in the centrifuge tubes. Stopper and shake for 20 minutes. Centrifuge and decant through filter paper (alternatively millipore) into 100 ml volumetric flasks. Repeat twice and bring volumetric flasks up to volume.

(3) *Hydroxylamine hydrochloride extraction.* Add 10 ml of 0.1*M* hydroxylamine hydrochloride in 1*M* ammonium acetate solution buffered to soil pH of 4.5. Shake 20 minutes as before and decant through filter paper into 100 ml volumetric flasks. Repeat twice and bring the flasks up to volume.

(4) *Hydrogen peroxide extraction.* Add to each sample 4 drops of 30% (= 100 volume) hydrogen peroxide. The initial reaction may froth violently but this can be prevented by adding some drops of ethanol. Heat the centrifuge tubes in an aluminium block (60° C) until the initial reaction has ceased. Carefully add some more hydrogen peroxide in 0.5-ml lots until the initial reaction has ceased. This takes between 8 and 12 hours. Add then 5 ml of hydrogen peroxide and leave the solutions overnight. Mix and heat to 80° C until all the hydrogen peroxide is destroyed (approximately 10 minutes). Add 10 ml of 1*M* ammonium acetate, centrifuge till clear and decant into 100 ml volumetric flasks. Extract two more times with 10 ml ammonium acetate. Bring the volumetric flask up to volume of 100 ml with distilled water.

(5) *Hydrazine chloride extraction.* Add 100 g 99% hydrazine hydrate to 1.8 litres of distilled water. Bring to pH (4.5) with concentrated hydrochloric acid and then bring the volume to 2 litres. Add 10 ml of the hydrazine chloride stock solution to the residues in the centrifuge tubes. Heat samples to 90° C in the aluminium block. Mix periodically. Bleaching characterizes the end of the reaction, which takes several hours. Add another 10 ml of the stock solution, and heat to ensure complete reaction, and leave overnight. Centrifuge, decant into 100 ml volumetric flasks and extract three times with ammonium acetate as in step 2.

(6) *Separation of the clay fraction from the silt residue.* Fill up the residues from the last procedure to a depth of approximately 8 cm and centrifuge at appropriate speed and time to ensure settling of all material $> 2\ \mu m$. It is important to increase and decrease the speed of the centrifuge as soon a possible. Conditions vary from instrument to instrument and must be calculated (Jackson, 1956). Decant the clay suspension and repeat four times or until the suspension becomes clear. Air-dry the resulting clay fractions and weigh. Keep the silt residues for step 8.

(7) *Perchloric acid leach of the clay fraction.* Transfer the dry clays to another set of glass centrifuge tubes, add 10 ml perchloric acid and heat with

periodic mixing at 200°C for 2 hours in a fume cupboard. After cooling, the samples are balanced with distilled water and centrifuged and the supernatants transferred to 50 ml volumetric flasks. Two washings with 5 ml distilled water, followed by two washings with 5 ml 1 : 10 hydrochloric acid and two final washings with distilled water (2 × 5 ml) are necessary. Centrifuge each time. Make up to volumes of 50 ml with distilled water.

(8) *Perchloric acid leach of (silt) residue fraction.* Add 10 ml perchloric acid to the silt residue from step 6, leach as in step 7. Air-dry residue, and weigh.

Alternatively the clay fraction and the residue may be digested with a hydrofluoric-perchloric acid mixture.

The metals extracted in the various procedures were in each case analyzed by atomic absorption spectrometry, using matching standards and applying correction for molecular absorption if necessary.

In addition to the techniques outlined above, the carbon content was determined by dry combustion followed by volumetric gas analysis (van Moort and de Vries, 1970).

SELECTION OF TEST MATERIAL AND SELECTION OF ELEMENTS ANALYZED

Soils with widely varying Fe contents were chosen by Gatehouse (1973) in order to evaluate the scavenging effect of the iron oxides. The first two of these soils near Dundas, western Tasmania, (South Comet area) are shallow inceptisols near quartzitic ridges (G1 and G2). G3 is a deep soil developed in a saddle in a fault zone and G4 represents a gossan on a pyrite-bearing sphalerite-galena ore.

The four soil profiles near Williamsford, western Tasmania (West Hercules area), studied by Russell (1976) are less variable and are all developed on volcanogenic sediments. Soil WH1 is, like G1 and G2, a shallow soil high on a slope, the other three profiles represent deeper soils down slope, in virgin forest. Soils WH2 and WH4 are located in an area of general high Pb values (> 200 ppm Pb). Soil WH1 is above this zone and soil WH3 is below.

The samples come from areas with potential Cu, Pb and Zn mineralization, and analysis has been restricted to these elements and potential scavengers (Mn, Fe, C and clay). The South Comet area contains some sphalerite veins while the West Hercules area is near a zinc mine.

Pits were dug and all profiles were sampled at 10-cm intervals. Gatehouse (1973) studied both the minus 20-mesh (< 850 μm) and minus 80-mesh (< 180 μm) fractions of the soils. The analysis of the minus 80-mesh fraction gave more erratic trends and the results are not recorded here. Russell (1976) mechanically ground the (rather fine) soils down to minus 20-mesh using a Hobart Cadet soil grinder.

The analytical treatment of the samples has been essentially the same, as summarized above. The South Comet samples, however, have not been extracted with distilled water and hydroxylamine hydrochloride. The later ana-

TABLE I

Elemental concentration in various soil phases. South Comet area (minus 20-mesh fraction = < 850 μm; all results in ppm unless otherwise stated)

			AMMONIUM ACETATE EXTRACTABLE					HYDROGEN PEROXIDE EXTRACTABLE (1)					
PROFILE	CO.NO.	DEPTH (in cm)	Fe	Mn	Zn	Pb	Cu	Fe	Mn	Zn	Pb	Cu	CARB
G1	KH11	0-20	13	2	0.3	-	3.1	38	1	27	17	8.5	9.
		20-30	5	1	0.1	-	5.6	29	2	40	8	6.2	12.
		30-40	-	-	0.1	-	2.3	15	-	-	-	5.7	1.
		40-50	-	-	-	-	2.2	-	-	-	-	2.8	0.
		50-60	180	3	1.4	-	2.3	9	-	-	-	2.9	0.0
G2	KH2	0-10	62	15	23	26	2.4	25	3	0.1	-	5.0	2.
		10-20	59	4	3	15	1.6	25	11	12	-	5.9	5.
		20-30	28	2	0.4	17	4.0	27	3	-	-	4.8	5.
		30-40	10	2	-	24	2.3	18	1	2.4	-	4.7	4.
		40-50	12	4	-	-	0.7	17	-	-	8	3.7	2.
		50-60	3	4	0.2	16	1.5	9	-	-	-	2.9	2.
		60-70	16	1	-	17	0.7	6	-	62.7	-	2.9	0.
G3	KH10	0-10	43	17	69	-	1.5	20	66	8.7	-	10.4	5.7
		10-20	54	16	48	16	6.4	21	14	6.3	-	7.6	5.6
		20-30	99	9	-	-	0.7	29	1	0.6	-	2.9	4.4
		30-40	120	4	0.4	17	4.1	12	1	9.4	-	-	2.9
		40-50	28	1	60	-	1.5	24	1	1.3	-	1.0	2.
		50-60	39	20	52	9	3.2	18	-	0.3	-	4.9	0.9
		60-70	30	1.7	3	8	7.9	208	0.40%	266	259	8.6	1.
G4	Sc2	0-10	103	173	69	192	6.0	307	0.31%	97	154	15.3	13.8
		10-20	210	85	65	155	2.7	53	0.32%	87	70	5.7	10.3
		20-30	18	34	62	65	1.6	61	1.70%	13	-	4.9	2.
Limit of Detection			2	0.2	0.1	2	0.5	2	0.2	0.1	2	0.5	

1) Extractable from manganese oxides and organic material.

- is below limit of detection

TABLE II

Elemental concentration in various soil phases. West Hercules area (total soil; all results in ppm unless otherwise stated)

			WATER SOLUBLE					AMMONIUM ACETATE EXTRACTABLE (1)					HYDROXYLAMINE HYDRO-CHLORIDE EXTRACTABLE (2)					HYDROGEN PEROXIDE EXTRACTABLE (3)					
PROFILE	CO.NO.	DEPTH (in cm)	Fe	Mn	Zn	Pb	Cu	Fe	Mn	Zn	Pb	Cu	Fe	Mn	Zn	Pb	Cu	Fe	Mn	Zn	Pb	Cu	CARB
WH1	400N	0-10	-	-	3	2	15	24	12	-	30	-	-	7	-	6	-	15	4	-	19	-	3.
	225W	10-20	-	-	-	-	5	36	5	-	15	-	-	3	-	15	-	7	-	-	-	-	2,
		20-30	-	-	0	0	2	52	5	-	24	-	-	4	-			2	-	-	-	-	1.
		30-40	2		-	1	1	76	10	-	13	-	-	3	-	-	-	-	-	-	-	-	1.
WH2	400N	0-10	40	28	-	3	2	593	277	-	394	-	-	153	-	28	-	1757	137	-	95	-	8.
	700W	10-20	40	22	-	1	1	652	144	-	297	-	-	165	-	135	-	2643	118	-	156	-	8.
		20-30	62	12	-	0	0	465	141	-	455	3	-	381	-	171	-	2745	131	0	154	-	7.
		30-40	63	9	-	2	-	181	117	-	606	6	-	349	-	48	-	343	89	4	146	-	6.
		40-50	66	4	-	3	-	107	52	-	635	-	-	63	-	80	-	320	137	2	139	-	3.
		50-60	90	4	-	7	2	28	29	-	1361	-	-	140	-	6	-	147	101	12	274	-	3.
		60-70	40	5	-	2	2	3	23	27	1255	-	-	-	-	0	-	88	54	29	185	-	2.
		70-80	74	2	-	5	2	10	19	16	1064	-	-	-	-	36	-	138	57	37	776	-	2.
		80-90	30	3	0	4	2	74	17	57	538	-	-	-	-	0	-	60	28	35	38	-	4.
WH3	400N	0-10	88	10	5	-	7	586	21	22	-	-	-	-	-	-	-	189	12	-	-	-	15.
	1125W	10-20	100	5	-	-	3	523	11	35	-	-	-	-	-	-	-	137	11	-	-	-	13.
		20-30	40	0	2	-	2	147	2	38	-	-	-	-	-	-	-	28	6	-	-	-	6.
		30-40	30	0	-	-	-	35	1	36	-	-	-	-	-	-	-	6	9	-	-	-	4.
		40-50	42	0	5	-	3	75	2	30	-	-	-	-	-	-	-	6	10	3	-	-	1.
WH4	800N	0-10	82	98	-	0	3	862	521	73	95	-	-	-	-	-	-	2073	149	14	21	-	26.
	700W	10-20	87	99	2	-	4	905	403	51	140	-	-	-	-	23	-	6	123	3	-	-	12.
		20-30	86	114	2	-	5	857	307	39	181	-	-	38	-	22	-	219	81	-	-	-	8.
		30-40	55	234	11	1	4	710	628	3	223	-	-	89	-	115	-	168	173	-	14	-	6.
		40-50	40	261	0	3	-	643	554	54	242	-	-	179	-	196	-	120	123	-	-	-	5.
		50-60	35	273	5	0	4	735	837	35	260	-	-	441		393	-	57	172	-	-	-	5.
		60-70	26	56	-	0	4	715	463	3	307	-	-	849	-	556	-	79	201	-	19	-	4.
		70-80	18	208	3	-	26	576	582	63	435	-	-	680	-	537	-	34	185	-	16	-	2.
		80-90	13	155	2	4	4	357	278	14	358	-	-	333	-	198	-	41	121	-	15	-	1.
		90-100	38	65	4	-	4	352	101	14	143	-	-	505	-	66	-	17	41	7	-	-	1.

1) The ammonium acetate extraction has been made later on separate samples.

2) The original results of the hydroxyl hydrochloride acetate extraction have been corrected for the results of the later ammonium acetate extraction.

- is below limit of detection

3) Because of the earlier removal of manganese oxides the data listed represent only values related to organic material.

HYDRAZINE CHLORIDE EXTRACTABLE					PERCHLORIC EXTRACTABLE IN CLAY						PERCHLORIC ACID EXTRACTABLE ON SILT FRACTION					TOTAL ON SUMMATION				
Fe	Mn	Zn	Pb	Cu	Fe	Mn	Zn	Pb	Cu	CLAY%	Fe	Mn	Zn	Pb	Cu	Fe	Mn	Zn	Pb	Cu
750	0.6	18	33	6.6	192	-	4	3	1.0	3.4	483	0.2	15	1	0.3	0.15%	3.8	40	54	19.5
548	-	50	25	3.8	230	-	13	2	1.9	4.9	505	0.6	14	4	1.2	0.13%	3.8	117	29	18.7
579	0.4	9	14	1.7	513	-	2	1	-	6.5	0.11%	0.4	13	4	0.5	0.22%	3.6	25	19	10.2
510	1	5	7	0.7	543	0.8	6	-	0.3	5.9	0.11%	0.4	11	2	0.5	0.22%	2.2	21	9	7.2
815	2	7	7	3.8	851	0.3	4	1	0.3	8.9	0.14%	0.5	16	4	1.2	0.33%	5.8	28	12	10.5
0.55%	7.7	21	50	8.9	743	1.3	9	3	1.7	12.4	934	0.1	11	4	0.5	0.73%	27.0	69	83	18.5
0.61%	12	13	59	7.7	744	1.6	4	4	2.0	12.1	560	0.3	9	4	2.1	0.75%	29.9	41	92	19.3
0.62%	1	17	39	10.5	467	0.3	21	2	2.3	10.0	481	0.1	5	1	1.8	0.72%	6.4	43	59	23.4
1.42%	7.3	18	30	2.7	494	0.3	0.3	4	0.6	10.0	939	0.6	3	6	2.4	1.57%	11	24	64	12.7
2.88%	17	37	38	4.9	732	1.2	6	2	-	17.6	678	0.3	11	4	0.3	3.02%	33	54	52	9.6
2.21%	7.4	58	38	8.3	845	1.2	2	1	1.9	18.9	758	0.3	3	2	5.8	2.37%	13	64	57	20.4
2.31%	12	72	48	6.3	954	1.8	2	4	1.6	ND	752	0.3	3	6	0.7	2.48%	15	139	75	12.2
0.75%	180	97	104	8.2	152	0.2	48	2	0.4	2.8	3.27%	100	302	2	6.1	4.04%	363	524	108	26.6
0.87%	57	90	66	11.9	131	0.8	66	1	0.6	3.0	3.15%	97	340	2	3.1	4.04%	185	550	85	29.6
1.73%	12	74	56	11.8	126	0.2	67	1	0.6	3.8	2.73%	96	352	6	4.2	4.49%	188	494	63	20.7
4.94%	23	37	89	43.9	145	0.5	91	1	1.0	3.7	570	101	372	6	6.0	5.03%	129	510	113	55.0
12.0 %	41	539	172	29.0	415	0.6	95	2	2.3	4.9	2.60%	68	569	6	5.5	14.7 %	112	0.13%	180	39.3
7.6 %	24	275	109	50.3	536	0.8	93	7	0.6	5.2	3.24%	95	536	28	28.1	10.9 %	139	956	153	87.1
16.2 %	94	697	251	12.7	120	0.4	33	2	-	1.3	3.73%	109	0.10%	14	10.6	19.9 %	0.42%	0.20%	524	39.8
19.0 %	0.93%	0.26%	0.59%	85.9	155	1	3	5	2	8.2	364	0.3	9	5	4.3	19.1 %	1.26%	0.28%	0.63%	113.5
30.7 %	1.22%	0.36%	0.78%	85.2	99	-	0.5	8	-	0.8	0.18%	6.6	8	180	1.4	30.9 %	1.55%	0.37%	0.82%	95.0
52.3 %	1.32%	0.47%	0.93%	99.6	139	-	10.1	5	1.9	0.7	0.23%	5.1	9	237	1.5	52.5 %	3.02%	0.48%	0.96%	109.5
2	0.2	0.1	2	0.5	1	0.1	0.1	1	0.2	0.2	1	0.1	0.1	1	0.2					

Analyst: S. Gatehouse, 1973.

HYDRAZINE CHLORIDE EXTRACTABLE					PERCHLORIC EXTRACTABLE IN CLAY						PERCHLORIC ACID EXTRACTABLE ON SILT FRACTION					TOTAL ON SUMMATION				
Fe	Mn	Zn	Pb	Cu	Fe	Mn	Zn	Pb	Cu	CLAY%	Fe	Mn	Zn	Pb	Cu	Fe	Mn	Zn	Pb	Cu
150	-	-	-	-	410	3	3	30	-	6.6	2060	31	50	91	-	0.26%	57	56	178	15
134	-	-	21	-	400	3	3	30	-	6.34	2350	34	13	97	-	0.29%	45	100	176	5
135	-	-	-	-	370	3	3	30	-	5.6	2760	37	12	123	12	0.33%	49	126	175	14
169	-	-	-	-	410	3	3	30	-	5.54	2800	44	19	109	13	0.34%	60	112	153	14
2.35%	245	-	137	-	1.16%	33	30	250	12	28.1	4.09%	128	73	613	-	7.84%	1000	103	1500	14
3.52%	370	13	165	-	1.23%	37	28	250	6	26.5	5.37%	173	68	619*	-	10.45%	1000	114	1600	7
3.71%	449	0	293	-	1.28%	38	29	240	6	23.4	7.63%	333	90	1830	-	13.02%	1500	119	3140	9
2.86%	366	2	274	-	1.46%	43	34	260	9	31.8	7.32%	543	98	1400	12	11.70%	1500	138	2800	27
2.16%	218	3	292	-	1.52%	40	37	260	9	34.6	6.71%	195	98	976	12	10.44%	709	140	2400	26
1.45%	91	4	475	-	1.94%	40	42	300	12	36.5	5.37%	117	142	988	12	8.79%	522	200	3500	26
0.93%	52	72	369	-	1.85%	40	62	300	15	39.9	4.51%	93	148	778	12	7.31%	262	338	2000	29
1.21%	83	34	395	-	1.79%	48	61	330	15	39.2	4.59%	121	194	461	18	7.62%	329	342	2600	35
0.37%	27	37	166	-	0.71%	26	38	230	13	25.9	2.56%	90	179	308	26	3.59%	179	346	1300	41
3230	-	36	-	-	8100	9	28	30	12	16.3	2.21%	37	68	50	-	3.43%	89	159	80	19
4444	-	29	-	-	2900	-	10	20	10	6.3	3.53%	59	85	68	13	4.34%	86	159	88	26
4434	-	42	-	-	1300	-	6	10	9	2.6	4.68%	57	101	63	19	5.28%	65	189	73	30
2830	-	40	-	-	3500	-	9	15	3	5.5	5.53%	63	119	63	19	6.17%	73	204	78	22
3442	-	24	-	-	2400	-	6	15	3	3.3	9.74%	84	175	58	39	10.34%	96	243	73	45
0.54%	71	25	70	-	7700	21	42	130	6	23.2	1.85%	48	28	155	12	3.46%	908	182	471	21
1.21%	168	27	54	-	9800	28	28	170	9	29.9	2.97%	89	44	228	13	5.26%	960	155	615	26
1.38%	135	20	61	-	1.10%	29	29	170	19	32.0	3.16%	84	39	239	13	7.6 %	788	129	683	37
1.60%	184	25	61	-	1.32%	36	30	210	6	34.6	4.01%	120	48	211	18	7.03%	1400	107	934	28
1.69%	192	3	74	-	1.78%	35	28	180	9	31.7	4.15%	119	44	396	13	7.11%	1463	129	1100	22
1.75%	247	35	87	-	1.37%	39	29	190	10	33.1	4.84%	163	59	386	20	8.04%	2200	163	1300	34
2.14%	400	19	116	-	1.81%	55	39	250	10	37.5	5.61%	323	77	645	13	9.64%	2400	138	1900	27
2.18%	343	14	112	-	1.37%	42	33	210	7	31.4	5.56%	294	216	654	20	9.18%	2300	329	2000	53
1.66%	261	23	73	-	1.30%	40	31	190	6	30.1	5.78%	329	143	609	19	8.79%	1517	213	1500	29
1.12%	116	23	48	-	0.71%	48	21	110	3	17.1	3.87%	117	110	196	25	5.75%	993	179	563	32

Analyst: D.W. Russell, 1976.

lyzed West Hercules samples have been subjected to a complete extraction sequence. The ammonium acetate extraction of this material, however, has been inserted later and the data of the following hydroxylamine hydrochloride extraction have been corrected accordingly.

DISCUSSION

The analytical data are tabled in Tables I and II. The elemental distributions over the various soil phases (relative to total amount present) are represented by bar diagrams (Figs. 1 and 2). It should be kept in mind that the West Hercules analyses represent the entire soil, and not just the minus 20-mesh fraction.

The West Hercules samples indicate that an appreciable proportion of Cu is water-soluble, particularly in the organic-rich top soils. The extracts which proved to be richer in Cu were also browner in colour, indicating the presence of soluble organic material. Zn and Pb compounds are apparently almost insoluble in water. Water-soluble Mn is of some importance in the last West Hercules profile. Although high values of water-soluble Fe are recorded, these are a negligible proportion of the total Fe present.

The exchangeable Fe and Mn correlate more or less with the presence of organic material in the profile. However, no particular correlation can be noticed between these elements and the abundance of clay. Exchangeable Zn, Pb and Cu do not correlate with either organic matter or clay. The exchange process is indiscriminate and it is impossible to say which soil phase shows the strongest exchange capacity for the particular elements.

Reducible Mn is an important soil component in the two deep manganese-rich profiles WH2 and WH4 and was observed in the deep South Comet soil G3 and gossan G4. The manganese oxides occur apparently only in the well-developed soils. Table I and Fig. 2 indicate that scavenging of Pb has occurred as can be expected from the literature (e.g. Taylor and McKenzie, 1966; Levinson, 1974).There is no evidence of Zn being associated with the manganese oxides.

Lower amounts of Fe, Mn, Pb and Zn than expected, appear to be complexed by the organic material present (cf. Baker, 1973). In the case of Mn and Zn, and, to a lesser extent, Pb, this complexing is of some importance. Cu is present in the organic phase in the South Comet area.

Both quantitatively and proportionally large amounts of Mn, Zn and Pb are extracted together with the Fe. This trend is clearly developed in the deeper soils of the West Hercules area. In both areas studied, these metals are absolutely and proportionally more abundant in the hydrazine chloride reducible iron oxides than in the manganese oxides and organic matter, with the exception of WH4, where manganese oxides and associated Pb are of some importance. In the South Comet area, more Cu is associated with iron oxides than with organic material and manganese oxides (combined in the

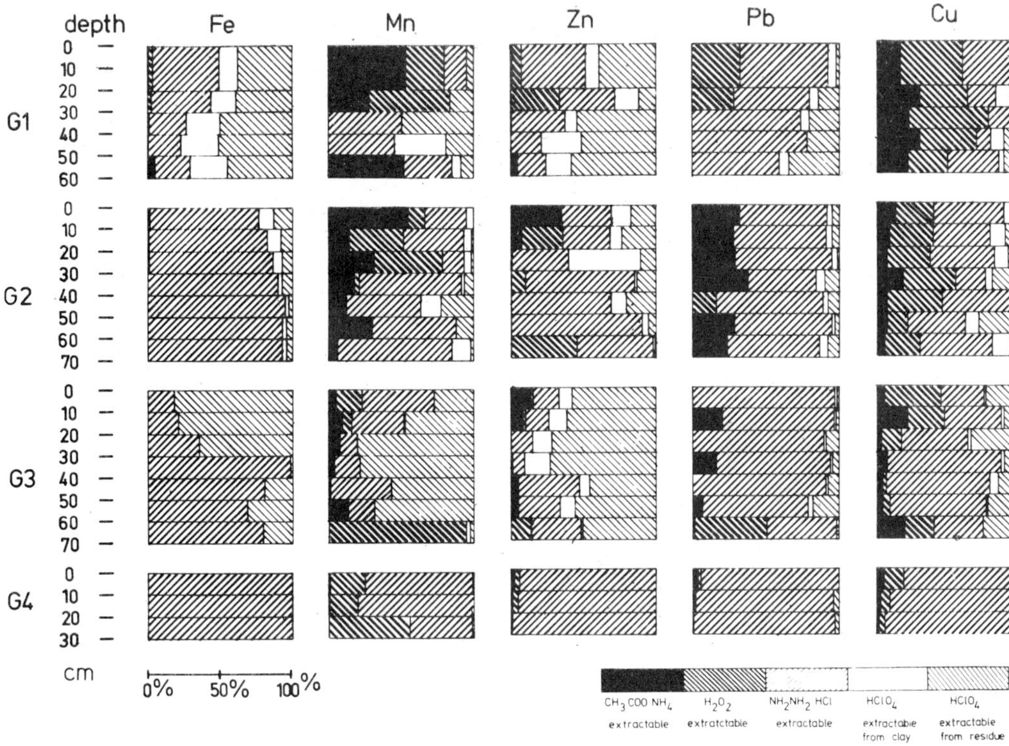

Fig. 1. Distribution of the total amount of Fe, Mn, Zn, Pb and Cu available in each 10-cm interval of South Comet soils, western Tasmania. The sequence of extraction is from left to right. The ammonium acetate extraction includes water-soluble as well as exchangeable ions (from clay, organic material, etc.). The hydrogen peroxide extraction includes metals complexed by organic material; it also reduces and extracts manganese oxides. Hydrazine chloride extractable metals are released after reduction of iron oxides. The perchloric acid extractions represent the metals in the clay fraction and coarser residue. (After Gatehouse, 1973.)

hydrogen peroxide extraction). The importance of scavenging by hydrous iron oxides has been underestimated in the past.

It is difficult to comment on the metal extraction from the clay fractions, particularly in the South Comet area, where the clay contents are very low. In the West Hercules area, the clay contents are rather constant to near the bottom of the profiles. Higher in the profiles there is some evidence of the absolute and relatively higher amounts of metals in the clay fractions.

The increase in Mn, Pb and Cu in the residue fraction of many profiles reflects the less weathered state of the deeper levels of the profiles. The same trend of elemental concentrations increasing with depth applies also to the total values. The higher proportion of residue in the West Hercules soils is the result of analyzing the entire sample, rather than the minus 20-mesh fraction.

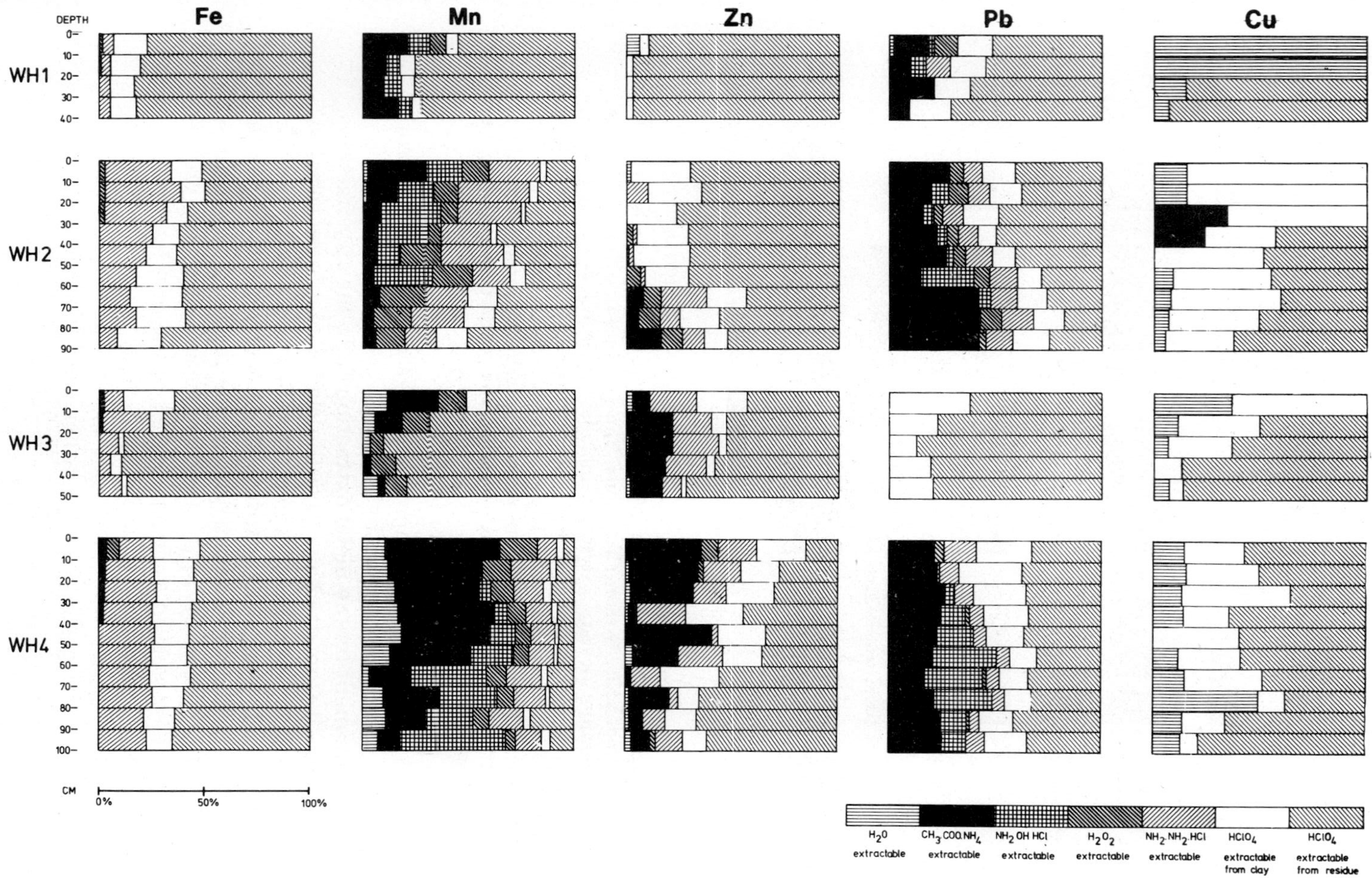

Fig. 2. Distribution of the total amount available of Fe, Mn, Zn, Pb and Cu in each 10-cm interval of the West Hercules soils, western Tasmania. The sequence is from left to right. The water extraction also extracts some organic material. The ammonium acetate extraction does not discriminate between the origin of the exchangeable cations (e.g. clay, organic material). The hydroxylamine hydrochloride extraction reduces and extracts manganese oxides. The hydrogen peroxide extraction destroys here only organic material and not manganese oxides (Fig. 1). Hydrazine chloride extractable metals are released after reduction of iron oxides. The perchloric acid extractions represent the metals in the clay fraction and coarser residue. (Modified after Russell, 1976.)

Study of the amount of both clay and residue phases appear to have little value as possible sulphides would have been extracted earlier by the hydrogen peroxide treatment. Analyses of fresh parent rock material would be of more value.

The low concentration of all elements analyzed in the immature soils G1 and G2 and also in WH1 is caused by the absence of secondary soil components able to scavenge.

The high concentrations in the deep soils G3, WH2 and WH4 are presumably of secondary nature and related to the influx of Mn and Fe and the accumulation of organic material.

The rather low values in soil WH3, situated in the middle of virgin forest, can be best explained by assuming that no metal enrichment by groundwater movement has occurred so far down slope. The soil is of moderate depth, like all soils in the lower part of the slope, and is apparently strongly leached.

PRACTICAL IMPORTANCE OF SEQUENTIAL ANALYSIS

In the soils studied, an association appears to exist between the amount of reducible Mn and Fe and the level of Pb and Zn. Cu appears to be organically bound, either in water-soluble form or not.

Considerable proportions of the total Zn, amounts of Pb, Cu, Mn and Fe in the soils are present in cationic form or are complexed or sorbed by manganese oxides, organic material and iron oxides.

These relations will probably hold in general, and no geochemical sampling programme should be undertaken without analysis of Mn and Fe. As the nature of soil will vary in different climates and different localities, an initial complete sequential analysis of some profiles will be valuable in early stages of exploration.

In the West Hercules area, the Pb anomaly may reflect mineralization. However, evidence from limited drilling is negative and it is more likely that the high Pb values, occurring at a distinct break of slope, reflect a hydromorphic anomaly formed by seepages of groundwater from the upslope Hercules host rocks.

REFERENCES

Asami, T. and Kumada, K., 1960. Comparison of several methods for determining free iron in soils. Soil Plant Food (Tokyo), 5: 179—183.

Baker, W.E., 1973. The role of humic acid from Tasmanian podzolic soils in mineral degradation and metal mobilization. Geochim. Cosmochim. Acta, 37: 269—281.

Band, R.B., 1969. Dispersion of nickel and molybdenum from mineralization in glaciated terrain, S. Norway. Ph.D. Thesis, Imperial College, London (unpublished).

Bradshaw, P.M., Clews, D.R. and Walker, J.L., 1973. Exploration Geochemistry. Barringer Research Ltd., Rexdale, Ont., 49 pp.

Chao, T.T., 1972. Selective dissolution of manganese oxides from soils and sediments. Soil Sci. Soc. Am., Proc., 36: 764—768.

Galabutskaya, E. and Govarova, R., 1934. Bleaching of kaolin. Miner. Syr'e, 9: 27—32.

Gatehouse, S., 1973. The primary and secondary copper, lead, zinc, iron and manganese dispersion geochemistry of the South Comet area, Western Tasmania. Honours Thesis, Geology Department, University of Tasmania, Hobart, Tasmania, 149 pp. (unpublished).

Gedeon, A.Z., Butt, C.R.M., Gardner, K.A. and Hart, M.K., 1977. The applicability of some geochemical analytical techniques in determining "total" composition of some lateritized rocks. J. Geochem. Explor., 8: 283—303.

Jackson, M.L., 1956. Soil Chemical Analysis — Advanced Course. Published by the author, Department of Soil Science, University of Wisconsin, Madison, Wisc., 894 pp. (4th printing 1968).

Latimer, W.M., 1964. Oxidation Potentials. Prentice Hall, Englewood Cliffs, N.J., 2nd ed., 392 pp.

Leeper, G.W., 1947. The forms and reactions of manganese in the soil. Soil Sci., 63: 79—94.

Levinson, A.A., 1974. Introduction to Exploration Geochemistry. Applied Publishing Ltd., Calgary, Alta., 612 pp.

Lynch, J.J., 1971. The determination of copper, nickel and cobalt in rocks by atomic absorption spectroscopy using a cold leach. In: R.W. Boyle and J.I. McGerrigle (Editors), Geochemical Exploration. Can. Inst. Min. Metall., Spec. Vol., 11: 313—314.

Rainwater, F.H. and Thatcher, L.L., 1960. Methods for collection and analysis of water samples. U.S. Geol. Surv., Water-Supply Paper, 1454: 301 pp.

Russell, D.W., 1976. The biogeochemistry and pedogeochemistry of the West Hercules area, Rosebery, Western Tasmania. Honours Thesis, Geology Department, University of Tasmania, Hobart, Tasmania, 315 pp. (unpublished).

Stecher, P.G. (Editor), 1968. The Merck Index. Merck, Rahway, N.J., 8th ed., 171 pp.

Taylor, R.M. and McKenzie, R.M., 1966. The association of trace elements with manganese minerals in Australian soils. Aust. J. Soil. Res., 4: 25—39.

Van Moort, J.C. and De Vries, D., 1970. Rapid carbon determination by dry combustion in soil science and geochemistry. Geoderma, 4: 109—118.